T0250413

CRC
Handbook
of
Microalgal
Mass Culture

Editor

Amos Richmond, Ph.D.
Cahn Professor in Economic Botany in Arid Zones
The Microalgal Biotechnology Lab
The Jacob Blaustein Institute for Desert Research
Ben-Gurion University of the Negev
Sede Boqer, Israel

CRC Press
Taylor & Francis Group
Boca Raton London New York

CRC Press is an imprint of the
Taylor & Francis Group, an **Informa** business

PREFACE

Two groups of algae may be readily distinguished from the standpoint of the technology involved in their cultivation, harvesting, and processing. One group is the microalgae, the morphological features of which may be resolved only with the aid of a microscope. The other group consists of macroalgae, macroscopic seaweeds. In general, seaweeds are attached to rocky substrates on the sea bottom and may reach several meters in length, e.g., the *Laminaria* species. In contrast, minute planktonic organisms, measuring only a few microns in diameter, like *Synechococcus* or *Chlorella* or the filamentous *Spirulina* are typical microalgae. These are the primary basis of aquatic food chains and account for ca. 40% of the total photosynthesis on earth. In nature, planktonic microalgae are so dispersed that they may be seen by the naked eye only if blooms, very dense populations of cells, are formed. The occurence of blooms, however, is unpredictable and their harvesting thus cannot support industrial production, which can be achieved only in artificial systems.

This handbook is devoted to the mass production of microalgae, and in my part, is based on some 10 years of experience in growing and studying microalgal cultures maintained at high population densities under laboratory conditions and in outdoor ponds. I was attracted to this field because of the exciting promise in growing unicellular algae on local resources as an alternative source of food and feeds for hungry or malnourished populations. At that time, the report of the United Nations' Advisory Committee on International Action to Avert an Impending Protein Crisis had just been published (see Chapter 2). The message of that report was clear — the world's population was increasing faster than the expansion of conventional agricultural production and a protein crisis affecting hundreds of millions of people the world over was looming. According to another report much referred to at that time, by the year 2000, the world supply of protein would fall short by some 39×10^6 tons.

The United Nations' committee recommended that conventional agricultural crops be supplemented with high-protein foods of unconventional origin, such as microorganisms. On this context, microalgae were natural candidates for several reasons:

1. Essentially the entire plant body has nutritional value since only a minor portion of the cell is indigestible.
2. These single-celled organisms contain a substantial amount of protein, up to 65 or even 70%, i.e., some 2 to 4 times that of seeds and leaves of vascular plants.
3. Microalgae have a very high yield potential, since the conditions for the culture may be readily controlled and the population density adjusted to that at which the highest efficiency of solar energy conversion per unit area is obtained.
4. Perhaps the most attractive aspect of algal mass culture as a source of food is that brackish and sea water can be used for their production.

This may be the key for establishing high bioproductivity in regions which lack sweet water and would otherwise remain poorly productive.

Today however, algaculture is still far from providing a source of inexpensive food on chemicals. Mass production of microalgae outdoors is a formidable task. Much more remains to be learned about the biology involved in this biotechnology and many technological details must be improved. Production procedures have to be simplified and the average annual yields have to increase severalfold before algaculture can become a significant agricultural endeavor. Nevertheless, the promise of cultured algae, particularly as a salt-tolerant crop in warm arid lands, is real. This is amplified by the growing realization that the deprivation and hunger of many hundreds of millions of people in developing areas of the world cannot be correctly and permanently relieved by importing food and materials from the developed countries. This is the consensus of all who are involved in analyzing the problems of hunger and poverty in these lands. The only meaningful solution rests in local development, which would depend on imported knowhow and capital at first, but which would aim at promoting economic independence. In this frame of reference, efficient utilization of saline-water for agricultural production is of particular importance. Indeed, the resources of sweet-water are dwindling the world over and shall often have valuable alternative uses. The aim of developing local resources therefore would include in many cases the production of salt-tolerant crops, such as selected species of algae that could be grown for various economic purposes.

Thus, even though the concept of mass production of microalgae to feed a hungry world has not yet been realized, it is still valid as a long-range goal. The concept should be expanded to include microalgae for biosynthesis of chemicals and special products, the emphasis being placed on production based on local resources and skills. Much more research will have to be carried out and detailed experience will have to be gained before cultured microalgae come of age as an industrial crop. I hope this handbook will be useful towards achievement of this goal.

It is a pleasant duty to thank Dr. Marj Tiefert for improving the English of Chapters 3, 6, 7, 9, 11, 14, and 17. Also, I am very much indebted to Ms. Ilana Brina for her most competent and devoted assistance in taking care of the many technical and administrative chores involved in preparing this book.

Amos Richmond
Sede Boqer, July 1984

THE EDITOR

Amos Richmond is Professor of Biology at the Jacob Blaustein Institute for Desert Research of Ben Gurion University of the Negev, the Sede-Boqer Campus. Dr. Richmond was born in 1931 in Tel-Aviv, Israel and received his B.Sc. in 1954 at the California State Polytechnic in St. Louis Obispo, California. He obtained his M.Sc. from the University of California at Los Angeles in 1956 and the Ph.D. in 1963 from Michigan State University at East Lansing, Michigan. He was thereafter a research-fellow at Purdue University in 1964 and at the University of California at Los Angeles in 1965-1966. He served as lecturer and senior lecturer in Botany at the Institute for Higher Education in the Negev from 1966 until 1971, and as chairman of the department of Biology of the Ben Gurion University of the Negev in 1970 to 1971. From 1971 to 1974 he was associate professor of Biology and Dean of the School of Natural Sciences at that university. He became Professor of Biology in 1975 and served as the first director of the Jacob Blaustein Institute for Desert Research at Sede-Boker from its inception in 1974 to 1983. In 1974 Dr. Richmond established at the Desert Research Institute the Micro-Algal Biotechnology Laboratory, which he had been directing since. Dr. Richmond received the Bergmann prize for distinction in applied research in 1984 and is the incumbent of the Miles and Lillian Cahn Chair in Economic Botany at B.G. University. He is the recipient of many research grants, among which from the German government, from the Israel Council for Research and Development, the Bi-national Agricultural Research Fund and the Solar Energy Research Institute in the U.S. He is the author of more than 70 publications and his current research interests are forcused on the growth physiology of salt-tolerant microalgae with particular reference to the physiology of growth in mass cultures aimed to yield products of economic potential.

CONTRIBUTORS

Sheldon Aaronson, Ph.D.
Professor
Department of Biology
Queens College
Flushing, New York

Aharon Abeliovich, Ph.D.
Senior Lecturer
The Laboratory of Environmental
 Microbiology
The Jacob Blaustein Institute for Desert
 Research
Ben-Gurion University of the Negev
Sede Boqer, Israel

Yael J. Avissar, Ph.D.
Lecturer
Department of Biology
Ben-Gurion University
Beer-Sheva, Israel

E. W. Becker, Ph.D.
Senior Scientist
Institute of Chemical Plant Physiology
University of Tubingen
Tubingen, Federal Republic of Germany

Zvi Cohen, Ph.D.
Lecturer
Microalgal Biotechnology Laboratory
The Jacob Blaustein Institute for Desert
 Research
Ben-Gurion University
Sede Boqer, Israel

Joseph C. Dodd, Ph.D.
Consulting Engineer
Microalgae Technology
Brawley, California

Zvi Dubinsky, Ph.D.
Senior Lecturer
Department of Life Sciences
Bar-Ilan University
Ramat-Gan, Israel

David O. Hall, Ph.D.
Professor
Department of Plant Sciences
King's College London
London, United Kingdom

Eithan Hochman, Ph.D.
Professor
Department of Economics
Ben-Gurion University of the Negev
Beer-Sheva, Israel

Drora Kaplan, Ph.D.
Lecturer
The Laboratory of Environmental
 Microbiology
The Jacob Blaustein Institute for Desert
 Research
Ben-Gurion University of the Negev
Sede Boqer, Israel

S. Herman Lips, Ph.D.
Professor
Department of Biology
Beer-Sheva and the Salinity and Plant
 Physiology Lab at Sede-Boqer
Ben-Gurion University of the Negev
Israel

Amos Richmond, Ph.D.
Professor
Microalgal Biotechnology Laboratory
The Jacob Blaustein Institute for Desert
 Research
Ben-Gurion University of the Negev
Sede Boqer, Israel

Carl J. Soeder, Ph.D.
Professor
Institute for Biotechnology
Kernforschungsanlage GmbH
Julich, Federal Republic of Germany

Y. Tsur, Ph.D.
Professor
Department of Economics
Ben-Gurion University of the Negev
Beer-Sheva, Israel

L. V. Venkataraman, Ph.D.
Scientist and Area Coordinator
Autotrophic Cell Culture Discipline
Central Food Technology Research
 Institute
Mysore, India

Avigad Vonshak, Ph.D.
Senior Scientist
Microalgal Biotechnology Laboratory
The Jacob Blaustein Institute for Desert
 Research
Ben-Gurion University of the Negev
Sede Boqer, Israel

TABLE OF CONTENTS

THE PRODUCTION OF BIOMASS: A CHALLENGE TO OUR SOCIETY

D. O. Hall

SUMMARY

The overuse and undersupply of biomass is currently a serious problem and potentially a greater long-term danger than lack of food. Today 14% of the world's primary energy is derived from biomass — equivalent to 20 million barrels of oil per day. Its predominant use is in the rural areas of developing countries where half the world's population lives, e.g., Nepal derives nearly 100%; Kenya, 75%; India, 50%; China, 33%; Brazil, 25%; and Egypt and Morocco 20% of their total energy from biomass. A number of developed countries also derive a considerable amount of energy from biomass: Sweden, 9%; Canada, 7%; and the U.S. and Australia 3% each. A number of European-wide studies have shown that about 5 to 10% of Europe's energy requirements could be met from biomass by the end of the 20th century. An especially valuable contribution could be in the form of liquid fuels which have become prone to fluctuating price and supply. The resources available, the effect of large agricultural surpluses (especially in North America and Europe), and the factors which will influence biomass energy schemes around the world are issues which, at present, are hotly debated.

Worldwide government expenditure on biomass energy systems is over $2 billion a year while the costs of surplus food production is over $60 billion a year. However, biomass energy is not necessarily the panacea for any country's energy problems, though currently the process of photosynthesis produces an amount of stored energy in the form of biomass which is almost 10 times the world's annual use of energy. Additionally, the productivity of biomass-for-energy species can be dramatically increased as has already been shown. Such improvements have been accomplished with a number of agricultural species which are well known such as maize, wheat, and rice.

The world produces 10 to 20% more food than is required to feed its 4.5 billion people an adequate diet. In North America and Europe the main problem with food is its easy overproduction and general over-consumption; however, there are an estimated 450 million undernourished people, mostly in Asia and Africa. Simplistically, if available food production was increased by 1.5% (equivalent to about 25 t of grains), and if this food was distributed equitably to those who need it, there would be no undernourished people in the world. The same argument applies if only 10% of the developed countries' grain production was diverted away from animals to humans. Health authorities have recommended lower meat and sugar consumption in the U.S. and U.K., and such changes are already occurring in some developed countries. These diet and other biotechnological changes will have long-term socioeconomic consequences.

The question is how to achieve both food *and* biomass fuel production *locally* on a sustainable basis. Both are required, thus planning and provision of the appropriate infrastructure and incentives must be provided. Increased support of research and development, training, and firm establishment of top priority to agriculture and forestry are essential in many countries of the world, if necessary, with significant help from abroad.

INTRODUCTION[1-24]

Biomass contributes a significant part of the world's energy, being an important provider of energy to a great number of people. Decisions that are made over the next few years will significantly influence the level of biomass energy use in the future. How much biomass will contribute over the long term will depend very much on decisions that are made both at the local level and at the national level, in addition to international policy making, especially for energy and food.

It is not within the scope of this chapter to cover any of the numerous contributions that algal biomass can make to food and energy chains. Nevertheless, it should always be borne in mind that when dealing with aquatic compared to terrestrial species there are usually totally different problems to be solved, whether it is for food, energy, chemicals, etc. Terrestrial plants are easier to grow and harvest but they often use valuable land.

"Food vs. fuel?" This is the wrong question. It should rather be, "How can we equitably distribute the existing ample supply of food?" The problem of uneven distribution of the world's adequate supply of food is well known and has been the subject of many reports over the last 50 years. This author is not competent to extend or improve on these studies. What I do, however, wish to point out is that both food *and* fuel are crucial limiting factors in development and must be *locally* available on a *sustainable* basis. Appropriate infrastructure incentives and planning related to agriculture and forestry enterprises certainly help alleviate shortages in order to allow socioeconomic development to proceed. Again this is not novel, but unless real recognition of the immediate importance of agriculture and forestry in many countries of the world is forthcoming, these developing nations will continue to suffer national shortages of food and/or fuel which may have a debilitating effect on their growth (both personal and political).

About 10% of the world's people are undernourished, but world food production is more than 10% greater than is required overall and per capita food production has been increasing at a compound rate of 0.5% in developing countries and 0.8% worldwide since 1950. Although the world population has increased by $^2/_3$ in the last 30 years (increased from 2.5 to 4.5 billion), the production of food has doubled. The amount of grains needed to solve undernourishment in the developing countries is only about 25 t annually — this represents only 1.7% of the total production worldwide and only 12% of the world's trade in grains. Biomass energy produced annually is about 10 times greater than the total amount of energy used worldwide; still many people in developing countries suffer chronic shortages of fuelwood and its overuse is creating serious environmental consequences.

Today, $^2/_3$ of the world's people depend on plants for nearly all their food and for the majority of their energy. The question should again be not "food vs. fuel?", but "how can we increase the productivity of plant-based agriculture (and forestry) in order to provide the required levels of both food and fuel at the national or regional level?" Since only less than 10% of the world's food and hardly any of the world's biomass fuel enters transnational trade it is local production of plant products which is all-important.

The average person in the rural areas of the developing world uses the total equivalent of about 1 t of wood per annum. This is used mainly for cooking and heating, but also for small-scale industry, agriculture, food processing, etc. The use of wood and charcoal in urban areas and for industry is often much greater than is realized. This author does not think many people appreciate the importance of biomass energy use because the statistics are not available to show this significance, and the consequences of biomass overuse are not readily evident. Until a few years ago world energy supply statistics listed biomass at 3 to 5%, if at all. It is now known that over half of all the trees cut down in the world today are used for cooking and heating. The problem of deforestation with its consequent flooding, desertification, and agricultural problems is not solely due to over-cutting of trees for cooking and heating. There are obviously other factors involved such as land clearing for agriculture, commercial and illegal cutting, absence of replanting, and so forth.

Recently there have been several good papers published on studies in Southern and East Africa which show that in an average family of six or seven, one person's sole job is to collect firewood and they will often have to walk great distances; this, of course, has other deleterious consequences. In urban environments, households can spend up to 40% of their income on fuelwood and charcoal. Another aspect which has been highlighted in Tanzania is the curing of tobacco; for each hectare of tobacco the wood from 1 ha of savannah woodland needs to be incinerated. There are many examples showing that it is not only domestic, but also agricultural, urban, and small-scale industrial fuelwood overuse which is having serious long-term consequences. Several attempts are being made by a number of international and national groups to try and reverse the deforestation problem by vigorously promoting reforestation, village fuelwood lots, or community forestry and agro-forestry. The World Bank in 1980 concluded that if one was to reverse the deforestation problem one would need to spend $6.75 billion over the next 5 years in order to start reforesting 50 million ha. There is little hope that this will happen, but it was what realistically was thought to be needed. There are a number of reasons why this will not be possible, but only one shall be mentioned here. It is the very low status that foresters have in developing (and also developed) countries. Consequently, it is all very well to promulgate reforestation schemes to help solve the energy crisis in various parts of the world, but unless one has the people on the ground with the experience and knowledge, implementation of these schemes is impossible.

The majority of people in the world exist by growing plants and processing their products. The main issue in developing countries is that of scarcity and the problem of trying to maintain, or possibly even to increase, the present level of use without harming agricultural or forestry and ecological systems. More efficient use of existing biomass and possible substitutes for biomass use, e.g., solar and wind-based technology and indigenous fossil fuels, should be considered and implemented as quickly as possible to reverse the trend of excessive biomass use, as is already occurring in many countries. In the developed world the expertise exists and is already being used to implement biomass energy programs from the standpoint of potential technology and economics. Biomass can provide a source of energy now and in the future; just how much it can contribute to the overall provision of energy will depend on existing local and national circumstances. It is imperative that each country establish its energy use patterns and the potential of biomass energy. This is not easy to accomplish quickly, but needs to be done as soon as possible.

How much photosynthetic production of biomass actually occurs on the earth? It has been shown that the world's total annual use of energy is only one tenth of the annual photosynthetic energy storage, i.e., photosynthesis already stores ten times as much energy as the world needs. The problem is getting it to the people who need it. Second, the energy content of stored biomass on the earth's surface today, which is about 90% in trees, is equivalent to our proven fossil fuel reserves. In other words the energy content of trees is equivalent to the commercially extractable oil, coal, and gas. Third, during the Carboniferous Era quite large quantities of photosynthetic products were stored, but in fact they only represent 100 years of net photosynthesis. The overall photosynthetic efficiency during the Carboniferous Era was less than 0.002%, thus, our total possible fossil fuel resources only represent a few years of net photosynthesis. Fourth, there is the problem of CO_2 cycling in the atmosphere. Many people are rightly concerned about this buildup of CO_2 if we continue to burn fossil fuels. It is a problem of cycling between two or three pools of carbon. The amount of carbon stored in the biomass is approximately the same as the atmospheric CO_2 and the same as the CO_2 in the ocean surface layers; there are three equivalent pools. The concern is over how the CO_2 is distributed between these pools and how fast it equilibrates into the deep ocean layers. However, we should appreciate that increasing CO_2 concentrations in the atmosphere may be good for plants (since CO_2 is a limiting factor in photosynthesis and plants have better water and fertilizer use efficiency at higher CO_2 concentrations). Plants could also act as CO_2 sinks if photochemical means for fixing CO_2 were not available to alleviate the problem in the future.

Table 1

**SOME ADVANTAGES AND PROBLEMS FORESEEN IN BIOMASS FOR
ENERGY SCHEMES**

Advantages	Problems
1. Stores energy	1. Land and water use competition
2. Renewable	2. Land areas required
3. Versatile conversion and products; some products with high energy content	3. Supply uncertainty in initial phases
4. Dependent on technology already available with minimum capital input; available to all income levels	4. Costs often uncertain
5. Can be developed with present manpower and material resources	5. Fertilizer, soil, and water requirements
6. Large biological and engineering development potential	6. Existing agricultural, forestry, and social practices
7. Creates employment and develops skills	7. Bulky resource; transport and storage can be a problem
8. Reasonably priced in many instances	8. Subject to climatic variability
9. Ecologically inoffensive and safe	9. Low conversion efficiencies
10. Does not increase atmospheric CO_2	10. Seasonal (sometimes)

The oil/energy problem of the 1970s had three clear effects on biomass energy and development. First, in a number of developed countries large research and development programs have been instituted which have sought to establish the potential, costs, and methods of implementation of energy from biomass. The prospects look far more promising than was thought even 3 years ago. Demonstrations, commercial trials, and industrial projects are being implemented. Estimated current expenditure is over \$1 billion per annum in North America and Europe. Second, in at least two countries, Brazil (which currently spends over half of its foreign currency on oil imports) and China (with over 7 million biogas digesters), large-scale biomass energy schemes are being implemented. The current investment is over \$1 billion per annum in Brazil. Third, in the developing countries as a whole there has been an accelerating use of biomass because oil products have become too expensive and/or unavailable.

Biomass as a source of energy has its pros and cons. Like all other energy sources one must realize that it is not the universal panacea. Some advantages and disadvantages are listed in Table 1. This author wishes to emphasize the large biological and engineering development potential which is available for biomass. Because no advances in research have been made for many years, the efficiency of production and use of biomass as a source of energy has not progressed in the way agricultural yields for food have increased.

Agricultural research has paid off very well (Table 2) and this may also be the case for biomass research and development as has been shown in recent fuelwood schemes. Thus there is an undoubted potential to increase biomass energy yields. Another advantage is the versatility of the biomass production and conversion technologies such that anyone involved in a particular project can select the routes most suited to the prevailing conditions and requirements of that area. The most obvious problems that immediately come to mind are land use in competition with food production. Existing agricultural, forestry, and social practices are also certainly a hindrance to promoting biomass as a source of energy, whether in a developing or an already developed country.

In North America and Europe the problem in agriculture and nutrition is overproduction, excessive consumption of animal products via feeding of grains, and surpluses affecting world trade, especially in relation to commodity prices for many developing countries' products. Obviously this is too simplistic, but the medium-term trends are important as they are likely to be aggravated by increased productivity, influence of new biotechnological processes, and changes in diet. Thus, for many countries the question should be how to

Table 2
STUDIES OF AGRICULTURAL RESEARCH PRODUCTIVITY — DIRECT COST-BENEFIT TYPE STUDIES[25]

Commodity	Country	Time period	Annual internal rate of return (%)
Hybrid corn	U.S.	1940—55	35—40
Hybrid sorghum	U.S.	1940—57	20
Poultry	U.S.	1915—60	21—25
Sugar cane	South Africa	1945—62	40
Wheat	Mexico	1943—63	90
Maize	Mexico	1943—63	35
Cotton	Brazil	1924—67	77 +
Tomato harvester	U.S.	1958—69	37—46
Maize	Peru	1954—67	35—40
Rice	Japan	1915—50	25—27
Rice	Japan	1930—61	73—75
Rice	Colombia	1957—72	60—82
Soybeans	Colombia	1960—71	79—96
Wheat	Colombia	1953—73	11—12
Cotton	Colombia	1953—72	None
Aggregate	U.S.	1937—42	50
		1947—52	51
		1957—62	49

achieve both food and biomass fuel production locally on a *sustainable* basis. Both are required, thus, planning and provision of the appropriate infrastructure and incentives must be provided. Increased support of research and development, training, and firm establishment of top priority to agriculture and forestry are essential in many countries of the world, if necessary, with significant help from abroad. Overall, biomass energy will not be a simple solution to the energy problems of developing countries. Biomass systems will not necessarily be inexpensive, nor will they be implemented easily, without a major commitment from governments and a considerable amount of political will at the national and international levels.

Only a relatively small amount of food (25 t of cereal out of a world production of 1500 tonnes at present) is needed to solve the world's hunger problem which condemns about 450 million people to suffer nearly perpetual undernutrition. How can this amount of food be produced and distributed so that those that need it actually receive it? This problem has vexed many and has been the subject of lengthy reports and will no doubt continue to do so in the future if, or until, the problem is solved. The gist of my message, which does not claim to be novel, is that it is possible to produce sufficient food and some fuel on the land using good agricultural and forestry practices, today and in the future. To what degree of self-sufficiency a country or region decides to achieve in food and fuel depends very much on local conditions, which are not easy to accurately ascertain from superficial surveys.

If we really want to solve the undernourishment/poverty problem at the peasant and village level in developing countries (where most but by no means all of the world's poor and malnourished people live) we must understand what the limiting factors are in food and biomass production at this level. Is it lack of water, fertilizers, pesticides, credit facilities, extension services, transport, storage, etc., or factors such as lack of human and animal power at crucial times of the year, tax incentives, high energy costs, too low product costs, and so on?

My contention is that with a reasonable degree of management, combined with the nec-

Table 3
SOURCES OF BIOMASS FOR CONVERSION TO FUELS

Wastes
  Manures
  Slurry
  Domestic rubbish
  Food wastes
  Sewage

Starch crops
  Maize
  Cassava

Sugar crops
  Cane
  Beet

Residues
  Wood residues
  Cane tops
  Straw
  Husks
  Citrus peel
  Bagasse
  Molasses

Aquatic plants
  Algae
  Chlorella
  Scenedesmus
  Navicula
  Multicellular
  Kelp

Land crops (ligno-cellulose)
  Trees
    Eucalyptus
    Poplar
    Luceana
    Casuarina

Water weed
  Water hyacinth
  Water reeds/rushes

essary incentives and infrastructures many regions of the world could be self-sufficient in their basic food requirements, and in addition provide reasonable amounts of energy from natural or specifically grown biomass.

BIOMASS[26-33]

Biomass is a jargon term used in the context of energy for a range of products which have been derived from photosynthesis; the products can be recognized as waste from urban areas and from forestry and agricultural processes, specifically grown crops such as trees, starch and sugar crops, hydrocarbon plants and oils, and aquatic plants such as water weeds and algae (Table 3). Thus, everything which is derived from photosynthesis is a potential source of energy since it is a solar energy conversion system. The problem with solar radiation is that it is diffuse and intermittent. Therefore if we are going to use it we must capture a diffuse source of energy and store it; plants accomplished this long ago.

The process of photosynthesis embodies the two most important reactions in life. The first is the water-splitting reaction which evolves oxygen as a by-product. All life depends on this reaction. Second is the fixation of CO_2 to organic compounds. All food and fuel is derived from CO_2 fixation in the atmosphere. When looking at an energy process we need to have some understanding of what the efficiency of this process will be; one needs to look at the efficiency over the entire cycle of the system, and for the process of photosynthesis we mean incoming solar radiation converted to a stored end-product. Most people agree that the practical maximum efficiency of photosynthesis is 5 to 6%. This may not seem like very much, but one should be mindful that this represents stored energy (Table 4).

Photosynthetic efficiency will determine biomass dry weight yields. For example, in the U.K. at 100 Wm^{-2} incoming radiation a good potato crop growing at 1% efficiency (usually not higher than this in temperate regions), will yield about 20 t dry weight per hectare per annum. Obviously if we can grow and adapt plants to increase photosynthetic efficiency the dry weight yields will increase and alter the economics of the crop. One of the more interesting

Table 4
PHOTOSYNTHETIC EFFICIENCY AND ENERGY LOSSES[34]

	Available light energy (%)
At sea level	100
50% loss as a result of 400—700 nm light being photosynthetically usable	50
20% loss due to reflection, inactive absorption, and transmission by leaves	40
77% loss representing quantum efficiency requirements for CO_2 fixation in 680 nm light (assuming 10 quanta/CO_2),[a] and remembering that the energy content of 575 nm red light is the radiation peak of visible light	9.2
40% loss due to respiration	<u>5.5</u> (Overall photosynthetic efficiency)

[a] If the minimum quantum requirement is 8 quanta/CO_2, then this loss factor becomes 72% instead of 77%, giving the final photosynthetic efficiency of 6.7% instead of 5.5%.

areas of research is understanding what the limiting factors are in photosynthetic efficiency in plants both for agriculture and biomass energy.

Currently the production of liquid fuels from biomass is of great interest, but yield limitations are becoming paramount in the economics of the overall processes. Table 5 highlights some important parameters.

There is another aspect of photosynthesis that we should all appreciate, i.e., the health of our biosphere and our atmosphere is totally dependent on the process of photosynthesis. Every 300 years all the CO_2 in the atmosphere is cycled through plants; this occurs every 2000 years to all the oxygen, and every 2 million years to all the water. Thus, 3 key ingredients in our atmosphere are dependent on cycling through the process of photosynthesis.

FOOD PRODUCTION, USE, AND PRODUCTIVITY[25,36-60]

World agriculture is already producing enough to feed everyone on a Western standard of diet, however, inequalities and inefficiencies in distribution seem to make it impossible for the poorest to ensure their share. The current food stock is sufficient to feed the 1977 population of 4.2 billion at a level of 2570 kcal/day (about 20% above requirements) and supply 69 g/day protein. The disparity between developed ($^1/_4$ world population) and developing ($^3/_4$ world population) countries is seen in Table 6, both in the amount of calorie intake and the contribution from animal sources (5 times lower). Miller[50] has estimated that of the total primary food production (62% grains) equivalent to 4514 kcal/person/day, only 2149 kcal is directly consumed by man and 2365 kcal is fed to animals, who in turn contribute 414 kcal to man to give an average worldwide total daily intake of 2563 kcal per capita. Over one third of the world's cereal production (about 1500 t/year of which only 12% enters world trade) and over one half of the soybean production is fed directly to animals. If only $^1/_4$ of the world soybean harvest was used for direct human consumption instead of animal feed it would provide 5 kg/year of a high protein food (750 kcal energy per day) for everyone in the world.[2]

The problem with using these facts to help solve the world's food problem are well discussed by FAO in its recent *Agriculture: Toward 2000*.[47] These facts, which are inconsistent with what is morally right, are often put forward to support the case for giving less cereals to animals and more to people. Unfortunately, neither the facts nor the solutions are simple.

The great bulk of the feed use of cereals occurs in the developed countries, followed at a great distance by the more prosperous developing countries; the low-income developing countries use less than 2% of their total cereal supplies for animal feed (Table 7). In the

Table 5
ENERGY CROP YIELDS[35]

Energy crop	Annual crop yield (t/ha)			Fuel type	Conversion yield[c] (% by weight)	Fuel yield (t/ha)			Energy content of fuel[d] GJ/t	Fuel yield (GJ/ha)		
	Average[a]	Good[b]	Best[b]			Average	Good	Best		Average	Good	Best
Sugar cane	54	75	125	Ethanol	5.5	3.0	4.1	6.9	26.6	80	109	184
Sweet sorghum[f]		35	70		6.8		2.4	4.8	26.6		64	128
Cassava[g]	8.7	20	50		14	1.2	2.8	7.0	26.6	32	74	186
Sweet potato		20	50		10		2.0	5.0	26.6		53	133
Trees	10	20	40	Methanol	38	3.9	7.7	19.3	19.8	77	153	382
Oil palm[h]	17	25	38	Vegetable oils	20	3.4	5.0	7.6	38.5	131	193	293
Coconut[i]		2.2	5.0		57		1.5	3.4	38.6		58	49
Soybean	1.1	2.7	4.0		20	0.2	0.5	0.8	38.5	8	21	31
Groundnut	0.9	1.3	3.8		45	0.4	0.6	1.7	38.5	16	22	66
Rapeseed	0.6	1.0	1.6		40	0.2	0.4	0.6	38.5	9	15	25
Sesame	0.3	0.8	2.0		50	0.2	0.4	1.0	38.5	6	15	39

a   Assumes one crop per year unless otherwise stated.

b   Conversion yield figures are averages. Variations can occur between crop varieties due to different sugar, starch or oil contents, and different conversion plant efficiencies. Ethanol yields are more commonly treated on an ℓ/t of feedstock basis. Expressed in this way the conversion yields for the four ethanol crops are as follows: sugar cane, 70 ℓ/t; sweet sorghum, 86 ℓ/t; cassava, 180 ℓ/t; sweet potato, 125 ℓ/t. Methanol yields at 386 kg (483 ℓ) from 1 t oven dry wood.

c   The energy contents of vegetable oils do show some variation. For example: sunflower, 35.3 GJ/t; coconut oil, 36.6 GJ/t; soya oil, 39.3 GJ/t. In the absence of comprehensive figures, an average has been taken of 38.5 GJ/t.

d   Average crop yield for developing countries.

e   Good and best yields taken from various sources. Figures intended to be indicative rather than absolute. The ease with which good and best yields can be obtained is not the same.

f   Best yields for sweet sorghum refer to double cropping.

g   No allowance has been made for the fuel that is needed to run the cassava distillery. If a fast growing tree plantation was used to supply wood fuel the total plantation area required would be increased by approximately 28% (assuming 20 t/ha for cassava and 10 ODT/ha for wood).

h   Crop yields refer to tonnes of fresh fruit bunches per hectare. For every tonne of palm oil, roughly 200 kg of palm kernels are also produced.

i   Crop yields refer to t/ha of copra.

Table 6
GLOBAL DISTRIBUTION OF ENERGY USE, POPULATION, AND FOOD SUPPLY (1978)

	Developed countries	Developing countries	World
Population	1.1 billion (26%)	3.1 billion (74%)	4.2 billion (100%)
Total energy use ($\times 10^9$)	208 GJ (69%)	92 GJ (31%)	300 GJ (100%)
per capita	189 GJ (6.3 tce)	30 GJ (1.0 tce)	71 GJ (2.4 tce)
Commercial energy use ($\times 10^9$)	206 GJ (80%)	52 GJ (20%)	258 GJ (100%)
per capita	187 GJ (6.2 tce)	17 GJ (0.6 tce)	61 GJ (2.0 tce)
Biomass energy use ($\times 10^9$)	2 GJ (5%)	40 GJ (95%)	42 GJ (100%)
per capita	1 GJ (0.03 tce)	13 GJ (0.4 tce)	10 GJ (0.3 tce)
Commercial energy in total energy	98%	57%	86%
Biomass energy in total energy	2%	43%	14%
Food supply[a]			
per capita (kcal)	3353	2203	2571
% of daily requirement	(129%)	(96%)	(106%)

Note: tce = Tonnes coal equivalent.

[a] Average requirement is calculated at 2600 kcal in developed countries and 2300 kcal in developing countries; 2370 kcal is a world average.

Table 7
PLANT AND ANIMAL PROTEIN PRODUCTION AND CONSUMPTION BY MAN AND LIVESTOCK IN 1975 (ESTIMATED, IN MILLION T)[18]

	U.S.		World	
Total cereal protein produced	17.0	100%	95	100%
Fed to livestock	15.5	91%	38	40%
Available to man	1.5	9%	57	60%
Total legume protein produced	9.3	100%	30	100%
Fed to livestock	9.0	97%	6	20%
Available to man	0.3	3%	24	80%
Total livestock protein produced	6.0	100%	33	100%
Fed to livestock	0.7	12%	3	9%
Available to man	5.3	88%	30	91%
Total protein produced[a]	34.1	100%	173	100%
Fed to livestock	26.1	77%	51	29%
Available to man	8.0	23%	122	71%

[a] Fish protein and other types of vegetable protein are included in this total.

short run, when total cereal supplies cannot be changed, saving a part of current supplies from going to livestock feeding in the more industrialized countries might make the quantities involved available as additional food. However, even if many people are desperately short of food, saving these cereals from being used as livestock feed would by no means automatically ensure that needy people would receive them. Supplies must be financed, generally transported to other countries, and then distributed.

The actual situation is very complex, but the issue is essentially one of sufficient income and demand on the part of the population groups which need to increase their cereal consumption. An "imposed" reduction in the feed use of cereals (perhaps in another part of

Table 8
TOTAL GRAIN PRODUCTION AND FOOD GAPS
(EXCESS OF PRODUCTION TO CONSUMPTION
PROJECTED TO 1990; $\times$ 10^6)[37]

	Population		
	Distribution (%)	Production	Gap
Developed countries (excluding U.S.S.R. and Eastern Europe)	16	640	+ 152
U.S.S.R. and Eastern Europe	8	370	− 61
China	20	290	− 10
Developing countries	8	205	− 24
Developing countries (low income)	48	260	− 47
World	100	1765	+ 10

the world) can do little by itself to solve this basic problem. Cereals provide 59% of all food calories in the developing countries and no less than 83% of the calorie content of their food imports. The "most vulnerable cereal importing countries which contain two thirds of the developing countries' population had a net cereal deficiency of 22 t in 1978 to 1979. Thus it would be facile to conclude that only a small increase in production or modest switch from feed to food use is called for. As is emphasized repeatedly in this study a lasting solution to the problem of chronic undernourishment involves production and distribution and employment/income. However at present and in the future the quantities of additional food intakes required . . . are almost minuscule compared with what the world produces and consumes".

Thus, although in all developing countries the per capita calorie consumption is 2203 kcal (96% of an "average daily requirement" of 2300 kcal), the situation varies from country to country, and within any given country can vary tremendously depending on individual income. In India, for example, 38% of the population in the lowest four income classes (out of nine) have calorie deficits varying from 306 to 1108 kcal/day whereas the national average intake in 1976 was 2217 kcal/person/day. In Kenya where there is only a marginal calorie deficit (as of this writing), 40% of the rural population have a deficit of 640 cal and in urban areas 40% have a deficit of 340 cal/day.

The world's grain production since 1950 and projected trends until 1990 are well summarized by Barr[37] (see Table 8). From 1950 to 1980 food production doubled while the population increased by two thirds, thus, per capita food production worldwide increased at 0.8% compound annual rate. The compound annual rates of increase in per capita food production have, however, been decreasing over the three decades from 1.6 to 0.6 to 0.3%. Developed countries have shown strong upward trends while most developing countries (with the exception of Africa) have experienced improved per capita output. The increase in output both in developed and developing countries came mostly (70 to 90%) from gains in yields rather than from expansions in area. In developing countries population increases have absorbed more than four fifths of the gain in food production, while in developed countries food production has increased twice as fast as the population. A disquieting occurrence has been the fact that in the 1970s food production declined worldwide in three of the years, while in the previous two decades (1950 to 1970) production did not decline in any one year. A trend analysis to 1990 from production and consumption patterns for the period 1960 to 1980 suggests that the "world can indeed produce sufficient grain to permit some limited improvement in per capita consumption". (Table 8 shows a surplus of 10 t in 1990, although it is likely that the surpluses could be even larger if North American and European production trends continue.)

The changing pattern of world grain trade shows the overwhelming reliance on North America as a net exporter in 1983 of 122 t to over 100 countries.[2] The great dependence of the U.S.S.R. on imported grains (39 t in 1983) is a quite recent and important development in the world food question. There is natural concern over this reliance on one area, but the productivity of North American farmers (with all their back-up support) is a reassuring factor in the world food equation. There is concern about the possibility of serious shortfalls in grain production; in the last 20 years the largest deficit in world production occurred in 1972 and resulted in a decrease in world stocks of 42 t. However, during this period the world's grain stocks have never fallen below 140 t which represents about 40 days world consumption.

There is concern as to whether increasing grain yields can be sustained to match and even improve on the world's increase in population (estimated at 6.2 billion by 2000 and the optimistic view is that it will level off at 10 billion). The factors here are soil erosion, loss of agricultural land due to water logging, salinity, urbanization, water shortages, environmental pollution problems, political upheavals, among others. There is cause for optimism and one must not always receive bad news uncritically.

Since the population will be increasing and high productivity land will be lost (an estimated net loss of 55 million ha by 2000; see Table 11 for land use patterns according to Buringh), it is imperative that crop productivity be increased on existing land, that new land is brought into production, and that the loss of good land is curbed. One should consider that higher productivity will account for over 80% of the increased production by the year 2000. Thus, rapid implementation of current knowledge on high-productivity agriculture is necessary if food consumption per capita is to hold its own, let alone increase. This does not imply a massive change in agricultural practice in the developing countries, but it does require that the gains of the ''green revolution'' be consolidated and be more widely applied than at present. The rapid adoption of high-yielding varieties of wheat and rice do indeed show that agriculture in developing countries can be ''modernized''. Production of rice in developing countries has grown by about 2.4%/year since the advent of the HYVs as compared with 0.9% in the early 1950s. Wheat output has grown by about 4% annually in recent years as compared with 2.4% earlier. During the same period, production in developing countries of maize, sorghum, millet, and other coarse grains fell. Overall, it augurs well for the future, particularly when suitable farm management practices are found and applied to medium- and small-scale farmers and to the more marginal regions.

On a global basis Buringh[44,45] has estimated the world's maximum food production capacity and has pointed out the advantages of ''modern'' agriculture as opposed to ''labor-oriented'' (no machinery or chemical fertilizers) agriculture. He calculates that the total potential agricultural land of the world is 3419 million ha (equal to one fourth of the world's total land area) compared to the 1405 million hectares presently cultivated (11% of total); at present two thirds of this land is used for cereal crop production. Buringh makes the staggering claim that 30 times the present grain production could be achieved in the world as compared to present production. He believes that it makes far more sense to increase the productivity of existing cultivated land by using modern techniques instead of opening up new land with low productivity. Tables 9 and 10 show potential agricultural land and the population possibilities.

Only one fifth of the presently cultivated land is used for improved or ''modern'' agriculture, while the remaining four fifths has low productivity levels. Reclaiming new land is seen as poor agricultural practice and also uneconomical. Productive agriculture on ''one-third of the presently cultivated and grazing land can produce enough food for the present population and consequently even more land is available for forest and wildlife''. Of course, there are many problems in introducing ''modern'' agriculture all over the world, e.g.,

Table 9
WORLD LAND USE (Mha) AND PRODUCTIVITY CLASS:
TRENDS TO 2000[13]

	Land use (1975)	Land losses	Land reclamation	Land use (2000)	Net change 1975/2000
Cropland					
High	400	100	45	345	−55
Medium	500	80	325	745	+245
Low	600	40	150	710	+110
	1,500	220	520	1,800	+300
Grassland					
High	200	30	0	170	−30
Medium	300	20	30	320	+20
Low	500	90	40	510	+10
Zero	2,000	0	0	2,000	0
	3,000	140	70	3,000	0
Forest					
High	100	25	45	30	−70
Medium	300	90	180	100	−200
Low	400	75	205	230	−170
Zero	3,300	0	0	3,140	−160
	4,100	190	590	3,500	1,600
Nonagricultural	400	200	0	600	+200
Other land	4,400	100	0	4,500	+100

water, fertilizers, energy, soil erosion, technical expertise, social practices, etc. However, the advantages seem very great, especially since it would free, or leave untouched, large areas which could be reforested for other uses, such as fuelwood provision, maintaining ecological and climatological balances, energy farming operations, etc.

Looking at Europe and the U.S. it should be noted that the land areas devoted to arable (cultivated cropland) agriculture are really rather small — in the U.K. and the U.S. this is only about one fifth of their total land areas. In the U.S. about 60% of this cropland is used for growing animal feed, while in the U.K. 92% of all farm land or 87% of the arable and pasture land is devoted to feeding livestock; in addition the U.K. imports 15 million t of feedstuffs per year. A small change in the modern diet toward eating less animal products, as has been most recently suggested in the U.K. and the U.S. to alleviate heart and cancer problems, would release large amounts of plant material and areas for other uses.[36,39,49] In the U.S. 91% of the vegetable protein and in the U.K. 70% of the primary products of agriculture are fed to animals. In both Europe and the U.S. there are surpluses of animal and plant products which can be difficult to store and economically subsidize. Worldwide 40% of the cereal production and 30% of the total protein production is fed to animals (Table 7).

In conclusion, maintaining and even increasing agricultural productivity in the developed countries should not be an insurmountable problem if past history is any criterion. In fact, the problem in the U.S. and the European Economic Council (EEC) at present is of increasing surpluses which are economically and politically troublesome, both to the countries themselves and to those involved in world food exports. The U.S. agricultural productivity is now considered to be in the ''science power'' phase with an annual growth rate of 1.6% since 1945 (total farm output per unit has doubled); the ''mechanical power phase'' occurred between 1920 and 1945 with a 1.2% annual growth rate.

Table 10
AGRICULTURAL PRODUCTION POTENTIALS (EXPRESSED IN CONSUMABLE GRAIN EQUIVALENTS, × 10⁶ t), SUSTAINABLE POPULATIONS FOR DIFFERENT AGRICULTURAL SYSTEMS, AND AGRICULTURAL LAND AREAS[44,45]

	South America	Australia & Oceania	Africa	Asia	North & Central America	Europe	World
Production							
Present (1975)	50	40	100	450	310	340	1,290
Modern agriculture on all potential agricultural land	2,932	623	2,863	3,770	1,870	1,100	13,156
Modern agriculture on present agricultural land	379	100	636	2,929	712	582	5,338
Labor-oriented agriculture on maximum agricultural land	241	57	236	426	391	256	1,606
Absolute maximum production							
Total	11,106	2,358	10,845	14,281	7,072	4,168	49,830
Per ha	18.0	10.5	14.3	13.2	11.3	10.4	13.4
Population							
Present	230	20	410	2,400	390	750	4,200
Using modern agriculture on present agricultural lands	474	235	795	3,661	890	728	6,673
Using labor-oriented agric. on maximum agricultural land	803	190	787	1,420	1,303	853	5,356
Land							
Land area (10⁶ ha)	1,780	878	3,030	4,390	2,420	1,050	13,548
Presently cultivated (10⁶ ha)	77 (4.4%)	32 (3.9%)	158 (5.2%)	689 (15.6%)	239 (11.3%)	211 (20.2%)	1,406 (10.7%)
Potential agricultural land (10⁶ ha)	596 (33.5%)	199 (23.2%)	711 (23.5%)	887 (20.2%)	628 (25.9%)	399 (37.9%)	3,419 (26.0%)
Maximum agricultural land (10⁶ ha)	383 (21.5%)	99 (11.3%)	478 (15.8%)	610 (13.9%)	526 (21.7%)	367 (35.0%)	2,463 (18.2%)

The world's main thrust should therefore be focused on improving agriculture in the developing countries themselves; as the FAO says "there is no other truly realistic option".[47] Soedjamoko of the United Nations University says, "The only way out is for developing countries to grow more of their own food . . . Additional food will have to be produced through higher yields per hectare — but without high-cost and energy intensive inputs".[59] Carruthers of Wye College (U.K.) has written, "Only by strengthening domestic agriculture will the impact of the food weapon be blunted. Furthermore, it is not only good politics to seek food security but also good economics. However, such a strategy will require many governments to change their development philosophy, their investment and research priorities and new levels and forms of aid to agriculture will need to be agreed". "The political and financial shocks of the 1970s may have some beneficial long-term effects for agriculture. It may be the final lesson to show developing nations that the vision of an industry-led, or export cash crop-led, development policy is illusory. Perhaps urban elites, who generally control development policy in developing countries, will recognise that it is now vitally important for national independence and integrity to commit scarce resources to agricultural investment on an unprecedented, massive scale".[46] Ruttan[57] of the University of Minnesota has stated that "Countries that cannot take advantage of yield-increasing biological and chemical technology will find it increasingly difficult to maintain their export earnings from agriculture or even to meet their domestic food needs. Only a country that establishes its own research capacity in agriculture can gain access to the advances in knowledge . . . ". In addition, a number of observers have recently emphasized that the problems associated with very high military expenditures in pre-empting funds and armed conflicts that disrupt agriculture must not be underestimated.[2]

However, the problems of implementing new agricultural policies and research and development to stimulate agricultural productivity are well known. Not the least problem is the lack of money — of the $150 billion spent annually worldwide on research and development only 3% is spent on agriculture with most of this being used in developed countries. Cost benefit analysis suggests that 2% of the value of agricultural products should be spent on research and development; among developing countries, particularly in Asia, less than 0.5% is currently being allocated.

It is worth considering the World Bank's role in agricultural research and development as it shows interesting changes.[25,56] The Bank's support for agricultural research activities is relatively recent, dating only from the late 1960s and early 1970s. Some gains in terms of increases in production benefits are beginning to emerge from the earliest Bank-supported research projects, although the full impact is likely to be felt a few years hence. The World Bank's experience in the field of agricultural research on the whole has been positive. More emphasis by the Bank on research has resulted in an increase in the flow of local resources for research purposes and has helped make national policymakers and research administrators more aware of the developmental potential from investing in research. Support of agricultural research and extension takes several forms: (1) agricultural and rural development projects that contain adaptive research and extension components, (2) national or statewide adaptive research and extension projects, (3) research components in educational projects, and (4) financial and administrative support of the CGIAR (Consultative Group on International Agricultural Research), which includes the rice research station, IRRI, and the wheat and maize station, CIMMYT, among 11 others with an annual budget of only $140 million. From 1977 to 1979, lending for research and extension, as embodied in the first three categories, constituted almost 9% of total Bank lending for agriculture and rural development. In recent years, about one third of this proportion has been allocated to research alone.

The time frame of a project has emerged as a key consideration in design and implementation. The World Bank's experience suggests that in several projects there were unrealistic expectations in terms of strengthening research institutions and in achieving practical research results. One element contributing to the lower-than-expected results has been the lack of adequate government commitment to agricultural research in the overall development of the rural sector.

Research is successful only if the improved technologies are adopted by farmers. Where adoption is slow, the explanation may well be that price incentives to producers are inadequate. An attractive economic environment for agriculture, and in particular, renumerative prices for producers, must be a part of the effort made to strengthen the national agricultural research program. The potential benefits of some of the research have sometimes not been realized because of the failure of governments to provide a suitable economic environment that can encourage the adoption of a technology. Few developing countries can afford to bypass the opportunity to maximize their agricultural (and forestry) productivity.

There are many factors involved in optimizing plant productivity which are just as applicable to biomass energy systems. Some are yield stability and resistance to stresses, water and nutrient use efficiencies, resistance to diseases and pests, etc. These must be combined with farming systems which provides plant breeding and seed supply systems, adequate extension and credit services, and applied and adaptive research to support the farmer whether farming on a large, medium, or small scale. A decrease in post-harvest losses which can easily be 10 to 20% of the harvested crop are also imperative. Research on growing plants in marginal areas should not be neglected as they are already important environments for large numbers of subsistence farmers and are far more extensive than the fertile areas. The ability of plants to produce worthwhile yields in the areas of water deficiency or excess nutrient deficiency or toxicity, salinity, and high or low temperatures will be increasingly important.

BIOMASS PRODUCTION AND INCREASED PRODUCTIVITY[9,33,35,61-71]

The developed countries constitute about one fourth of the world's population but consume four fifths of the world's commercial energy. The average per capita consumption of energy (including biomass) in the developed countries is about 208 GJ (equal to 6.3 t coal) and about 30 GJ (equal to 1.0 t coal) in developing countries where noncommercial, mostly biomass, energy use is equivalent to 0.4 t coal or 1 t wood per person per year (see Table 5). Reddy[20] has calculated that the equivalent of 0.3 to 0.4 t coal per person per year is the energy requirement for minimum provision of food and habitat for survival in a rural environment. Reddy also estimates that about 1.5 t coal equivalent is necessary to provide the annual minimum requirement for a satisfactory life, including food, habitat, and transportation.

The importance of biomass energy in the developing countries today is indisputable. In many countries, biomass supplies more than one half of the total amount of energy used; in some countries this reliance is as great as 95%. These startlingly high figures result from the fact that the traditional biomass fuels, wood, charcoal, crop residues, and dung, are still the primary cooking and heating fuels for the vast majority of the population of the developing countries, especially the rural and urban poor. The role of biomass energy is in jeopardy, however. In many countries, biomass fuel supplies can no longer meet demand and for a variety of reasons, severe shortages are developing. These problems, sometimes known as the "second energy crisis", are superimposed on those caused by the rising price of oil. Despite their modest consumption levels, many of the developing countries have been hard hit by the oil crisis and are now facing acute balance of payments difficulties as a consequence. Palmedo, in his study of 88 developing countries, showed that over one half of them depend on liquid fuels for at least 90% of their commercial energy; in only four of the oil importing countries did liquid fuels represent less than 50% of the total commercial energy use.[68] As a consequence, many developing countries are obliged to devote over one half their export earnings to provide their liquid fuel requirements which so often represent a relatively small percentage of their total energy use. Because of these dual energy problems, there is a clear need to develop indigenous energy sources in the developing countries, renewable as well as nonrenewable. This is particularly urgent in view of the fact that expanding energy supplies will be necessary simply to maintain present levels of consumption for a growing population, let alone providing for the major increase in energy consumption that will be essential if economic development is to proceed. Among the various options, there is significant scope for the development of biomass energy systems. Present biomass resources can be utilized more efficiently and a variety of approaches, new and old, are available for growing and converting biomass for energy with higher productivities.

For supplying domestic energy needs, there are very few viable alternatives to biomass. For the vast majority of the poorer sectors of the developing countries, locally grown biomass fuels are the predominant source of energy within their financial reach. This situation is not likely to change in the near future since, without heavy subsidies, virtually all nonbiomass alternatives — kerosene, bottled gas, etc. — are much too expensive to have any widespread impact, other than perhaps for towns. Developing low-cost methods of producing and using biomass energy is therefore of the highest priority if the future domestic energy requirements of the poor are to be met.

Because of the heterogeneous nature of biomass energy systems it is difficult to make generalizations about their properties or their potential. One feature that they do have in common, however, is the high degree to which they are interlinked with other factors, environmental, agricultural, socioeconomic, and political. It is vital that these interactions are understood from the outset since they are of crucial importance in determining the limitations on biomass energy systems, as well as their potential impact. Limitations on the use of biomass energy systems, for example, can occur for a number of technical and economic reasons. Thus, one of the most important of these constraints is the competition for land between food and fuel production. In many cases, the introduction of new biomass systems will also be restricted by the lack of financial and other resources, and the absence of the necessary infrastructures, extension services, manufacturing capabilities, and technical know-how.

Assessing the potential of biomass energy in the developing countries and arriving at effective implementation policies is undoubtedly a very complex task. There are many limitations and constraints to take into account and many decisions to be made concerning the choice of systems, scale options, and the implementation approaches to be taken. Furthermore, there are a large number of unknowns involved in biomass energy development. While some biomass systems have been tried and found to be successful under some conditions, others have yet to be tested on a large scale. In almost all cases, detailed local surveys and field testing will be essential since many of the relevant variables — social, economic, and technical — are highly site-specific and because very little information generally exists regarding current patterns of biomass energy use. Adaptation of systems to fit local conditions is also likely to play a crucial part in ensuring the success of biomass energy schemes.

Overall, biomass energy will not be a simple solution to the energy problems of the developing countries. Biomass systems will not necessarily be cheap, nor will they be implemented easily, without a major commitment from governments and a considerable amount of political will at the national and international level.

An examination of the presently available biomass energy and its current use on a regional or a country basis reveals a similar dichotomy to that seen in the case of food production and consumption. In North America and Europe the maximum calculated energy potential from biomass is only one half the present energy use; in Africa and Latin America biomass could provide many times more energy than required; while in Asia an excess is also calculated. However, the Food and Agriculture Organization has calculated that at present about 100 million people are suffering from "acute scarcity" of fuelwood while about 1 billion people have a "deficit".[47] In addition, if present trends continue, these figures may reach 35 million and 2.77 billion, respectively, in the year 2000 — a truly frightening proportion of the estimated population of 6.2 billion.

If individual countries are studied with respect to their energy use and the possible sustainable yields of biomass (or wood only), quite widely varying estimates can be derived and indeed have been published, depending on the assumptions of yields, land use, conversion routes, etc. What these estimates show is that in many developing countries of the world it is theoretically possible for biomass to be a significant and sustainable source of energy, if the appropriate actions could be and were implemented. We realize, however, that this is not often possible under the prevailing circumstances.

Against this background a number of national and international agencies instituted aid programs which stress reforestation, social forestry, agroforestry, and other schemes aimed at increasing the production and supply of fuelwood.

Despite the well-known problems of overcutting for domestic, agricultural, and industrial use in many countries there are some limited success stories of reforestation schemes. For example, China is reported to have increased its proportion of forests from 5% in 1949 to 13% in 1978, an increase of 72 million ha. South Korea in the 1970s gave priority to rural community reforestation and planted 0.6 million ha of village woodlots (this is equal to one half the area planted to rice). The social forestry scheme in Gujarat (India) is known for its successful implementation and may be emulated by other regions in India. The scheme began with plantings along roads, railways, and canals and then spread to village woodlots which were planned in consultation with the villagers in order to provide good maintenance and acceptable returns to the villagers. Other countries such as Indonesia, Philippines, Bangladesh, Thailand, Nepal, Kenya, Malawi, Mali, and Jamaica, among many, are now known to have embarked on fuelwood and other biomass projects.

Of course not all "biomass" schemes are forestry-based and not all of them occur in developing countries. The following lists some examples of biomass programs around the world.

Brazil[72-76] — The largest program is in Brazil, which is currently spending over $1 billion of government monies on subsidizing the production of alcohol made from sugar cane. The reason that so much money is being spent (the new program was established in 1975, although they have been blending alcohol in gasoline since the 1930s) is due to the fact that currently Brazil spends over one half of its total foreign income, about $11 billion on importing oil in order to provide the petroleum to run its transport system. Interestingly enough, it still derives about 25% of its energy from biomass. What Brazil is trying to do is to reverse this great dependence on imported petroleum by the production of alcohol; currently they are producing over 9 billion ℓ; it has been proposed that it be expanded to approximately 11 billion ℓ by 1987, which is 10% of the total forecast oil consumption. At present all the petrol sold in Brazil is blended to 20% with alcohol to run 7 million cars. In addition about 2 million cars now run on hydrated alcohol which is 95% by volume.

This program is not perfect, however. One of the problems which was immediately evident is water pollution (mainly in rivers). For every liter of alcohol produced about 10 to 15 liters of stillage is produced and this has a very high chemical and biological oxygen demand. Stillage has a high protein content and it can be a valuable food when dried. It can also be fermented to methane or be put into lakes to grow water hyacinths and algae to be fed to cattle or even directly fermented.

Competition between biofuels and food production is recognized. Sugar cane production (1981 study) occupies 2.6 million of 40 million ha cultivated with another 88 million ha in planted forests and pastures; this is out of a total land area of 884 million ha. There is ample land available for energy crops in Brazil, however, the regional location and the quality of land dedicated to alcohol crops are important questions. Increased productivity and possibly intercropping are required improvements which are widely recognized.

U.S.[26,27,30,77,78] — The biomass resources of the U.S. are indeed quite large. Currently, Americans derive about 3% of their total energy requirements from biomass, which is equivalent to 1 million barrels of oil per day. The energy content of the standing forests of the U.S. are at least 50% greater than the oil reserves and about equivalent to the gas reserves. In 1980, the Office of Technology Assessment (OTA) of the U.S. Congress estimated that "depending on a variety of factors, including the availability of cropland, improved crop yields, the development of efficient conversion processes, proper resource management, and the level of policy support, bioenergy could supply as few as 4 to 6 Quads/

year, or as many as 12 to 17 Quads/year by 2000 (or up to 15 to 20% of current U.S. energy consumption (about 74 Quads 74 × 10^{18}J). Superimposed on these calculations are the great price support and other programs to help American farmers with surplus food production. In 1983/1984 these amounted to over $40 billion and thus considerably affect the ability of biomass energy schemes to be acceptable and/or profitable.

The U.S. has a gasohol program which blends 10% alcohol, primarily derived from corn (maize), with unleaded gasoline to provide a high octane fuel. Currently about 500 million gallons of ethanol are produced per annum (about $^1/_4$ of total production) at a cost which is competitive with fossil fuel derived ethanol. One of the serious discussions about using biological material for producing alcohol is whether more energy output is obtained than put into the system — new fermentation and distillation systems use only a quarter the amount of energy as pre-1980 systems, and secondly whether food is diverted from the food market and from world trade.

Flaim and Hertzmark[78] have looked carefully into these issues and conclude: "The impacts on domestic food and fiber markets appear relatively modest for production levels of 1,000 million gallons/year. Doubling or tripling this amount may not have serious effects on domestic markets, but export markets will be sensitive to these higher levels of ethanol production". The major impact of ethanol production from grain will be on prices of feed grains. European consumers will probably bear most of the costs of a U.S. ethanol program through higher meat prices. This is because the joint by-products of ethanol production exert a downward pressure on soybean prices, which is a result of gluten meal and distillers' grains being substitutes for soybeans in feed markets. This effect is largely responsible for the switching of acreage from soybeans to corn and thus feed-grain exports will less often be substituted with more expensive gluten and distillers' grains. Using the national agricultural policy model Flaim and Hertzmark[78] show that "The changes in acreage due to ethanol production are approximately linear up to 1000 million gallons of annual production. The most striking feature of these results is the relatively small increment of land required for ethanol production".

There are numerous biomass energy schemes underway or being considered, including the use of agricultural and forestry residues as a source of liquid fuels and/or heat. There is considerable discussion as to the net energy realizable and also the costs and environmental effects of such collection. Costs of using such generally low density material reflect its collection, transportation and processing. However, adverse effects on the soil (erosion, water and nutrient loss, etc.) are the major concern. Nevertheless there are definite opportunities for use of residues especially with changed agricultural practices such as minimal tillage.

Europe[79-81] — Nearly all the countries in Europe have energy from biomass schemes. Why should Europeans look seriously at biomass as a source of energy? Currently the EEC derives over one half its total energy requirements from imported oil. For Europe as a whole some substitution of imported liquid fuels is what makes biomass so appealing. The EEC program examines the use of agricultural resources, the use of forestry resources, the use of algae, the digestion of biological materials to produce methane, and thermochemical routes such as gasification to produce methanol. A recent study indicated that the EEC could produce about 85 million t of oil equivalent, which is nearly 2 million barrels of oil per day, and would provide about 7% of our estimated 1985 energy demand. This is twice what we use for agriculture. The study also showed that if there was a maximum disturbance to agriculture and forestry where a crash program was required, we could possibly achieve a 20% provision of our energy requirement in the EEC. It is highly unlikely that this will occur.

The EEC currently spends about three fourths (equal to $14 million) of its budget on the Common Agricultural Policy and much of this goes to subsidizing animals directly or

Table 11
COSTS OF PRODUCTION OF BIOMASS
FUELS (USEFUL ENERGY ESTIMATED FOR
EEC)[81]

Relative to Oil at 100 Units in 1980

	Relative cost	Actual cost (1982)
1. Heat from dry bio- mass, gasification, etc.	80 —100	
2. Methanol from wood	120 —150	350— 700/t
3. Anaerobic digestion of wastes[a] and catch crops		
Gas	150 —300	
Electricity	More than 300	
4. Ethanol		
Sugar crops	150 —300	700—1000/t
Grains	100[b]—200	

[a] No credits for disposal.
[b] Assuming by-product credits.

Note: Cost in American dollars (1985).

indirectly. To provide the food for these animals we devote a very large percentage of our land area for growing grains, we also import large quantities of grains from the United States and other countries. We produce milk, butter and cheese — it is a problem getting rid of these mountains, so very often they are fed back to animals. Since 90% of the energy in the food is lost every time you go through the animal, it seems to be rather unusual behavior. The EEC is also the world's second largest exporter of sugar. We produce about 4.5 million tonnes of surplus sugar every year. We also are the second biggest meat exporters in the world, have the biggest wine lakes, and have the biggest olive oil lakes in the world. The problem in Europe is not food shortages, it is over-production. Even 2 years ago it was unthinkable to mention the possibility that we could use some of our agricultural land and surpluses to produce something different from them, such as energy at a reasonable cost and at the same time provide regional employment (Table 11).

OIL AND HYDROCARBON PLANTS AND ALGAE[82-89]

It has long been known that all types of vegetable oils can be used in diesel engines. Numerous studies show that in sunny countries, if a maize farmer devoted 10% of his land area to growing sunflowers, he could run all the diesel-powered machines on the farm. Probably a blend of from 10 to 30% is preferable. There is work on the esterification of sunflower oil as methyl or ethyl esters; the esterified oil has fuel properties very close to those of diesel and the esterification can be done on the farm. The Brazilians are examining the extraction of oil from peanuts and carefully ascertaining the use of palm oil (which can have high yields) as a 6% blend into diesel. Oils from soybean, casto seeds, and indigenous plants such as malmeleiro and babassu nut are also being investigated. The main problem to further commercial development is the present low bioproductivity of the plants now available. The recent success of a UK/Malaysian R & D programme in selecting high yielding clones of oil palm (up to 14 t oil/ha/year — over 2 to 3 × current average yields) and their rapid propagation using tissue culture techniques is a dramatic example of what can be done to exploit oil producing (and other) plants once such a programme is initiated and fully supported. There have been a great many proposals recently to use plants directly to produce petroleum. One of the main proponents was Calvin, who advocated growing *Euphorbia lathyrus* for the extraction of hydrocarbons. These have molecular weights close to that of petroleum. There have been a number of trials, mostly in Arizona, to establish whether this is economically viable. Initial hopes of high yields were not substantiated, but the studies showed yields of up to about 10 barrels (1.5 t) of oil per hectare per annum under irrigation. The question is whether such yields are sustainable in arid environments. Dryland studies in Kenya with *E. tirucalli* have shown that quite high yields of biomass can be obtained without fertilization in the first years growth, but is this sustainable over many years? Intercropping with leguminous trees adapted to such environments (fixing nitrogen and producing fodder) may possibly be advantageous.

There is another requirement from oil, and that is for the manufacture of synthetic rubbers. Guayule, *Parthenium argentatum*, which grows naturally in northern Mexico and the southern U.S., can be used as a source of rubber because it has properties similar to those of the rubber tree. Natural stands have been harvested widely in the past but selection is necessary for high yields. There are pilot plants of 1 t/day in Saltillo, Mexico and in Texas, and a 50 t/day demonstration plant has been proposed for Mexico. Research conducted in Australia has shown that guayule is the most commercially promising of all the hydrocarbon plants for their conditions.

The alga *Botrycoccus braunii* has been shown to yield 70% of its extract as a hydrocarbon liquid closely resembling crude oil. This has led to work on controlled growth cultures and on immobilizing these algae in solid matrices such as alginates and polyurethane, thereby allowing the use of a flow-through system to produce hydrocarbons. The green alga *Dunaliella* discovered in the Dead Sea produces glycerol, β-carotene, and protein. This alga does not have a cell wall and grows in very high salt concentrations such as exist in the Dead Sea; thus, to compensate for the high salt externally it produces glycerol internally which can be easily extracted. There are numerous examples of other quite valuable chemicals being produced and/or extracted by algae such as polysaccharides, detergents, and ammonia, all of which may be practically collected in the future.

CONCLUSIONS

The tragedies of hunger, cold, and excessive labor are too commonplace for anyone not to feel compassion when considering the problems of food and fuel shortages for so many of the world's population. There is excessive food *and* energy in the world as seen from the produce "mountains" in many developed countries and from the surplus oil and coal presently available. However, it still is very expensive to import fuel and it can consume a large percentage of a country's external income. This often is not the case for food imports since these usually only provide the top few percent of total consumption. There are, of course, exceptions, however, note that only about one tenth of the world's grain is traded internationally, while well over half of the world's commercial energy is traded abroad.

It is essential that countries with food shortages develop their agriculture to the utmost. As has been said many times this will not occur until agriculture is given its due recognition financially, administratively, and politically. It is easy to provide "shopping lists" of what is required, e.g., research and development institutions, extension services, incentives, etc., but sustained food production on a local basis (which can tolerate weather fluctuations) necessitates commitment to long-term planning. Only recently has there been recognition of the need for this type of commitment nationally and internationally, but it will be some time before the neglect of the last two to three decades can be reversed. An overreliance on the world agricultural commodity markets for export earnings can be disastrous as evidenced by fluctuation in earnings potential and diversion of scarce resources away from local food production; obviously a correct balance between local and export-oriented production is necessary.

There is an optimism that food production can keep ahead of population growth (within limits, of course) since crop productivity can be increased from the favorable base of understanding which we now have in agriculture. What is needed most is a pragmatic application of all this knowledge with a substantial input of funds rather than the meager percentage (usually much less than 10%) of most countries' investment in research and development, infrastructure, etc., which now occurs. There is quite sufficient evidence that such an investment pays off in economic terms if the programs are reasonably well managed. Actually, what is most impressive is how much is achieved with such low levels of investment and infrastructure (both in production and delivery systems). With substantial and well-directed programs crop production could increase dramatically in many parts of the world.

Of greater concern, in the author's opinion, are the more serious consequences of overuse of biomass as a source of energy (but also for other commercial purposes) which is occurring in so many countries. Excessive costs and shortages of fossil fuels have forced many people back to a reliance on biomass as their main source of energy. This is not only a problem for rural people but urban dwellers in many parts of the world use wood, charcoal and dung for cooking and heating; increasing denudation of vegetation for large distances around cities and towns are ready manifestations of this increasingly serious problem. The quality of planning and control required to manage and/or change biomass use is generally lacking where it is most needed. It is also likely to become even more serious as urbanization increases and fossil fuel availability and price become problematical.

There is however now a widespread recognition of the biomass energy/deforestation problem by national and international agencies. For example, foresters and their research efforts are now being encouraged from previously abysmally low levels of incentives and finance by nearly all the national and international agencies. The reawakened realization that forestry can also involve a contribution of trees to agriculture, energy, rural welfare and development, environmental protection, etc., augers well for the future. Nevertheless, there is great reason for optimism, in that the potential yield increases which are possible for biomass energy production (and other products) are very great and are only now being more widely recognized. Techniques applied to agriculture such as advanced breeding, genetics, plant tissue cultures, micropropagation, root mycohrriza and N_2 fixation, physiological control, etc., can be, and are being, applied in the more modern approaches to tree production as forests, energy plantations, agroforestry cropping, etc. These ideas and techniques need to be much more widespread and incorporated with advanced techniques of general forestry management, which must also recognize the sociological dimensions of forestry. The greatest impediment now to the implementation of such techniques is the lack of trained personnel due to the previous near-total neglect of "modern" forestry by planning authorities.

Both food and biomass energy production in all its various forms must be interlinked with the problems of rural development and poverty. A knowledge of such linkages at village and regional levels is very important. However, because the interrelationships of agricultural systems are often difficult to comprehend, the sustained development of adequate food and fuel production must not be neglected as being too difficult to solve singly or together. Again, not novel thoughts, but probably worth repeating if only to encourage further serious work in trying to understand the limiting factors in food (and biomass) production in any given area.

Given the right commitments, the author believes that the evidence shows that food production and distribution can be easily improved so as to provide adequate diets to the world's current and foreseeable population into the beginning of the 21st century. The surplus food producing countries of the developed world could feed the world if political and economic policies were so decreed. However, the developing countries need to improve their own food producing and distribution capabilities. The author believes this is feasible, given the proper incentives. The more serious problem is how to stimulate biomass production and its more efficient use and management. This problem can also be overcome given time, but because of the longer time lags in biomass production compared to agriculture and the very inadequate base from which biomass energy systems are launched, the overuse of biomass is the more serious problem of the two mentioned. However, it is very encouraging to note that in the recent agreement for cooperation between India and the U.S. improvement of food production and biomass production were the top two priorities in the program.

REFERENCES

1. **Arnold, J. E. M.,** Wood energy and rural communities, *Natl. Resour. Forum,* 3, 229, 1979.
2. **Brown, L. R., Ed.,** *State of the World 1984,* Worldwatch Institute, Washington, D.C., 1984.
3. **Burley, J.,** Wood scarcity forces poor to change basic life patterns, *UN Univ. Newsl.,* 6(6), 3, 1982.
4. **Deudney, D. and Flavin, C.,** *Renewable Energy,* Worldwatch Institute, Washington, D.C., 1983, chap. 11.
5. **Earl, D. E.,** *Forest Energy and Economic Development,* Clarendon Press, Oxford, England, 1975.
6. **Eckholm, E.,** *Down to Earth, Environment and Human Needs,* Pluto Press, London, 1982.
7. Food and Agriculture Organization, *Energy Cropping Versus Food Production,* FAO, Rome, 1981.
8. **Foley, G., Barnard, G., and Eckholm, E.,** *Fuelwood: Which Way Out?* Earthscan, London, 1984.
9. **Hall, D. O., Barnard, G. W., and Moss, P. A.,** *Biomass for Energy in the Developing Countries,* Pergamon Press, Oxford, 1982.
10. **Hall, D. O.,** Biomass for energy — fuels now and in the future, *J. R. Soc. Arts,* 130, 457, 1982.
11. **Hall, D. O.,** Food versus fuel: a world problem? in *Economics of Ecosystem Management,* Hall, D. O., Myers, N., and Margaris, N., Eds., Dr. W. Junk Publications, The Hague, 1984, 207.
12. **Hall, D. O. and Coombs, J.,** Biomass production in agroforestry for fuels and food, in *Plant Research and Agroforestry,* Huxley, P. A., Ed., International Council on Research in Agroforestry, Nairobi, Kenya, 1983, chap. 12.
13. **Holdgate, M. W., Kassas, M., and White, G. F., Eds.,** *The World Environment 1972—1982. A Report by the United Nations Environment Programme,* Natural Resources and the Environment Series, Vol. 8, Tycooly International Publications, Dublin, 1982.
14. **Lanly, J. P.,** Assessment of the forest resources of the tropics, *For. Prod. Abstr.,* 6, 137, 1983.
15. **Lewis, C.,** *Biological Fuels,* Edward Arnold, London, 1983.
16. National Academy of Sciences, *Food, Fuel and Fertilizer from Organic Wastes,* NAS, Washington, D.C., 1981.
17. **Overend, R. P.,** Biomass energy, current status and promise, in *Energy Options 4th Int. Conf.,* Institute of Electrical Engineers, London, 1984, 62.
18. **Pimentel, D. and Pimentel, M.,** *Food, Energy and Society,* Edward Arnold, London, 1979.
19. **Reddy, A. K. N.,** An Indian village and agricultural ecosystem — case study of Ungra village, *Biomass,* 1, 77, 1981; *Biomass,* 2, 255, 1982.
20. **Reddy, A. K. N.,** Alternative energy policies for developing countries: a case study of India, in *World Energy and Productivity,* Bohm, R. A., Clinard, L. A., and English, M. R., Eds., Ballinger Press, Cambridge, Mass., 1981, 289.
21. **Sen, A.,** *Poverty and Famines: An Essay on Entitlement and Deprivation,* Clarendon Press, Oxford, 1982.
22. **Smil, V. and Knowland, W., Eds.,** *Energy in the Developing World,* Oxford University Press, London, 1979.
23. **Tiwari, K. M.,** Fuelwood — present and future with special reference to developing countries, in *Energy from Biomass,* Vol. 2, Strub, A., Chartier, P., and Schleser, G., Eds., Applied Science, London, 1983, 682.
24. World Bank, Forestry Sector Policy Paper, World Bank, Washington, D.C., 1978.
25. World Bank, Agricultural Research: Sector Policy Paper, World Bank, Washington, D.C., 1981.
26. Bio-Energy Council, International Bio-Energy Directory and Handbook, The Bio-Energy Council, Washington, D.C., 1984.

27. **Cote, W. A., Ed.,** *Biomass Utilization,* Plenum Press, New York, 1983.
28. **Coombs, J., Hall, D. O., and Chartier, P.,** *Plants as Solar Collectors,* D. Reidel, Dordrecht, West Germany, 1983.
29. **Hall, D. O.,** Photosynthesis for energy, in *Proc. 6th Int. Congr. Photosynthesis,* Vol. 2, Sybesma, C., Ed., Martinus Nijhoff, The Hague, 1984, 727.
30. Office of Technology Assessment, *Energy from Biological Processes,* OTA, U.S. Government Printing Office, Washington, D.C., 1980.
31. **Sheppard, W. J. and Young, B.,** Biomass — the agricultural perspective, in *Biomass Utilization,* Cote, W. A., Ed., Plenum Press, New York, 1983, 117.
32. **Smil, V.,** *Biomass Energies,* Plenum Press, New York, 1983.
33. **Zaborsky, O. R., Ed.,** *Handbook of Biosolar Resources,* Vols. 1 and 2, CRC Press, Boca Raton, Fla., 1982.
34. UK-ISES, *Solar Energy: a UK Assessment,* U.K. International Solar Energy Society, London, 1976, chap. 9.
35. **Barnard, G.,** Liquid fuel production from biomass in the developing countries, in *Bioconversion Systems,* Wise, D. L., Ed., CRC Press, Boca Raton, Fla., 1983, chap. 5 to 8.
36. **Beardsley, T.,** UK nutrition, *Nature (London),* 304, 103, 1983.
37. **Barr, T. N.,** The world food situation and global grain prospects, *Science,* 214, 1087, 1981.
38. **Berg, A.,** *Malnourished People: A Policy Review,* Poverty and Basic Needs Series, World Bank, Washington, D.C., 1981.
39. **Blaxter, K. and Fowden, L., Eds.,** *Food, Chains and Human Nutrition,* Applied Science, London, 1982.
40. **Blaxter, K. and Fowden, L., Eds.,** *Food, Nutrition and Climate,* Applied Science, London, 1982.
41. **Boyer, J. S.,** Plant productivity and environment, *Science,* 218, 443, 1982.
42. **Brady, N. C.,** Chemistry and world food supplies, *Science,* 218, 847, 1982.
43. **Bunting, A. H.,** Changing perspectives in agriculture in developing countries, an agronomist's view, *Agric. Econ.,* 32, 287, 1981.
44. **Buringh, P., van Heemst, H. D. J., and Staring, G. J.,** *Computation of the Absolute Maximum Food Production of the World,* Agricultural University, Wageningen, The Netherlands, 1975.
45. **Buringh, P. and van Heemst, H. D. J.,** *An Estimation of World Food Production Based on Labour-Oriented Agriculture,* Centre for World Food Market Research, Wageningen, The Netherlands, 1977.
46. **Carruthers, I.,** Where should the world's food be grown? in *Food, Nutrition and Climate,* Blaxter, K. L. and Fowden, L., Eds., Applied Science, London, 1982, 392.
47. Food and Agriculture Organization, *Agriculture Toward 2000,* FAO, Rome, 1982.
48. **Hulse, J. H.,** Food science and nutrition. The gulf between rich and poor, *Science,* 216, 1291, 1982.
49. **Maugh, T. M.,** Cancer is not inevitable — diet, nutrition and cancer, *Science,* 217, 36, 1982.
50. **Miller, D. S.,** Man's demand for energy, in *Food Chains and Human Nutrition,* Blaxter, K. L., Ed., Applied Science, London, 1980, 23.
51. National Research Council, *World Food and Nutrition Study, The Potential Contributions of Research,* National Academy of Sciences, Washington, D.C., 1977.
52. National Research Council, *Priorities in Biotechnology Research for International Development,* National Academy of Sciences, Washington, D.C., 1982.
53. **Nugent, J. and O'Connor, M., Eds.,** *Better Crops for Food,* Pitman, London, 1983.
54. **Parikh, K. and Rabar, F., Eds.,** *Food for All in a Sustainable World: the Int. Inst. Appl. Sys. Anal., Food and Agriculture Program,* I.I.A.S.A., Laxenburg, Austria, 1981.
55. **Pinstrup-Andersen, P.,** *Agricultural Research and Technology in Economic Development,* Longmans, London, 1982.
56. **Plunkett, D. L. and Smith, N. J. H.,** Agricultural research and third world food production, *Science,* 217, 215, 1982.
57. **Ruttan, V. W.,** The global agricultural support system, *Science,* 222, 11, 1983.
58. **Swaminathan, M. S.,** Biotechnology research and third world agriculture, *Science,* 218, 967, 1982.
59. **Anon.,** The interlinked futures of food and energy, *UN Univ. Newslett.,* 5(2), 1, 1981.
60. **Wortman, S.,** World food and nutrition, the scientific and technological base, *Science,* 209, 157, 1980.
61. **Barnard, G. W. and Hall, D. O.,** Energy from renewable resources: ethanol fermentation and anaerobic digestion, in *Biotechnology,* Vol. 3, Dellweg, H., Ed., Verlag Chemie, Weinheim, West Germany, 1982, chap. 4.
62. **Hall, D. O., Coombs, J., and Higgins, I. J.,** Energy and biotechnology, in *Biotechnology — Principles and Applications,* Higgins, I. J., Ed., Blackwell Scientific, Oxford, 1985, chap. 7.
63. **Dunkerley, J., Ramsay, W., Gordon, L., and Cecelski, E.,** *Energy Strategies for Developing Countries,* John Hopkins University Press, Baltimore, 1981.
64. *New Trends in Research and Utilization of Solar Energy Through Biological Systems, Experientia,* Spec. Issue, 38, 1—66, 145—228, 1982.

65. **Huxley, P. A., Ed.,** *Plant Research and Agroforestry,* International Council on Research in Agroforestry, Nairobi, Kenya, 1983.
66. **Moss, R. P. and Morgan, W. B.,** *Fuelwood and Rural Energy Production and Supply in the Humid Tropics,* Natural Resources and Environment Series, Vol. 4, UN University Press, Tokyo, 1981.
67. **Nair, P. K. R.,** *Agroforestry Species — A Crop Sheets Manual,* International Council on Research in Agroforestry, Nairobi, Kenya, 1980.
68. **Beardsworth, E. and Hale, S.,** *Energy Needs, Uses and Resources in Developing Countries,* Brookhaven National Laboratory, Upton, New York, 1978.
69. **Rabson, R. and Roberts, P.,** The role of fundamental biological research in developing future biomass technologies, *Biomass,* 1, 17, 1981.
70. **Spears, J. S.,** *Overcoming Constraints to Increased Investment in Forestry,* World Bank, Washington, D.C., 1980.
71. World Bank/Food and Agriculture Organization. *Forestry Research Needs in Developing Countries — Time for a Reappraisal,* World Bank, Washington, D.C., 1981.
72. **Goldemberg, J.,** Energy problems in Latin America, *Science,* 223, 1357, 1984.
73. **Hall, D. O.,** Home-grown fuel: rise of the gasohol challenge, *South Mag.,* p. 77, December 1983.
74. **Morgan, W. B., Ed.,** Renewable energy and systems thinking: the Brazilian case, *Resource Manage. Optimization,* 3(1), 1, 1983.
75. **Saint, W. S.,** Farming for energy: social options under Brazil's national alcohol programme, *World Dev.,* 10, 223, 1982.
76. **Trinidade, S. G.,** Energy crops — the case of Brazil, in *Energy from Biomass,* Vol. 1, Palz, W., Chartier, P., and Hall, D. O., Eds., Applied Science, London, 1981, 59.
77. Biomass Panel of ERAB, Biomass energy (in U.S.A.), *Solar Energy,* 30, 1, 1983.
78. **Flaim, S. and Hertzmark, D.,** Agricultural policies and biomass fuels, *Ann. Rev. Energy,* 6, 89, 1981.
79. **Hall, D. O. and Scurlock, J. M. O.,** Biomass for energy in Europe, in *5th Canadian Bioenergy R and D Seminar,* Hasnain, S., Ed., Applied Science, London, 1985, 56.
80. **Pankhurst, E.,** The prospects for biogas — a European point of view, *Biomass,* 3, 1, 1983.
81. **Strub, A. S., Chartier, P., and Schleser, G.,** *Energy from Biomass,* Vol. 2, Applied Science, London, 1983.
82. **Foster, K. E. and Karpiscak, M. M.,** Arid lands plants for fuel, *Biomass,* 3, 269, 1983.
83. **Gudin, C. and Chaumont, D.,** Solar biotechnology study, in *Energy from Biomass,* Ser. E, Vol. 5, Palz, W. and Pirrwitz, D., Eds., D. Reidel, Dordrecht, West Germany, 1984, 184.
84. **Hall, D. O.,** Renewable resources (hydrocarbons), *Outlook Agric.,* 10, 246, 1980.
85. **Johnson, J. and Hinman, H. E.,** Oil and rubber from arid lands, *Science,* 208, 460, 1980.
86. **Lipinsky, E. S.,** Chemicals from biomass: petrochemical substitution options, *Science,* 212, 1665, 1981.
87. **McLaughlin, S. P. and Hoffman, J. J.,** Survey of biocrude-producing plants from the South West, *Econ. Bot.,* 36, 323, 1982.
88. **Orion, G. and Bodo, L. B.,** Oxygenate fuels for diesel engines: a survey of worldwide activities, *Biomass,* 3, 77, 1983.
89. **Raymond, L. P.,** Aquatic biomass as a source of fuels and chemicals, in *Energy, Resources and Environment,* Yuan, S. W., Ed., Pergamon Press, Oxford, 1982, 75.

AN HISTORICAL OUTLINE OF APPLIED ALGOLOGY

C. J. Soeder

INTRODUCTION

A brief account of the historical development of applied algology is given, providing references to basic algological literature. A shorter section on seaweeds and seaweed farming is followed by a more detailed description of the onset of microalgal uses for nutritional purposes, sanitary engineering schemes, production of chemicals, and energy farming.

Seaweeds and Microalgae

A person living near a rocky shoreline of the Atlantic Ocean would probably refer the keyword "algae"[1,2] to seaweeds and at low tide he might be able to distinguish between red, brown, and green algae by color and shape. He might perhaps be surprised to learn that the open sea is inhabited by numerous species of planktonic algae, mostly of so minute a size that a microscope is needed to watch and distinguish them. Although aware of the dangers of a red tide, he might not know that this phenomenon is due to poisonous dino-flagellates like *Gonyaulax catenata*.[3] Would he be enough of a naturalist to appreciate that other dinoflagellates are causing the phosphorescence of the sea?

Now consider the proverbial landlubber. He would know of algae as "the green stuff in our swimming pool", and try a new product every year to rid the pool of this aquatic vegetation. Would he not look doubtful, if told that the macroscopic seaweeds are algae, too?

Indeed, the heterogeneity of the algae as the most diverse group of lower plants is tremendous. If we take giant kelp (*Macrocystis pyrifera*), measuring some 50 m in length, as a typical macroalga and the unicellular *Chlorella vulgaris* of 3 μm in diameter as a paragon of a microalga, it is hard to believe that systematic botany puts both of these plants into the same category: "ALGAE".

The Latin language designates seaweeds in general by the term "alga". Obviously unaware of the evolution of the flowering plants from green-algal ancestors as postulated currently, a Roman poet may have disliked the smell of seaweeds decaying on a beach when writing "nihil vilior alga" ("there is nothing worse than algae", Horatius, 30 B.C.). Today, about 30,000 algal species are known, and their classification has changed significantly in recent years due to the recognition of chemotaxonomic and submicroscopic characters previously unknown.[1] A matter of unresolved debate is the position of the blue-green algae,[4] which are structurally so similar to bacteria that microbiologists called them *Cyanobacteria*, whereas botanists maintain that the blue-greens are too distinct a group and should be called *Cyanophyta*.

Global net primary production amounts to about 2×10^{11} t of organic matter per annum, produced by photosynthetic fixation of 8×10^{10} tons of carbon.[5] Of these enormous quantities some 40% can be attributed to algal photosynthesis, seen predominantly in the oceans.[6] (For basic information on algal ecology, see Round[7] and Wetzel;[8] for algal physiology and biochemistry, see Stewart[9] and Lobban and Wynne.[10]

With emphasis on microalgae, this chapter describes major trends in the historical development of applied algology. Elaborate reviews on the same topic have already been published by several authors.[11-15]

There is a good number of white patches on the present world map of applied algology. Language barriers and a certain lack of actual publications cause the Western observer of the scene to be unaware of recent progress in the U.S.S.R. and its satellite countries. Since the 1960s research by Japanese companies on the mass production of *Chlorella* and *Spirulina*

Table 1
ANNUAL AMOUNTS OF SEAWEEDS HARVESTED IN
JAPAN FOR HUMAN CONSUMPTION

Seaweed	Japanese name	Annual crop × 1000 t on freshweight basis	
		From nature	From seaweed farms
Laminaria japonica	Kombu	159	22
Undaria pinnatifida	Wakame	19	127
Porphyra species	Nori	—	291
Other seaweeds		36	

Note: Data compiled from Chapman and Chapman.[19] Present production of *Porphyra* alone represents an annual income of about 10[6] U.S. dollars for the seaweed-farming enterprises.

remained mainly undisclosed, except for a few short communications[16] and patents. There is also no survey of microalgal processes in the People's Republic of China. Since the more recent achievements and projects in the utilization of microalgae are dealt with more competently and in sufficient detail by other authors of this volume, I shall attempt to consider here the earlier stages of the subject. However, no attempt is made to speculate which of the potential algal uses[17] will become practically relevant, because the selection by which economic realities emerge from technical possibilities goes too often its own way, and depends on market acceptance and readiness for venture at the most opportune time.

Uses of Seaweeds

Sea Vegetables

Since prehistoric times coastal human populations must have collected and eaten a variety of seaweeds. What is known today about previous and still existent uses of algae as sea vegetables has been compiled by Levring et al.[18] and by Chapman and Chapman;[19] even a sea vegetable cookbook is available.[20] For example, a green sea lettuce (*Ulva lactuca*) was consumed in Brittany and elsewhere and esteemed for curing scurvy. The red alga, *Chondrus crispus* (Irish moss), is still commercially available as food in Ireland and used in folk medicine to treat bronchitis and diarrhea, and it is used as a gland ointment. Along the coasts of Latin America it is not uncommon to find red algae such as *Gracilaria* as "cushuro" on the menu. This Indio term for algae in general is also used for edible freshwater algae.[21] Only in Japan, however, has human consumption of seaweeds developed into a real cult. "Seaweed fishing" has been described in the literature since 274 A.D.[19] The consumption of sea vegetables and of the various sea vegetable products in Japan is still impressive, not only in terms of commercial values but also in quantities involved (Table 1). Approximately 20 kinds of sea vegetables are normal items on the Japanese shopping list. They cover, on the average, 10% of Japanese food requirements.

Seaweed Farming

When exploitation of natural stands of seaweeds was no longer sufficient to meet the steadily increasing demand of the Japanese for sea vegetables, earlier techniques developed in the Imperial Algae Gardens were transformed into regular seaweed farming in Tokyo Bay around 1624. This concerned first the laver (*Porphyra laciniata* and other *Porphyra* species) which was originally propagated by cuttings and cultivated on horizontal bamboo rods and strings. This technique has been improved recently after elucidation of the complicated life cycle of laver. Today Japanese laver farms cover more than 50 km² of shallow bay areas and engage over 70,000 workers. Laver from natural stands is still eaten in South Wales and elsewhere as well.[19]

During the last three decades, commercial farming of marine brown algae (*Undaria* since 1962, *Laminaria* since about 1965) has become important in Japan and also serves the increasing demand of the seaweed industry for raw materials from which alginic acid and other phycocolloids can be extracted. After the initiation of kelp farming in 1952, southern China produced 275,000 t dry matter of *Laminaria* on a culture surface of 15 km^2 in 1979.[22]

In the 1960s Japanese technology for farming marine red algae was adopted for mass cultivation of *Eucheuma* in the Philippines. Doty[23] impressively described how this innovation was introduced by giving local fishermen the opportunity to actually steal methods and starter algae, thereby converting themselves into seaweed farmers. In the meantime the Philippinian *Eucheuma* harvests have reached 16,500 t in 1979[24] and are used mainly for the extraction of carrageenans. *Eucheuma* is also being produced in China, in addition to *Gracilaria*.[22]

Surprisingly enough, seaweed farming has been established in only a few areas, although the actual potential is definitely larger. Recent attempts to produce *Chondrus* at the industrial scale in Canada have not yet led to commercial success despite conspicuous research inputs by top algologists. The reasons behind this apparently deadlocked situation have been analyzed and interpreted by Gellenbeck and Chapman[25] and are common to many other algal biotechnologies: because of the limited scale of accomplished testing programs, cost estimates and operations research are still so unreliable that industry hesitates to finance testing and development programs of the caliber required for an actual breakthrough.

A giant project for offshore farming of giant kelp has also been designed. Supporting structures resembling spider webs of 2 mi in diameter are envisaged to carry the growing *Macrocystis* plants to be harvested at preset intervals from special crafts. The prospects of this ambitious project are described by Wilcox[26] as follows: "The Ocean Energy and Food Farm Project promises by 1985—1990 to be able to demonstrate 100,000 acre ocean farm yielding some 16 million Btu per year of food plus some 160 million Btu of methane per year for each acre of cultivated ocean. On this basis, the total food and natural gas energy presently being consumed by the USA each year could be produced from a square of ocean approximately 470 mi on each side." Moreover, this project is expected to convert *Macrocystis* biomass into synthetic liquid fuels, lubricants, plastics, etc.

Seaweed Industry

Among the earliest historical records of the development of seaweed industry[19] are 17th century regulations of the French Government which set a number of rules for the harvesting and processing of seaweeds. At that time dried algae were incinerated in kelp kilns on the coasts of Normandy and Brittany in order to obtain a mixture of soda and potash for glass manufacture. From 1720 onward, kelp kilns were also worked in Great Britain and by the end of the 18th century, Scotland alone produced more than 20,000 tons of alkaline carbonates from about 400,000 tons of fresh seaweeds per annum.

After the invention of a less expensive way to produce soda by Leblanc in 1791, the objective of the European kelp industry was switched to producing iodine after its discovery by Courtois in 1811. Industrial extraction of iodine from seaweeds in Japan developed much later and peaked around 1920 with an annual output of some 250 tons. Today, iodine is still a valuable by-product of processes employed in the recovery of phycocolloids from various seaweeds.

The present basis of seaweed industries flourishing in several countries is the production of phycocolloids. These unique acid polysaccharides serve as gelling agents (especially agar, carrageenan, etc.), while alginates are mainly applied as emulsifiers, thickeners and stabilizers in foodstuffs, and in the manufacture or modification of pharmaceutical and chemical products. The uses of phycocolloids as water-holding agents in paper industry and textile printing are also important. For details and for information on algal "drugs from the sea", see Levring et al.,[18] Hoppe et al.,[27] Hoppe and Levring,[28] and Chapman and Chapman.[19]

Although phycocolloids may be partially replaceable by bacterial gums (xanthans) they appear to be indispensable for many purposes. Current prices for clean phycocolloids range between $30 and $880/kg.[25]

Uses of Microalgae
Lake Plums, Pond Scums, and Associates
If we project our imagination from present knowledge back to prehistoric times, we may perceive tribes of hunters collecting flat gelatinous masses of the blue-green alga, *Nostoc commune*, from soil, then eating raw or cooked what the English would call ages later "star jelly" or "witches' butter". To enrich their diets, early hunters may also have consumed filamentous freshwater algae occurring in larger masses and collected the scums of *Spirulina platensis* from the shores of warm alkaline lakes.

When searching (generally in vain) algological literature for records of the uses of microalgae, one is left with the impression that this topic has been taboo for quite some time. Nevertheless, Johnston[29] was able to excavate an early report which he summarizes as follows: "Dr. Hooker, in a paper read before the Linnean Society of London, January 20, 1852, mentioned that *Nostoc edule* (= *N. pruniforme*, the lake plum) was found abundantly in streams in Tartary. It was highly esteemed as an ingredient in soups. This form of *Nostoc* was well known and eaten in Mongolia and China in which countries it was used extensively as an article of commerce, generally sold in dried form." Johnston[29] also points out the actually widespread consumption of species of *Nostoc*, a habit also observed in the Andes Mountains of Peru.[21] One of the most exciting discoveries of the Belgian Sahara Expedition of 1964/1965 was the finding that blue-green *Spirulina platensis*, forming dense blooms in brackish warm waters, is collected and regularly eaten by natives around Lake Tchad. In his report on microalgae that are or have been eaten by man, Johnston[29] not only refers to the dietary significance of *Spirulina* in central Africa, but also to its former use in Aztec Mexico.[30]

According to Johnston,[29] species of filamentous green algae, *Oedogonium* and *Spirogyra*, are collected, sold, and eaten in Burma, Thailand, Vietnam, and India. The same holds for the flat thalluses of *Prasiola yunnanica* (resembling thin salad leaves) in China and *P. japonica* in Japan, both of which have an appreciably high protein content. The list of these freshwater algae which have been mostly unrecognized by botanists as being edible and eaten, is probably incomplete. It is, therefore, perhaps a legitimate question to ask how unconventional a group of protein sources microalgae are, among the so-called novel sources of protein. Perhaps the real unconventionality is their production at the technical scale.

Cultivation of Microalgae: From Test Tube to Acres
When a tiny spherical green alga was given the name *Chlorella vulgaris* by the Dutch bacteriologist Beijerinck,[31] he simultaneously introduced a technique of cultivating microalgae in test tubes and petri dishes on solid medium and provided the first information on growth physiology. In the years thereafter growth of *Chlorella* and *Scenedesmus* was studied by an increasing number of botanists with respect to mineral nutrient demand, organotrophic growth, etc. A second hallmark was the introduction of cultivated *Chlorella* by Warburg[32] as a key to elucidating algal photosynthesis and respiration. He began to cultivate representatives of other algal groups, as well; Pringsheim[33] considerably improved the techniques of monospecific or bacteria-free cultivation of microalgae.

The idea of producing microalgae at the technical scale first occurred to German scientists who thought about the means to abating the acute shortage of food supply during World

War II. The primary idea was to produce lipids from nitrogen-starved diatoms which were shown to accumulate appreciable amounts of fat under laboratory conditions.[34] Since the productivity of nutrient-deficient algae is rather low, technological interest soon after switched to the production of proteinaceous microalgal matter, and thus scientists began to intensively investigate the potential of *Chlorella* which was known to double its biomass a few times per day in highly illuminated laboratory cultures. If such a photosynthetic organism with a crude protein content of 50 to 60 or even 88%, as found by Spoehr and Milner[35] could be cultivated at the industrial scale, would this not lead to a revolution in plant production?

This question began to be studied from 1947 onward in several countries. It was learned quickly that artificial illumination by lamps is by far too costly for true mass production of algae. Algae has, therefore, to rely on natural sunlight. As predictable from laboratory experiments, microalgae grow in outdoor mass cultures at the expense of light energy, mineral nutrients, carbon dioxide, and mechanical energy. Provided that nutrient supply is adequate, productivity depends primarily on irradiance and temperature. A conversion of these basic facts into reliable industrial processes led stepwise to a technological development which has still not come to an end. Investigators had to learn how to prevent the invasion of cultures by predators to which suspensions of *Chlorella* are quite susceptible. As in other types of monocultures, phycopathogenic viruses, bacteria, fungi, and amoebae turned up and had to be controlled. As may be noted already at this point, *Coelastrum sphaericum* and *Spirulina* proved to be particularly resistant in that respect. Economically feasible ways had to be found to solve the problems encountered in reactor construction, agitation of culture liquid, carbon supply, and harvesting. This was altogether more difficult than originally envisaged.

The first systematic efforts for translating laboratory methods into engineering specifications for a large-scale culture plant were conducted in Stanford (California) at the Department of Plant Biology of the Carnegie Institution of Washington from 1948 to 1950.[36] At that time the Institute was headed by C. S. French, one of the leading photosynthesis researchers under whom the first fully satisfying apparatus for continuous culture of *Chlorella* was developed.[37] At Berkeley (not far from Stanford), Calvin and colleagues had just adopted ^{14}Carbon tracer techniques for elucidating the path of carbon in photosynthesis,[38] experimenting also with *Chlorella*. It was apparently not accidental that W. J. Oswald, a young sanitary engineer at the University of California at Berkeley, became deeply interested in the role of microalgal photosynthesis in algal ponds and developed the high-rate algal pond for "photosynthetic wastewater treatment".[39]

A research and development program for large-scale production of green microalgae in ponds was started under Gummert in 1949 at Essen (Germany), following Kraut's suggestion that investigatory effort should be focused on production of *Scenedesmus obliquus*.[40]

In 1951 the first *Chlorella* pilot plant was constructed and operated for the Carnegie Institution by Arthur D. Little, Inc. at Cambridge, Mass. Another series of laboratory and pilot plant studies followed in Japan under the guidance of Tamiya after his stay at Stanford. His strong working group at the famous Tokugawa Institute in Tokyo concentrated again on *Chlorella* research, as was also the case for a study conducted in Israel. The results of this first boom of applied research on microalgae were published in the classic and still fascinating report edited by Burlew.[41] This book (*Algal Culture*) contains an interesting article on the use of a "natural" algal suspension consisting mainly of *Chlorella* for the nutrition and apparently successful treatment of leprosy patients by Jorgensen and Convit.[42]

From the early studies it was concluded by Tamiya[11] that the production of 1 ton of dry *Chlorella* meal would cost about $520 (U.S.) and could, therefore, not compete with inexpensive proteinaceous plant materials such as soybean meal. This conclusion is still valid as far as the production of pure microalgal meal is concerned.[43] Nonetheless, since the 1960s, *Chlorella* production has become a commercial success in Japan and Taiwan, because *Chlorella* products were favorably accepted as novel health food items in the Far East.

At Stanford, the Department of Plant Biology of the Carnegie Institution of Washington had been from 1948 to 1950 the very cradle of applied research on microalgae, before returning to exciting basic research. This primary spark ignited activities in so many other laboratories that it is virtually impossible to give full credit to all of them. Before elaborating on selected examples of further development, some types of culture units shall be described in general terms.

Development of Production Units

Among production units for microalgae, one has to distinguish between the industrial approaches and simplified devices meant for rural application in less developed countries. For example, a simple ditch in India, in which a dense blue-green bloom of *Spirulina* thrives on the residues of fermented cow dung, stirred occasionally by means of a broom, and harvested by cloth filtration to be dried with solar energy[44,45] may be looked upon as an almost natural branch of tropical agriculture. How to convert a cylindrical metal container by a few cuts and with the aid of some bamboo rods into a little windmill stirring an algal pond has been described in a brochure by Seshadri et al. In southern Asia there is no distinct borderline between conventional pond management and microalgal biotechnology, and some of the achievements made in that area are applied to such an extent that they are not described anywhere in scientific or technical publications. This might also hold for the cultivation of *Chlorella* on hog manure in the People's Republic of China. Another example known from occasional newspaper reports rather than from scientific articles, concerns the "*Chlorella* eaters" of Auroville in India.[46]

Returning to the industrial approach, the first *Chlorella* pilot plant of Arthur D. Little, Inc. consisted of flat thin-walled plastic tubings (1.2-m wide, 20-m long) through which the algal suspension (volume up to 4.5 m^3) was circulated. As described in Burlew[41] this arrangement served to keep the cultures pure and to lose no CO_2, which was recognized from the beginning as an important cost factor. The concept of growing algae in closed ponds was studied again quite recently in order to minimize water loss in arid climates.[47]

Coiled tubular reactors with a smaller diameter for plug-flow operation may be used for the production of pure or axenic microalgal biomass, but is used mainly for biochemical research purposes.[48] The applicability of similar systems under field conditions has been discussed by Setlik et al.[49] Small culture units such as bubbled columns[34] or lollipop flasks[50] (commonly employed for bench scale work) are generally important for providing inoculum for pure cultures of microalgae in ponds.[16,36,51]

When cultivating *Scenedesmus* outdoors in long open troughs, Meffert and Stratmann[52] discovered that their cultures of *Scenedesmus obliquus* remained essentially free from alien algae under completely unsterile conditions for extended periods of time. The degree of purity of the cultures was usually >98%, and commensalic bacteria comprised <1% of total biomass. Since the startling stability of microalgal monocultures appears to require a comparatively high areal density (biomass per square meter), the phenomenon can be understood as being due to pronounced light limitation leading to doubling times of <0.5/day.[53] The chance which potentially fast-growing algae would find for competing successfully with the established population is as small as speeding a race car in a traffic jam.

Since experience with the closed system of the first American pilot plant was not encouraging, almost all of the mass culture systems developed later were open to the atmosphere. After the experiments with the first pilot plant of the Tokugawa Institute at Tokyo[50] and having worked for some years with "bubbling cultures" in rectangular troughs,[54] the "*Chlorella* factories", established first in Japan and later in Taiwan (between 1960 and 1970), constructed circular cultivation ponds of up to 45 m in diameter, some of them even covered with glass domes.[51] The ponds are operated according to the "open circulation method''[55] or mixed by a stirrer resembling a spokeswheel;[51] however, circular ponds have the following disadvantages: expensive construction in heavily reinforced concrete, high energy consumption for stirring, great difficulty in overcoming the lack of turbulence in the center, and inefficient land use. Among the horizontal culture systems, the raceway type[51,56] is preferred, either in its simple version or as a meandering system. Raceway ponds can be built from prefabricated concrete parts and/or large sheets of plastic;[17] one can even do without solid lining.[57] Generation of turbulent flow in closed horizontal channels is accomplished mostly by paddlewheels, but propellers, immersed pumps, or an airlift device[58] have also been used. A very simple stirring device is the mixing board introduced recently by Wagener for rectangular ponds.[56]

Microalgae can also be cultivated in shallow layers on inclined surfaces. The first set of sloping culture units was designed and operated by Setlik et al.[49] in Czechoslovakia. Their cascade system of which the culture unit of Mituya et al.[50] was a forerunner, consists of a sloping plane whereon high turbulence of the downflowing algal suspension is created by small baffles. The suspension is collected at the lower rim of the slope, flows into a sink, and is pumped back to the origin. Pumping is restricted to daytime, the suspension being stored overnight in a reservoir. Following the same principle, less expensive versions were constructed with Czechoslovakian aid in southern Bulgaria,[59] where CO_2 of volcanic origin was amply available. Specific difficulties encountered in the operation of the Czechoslovakian sloping system consisted in uneven CO_2 supply and cleaning problems. A very promising type of a sloping culture unit was designed and tested by Heussler et al.[60] in Peru. Meandering channels are built on a slope perpendicular to the length of the channels. A similar design was proposed by Berend et al.[61] The most recent unit was developed by A.E.I.C., Inc. and tested near Johannesburg.[62] Their pilot plant has the special feature of sequencing ponds with stepwise decrease in water depth and the gradual increase of suspension density. This leads to comparatively high final biomass concentrations and reduced energy expenses for harvesting.

As a device for mass cultivation of attached microalgae (including filamentous species) Salageanu[63] used a rolling plastic-film system. The microalgae was held in a vertical position and dipped into the nutrient solution at the lower turning point of the sheet.

The cultivation of microalgae in illuminated fermenters is, of course, the method of choice for producing isotope-labeled biochemicals. For some time, however, large fermenters of 5 m³ have been used by the Nihon Chlorella Company in Japan for heterotrophic *Chlorella* production on glucose.[51] With regard to developments in CO_2 supply and methods for harvesting and processing of microalgae, refer to the remaining chapters of this volume. It should be noted at this point, however, that most of the Japanese *Chlorella* factories are or have been working with acetic acid as the carbon source for microalgal growth and a batchwise operation is preferred there.[16]

The Perennials in Research

Applied research on true mass production of microalgae did and still does rely mainly on short-term grants. Besides these biannual to quadrannual projects, institutionalization of applied research on microalgae is comparatively rare. Of the traditional perennials in the field one should mention institutions financed by public funds: the Centro di Studio dei Microorganismi Autotrofi, Universita di Firenze (Italy); the Laboratory of Biotechnology, Czechoslovakian Academy of Sciences, Trebon; the Biologicheskij Institut, Sect. Massovoho Kultivironvaniya Vodorosley, University of Leningrad (U.S.S.R.);[64] and the Research and Production Laboratory of Algology, Bulgarian Academy of Sciences, Sofia. Some other perennials have ceased to exist in their original form as algal research institutes. After having been subsidized for some time by American foundations, the Tokugawa Institute for Biological Research had to be dissolved in 1965; after almost 30 years of applied research on microalgae, the Kohlenstoffbiologische Forschungsstation (formerly at Essen) terminated its activities at Dortmund in 1979. In the case of the Tokugawa Institute, the authorities were apparently satisfied that the degree of industrialization *Chlorella* production had reached was such that specific research tasks could be taken over by profitable *Chlorella* companies. In Germany industry was not satisfied as to the application of microalgal biotechnologies. In addition, it appeared unwise to work on apparently expensive methods for protein production in a country which is supersaturated with agricultural products.

Among the perennials of the scene already considered here in retrospect, several industrial laboratories must be mentioned as well. Besides the research divisions of Japanese companies such as *Chlorella* Industries, there is the Biosciences Division of the Research Institute for Advanced Studies of the Martin Marietta Corporation at Baltimore and the Algological Laboratory of the Institut Francais du Pétrole at Lavera near Marseille. While the former excelled in photosynthesis research and contributions to algal space technology, the latter greatly advanced experience and information on the mass culture and usefulness of *Spirulina*.

Since at least the appearance of the Burlew book,[41] the following advantages in the mass culture of microalgae have been uncovered:

- Appreciably higher protein yields obtained with higher plants per unit surface and unit time.
- Absence of leaves, stems or roots, i.e., there are no useless parts of the biomass.
- Independence of soil quality, i.e., no competition with agriculture for land.
- Daily harvest of regrown biomass.
- The whole process can be automated.

On the other hand, the following disadvantages have been encountered:

- Comparatively high investments.
- Costliness of the pure CO_2 or organics (acetate, glucose) required.
- Demand for chemicals as sources of mineral nutrients.
- Danger of accumulation of environmental pollutants from water, chemicals, and/or atmospheric emissions. This potential hazard is common to all microbial biomass production systems and has, in the case of microalgae, especially, been investigated by Payer et al.[65]

The initial work on mass cultures of microalgae in the U.S.S.R. were reviewed by Gromov.[64] About the later Russian work we are not well informed and a summary would be welcome. In Moscow, Gajevskaja[66] studied the production of *Chlorella* in artificially illuminated tanks with the aim of using the biomass produced for rearing larvae of aquatic animals. When reviewing her own work, the same author also gave a detailed account of the achievements in the U.S.[67] She mentions invasions by zooplankton organisms as a major management problem of open outdoor cultures and points out the potential advantages of the production of microalgae in the arctic and antarctic summer. At the Algological Laboratory of the University of Leningrad, the potential of microalgae production was first studied under field conditions in 1957 and the productivity of various strains of *Chlorella*, *Scenedesmus*, and *Ankistrodesmus* was compared. The work at Leningrad concentrated later on the potential of *Dunaliella* as a source of carotenoids, but this development apparently did not lead to direct practical applications.

In the 1960s, closed culture systems were intensely studied under the direction of Nichiporovich and Semenenko[68] for extraterrestrial life support during prolonged space missions. By optimizing intensely illuminated laboratory cultures of *Chlorella*, daily yields of up to 150 g algal dry matter per m² were obtained as an average of 40 days.[69] This enormous value is still a challenge to researchers working with other microalgae production systems. The realization of a capsule atmosphere as a two-species microcosm (man + microalgal suspension) was consequently developed and based on careful measurements and the operational control of photosynthetic and respiratory quotients. The first full-size experiment lasted 30 days.[70] This work, probably continuing even today at Novosibirsk, also included the nutritional evaluation of *Chlorella* as a protein source for humans[71] and actual regeneration of human excrement by means of mixed algae-bacteria cultures.[72] Other studies concerned the effects of actual orbital flight and cosmic radiation on photosynthetic and mutation rates.

In parallel to the Russian work, the application of intensive algal cultures for bio-regenerative space and extraterrestrial life support systems has also received considerable attention by the space agencies of the U.S. Biological problems involved and technological details required were studied by numerous physiologists and engineers. The famous high-temperature *Chlorella* TX 71105[73] provided an unusually high photosynthetic potential for these studies. In parallel to the successful nutritional evaluation of *Chlorella* as a foodstuff for humans,[74] a number of estimates were worked out concerning the "one-man unit" (i.e., a continuously illuminated algae-bacteria culture compensating the entire metabolism of one man, which seems to require a minimal volume on the order of 100 ℓ).[75] One of the bioreactors developed for operation under gravity-free conditions consisted of a spinning cylinder, called the algatron, to whose inner surface the suspension adhered by centrifugal force.[76,77]

The applied algological work at Firenze (Italy) was launched in 1956 and led to the installment of a small pilot plant in 1957.[78] The algal suspensions in simple ditches with a plastic lining were mixed by sparging a CO_2-in-air mixture. Over the years extensive productivity measurements were carried out with a variety of chlorococcal and blue-green algae, including nitrogen-fixing strains. The efficiency of CO_2 utilization was optimized and the effects of light and temperature on fatty and amino acid ratios were thoroughly investigated.[79] More recently, the Italian group studied mariculture on land for energy farming in coastal deserts. The system implies biotransformation of solar energy into algal biomass which is harvested and subjected to anaerobic digestion. Although the energy balance was definitely positive, the economic feasibility of this interesting system still remains to be proven at the technical scale.[56]

The Laboratory of Experimental Algology and the Department of Applied Algology of the Czechoslovakian Academy of Sciences began its work at Trebon in 1960 and quickly attained an international reputation. After important contributions to the understanding of microalgal growth physiology, the group developed a unique system for mass cultivation of microalgae on an inclined baffled slope (see section entitled "Development of Production Units"). The original idea behind this sophisticated set-up was dual-purpose: in summer the algal culture would shade and cool a greenhouse, and during winter, when microalgae production was unfeasible, the greenhouse would only be used as such. According to an average water depth of only 3 to 5 cm, higher biomass concentrations than usual (up to 3 g/ℓ of algal dry matter) were optimal. The clean algal biomass produced at Trebon was obviously quite expensive and applications were therefore sought mainly in pharmacy.[80] Reduced financial support shifted the activities at Trebon more to basic research after 1969.

The German pilot plant, operated since 1958 at Dortmund, first served for the production of *Scenedesmus* biomass for nutritional evaluation. Beginning with tests in rats[81,82] the nutritional value of *Scenedesmus obliquus* was thoroughly tested in other animal species and finally supplied successfully to human adults. Later tests also included *Coelastrum sphaericum*.[83] After improvements in reactor design, culture management, and processing technology,[51] the West German government supported major microalgae research projects in India, Peru, and Thailand aiming to further develop the use of microalgae as a proteinaceous and vitamin-rich component of human diets. The results of these projects are summarized by Shelef and Soeder[84] and Becker and Venkataraman.[45] Besides the apparent impossibility of producing pure microalgal biomass at sufficiently low costs, the lack of a full toxicological evaluation of *Scenedesmus obliquus* for human consumption, according to the PAG Guidelines[85,86] (published after the start of the projects) prevented straight success.[45] The same holds for the more recent Egyptian-German Project at the National Research Centre in Cairo-Dokki guided by El-Fouly. It is still fascinating to read the recorded disputes between toxicologists and algologists in Soeder and Binsack[87] and to realize how high the hurdle of toxicological safety appears.

Algological work at the Woods Hole Oceanographic Institution, Mass. has a long tradition. In 1971, Ryther started intensive work on photosynthetic wastewater treatment in seawater ponds. The objective was in fact to remove residual nutrients from the effluent of a sewage plant. The effluent flowed at first into large unstirred ponds of 150 m² net surface in which dense blooms of planktonic diatoms developed.[88,89] The phytoplankton was then led to raceways with stacked molluscs harvesting the microalgae and initiating a complex food chain.[90] At the end of the foodchain seaweed cultures with stocking densities of 4 to 5 kg × m⁻² (!) removed residual nitrogen.[91] These interesting studies were accompanied by important basic research,[92] but did not lead to direct application; however the work at Woods Hole stimulated the entire development of systems for mass cultivation of marine microalgae and was certainly influential on the installment of pond systems for technical mass production of phytoplankton with an "artificial upwelling method" in the Virgin Islands.[93]

Microalgal Biomass for Sale

In the course of the various research and development projects which have been briefly discussed in the preceding section, the productivity of microalgal cultures increased considerably. For example, the first Japanese pilot plant yielded only 3.5 g/m²/day of *Chlorella* dry matter; a few years later comparative values averaged 16 to 17 g/m²/day. Provided the nutrient supply is adequate, algal productivity depends primarily on irradiance and temperature.[94] Sizeable autotrophic yields can, therefore, only be obtained in suitable areas of warmer countries. The highest productivities measured to the present for autotrophic mass cultures of microalgae under field conditions amount to 55 g/m²/day and were obtained under favorable conditions in Peru[95] and South Africa.[94] Extrapolation to annual yields amounted to about 100 t/ha.

The initial hope that microalgae might become a potent weapon in the fight against protein malnutrition has not come to pass,[43] although mass cultures of *Chlorella*, *Scenedesmus*, and *Spirulina* definitely yield appreciably higher amounts of protein per unit surface per year than conventional crop plants, including the soybean. However, pure microalgal biomass is simply too expensive to compete successfully with soya or other legumes.[96] Besides the price factor, one must consider potential quantities, especially in the case of freshwater algae production. Since inland waters cover barely 2% of the surface of any continent and approximately 0.7% of the globe, the production of vast quantities of freshwater algae is physically impossible. However, *Chlorella* and its products have been commercially successful as health food commodities in Japan since 1960. According to Tsukada et al.,[97] Japanese *Chlorella* production exceeded 350 t in 1976. Production costs ran at $5 to 11/kg. Since *Chlorella* tablets or granulates are sold in health food stores for about $100/kg, the total retail sales volume must be at least $3.5 × 10⁷ per annum. Another source of information is Kawaguchi,[16] who states that eight larger *Chlorella* factories were operated by the end of 1977 in Japan and production costs amounted to at least $11/kg.

Kawaguchi[16] also lists one *Chlorella* factory for Korea and one for Malaysia; however, 34 *Chlorella* companies are listed for Taiwan, where commercial production began in 1970. For the 756 t produced in 1976, the Taiwanese manufacturers earned over $1.6 × 10⁷ in 1976.[98] Considering the fact that part of the *Chlorella* biomass is used for extraction of the very expensive "*Chlorella* Growth Factor", total retail sales of *Chlorella* products must have reached $1.2 to 1.3 × 10⁸ per annum in the Far East.

The year 1977, however, brought a serious loss of market volume for *Chlorella* manufacturers. The number of persons consuming *Chlorella* tablets dropped significantly below its peak value of 2 million, when it became known that a batch of spoiled *Chlorella* tablets in which photoconversion of chlorophyll to pheophorbide had taken place. This caused heavy skin lesions in a few consumers. The author is not informed to what extent the *Chlorella* market in the Far East has regained its former volume.

After the discovery of *Spirulina* as a local food around Lake Tchad, the French Petroleum Institute developed a large-scale production of this attractive cyanophyte near Marseille. Attempts to introduce this technology in Algeria, Egypt, the West Indies, and Mexico have apparently been unsuccessful, although they initiated the harvesting of the seminatural constant bloom of *Spirulina maxima* from the soda brines of Lake Texcoco near Mexico City. Relying on this unique source, Sosa Texcoco Ltd. became the largest manufacturer of *Spirulina* in the world, selling presently about 500 t of *Spirulina* meal for which wholesale prices in Europe are about \$10/kg. Because of the international sales boom of *Spirulina* tablets, advertised as a natural appetite control, retail prices for these products peaked at \$140 to 160/kg in 1983. *Spirulina* meal was also successful as a petfish-feed ingredient in Japan where local *Spirulina* production started in 1977 on Okinawa. Another Japanese *Spirulina* plant with an annual capacity of 200 t is being operated in Thailand and the production of the plant in Israel is promising (described elsewhere in this volume).

For some years an American company produced the green microalga *Spongiococcum* as a source of carotenoids for commercial chicken feed, after it had become known that this microalgal product is superior to alfalfa meal in improving the color of egg yolk. The production of *Spongiococcum* was, however, terminated in 1965.[12]

Without going into further detail here, it should be pointed out that production of microalgae is an established method for commercial rearing of larvae of shrimp, molluscs, and certain fish. The microalgal biomass is either used directly or as a feed for zooplankton such as *Brachionus plicatilis*.[99] Although some Japanese companies are operating large culture units in order to obtain sufficient amounts of live phytoplankton,[50] the amounts of *Isochrysis, Chlorella, Nannochloris, Phaeodactylum, Skeletonema*, etc., are usually small, however indispensable in modern marine aquaculture.[100,101] Among the various future options for commercial applications of microalgae, those suited for multipurpose uses appear to be the most attractive. A paradigmatic example of such an approach is the Israeli project to produce glycerol from the halophilic microalgae *Dunaliella*. Under salt stress, *Dunaliella bardawil* may contain as much as 40 to 50% of glycerol in the dry matter and up to 10% of β-carotene, which is therefore considered a future valuable by-product of the process which would also render a proteinaceous residue to be used in animal nutrition.[102]

Sewage, Algal Ponds, and Albazod

The application of high-rate algal ponds in wastewater treatment appears to bear the greatest potential of all biotechnologies based on microalgae, if exploited fully as a multipurpose system.

Contrary to the "clean processes" described in the preceding sections, the role of microalgae in photosynthetic wastewater treatment is tied to interaction with the bacteria responsible for eventual aerobic degradation of the wastewater organics. Photosynthesizing according to light energy input, microalgae make use of remineralized end products of bacterial metabolism (CO_2, ammonium, etc.). The algae in turn supply the bacteria with the oxygen required for full breakdown of degradable organics. This symbiotic relationship, which exists in all natural waters, is intensified in facultative and aerobic algal ponds,[103] and to a still larger extent in the mechanically mixed high-rate algal ponds ("HRAPs") developed by Oswald in California. Although the potential of the HRAP for combined water reclamation and nutrient recycling via the recovery of microbial biomass was envisaged from the beginning,[39] it took quite some effort[12,104-106] until such a multipurpose scheme was used at the technical scale as in Singapore.[107]

The term "albazod" has been introduced recently to make it clear that the particulate matter to be recovered from HRAPs consists of algae (i.e., microalgae), bacteria, zooplankton, and detritus.[108] Thus, imprecise expressions such as "algal biomass", "algal bacterial matter", etc., can be avoided.

With regard to algal ponds in general, they are now considered to be the most economic means for treatment of urban sewage or excess amounts of liquid agricultural wastes in warmer countries, wherever sufficient areas of comparatively low value are available.[108,109] There, the specific advantages of pond systems are related to the production of fish and, given the demand, of irrigation water as added values. By contrast, the attractivity of algal ponds for smaller communities of the temperate zones is more related to sanitation at low cost.

As a result of extended studies on smaller systems, Oswald and Golueke[110] presented the engineering and operational characteristics of the first large-scale HRAP designed for low-cost operation. According to the basic design of a large, shallow compacted dirt pond bordered by a low levee (about 1 to 1.5 m high) divided into long channels (10 to 30 m wide) by means of baffles, the still-operational HRAP plant at Richmond, California began operation in 1960. While the Richmond pond covers 2700 m² and has a working volume of about 1000 m³, an increasing number of larger HRAPs have been installed in California,[111,112] Florida,[113] and elsewhere.

Oswald and co-workers screened various methods for harvesting and processing of albazod, introducing alum flocculation and drum drying as important techniques.[12] Later, two of Oswald's former students were in charge of important large projects (McGarry in Thailand and Shelef in Israel), in which flocculation was followed by flotation.[84] Albazod meal produced by Oswald's group at Richmond was successfully tested as a proteinaceous feed component in pigs as well as in sheep.[114,115] The coarse chemical composition of albazod does not differ significantly from that of pure microalgal biomass. In the dry matter albazod contains 50 to 55% crude protein and was found to be a suitable protein source for rats,[116] carp and *Tilapia*,[117] chicken,[118] and pigs.[104] In chicken and carp raised on albazod-containing diets, heavy metals and organic pollutants were not accumulated in nonpermissible amounts.[119]

Neither in California nor in Israel is the utilization of albazod practiced at the industrial scale as yet. The important developments at Singapore, where treatment of piggery wastes is required by a recent law[107] will apparently also not reach full industrial application right there because of the limitation of land availability.[120] Intensive work in Florida[113] will also help lead to the eventual breakthrough in HRAP technology.

Current HRAP technology is characterized by Taiganides:[107] "The three resources which can be recovered from organic wastes such as pig wastes and wastewaters are: water, energy and nutrients. However, the recovery of these resources cannot be justified on economic terms but must be carried out in conjunction with pollution control or sanitation operations. Wastewater treatment to meet environmental pollution standards is sufficiently expensive to justify employing purification processes which yield useful products."

ACKNOWLEDGMENT

The author thanks Doris Schröder for competent processing of the manuscript and for able help with the bibliography.

REFERENCES

1. **Bold, H. C. and Wynne, M. J.,** *Introduction to the Algae,* Prentice Hall, Englewood Cliffs, N.J., 1978.
2. **Lee, R. E.,** *Phycology,* Cambridge University Press, Cambridge, 1982.
3. **Kneifel, H.,** Amines in algae, in *Marine Algae in Pharmaceutical Science,* Hoppe, H. A., Levring, T. and Tanaka, Y., Eds., Walter de Gruyter, Berlin, 1979, 365.
4. **Carr, N. G. and Whitton, B. A.,** *The Biology of Cyanobacteria,* Bot. Monogr. Vol. 17, Blackwell Scientific, Oxford, 1982.
5. **Hall, D. O.,** Solar energy through biology: fuels from biomass, *Experientia,* 38, 3, 1982.
6. **Lieth, H. and Whittaker, R. H.,** *Primary Productivity of the Biosphere,* Springer Verlag, Berlin, 1975.
7. **Round, F. E.,** *The Ecology of Algae,* Cambridge University Press, Cambridge, 1981.
8. **Wetzel, R. G.,** *Limnology,* 2nd ed., Saunders College Publications, Philadelphia, 1983.
9. **Stewart, W. D. P., Ed.,** *Algal Physiology and Biochemistry,* Bot. Monogr. Vol. 10, Blackwell Scientific, Oxford, 1974.
10. **Lobban, C. S. and Wynne, J. S.,** *The Biology of Seaweeds,* Bot. Monogr. Vol. 17, Blackwell Scientific, Oxford, 1981.
11. **Tamiya, H.,** Mass culture of algae, *Ann. Rev. Plant Physiol.,* 8, 309, 1957.
12. **Oswald, W. J. and Golueke, C. G.,** Harvesting and processing of waste grown algae, in *Algae, Man, and the Environment,* Jackson, D. J., Ed., Syracuse University Press, Syracuse, N.Y., 1968.
13. **Benemann, J. R., Weissmann, J. C., and Oswald, W. J.,** Algal biomass, in *Microbial Biomass,* Vol. 4, Rose, A. H., Ed., Academic Press, New York, 1979, 177.
14. **Goldman, J. C.,** Outdoor algal mass cultures. I. Applications, *Water Res.,* 13, 1, 1979a.
15. **Goldman, J. C.,** Outdoor algal mass cultures. II. Photosynthetic yield limitations, *Water Res.,* 13, 119, 1979b.
16. **Kawaguchi, K.,** Microalgae production systems in Asia, in *Algae Biomass: Production and Use,* Shelef, G. and Soeder, C. J., Eds., Elsevier/North-Holland, Amsterdam, 1980, 25.
17. **Richmond, A. and Preiss, K.,** The biotechnology of algaculture, *Interdiscipl. Sci. Rev.,* 5, 60, 1980.
18. **Levring, T., Hoppe, H. A., and Schmid, O. J.,** *Marine Algae,* de Gruyter, Hamburg, 1969.
19. **Chapman, V. J. and Chapman, D. J.,** *Seaweeds and Their Uses,* 3rd ed., Chapman & Hall, London, 1980.
20. **Madlener, J. C.,** *The Seavegetable Book,* Potter, New York, 1977.
21. **Aldave-Pajares, A.,** Cushuro, algas azul-verdes como alimento en la region alta andina Peruana, *Bol. Soc. Bot. Libertad Trujillo,* 1, 5, 1969.
22. **Tseng, C. K.,** Marine phycoculture in China, in *Proc. 10th Int. Seaweed Symp.,* Levring, T., Ed., Walter de Gruyter, Berlin, 1981, 123.
23. **Doty, M. S.,** *Eucheuma* — current marine agronomy, in *The Marine Plant Biomass of the Pacific Northwest Coast,* Krauss, R. W., Ed., Oregon State University Press, Corvallis, 1977, 203.
24. **Laite, P. and Ricohermoso, M.,** Revolutionary impact of *Eucheuma* production in the South China Sea on the carrageenan industry, in *Proc. 10th Int. Seaweed Symp.,* Levring, T., Ed., Walter de Gruyter, Berlin, 1981, 595.
25. **Gellenbeck, K. W. and Chapman, D. J.,** Seaweed uses: the outlook for mariculture, *Endeavour, New Ser.,* 7, 31, 1983.
26. **Wilcox, H.,** Ocean farming, in *Capturing the Sun through Bioconversion,* Washington Center for Metropolitan Studies, Washington, D.C., 1976, 276.
27. **Hoppe, H. A., Levring, T., and Tanaka, Y., Eds.,** *Marine Algae in Pharmaceutical Science,* Walter de Gruyter, Berlin, 1979.
28. **Hoppe, H. A. and Levring, T., Eds.,** *Marine Algae in Pharmaceutical Science,* Vol. 2, Walter de Gruyter, Berlin, 1982.
29. **Johnston, H. W.,** The biological and economic importance of algae. III. Edible algae of fresh and brackish waters, *Tuatara,* 18, 19, 1970.
30. **Durand-Chastel, H.,** Production and use of *Spirulina* in Mexico, in *Algae Biomass: Production and Use,* Shelef, G. and Soeder, C. J., Eds., Elsevier/North-Holland, Amsterdam, 1980, 51.
31. **Beijerinck, M. W.,** Kulturversuche mit Zoochlorellen, Lichenengonidien und anderen niederen Algen, *Bot. Zeitung.,* 48, 725, 1890.
32. **Warburg, O.,** Über die Geschwindigkeit der photochemischen Kohlensäurezersetzung in lebenden Zellen, *Biochem. Z.,* 100, 230, 1919.
33. **Pringsheim, E. G.,** *Pure Cultures of Algae,* Cambridge University Press, England, 1947.
34. **Harder, R. and von Witsch, H.,** Über Massenkultur von Diatomeen, *Ber. Dtsch. Bot. Ges.,* 60, 146, 1942.
35. **Spoehr, H. A. and Milner, H. W.,** The chemical composition of *Chlorella*: effects of environmental conditions, *Plant Physiol.,* 24, 120, 1949.

36. **Davis, E. A., Dedrick, J., French, C. S., Milner, H. W., Myers, J., Smith, J. H. C., and Spoehr, H. A.,** Laboratory experiments on *Chlorella* culture at the Carnegie Institution of Washington Dept. of Plant Biology, in *Algal Culture: From Laboratory to Pilot Plant,* Publ. no. 600, Burlew, J. S., Ed., The Carnegie Institution, Washington, D.C., 1953, 105.

37. **Cook, P. M.,** Some problems in the large-scale culture of *Chlorella,* in, *The Culture of Algae,* Brunel, J., Prescott, G. W., and Tiffany, L. H., Eds., Charles F. Kettering Foundation, Dayton, Ohio, 1950, 53.

38. **Calvin, M. and Benson, A. A.,** The path of carbon in photosynthesis. I., *Science,* 107, 476, 1948.

39. **Gotaas, H. B. and Oswald, W. J.,** Algal Symbiosis in Oxydation Ponds, 2nd Progr. Rep., Institute of Engineering, University of California, Berkeley, Ser. No. 3, 1, 1951.

40. **Gummert, F., Meffert, M.-E., and Stratmann, H.,** Nonsterile large-scale culture of *Chlorella* in greenhouse and open air, in *Algal Culture: From Laboratory to Pilot Plant,* Publ. no. 600, Burlew, J. S., Ed., The Carnegie Institution, Washington, D.C., 1953.

41. **Burlew, J. D., Ed.,** *Algal Culture: From Laboratory to Pilot Plant,* Publ. no. 600, The Carnegie Institution, Washington, D.C., 1953.

42. **Jorgensen, J. and Convit, J.,** Cultivation of complexes of algae with other freshwater organisms in the tropics, in *Algal Culture: From Laboratory to Pilot Plant,* Publ. no. 600, Burlew, J. S., Ed., The Carnegie Institution, Washington, D.C., 1953.

43. **Litchfield, J. H.,** Single-cell proteins, *Science,* 219, 740, 1983.

44. **Becker, E. W. and Venkataraman, L. V.,** Production and processing of algae in pilot plant scale experience of the Indo-German project, in *Algae Biomass: Production and Use,* Shelef, G. and Soeder, C. J., Eds., Elsevier/North-Holland, Amsterdam, 1980, 35.

45. **Becker, E. W. and Venkataraman, L. V.,** *Biotechnology and Exploitation of Algae: The Indian Approach,* GTZ Press, Eschborn, Germany, 1982.

46. **Venkataraman, L. V.,** New possibilities for microalgae production and utilization in India. *Ergeb. Limnol.,* 11, 199, 1978.

47. **Walmsley, R. D., Wurts, T., and Carr, L.,** Concepts and design considerations for the mass culture of algae in closed ponds, in *Wastewater for Aquaculture,* Grobbelaar, J. U., Soeder, C. J., and Toerien, D. F., Eds., Publ. Ser. C, University of the Orange Free State, Bloemfontein, Republic of South Africa, 1981, 136.

48. **Jüttner, F. et al.,** Massenzucht phototropher Organismen in einer automatischen Kulturanlage, *Arch. Mikrobiol.,* 77, 275, 1971.

49. **Setlik, I. et al.,** Dual purpose open circulation units for large scale culture of algae in temperate zones. I. Basic design considerations and scheme of pilot plant, *Algol. Studies,* 1, 111, 1970.

50. **Mituya, A., Nyunoya, T., and Tamiya, H.,** Pre-pilot plant experiments on algal mass culture, in *Algal Culture: From Laboratory to Pilot Plant,* Publ. no. 600, Burlew, J. S., Ed., The Carnegie Institution, Washington, D.C., 1953, 273.

51. **Stengel, E.,** Anlagentypen und Verfahren der technischen Algenmassenproduktion, *Ber. Dtsch. Bot. Ges.,* 83, 589, 1970.

52. **Meffert, M. E. and Stratmann, H.,** Algen-Großkulturen im Sommer 1951, *Forschungsber. Wirtsch. Verkehrsminist.,* 8(NRW), 3, 1954.

53. **Stengel, E. and Soeder, C. J.,** Control of photosynthetic production in aquatic ecosystems, in *Photosynthesis and Productivity in Different Environments,* Cooper, J. P., Ed., Cambridge University Press, Cambridge, England, 1975, 645.

54. **Morimura, Y., Nihei, T., and Sasa, T.,** Outdoor bubbling culture of some unicellular algae, *J. Gen. Appl. Microbiol.,* 1, 173, 1955.

55. **Kanazawa, T., Fujita, C., Yuhara, T., and Sasa, T.,** Mass culture of unicellular algae using ''open circulation methods'', *J. Gen. Appl. Microbiol.,* 4, 135, 1958.

56. **Balloni, W., Florenzano, G., Materassi, R., Tredici, M., Soeder, C. J., and Wagener, K.,** Mass cultures of algae for energy farming in coastal deserts, in *Energy From Biomass, 2nd E.C. Conference,* Strub, A., Chartier, P., and Schleser, G., Eds., Applied Science, London, 1983, 291.

57. **Shelef, G., Moraine, R., and Oron, G.,** Photosynthetic biomass production for sewage, *Ergeb. Limnol.,* 11, 3, 1978.

58. **Clement, G. and van Landeghem, H.,** *Spirulina* — ein günstiges Objekt für die Massenkultur von Mikroalgen, *Ber. Dtsch. Bot. Ges.,* 11, 559, 1970.

59. **Vendlova, J.,** Outdoor cultivation in Bulgaria, *Ann. Rep. Algol. Lab. Trebon,* 1968, 143, 1969.

60. **Heussler, P., Castillo, S., Merino, M. F., and Vasques, V.,** Improvements in pond construction and CO_2 supply, *Ergeb. Limnol.,* 11, 254, 1978.

61. **Berend, J., Simovitch, E., and Ollian, A.,** Economic aspects of algal animal food production, in *Algae Biomass: Production and Use,* Shelef, G. and Soeder, C. J., Eds., Elsevier/North-Holland, Amsterdam, 1980, 799.

62. **Bosman, J. and Hendricks, F.**, The development of an algal pond system for the removal of nitrogen from an inorganic industrial effluent, in *Proc. Symp. Aquaculture in Wastewater*, CSIR Press, Pretoria, 1980.

63. **Salageanu, N.**, Versuche zur Aero-Massenkultur einiger mikroskopischer Algenarten, *Ber. Dtsch. Bot. Ges.*, 83, 549, 1970.

64. **Gromov, B. V.**, Main trends in experimental work with algal cultures in the USSR, in *Algae, Man and Environment*, Jackson, D. F., Ed., Syracuse University Press, Syracuse, N.Y., 1968, 249.

65. **Payer, H. D., Runkel, K. H., Schramel, P., Stengel, E., Bhumiratana, A., and Soeder, C. J.**, Environmental influences on the accumulation of lead, cadmium, mercury, antimony, arsenic, selenium, bromine in unicellular algae cultivated in Thailand and in Germany, *Chemosphere*, 6, 413, 1976.

66. **Gajevskaja, N. S.**, On the cultivation of protococcus algae with fluorescent lamps, *Byull. Mosk. Ova. Ispyt. Prir.*, 57, 35, 1952.

67. **Gajevskaja, N. S.**, Probleme der Verwertung einzelliger Algen, *Priroda*, 4, 3, 1956.

68. **Nichiporovich, A. A., Semenenko, V. E., Vladimirova, M. G., and Spektorov, K. S.**, Some principles of intensification of photosynthetic productivity in unicellular algae, *Istvestia Acad. Sci. USSR Ser. Biol.*, 2, 163, 1962.

69. **Semenenko, V. E., Vladimirova, M. G., Soglin, L. N., Tauts, M. I., Phillipovskiy, Iu. N., Klyachko-Gurvich, G. L., Kuznetsov, E. D., Kovanova, E. S., and Raijkov, N. I.**, Prolonged continuous directed cultivation of algae and physiological and chemical characteristics of the productivity and efficiency of light energy utilization by *Chlorella, Upr. Biosynthez*, 128, 136, 1966.

70. **Kirensky, L. V., Terskov, I. A., Gitel'zon, I. I., Lisovsky, G. M., Kovrov, B. G., Sid'ko, F. Y., Okladnikov, Y. N., Antonyuk, M. P., Belyanin, V. N., and Rerberg, M. S.**, Gas exchange between man and a culture of microalgae in a 30-day experiment, *Kosm. Biol. Med.*, 1, 23, 1967.

71. **Kondralyew, Y. I., Bychkov, V. P., Ushakov, A. S., Boiko, N. N., and Klyushkina, N. S.**, Use of 50 and 100 g dry matter of unicellular algae in human diets, *Vopr. Pitan.*, 25, 9, 1966.

72. **Rerberg, M. S., Vorobieva, T. I., Kuzmina, R. I., and Barkhatova, I. M.**, Regeneration of human excretions by means of a naturally developing algae-bacteria community, *Probl. Kosm. Biiol., Acad. Sci. USSR, Moscow*, 598, 1965.

73. **Sorokin, C. and Myers, J.**, *Characteristics of a High-Temperature Strain of Chlorella*, Carnegie Institution, Washington, D.C., Yearbook No. 53, 147, 1954.

74. **Lee, S. K., Fox, H. M., Kies, C., and Dam, R.**, The supplementary value of algae in human diets, *J. Nutr.*, 92, 281, 1967.

75. **Lachance, P. A.**, Single-cell protein in space systems, in *Single-Cell Protein*, Mateles, R. T. and Tannenbaum, S. R., Eds., MIT Press, Cambridge, Mass., 1968, 122.

76. **Shelef, G., Oswald, W. J., and Golueke, C. G.**, The continuous culture of algal biomass on wastes, in *Continuous Cultivation of Microorganisms*, Malek, I., Ed., Academia, Prague, 1969, 601.

77. **Shelef, G., Oswald, W. J., and McGauhey, P. H.**, Algal reactor for life support systems, *J. Sanit. Eng. Div. Am. Soc. Civil Eng.*, 96(SA1), 91, 1970.

78. **Florenzano, G.**, Prime Ricerche in Italia. Nell'impianto sperimentale di Firenze, sulla cultura massiva non sterile de alghe, *N. Giorn. Bot. Ital.*, 65, 1, 1958.

79. **Materassi, R., Paoletti, C., Balloni, W., and Florenzano, G.**, Some considerations on the production of lipid substances by microalgae and cyanobacteria, in *Algae Biomass: Production and Use*, Shelef, G. and Soeder, C. J., Eds., Elsevier/North-Holland, Amsterdam, 1980, 619.

80. **Rydlo, O. and Maly, J.**, On the possibilities of using the fresh-water alga *Scenedesmus obliquus* in medicine preparation, *Farmaceut. Obzor (Bratislava)*, 39, 49, 1970.

81. **Meffert, M.-E.**, Die Wirkung der Substanz von *Scenedesmus obliquus* als Eiweißguelle in Fütterungsversuchen und die Beziehung zur Aminosäure. — *Eiwei βzusammensetzung. Forschungsber.*, Westd. Verlag, Cologne, 1961.

82. **Kraut, H., Jekat, F., and Pabst, W.**, Ausnutzungsgrad und biologischer Wert des Proteins der einzelligen Grünalge *Scenedesmus obliquus*, ermittelt im Ratten-Bilanz-Versuch, *Nutr. Dieta*, 8, 130, 1966.

83. **Kofranyi, E.**, The nutritional value of the green algae *Scenedesmus acutus* for humans, *Ergebn. Limnol.*, 11, 150, 1978.

84. **Shelef, G. and Soeder, C. J., Eds.**, *Algae Biomass: Production and Use*, Elsevier/North Holland, Amsterdam, 1980.

85. Protein Advisory Group, Guideline No. 6 for Preclinical Testing of Novel Sources of Protein, *PAG Bull.*, 4(3), 17, 1972.

86. Protein Advisory Group, Guideline No. 15 on Nutritional and Safety Aspects of Novel Sources of Protein for Animal Feeding, United Nations, New York, 1974.

87. **Soeder, C. J. and Binsack, R., Eds.**, *Microalgae for food and feed, Ergeb. Limnol.*, 11, 300, 1978.

88. **Goldman, J. C. and Ryther, J. H.**, Temperature-influenced species competition in mass cultures of marine phytoplankton, *Biotechnol. Bioeng.*, 18, 1125, 1976.

89. **Goldman, J. C.,** Physiological aspects in algal mass cultures, in *Algae Biomass: Production and Use,* Shelef. G. and Soeder. C. J.. Eds.. Elsevier/North-Holland. Amsterdam. 1980. 343.
90. **Ryther, J. H., Goldman, J. C., Gifford, C. E., Huguenin, J. E., Wing, A. S., Clarner, J. P., Williams, L. D., and Lapointe, B. E.,** Physical models of integrated waste recycling: marine polyculture systems, *Aquaculture,* 5, 163, 1975.
91. **Ryther, J. H., Corwin, N., and de Busk, T. A.,** Nitrogen uptake and storage by the red alga, *Gracilaria tikvahiae, Aquaculture,* 26, 107, 1981.
92. **Goldman, J. C.,** Temperature effects on the steady-state growth. phosphorus uptake and the chemical composition of marine phytoplankton, *Microbiol. Ecol.,* 5, 153, 1979c.
93. **Roels, O. A., Laurence, S., Farmer, M. W., and van Hemelryck, L.,** Organic production potential of artificial upwelling marine aquaculture, in *Microbial Energy Production,* Schlemgel. H. G. and Barnea. J.. Eds.. E. Goltze KG, Göttingen, 1976. 69.
94. **Grobbelaar, J. U.,** Deterministic model for describing algal growth in large outdoor mass algal cultures, in *Wastewater for Aquaculture,* Grobbelaar. J. U.. Soeder, C. J., and Toerien, D. F.. Eds.. University of the Orange Free State. Bloemfontein, Republic of South Africa, 1981.
95. **Castillo, S., Merino, M. F., and Heussler, P.,** Production and ecological implications of algae mass culture under Peruvian conditions, in *Algae Biomass: Production and Use,* Shelef, G. and Soeder, C. J.. Eds.. Elsevier/North-Holland, Amsterdam, 1980, 123.
96. **Behr, W. and Soeder, C. J.,** *Commercial Aspects of Utilizing Microalgae with Special Reference to Animal Feeds,* U.O.F.S. Publ. Ser. C, No. 3, University of the Orange Free State, Bloemfontein, Republic of South Africa, 1981. 63.
97. **Tsukada, O., Kawahara, T., and Miyachi, S.,** Mass cultures of *Chlorella* in Asian countries, in *Biological Solar Energy Conversion,* Mitsui. A.. et al.. Eds.. Academic Press, New York, 1977, 363.
98. **Soong, P.,** Production and development of *Chlorella* and *Spirulina* in Taiwan. in *Algae Biomass: Production and Use,* Shelef. G. and Soeder, C. J.. Eds.. Elsevier/North-Holland. Amsterdam. 1980. 97.
99. **Soeder, C. J.,** The technical production of microalgae and its prospects for marine aquaculture, in *Harvesting Polluted Waters,* Devik. O.. Ed.. Plenum Press, New York, 1976, 11.
100. **Persoone, G. and Claus, C.,** Mass culture of algae: a bottleneck in the nursery culturing of molluscs, in *Algae Biomass: Production and Use,* Shelef, G. and Soeder, C. J.. Eds., Elsevier/North-Holland. Amsterdam. 1980. 265.
101. **Ukeles, R.,** American experience in the mass culture of micro-algae for feeding larvae of the American oyster, *Crassostrea virginica.* in *Algae Biomass: Production and Use,* Shelef, G. and Soeder, C. J.. Eds.. Elsevier/North-Holland. Amsterdam, 1980, 287.
102. **Ben-Amotz, A., Sussman, I., and Avron, M.,** Glycerol production by *Dunaliella, Experientia,* 38, 49, 1982.
103. **Benefield, L. D. and Randall, C. W.,** *Biological Process Design for Wastewater Treatment,* Prentice-Hall. Englewood Cliffs, N.J., 1980.
104. **Oswald, W. J.,** Growth characteristics of microalgae cultured in domestic sewage: environmental effects on productivity, in *Prediction and Measurement of Photosynthetic Productivity,* Center for Agricultural Publishing, Wageningen, The Netherlands, 1970, 473.
105. **McGarry, M. G. and Tongkasame, Ch.,** Water reclamation and algal protein production in the tropical environment. *Water Pollut. Control Fed.,* 43, 824, 1971.
106. **Shelef, G., Azov, Y., Moraine, R., and Oron, G.,** Algal mass production as an integral part of a wastewater treatment and reclamation system, in *Algae Biomass: Production and Use,* Shelef, G. and Soeder. C. J.. Eds.. Elsevier/North-Holland, Amsterdam, 1980, 163.
107. **Taiganides, E. P.,** Biomass from the treatment of pig wastes, *Wissen. Umwelt,* 4, 256, 1982.
108. **Soeder, C. J.,** Aquatic bioconversion of excrements in ponds, in *Animals as Waste Converters,* Ketelaars, E. H. and Iwema, B., Eds., Pudoc Wageningen, Wageningen, The Netherlands, 1984, 130.
109. **Arthur, J. P.,** Notes on the Design and Operation of Waste Stabilization Ponds in Warm Climates of Developing Countries, Tech. Paper No. 6, World Bank, Urban Development Department, Washington, D.C.. 1982.
110. **Oswald, W. J. and Golueke, C. G.,** Biological transformations of solar energy, *Adv. Appl. Microbiol.,* 2, 223, 1960.
111. **Oswald, W. J.,** Experiences with new pond designs in California, in *Ponds as a Wastewater Treatment Alternative,* Water Resources Symp. No. 9, Gloyna, E. F., Malina, J. F., and Davis, E. M.. Eds.. University of Texas. Austin, 1975.
112. **Benemann, J. R., Koopman, B., Weissman, J., Eisenberg, D., and Goebel, D.,** Development of microalgae harvesting and high-rate pond technologies in California. in *Algae Biomass: Production and Use,* Shelef, G. and Soeder. C. J.. Eds.. Elsevier/North-Holland. Amsterdam. 1980. 457.
113. **Lincoln, A. P. and Hill, D. T.,** An integrated microalgae system, in *Algae Biomass: Production and Use,* Shelef. G. and Soeder. C. J.. Eds.. Elsevier/North-Holland. Amsterdam. 1980. 229.

114. **Hintz, H. F., Heitmann, H., Weir, W. C., Torell, D. T., and Meyer, J. H.,** Nutritive value of algae grown on sewage, *J. Anim. Sci.,* 25, 675, 1966.

115. **Hintz, H. F. and Heitmann, H.,** Sewage-grown algae as a protein supplement for swine, *Anim. Prod.,* 9(2), 135, 1967.

116. **Mokady, S., Yannai, S., Einav, P., and Berk, Z.,** Nutritional evaluation of the protein of several algae species for broilers, *Ergebn. Limnol.,* 11, 89, 1978.

117. **Sandbank, E. and Hepher, B.,** Microalgae grown in wastewater as an ingredient in the diet of warmwater fish, in *Algae Biomass: Production and Use,* Shelef, G. and Soeder, C. J., Eds., Elsevier/North-Holland, Amsterdam, 1980, 697.

118. **Lipstein, B. and Hurwitz, S.,** The nutritional and economic value of algae for poultry, in *Algae Biomass: Production and Use,* Shelef, G. and Soeder, C. J., Eds., Elsevier/North-Holland, Amsterdam, 1980, 667.

119. **Yannai, S., Mokady, S., Sachs, K., Kantorowitz, B., and Berk, Z.,** Certain contaminants in algae and in animals fed algae-containing diets, and secondary toxicity of the algae, in *Algae Biomass: Production and Use,* Shelef, G. and Soeder, C. J., Eds., Elsevier/North-Holland, Amsterdam, 1980, 757.

120. **Taiganides, E. P.,** Personal communication to F. H. Mohn, 1984.

PHOTOSYNTHESIS AND ULTRASTRUCTURE IN MICROALGAE

S. Herman Lips and Yael J. Avissar

LIGHT SPECTRA, QUANTUM THEORY, AND ENERGY LIMITS

The ultimate energy source for all organisms is the sun. Its nuclear mass is converted into energy according to the equation $E = mc^2$ formulated by Einstein. This energy is partially radiated into space, some of it in the form of light which reaches the earth's surface where it allows the photosynthetic assimilation of CO_2 by plants, algae, and some bacteria.

Light Radiation

Visible light is the radiation perceived by the human eye. It constitutes a relatively small fraction of the electromagnetic spectrum (Figure 1). Photosynthetic organisms can use only a fraction of the light spectrum, i.e., those wavelengths that are absorbed by their photosynthetic pigments.[1]

The wavelength of light (Figure 2) can be defined as the distance travelled by one complete cycle of the wave.[2] Wavelengths can be expressed in a variety of units. Table 1 summarizes some of the terms used and their values in relation to the meter.

All of these units have been used in the scientific literature to describe wavelengths, but nanometer is the most widely used today. The wavelengths of visible light are 400 to 740 nm and it can be subdivided into bands of individual colors (Table 2). Colors as we know them constitute a response of our mind to the perception of certain wavelengths of radiation by our eye pigments. The same radiation will be perceived by other organisms (animals, insects) in a different way due to their own eye pigments' absorption characteristics and the particular response of their brains. In other words, what looks red to humans could be invisible or gray to a fly.

The smallest "packet" of light energy is a photon, a concept equivalent to that of a molecule of chemical elements. While 6.023×10^{23} (Avogadro's number, N) molecules constitute 1 mol of a given element (e.g., NaCl), this number of photons is equivalent to 1 mol of light or an Einstein. Since a mole refers to a mass equal to the molecular weight in grams, a unit containing N molecules, this definition does not apply to light which has no mass. The concept of the Einstein as a mole of photons is, however, useful and its usage widespread.

A light wave, besides its wavelength (λ), is characterized by its frequency of oscillation (υ) and its velocity of propagation (V). These three quantities are related as described in the following formula:

$$\lambda\upsilon = V$$

The speed of light (c) in a vacuum is constant, independent of the wavelength and is 299,792 km/sec or 3×10^8 m/sec. Light passing through a medium other than vacuum (air, water, glass) has a speed lower than c. The frequency υ does not change in various media, so that the values of λ and V will be affected; this is the reason for the generalized use of λ to describe a wavelength, as shown in Table 2.

Energy of Light

Light may be characterized in two different ways: as a wave phenomenon (like sound) or as a particle phenomenon (like molecules in a chemical solution). For this reason light can be seen as composed of discrete particles called photons, which correspond conceptually to the idea of atoms, the smallest physical unit retaining elementary characteristics.

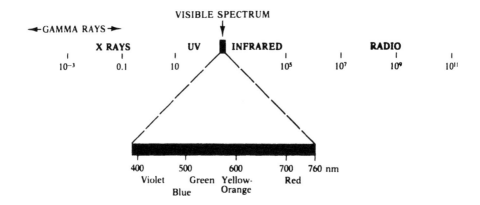

FIGURE 1.   The electromagnetic spectrum.[1]

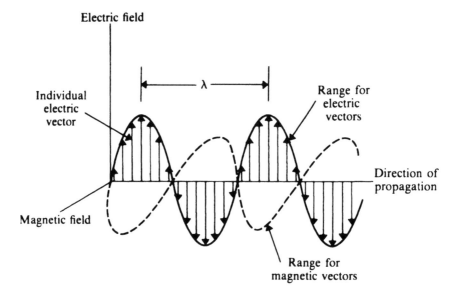

FIGURE 2.   Propagation of electromagnetic waves. (From Nobel, P. S., *Introduction to Biophysical Plant Physiology*, W. H. Freeman, San Francisco, 1974. With permission.)

The light energy (Eλ) carried by a photon is called a quantum and is directly proportional to its frequency and inversely proportional to its wavelength in a vacuum. These relationships can be expressed by the following equation:

$$E\lambda = h\upsilon = hc/\lambda$$

where h is Planck's constant which has the dimensions of energy times time. The units are selected according to the application required: 6.626×10^{-27} erg-sec, 0.4136×10^{-14} ev-sec, or 1.548×10^{-37} kcal-sec. Now we can calculate the energy of any wavelength of light. For instance, 680 nm (red light); it has a frequency of 4.41×10^{14} c/sec. A quantum of this light (680 nm) has an energy hυ of

$$(0.4136 \times 10^{-14} \text{ eV-sec})(4.41 \times 10^{14}/\text{sec}) = 1.82 \text{ eV (Table 2)}$$

Table 1
UNITS USED TO MEASURE THE WAVELENGTHS
OF LIGHT AND THEIR RELATION TO THE METER

Unit	Symbol	Length	Equivalent to
Micron	μm	10^{-6} m	10^4 Å, 10 nm, 10 mμ
Millimicron	mμ	10^{-9} m	10 Å, 0.001 μm, 1 nm
Nanometer	nm	10^{-9} m	10 Å, 0.001 μm, 1 nm
Angstrom	Å	10^{-10} m	10^{-4} μm, 0.1 nm, 0.1 mμ

Table 2
COLORS, WAVELENGTH, AND ENERGY OF LIGHT

Wavelength range (nm)	Color	Energy of representative wavelengths	
		kcal/Einstein	eV/photon
Less than 400	UV	112.5 (254 nm)	4.88
400—425	Violet	69.7 (410 nm)	3.02
425—490	Blue	62.2 (460 nm)	2.70
490—560	Green	55.0 (520 nm)	2.39
560—585	Yellow	49.3 (580 nm)	2.14
585—640	Orange	46.2 (620 nm)	2.00
640—740	Red	42.1 (680 nm)	1.82
Above 740	IR	20.4 (740 nm)	0.88

From Nobel, P. S., *Introduction to Biophysical Plant Physiology*, W. H. Freeman and Company, San Francisco, Copyright 1974. With permission.

The energy of 1 Einstein of the same light wavelength would be the energy of a photon times the number of photons in 1 Einstein or

$$E\lambda = Nh\upsilon$$

which, for the example chosen would be equal to

$$(6.023 \times 10^{23}/mol)(1.584 \times 10^{-37} \text{ kcal-sec})(4.41 \times 10^{14}/sec)$$

$$= 42.1 \text{ kcal/Einstein (Table 2)}$$

The metabolic implications of the amounts of energy carried by photons of different wavelengths, converted first into ATP, and eventually into new carbon to carbon bonds of carbohydrates will be discussed in a later section.

Radiant Flux and Illumination

How does one measure incident light? The following are two common ways of measuring light flux.

Photometric

Photometric is based on the available illuminating power and related to the wavelength sensitivity of the human eye; consequently, this method does not take into account other wavelengths, such as UV and IR, which may be significant for a given photosynthetic light receptor even if the human eye is insensitive to it. These measuring devices read generally

Table 3
CONVERSION FACTORS FOR DIFFERENT
RADIOMETRIC READINGS OF LIGHT

Units	cal/cm²-min	W/cm²	erg/cm²-sec
cal/cm²-min	—	0.0697	6.97×10^5
W/cm²	14.34	—	1×10^7
erg/cm²	1.434×10^{-6}	1×10^{-7}	—

in lux or footcandles. The candle is a unit of luminous intensity. The present international unit of luminescence is the "candela", 1/60 of the light intensity emitted from 1 cm² of a blackbody radiator at the melting temperature of pure platinum (2042 K). A blackbody is a convenient idealization describing an object which does not absorb or emit only at particular wavelengths, but rather at all wavelengths; it is uniformly "black" at all wavelengths. A source emitting 1 candela of light produces 4 pi lumens. One footcandle is 1 lumen/ft², one lux is 1 lumen/m². One footcandle equals, therefore, 10.76 (the number of ft²/m²) lux. The main limitation of this method of light flux measurement is that the wavelength composition of the light source is assessed on the basis of luminescence and not of photosynthesis. In other words, what the human eye appreciates is not necessarily what the photosynthetic pigments absorb. The usefulness of the method may be limited to relative estimations of the light flux emitted by a given source such as sunlight, neon, or incandescent lights, not for accurate quantitative estimations of different sources.

Radiometric

Radiometric is based on the total energy of the radiation. Blackened thermocouples, thermistors, or thermopiles respond to radiant energy and are consequently sensitive to radiation in the UV and IR as well as in the visible range. Readings are expressed in units of energy per unit area and time: cal/cm²-min, ergs/cm²-sec or W/cm². Calories are energy units used generally in biological systems due to their significance in biochemical reactions. Some useful conversion factors for radiometric readings are shown in Table 3.

Radiant flux at a specific wavelength measured in radiometric units can be converted to photon flux (photons/cm²-sec) by using the energy carried by an individual photon.

Sunlight

Electromagnetic radiation from the sun is the main source of energy for biological systems. The flux of solar radiation incident on the atmosphere of this planet (the solar constant) is about 1.98 ± 3 cal/cm²-min. Of this, only about 0.5 cal/cm²-min reaches the earth's surface on a clear summer day. This is an average value which varies with latitude, altitude, and atmospheric conditions. In photometric units, the light intensity on a clear day can reach about 100,000 lux on the surface of the earth, equivalent to about 0.6 cal/cm²-min. The maximum photon flux under these conditions will vary from between 0.9×10^{17} photons/cm²-sec in the blue region to 1.2×10^{17} in the yellow region of the spectrum.

About 5% of the photons incident on the earth's atmosphere are in the UV region, most of which is prevented from reaching the surface of the earth by a layer of ozone (O_3) in the stratosphere. The absorption of UV by ozone protects organisms from UV-induced mutations. It seems reasonable to suppose that there were evolutionary pressures for selecting photochemical systems effective between 400 and 900 nm such as photosynthesis and vision. The relative amount of visible light reaching the atmosphere is about 28% and of infrared light, 67%. Water absorbs strongly near 900 nm, 1100 nm, and above 1200 nm. Atmospheric CO_2 also absorbs infrared radiation. Due to the strong absorption of UV and IR radiation by atmospheric gases, the composition of solar radiations reaching the earth's surface is

Table 4
TONS OF CARBON INCORPORATED ANNUALLY INTO
ORGANIC MATTER THROUGH PHOTOSYNTHESIS

Type of vegetation	Area (millions of km^2)	Annual yield (tons of C/ km^2)	Total yield (billions of tons of C/year)
Forests	44	250	11.0
Grassland	31	35	1.1
Farmland	27	150	4.0
Desert	47	5	0.2
Total (land)	149	—	16.3
Ocean	361	62—375	19—135
Total	510	—	35—151

different of that incident in the atmosphere and is relatively richer in the visible wavelengths; 2% of the photons are in the UV, 45% are in the visible, and 53% are in the infrared.[2]

THE PROCESS OF PHOTOSYNTHESIS

The term photosynthesis means "synthesis with the help of light". This is generally ascribed to the synthesis of organic matter by plants, although it implies a variety of organic as well as inorganic and physical reactions. Photosynthesis is the basic process of life on earth. It creates living matter out of inert inorganic materials (CO_2 and H_2O), replenishes oxygen in the atmosphere, and traps and converts solar energy into chemical energy available to metabolic functions of living organisms.

By photosynthesis, plants store the energy of sunlight in the form of chemical energy. Julius Robert Mayer (1814 to 1878), one of the pioneer investigators of photosynthesis, saw in this energy conversion a particularly important illustration of the law of conservation of energy to whose formulation he contributed. Before Mayer, only the chemical functions of plants as "creators of organic matter" on earth was comprehended. After Mayer, their physical function as energy transducers and providers for life became clear as well.

Estimates of the total yield of photosynthesis on earth are only approximate. The largest item is the one of which we know least, the production of organic matter in the ocean (Table 4).

In recent years, new interest in the ocean as a potential source of food for the growing world population has led to extensive oceanographic studies. These studies have shown that the biological productivity of the ocean is far from uniform. According to a review published in 1966, the photosynthetic production varies in different parts of the ocean by a factor of 16. The reasons for these wide differences in biological productivity lie in variations in the supply of certain nutrient elements, such as nitrogen, phosphorous, iron, and manganese whose distributions depends strongly on vertical and horizontal water currents.

Photosynthesis in plants is superimposed on the reverse process, respiration, i.e., slow combustion of organic matter to form water and CO_2 with the release of chemical energy in the form of heat and "energy-rich" phosphate bonds. An excess of photosynthesis over respiration is what permits the growth of photosynthetic organisms and storage of food reserves such as starch, oil, and fat.

Hill Reaction

How does the absorption of light by the pigment molecules of the thylakoid membranes result in the conversion of light energy into chemical energy? The answer lies within a discovery made in 1937 by Robert Hill. He found that when leaf extracts containing chloroplasts were supplemented with a nonbiological hydrogen-accepting molecule and then illuminated, evolution of oxygen and simultaneous reduction of the hydrogen acceptor took place, according to the equation:

$$2H_2O + 2A \rightarrow 2AH_2 + O_2$$

in which A is the artificial hydrogen acceptor and AH_2 its reduced form. One of the hydrogen acceptors Hill used was the dye, 2,6-dichlorophenolindophenol, which is blue in its oxidized form (A) and colorless in its reduced form (AH_2). When the leaf extract supplemented with the dye was illuminated, the blue dye became colorless and oxygen was evolved. Neither oxygen evolution nor dye reduction took place in the dark. This constituted the first specific clue to how absorbed light energy is converted into chemical energy: it causes electrons to flow from H_2O to an electron-acceptor molecule. Moreover, Hill found that carbon dioxide was not required for this reaction, nor was it reduced to a stable form under these conditions. He therefore concluded that oxygen evolution can be dissociated from carbon dioxide reduction. The reaction summarized above is known as the Hill reaction and any artificial acceptor A as a Hill reagent.

At this stage a search was begun to identify naturally occurring, biologically active counterparts of the Hill reagent: the electron acceptor in the chloroplasts, that normally accepts hydrogen atoms from water during illumination. Several years later it was found that the coenzyme $NADP^+$ is the natural biological acceptor in chloroplasts, according to the equation

$$2H_2O + 2NADP^+ \xrightarrow{\text{light}} 2NADPH + 2H^+ + O_2$$

A very important characteristic of this reaction is that the net flow of electrons is from water to $NADP^+$, whereas in mitochondrial respiration the net flow is in the opposite direction, from NADPH or NADH to water, with the loss of free energy. Because net electron flow induced in chloroplasts by light is in the reverse or uphill direction (from H_2O to $NADP^+$), it cannot occur without the input of free energy: the energy required to push electrons uphill comes from the light absorbed when the chloroplasts are illuminated.

Utilization of Light Energy in the Hill Reaction

When a chlorophyll molecule in the thylakoid membrane is excited by light, the energy level of an electron in its structure is boosted by an amount equivalent to the energy of the absorbed light and the electron enters a high energy state. The "packet" of excitation energy (the exciton) now migrates rapidly through the cluster of light-harvesting pigment molecules to the reaction center of the photosystem, where it is transferred to an electron of the reaction-center chlorophyll.

This excited electron is expelled from the reaction center and is accepted by the first member of a chain of electron carriers. As a result, the first electron carrier of this chain becomes reduced (gains an electron), whereas the reaction center has become oxidized (has lost an electron). The reaction center in this oxidized condition is said to have an electron "hole". The energy-rich electron, which has a very high reducing potential, now passes from the first electron acceptor downhill along a chain of electron-carrier molecules to $NADP^+$, since the $NADP^+$-NADPH redox couple has a rather negative standard potential -0.32 V (Figure 3).

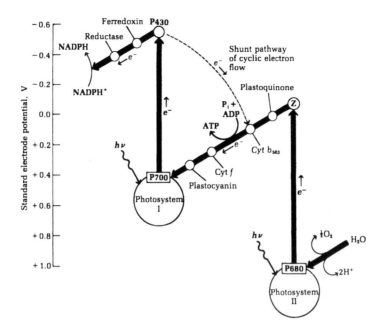

FIGURE 3. The "Z-scheme" of photosystems and electron transport in photosynthesis.[3]

Photosystems I and II

The set of light-harvesting or antenna pigments and its associated reaction center which supplies electrons for the reduction of $NADP^+$ is Photosystem I (PS I), which is maximally excited by light of wavelength 700 nm. It has been found, however, that chloroplasts must be illuminated not only at 700 nm but also at lower wavelengths for maximum rates of oxygen evolution. If chloroplasts are illuminated only at 600 nm and not at 700 nm, there is a large drop in oxygen evolution, called the *red drop*, since 700 nm is at the red end of the spectrum. These observations led to the conclusion that two photosystems absorbing light at different peak wavelengths function together in the oxygen-evolving light reactions of photosynthesis. The diagram in Figure 3, called the Z scheme, outlines the pathway of electron flow between the two photosystems, as well as the energy relationships in the light reactions.

When light quanta are absorbed by PS I, energy-rich electrons are expelled from the reaction center and flow down a chain of electron carriers to $NADP^+$ to reduce it to NADPH. This process leaves an empty electron hole in the PS I reaction center. This hole is in turn filled by an electron expelled by illumination of PS II, which arrives via a connecting chain of electron carriers. This leaves an electron hole in PS II, however, which is in turn filled by electrons from water. The water molecule is split to yield (1) electrons, which are donated to the holes in PS II; (2) H^+ ions, which are released to the medium; and (3) molecular oxygen, which is released to the gas phase. The equation for water cleavage is

$$2H_2O \rightarrow 4H^+ + 4e^- + O_2$$

The Z scheme thus accounts for the complete route by which electrons flow from H_2O to $NADP^+$ according to the equation

$$2H_2O + 2NADP^+ \xrightarrow{\text{light}} O_2 + 2NADPH + 2H^+$$

For each electron flowing from H_2O to $NADP^+$, two light quanta must be absorbed, one by each photosystem. To form one molecule of O_2, which requires the transfer of four electrons from two H_2O to two $NADP^+$, a total of eight quanta must be absorbed, four by each photosystem.

When the reaction center of PS I, a complex of a chlorophyll molecule with a specific protein, is excited by light quanta received from the antenna molecules, there is a decrease in the absorption of light by the chloroplasts at 700 nm. For this reason the reaction center of PS I is designated P700. The first electron-carrier molecule in the chain from P700 to $NADP^+$ is believed to be an iron-sulfur protein called P430. The next electron carrier is ferredoxin, which has been isolated and crystallized, has a molecular weight of about 10700, and contains two iron atoms bound to two acid-labile sulfur atoms. The iron atoms in P430 and ferredoxin transfer electrons via one-electron Fe(II)-Fe(III) valence changes.

The third electron carrier is a flavoprotein called ferredoxin-NADP oxidoreductase. It transfers electrons from reduced ferredoxin (Fd_{red}) to $NADP^+$, reducing the latter to NADPH:

$$2Fd_{red}2 + {} + 2H^+ + NADP^+ \rightarrow 2Fd_{ox}3 + {} + NADPH + H^-$$

Next we have the connecting chain of electron carriers that leads downhill from the excited reaction center of PS II to the empty holes in PS I. The oxidized reaction center of PS II absorbs at 680 nm and is designated P680. The first electron carrier in the chain has not been characterized clearly and is usually called Q. Reduced Q passes electrons downhill to plastoquinone or PQ, a fat-soluble quinone with a long isoprenoid side chain, which resembles ubiquinone of the mitochondrial respiratory chain. Reduced plastoquinone then donates electrons to a b-type cytochrome called cytochrome b_{563}, which in turn passes electrons on to cytochrome f. From cytochrome f electrons go on to plastocyanin, a blue copper protein. The copper atom of this protein undergoes Cu(I)-Cu(II) cycles. Plastocyanin is the immediate donor of electrons to empty holes in P700. The electron holes left in the reaction center P680 of PS II are refilled by electrons removed from water by a little understood Mn^{2+}-containing enzyme complex called H_2O-dehydrogenase.

We speak of the entire set of reactions of Figure 3 as the light reactions of photosynthesis. However, the term light reactions is not entirely accurate since the only points in the light reactions that really require light are the two steps in which the two photochemical reaction centers are excited. All the other steps in photosynthetic electron transport can occur in the dark, once the electrons have been boosted by absorption of light energy.

Photophosphorylation

In 1954, Arnon and colleagues discovered that phosphate is incorporated into ATP during photosynthetic electron transport in illuminated spinach chloroplasts. At the same time, a similar discovery was made by Frankel when he illuminated membranous pigment-containing structures called chromatophores, obtained from photosynthetic bacteria. They concluded that some of the light energy captured by the photosynthetic systems of these organisms is transformed into the phosphate-bond energy of ATP. The process was called photophosphorylation, to distinguish it from oxidative phosphorylation in respiring mitochondria.

Oxidative phosphorylation of ADP to ATP in mitochondria occurs at the expense of the free energy released as electrons flow downhill along the electron-transport chain from substrates to oxygen. Photophosphorylation of ADP to ATP is coupled to the energy released as electrons flow down the photosynthetic electron transport chain from excited PS II to the electron holes in PS I. It seems that only one ATP is formed per pair of electrons passing this connecting chain, although some scientists have suggested that two ATP molecules are formed (Figure 4).

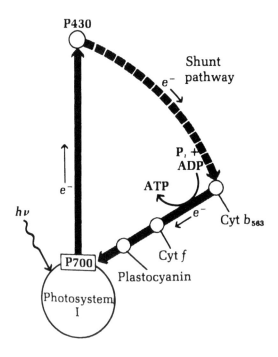

FIGURE 4. Pathway of electron transport in cyclic phosphorylation.[3]

The electron flow from water to PS II to PS I is unidirectional or noncyclic (Figure 3). Another type of light-induced electron flow that can take place in chloroplasts is cyclic electron flow. The latter path involves only PS I, oxygen is not evolved, and NADP$^+$ is not reduced. It is called cyclic because the electron boosted to the first electron acceptor P430 by illumination of P700 instead of reducing NADP$^+$ flows back into the electron hole of PS I by a shunt or bypass pathway. This shunt involves some of the electron carriers of the chain between PS I and II, including the segment that contains the phosphorylation step. Thus, continuous illumination will keep the electrons circulating and generating ATP. Electrons will be excited and ejected from the reaction center of PS I and return to it after flowing downhill around the remainder cycle. The overall reaction for cyclic electron flow and the associated photophosphorylation is

$$P_i + ADP + \text{light energy} \rightarrow ATP + H_2O$$

Cyclic electron flow and photophosphorylation are believed to occur when the photosynthetic cell is already amply supplied with NADPH but requires additional ATP for its metabolic needs.

The processes of photophosphorylation in chloroplasts and of oxidative phosphorylation in mitochondria have several similarities. Like the inner mitochondrial membrane, the thylakoid membrane has an asymmetric molecular organization. The electron transferring molecules in the connecting chain between PS II and PS I seem to be oriented in the thylakoid membrane so that electron flow results in the net movement of H$^+$ across the membrane, from the outside of the thylakoid membrane toward the inner compartment, so that the inside of the vesicle becomes more acidic than the outside. These properties are consistent with the chemiosmotic hypothesis originally proposed for oxidative phosphorylation.

Based on these characteristics of the ATP generating system, Jagendorf and Uribe (1966) performed an important experiment showing that an alkaline-outside pH gradient across the

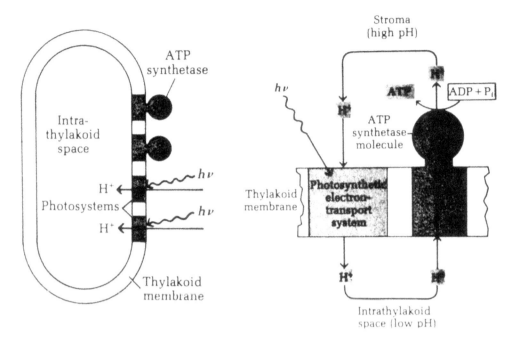

FIGURE 5.   H⁺ movement across the thylakoid membrane of the chloroplast.³

thylakoid membrane can furnish the driving force to generate ATP. They soaked chloroplasts in the dark in a buffer at pH 4, which slowly penetrated into the inside of the thylakoids, lowering their internal pH. ADP and phosphate were added to the dark suspension of chloroplasts and at the same time the pH of the outer medium was raised to 8 by adding an alkaline buffer. This created a large transient pH gradient across the membrane. These conditions were sufficient to allow ATP generation in the dark. The pH gradient required for ATP formation is normally generated in the chloroplast by the electron transport system energized by light (Figure 5).³

CO₂ Assimilation or Photosynthesis of Carbohydrates

The simplest equation expressing the utilization of light energy for the synthesis of carbohydrates is

$$CO_2 + 2H_2O \xrightarrow{8h\nu} [CH_2O] + H_2O + O_2$$

or

$$6CO_2 + 12H_2O \xrightarrow{48h\nu} [CH_2O]_6 + 6H_2O + 6O_2$$

where the term $[CH_2O]$ represents the basic formula of carbohydrates. Glucose would be $[CH_2O]_6$. To convert CO_2 to CH_2O it is necessary to separate H⁺ and electrons, which involves breaking H–O bonds. The energy required to do so is supplied by light quanta absorbed by PS II (see above). The standard free energy for the synthesis of glucose from CO_2 and H_2 is +686 kcal/mol. To generate 6 molecules of O_2, 48 (or 8 × 6) light quanta must be used. Since the energy of 1 Einstein of photons may range from 72 kcal at 400 nm to about 41 kcal at 700 nm (Table 2), the energy of the 48 quanta required to reduce 6 CO_2 molecules to 1 molecule of sugar will be between 1968 and 3456 kcal, depending on the wavelengths of absorbed light. This range is well above the 686 kcal required for the synthesis of 1 mol of glucose.

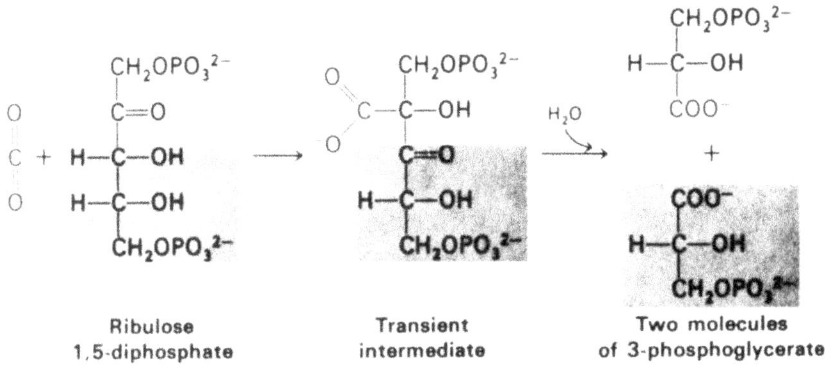

FIGURE 6. CO_2 fixation in the RBP carboxylase reaction. (From Stryer, L., *Biochemistry*, W. H. Freeman, San Francisco, 1974. With permission.)

The path of CO_2 assimilation seems to be essentially the same in all organisms. It is called the Calvin Cycle, the Benson-Calvin Cycle, the Photosynthetic Carbon Reduction Cycle, or the Reductive Pentose Phosphate Pathway. Some plants have, in addition to the central CO_2 reduction path, an auxiliary or preliminary fixation mechanism called C_4 and CAM which will be described later.[4]

The Reductive Pentose Phosphate pathway is characterized by four central features as follows.

Carboxylation of RBP (Ribulose 1,5-Bisphosphate) by Addition of CO_2 to Give 2 Molecules of PGA (3-Phosphoglycerate)

This reaction is catalyzed by RBP carboxylase, a complex and relatively large enzyme with high affinity for CO_2. Work leading to the discovery of this important enzyme was done by Melvin Calvin and his associates in the late 1940s. A suspension of green algae was illuminated in the presence of radioactive carbon dioxide ($^{14}CO_2$) for a few seconds and then quickly killed and extracted. The first compound that became labeled with ^{14}C was 3-phosphoglycerate. Most of the isotope was located in the carboxyl carbon atom of PGA. Further research led to the identification of the enzyme responsible for this reaction (Figure 6).

RBP carboxylase catalyzes the covalent insertion of CO_2 into RBP with the simultaneous cleavage of the resulting 6-carbon intermediate to form two molecules of PGA, only one of which bears the isotopic carbon introduced as $^{14}CO_2$ in its carboxyl group.

RBP carboxylase, which is exclusive to photosynthetic organisms, is a complex structure composed of eight large catalytic subunits and eight small regulatory subunits. It has a molecular weight of 550,000 and is located on the outer surface of the thylakoid membranes. RBP-carboxylase is the most abundant enzyme in the biosphere and one of the key enzymes in biomass production in plants.

Reduction of the PGA Resulting from the Carboxylation of RBP at the Expense of NADPH and ATP to Triose Phosphate

How is PGA converted into sugar? The Berkeley group proposed a complex cyclic mechanism for the total biosynthesis of all six carbon atoms of glucose from carbon dioxide. The series of reactions leading to the formation of glucose is given in the following equations:

(1a) $6CO_2 + 6RBP + 6H_2O \xrightarrow{\text{RBP-carboxylase}} 12\ 3\text{-PGA}$

(1b) $12\ 3\text{-PGA} + 12ATP \xrightarrow{\text{PGA kinase}}$

$12\ 3\text{-phosphoglyceroyl phosphate} + 12ADP$

(1c) $12\ 3\text{-phosphoglyceroyl phosphate}$

$+ 12NADPH + 12\ H^+ \xrightarrow{\text{GAP dehydrogenase}}$

$12GAP \text{ (glyceraldehyde 3-phosphate)} + 12NADP^+ + 12P_i$

(1d) $5GAP \xrightarrow{\text{Triose phosphate isomerase}} 5DHAP \text{ (dihydroxyacetone phosphate)}$

(1e) $3GAP + 3DHAP \xrightarrow{\text{aldolase}} 3FBP \text{ (fructose 1,6-biphosphate)}$

(1f) $3FBP + 3H_2O \xrightarrow{\text{Fructose diphosphatase}} 3F6P \text{ (fructose 6-phosphate)} + 3P_i$

(1g) $F6P \xrightarrow{\text{Phosphoglucose isomerase}} G6P \text{ (glucose 6-phosphate)}$

(1h) $G6P + H_2O \xrightarrow{\text{Glucose-6-phosphatase}} GLUCOSE + P_i$ (1)

To synthesize each molecule of glucose from 6 molecules of CO_2 — 18ATPs and 12NADPHs are utilized. Both ATP and NADPH are replenished by the ongoing light reactions of photosynthesis. This linear or sequential series of reactions presents a great problem: how to continue the supply of RBP, the molecule incorporating CO_2. Calvin and co-workers proposed an elegant solution to the problem, the regeneration of RBP in the carbon-reducing pathway known as the Calvin Cycle or the Pentose Phosphate Reduction Pathway.

Regeneration of RBP

This is a process in which five molecules of triose phosphate are rearranged into three molecules of RBP, the initial acceptor of CO_2 in Reaction 1.

This process of rearrangement of triose phosphate molecules into RBP molecules is not done in a single step. Several reactions are necessary for the gradual regeneration of CO_2-acceptors in the Calvin cycle:

(2a) $2F6P + 2GAP \xrightarrow{\text{transketolase}}$

$2X5P \text{ (xylulose 5-phosphate)} + 2E4P \text{ (erythrose 4-phosphate)}$

(2b) $2E4P + 2DHAP \xrightarrow{\text{aldolase}} 2SBP \text{ (sedoheptulose 1,7-bisphosphate)}$

(2c) $2SBP + 2H_2O \xrightarrow{\text{phosphatase}} 2S7P + 2P_i$

(2d) $2S7P + 2GAP \xrightarrow{\text{transketolase}} 2R5P \text{ (ribose 5-phosphate)} + 2X5P$

(2e) $2R5P \xrightarrow{\text{isomerase}} 2Ru5P \text{ (ribulose 5-phosphate)}$

(2f) $4X5P \xrightarrow{\text{epimerase}} 4Ru5P$

(2g) $6Ru5P + 6ATP \xrightarrow{\text{phosphoribulokinase}} 6RBP + 6ADP$ (2)

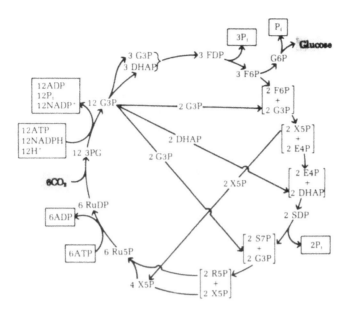

FIGURE 7.   The Calvin Cycle for the conversion of CO_2 into glucose and for the regeneration of RBP.[3]

The sum of reactions 1(a—h) + 2(a—g) is

$$6CO_2 + 18ATP + 12H_2 + 12NADPH + 12H^+ \rightarrow$$

$$[CH_2O]_6 + 18P_i + 18ADP + 12NADP^+$$

Plants and algae that use RBP-carboxylase as the first CO_2 fixation step are called C_3, because the first fixation product is PGA, a 3-carbon compound. The reactions leading from the initial incorporation of CO_2 into RBP and the regeneration of the latter in the Calvin Cycle are schematically outlined in Figure 7.

Autocatalysis and Self-Adjustment

For every three molecules of CO_2 assimilated, one molecule of triose phosphate is produced and serves as the basic element for further synthesis of glucose, starch, sucrose, lipids, amino acids, etc. The triose phosphate can also be converted into RBP, increasing the number of CO_2-acceptor molecules. This reaction constitutes one of the basic elements of dry matter production by photosynthetic organisms. The basic process of CO_2 fixation is entirely self-sufficient (autocatalytic) in the sense that it can produce and increase the amount of its own substrate. If an organism would regenerate precisely the same quantity of substrate as it utilized, then it would have no ability to adjust its rate of photosynthesis to changing conditions (light intensity, temperature, water status, and mineral nutrition). Plants must therefore have at least two mechanisms for the regulation of the pool size of RBP: (1) the capacity to invest photosynthetic products in the synthesis of additional RBP when conditions favor enhanced photosynthesis — the Calvin cycle is a mechanism which meets this essential criterion; (2) the capacity to oxidize or decrease the amount of RBP when conditions are less favorable for photosynthesis — this is achieved by the process of photorespiration, which will be discussed later.

C_4 Photosynthesis or the Hatch-Slack Pathway

Many plants in the tropics, as well as temperate-zone crop plants native to the tropics (corn, sugar cane, sorghum) assimilate CO_2 through a pathway different in its initial steps from the Calvin cycle, called the Hatch-Slack or C_4 pathway. At the outset both the C_3 and the C_4 plants ultimately use the Calvin Cycle reactions. The significant difference is that the C_4 plants possess additional steps in which CO_2 is initially incorporated into a carbon compound with four carbon atoms before it is passed on to PGA. The C_4 pathway would seem unnecessary to photosynthetic organism in water and, indeed, it is unknown in algae which operate exclusively on the basis of the Calvin Cycle.

Photorespiration

Respiration and oxidative phosphorylation are carried out at night in the mitochondria of plant leaves at the expense of photosynthetic products accumulated during the light periods. Measurements of the rates of oxygen and carbon dioxide exchange have shown that C_3 plants respire in the light and consume some oxygen while they carry out oxygen-generating photosynthesis. The respiratory consumption of oxygen in leaves, or photorespiration, is not entirely mitochondrial and is light-dependent.

The major substrate oxidized by photorespiration is glycolic acid. Glycolate is oxidized in the peroxisomes of leaf cells to glyoxylate which can be converted into the amino acid glycine. Glycolate is formed by the oxidative breakdown of RBP by an enzyme which adds O_2 to it and is called RBP-oxygenase. RBP-carboxylase and RBP-oxygenase are the same molecular entity, the names characterizing two different catalytic capacities of the same enzyme molecule. RBP can react with either CO_2 or O_2. When CO_2 concentration is low relative to the concentration of O_2, the addition of oxygen to RBP will increase. The products of the oxygenation of RBP are PGA and phosphoglycolate (Figure 8).[5] Photorespiration in microalgae is described in Chapter 11.

ULTRASTRUCTURE OF MICROALGAE

Prokaryotes (Cyanobacteria)

Cyanobacteria (blue-green algae) are phototrophic microorganisms that carry out oxygenic photosynthesis similar to that of higher plants. Of the principal subgroups the chroococcacean and the pleurocapsalean cyanobacteria have the simplest structural organization since they are unicellular rods or cocci existing singly or in aggregates. In the remaining subgroups the unit of structure is a filament of cells, or trichome. The trichomes of cyanobacteria may consist entirely of vegetative cells, or may also contain (under certain growth conditions) structurally and functionally differentiated cells called heterocysts and akinetes (spores).

The internal organization of the cyanobacterial cell is prokaryotic, yet it is considerably more complicated structurally than most bacteria. The structural features of the cell observable with the light microscope include a central region (centroplasm or nucleoplasm) rich in nucleic acid, a peripheral region (chromoplasm) containing the photosynthetic thylakoid membranes and various inclusions, and several enveloping layers consisting of the plasmalemma, a pellicular wall, and often, a layer of mucilage (Figures 9 and 10).[6,7]

Envelope

Transmission electron microscopy permits further resolution of the cellular components. The cell wall consists of four layers, designated as LI, LII, LIII, and LIV, starting with the innermost layer. The plasmalemma or plasma membrane surrounds the protoplast directly underneath the cell wall[8] and is in contact with the thylakoid membranes at several locations. The thylakoid membranes form an interconnecting network of concentric shells, merging only at the inner surface of the plasmalemma.[7]

Inclusions

Four classes of inclusions are universally present in vegetative cells of cyanobacteria:

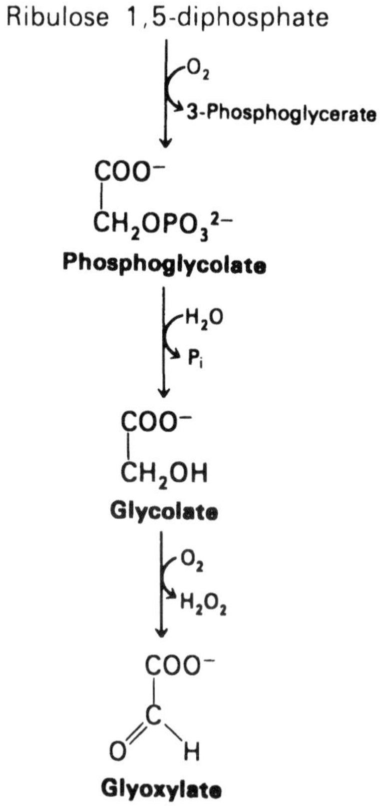

Ribulose 1,5-diphosphate

$$COO^-$$
$$CH_2OPO_3{}^{2-}$$
Phosphoglycolate

$$COO^-$$
$$CH_2OH$$
Glycolate

$$COO^-$$

Glyoxylate

FIGURE 8. Oxygenation of RBP and the products of the reaction. (From Bassham, J., in *Biochemistry*, Stryer, L. Ed. 1981. With permission.)

polyphosphate granules, glycogen granules, cyanophycin granules, and carboxysomes. The polyphosphate granules (volutin, metachromatin) are large metachromatic granules located adjacent to the centroplasm and are identified on the basis of their staining and solubility properties. These granules disappear upon culture in the absence of phosphate and reappear following the addition of phosphate to the culture medium.

The glycogen granules are found between the photosynthetic lamellae and can appear as crystals, rods, or discs with a special affinity for lead hydroxide stain. These granules are composed principally of glycogen as proven by iodine staining and chemical analysis. Glycogen granules disappear with prolonged darkness and reappear following illumination, a phenomenon that confirms their being composed of photosynthetic reserve material.

Cyanophycin granules (structured granules) are refractive granules present in vegetative cells and prominent in spores (akinetes), and they consist of high molecular weight copolymers of aspartic acid and arginine (1:1 ratio). These granules stain with certain protein-specific reagents but not with others and their synthesis is not inhibited by inhibitors of protein synthesis.

Carboxysomes are polyhedral granules surrounded by a membrane. They are composed of tightly packed molecules of RBP-carboxylase.[8,9]

Photosynthetic Apparatus

The plane of the photosynthetic membranes is usually parallel to the cell surface. Most often the thylakoid membranes appear compressed, but they separate considerably upon

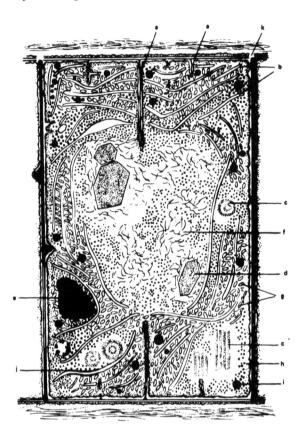

FIGURE 9.   Diagram of a section of a cell from a blue-green algal filament.[6] (a) Cross wall, (b) thylakoids or photosynthetic membranes, (c) cylindrical body seen in cross section (upper) and longitudinal section (lower), (d) polyhedral body, also found in bacteria, (e) structured granule, also called cyanophycin granule, (f) DNA standards in the "centroplasm", (g) glycogen granules, (h) ribosomes, (i) lipid globule, (j) phycobilosomes, and (k) pore. (From Trainor, F. R., *Introductory Phycology*, Wiley-Interscience, New York, 1978. With permission.)

growth at high light intensity. Macromolecular aggregates called phycobilisomes, consisting of phycobiliproteins are attached to the outer surface of the photosynthetic lamellae. Phycobilisomes are about 7 million daltons large in cyanobacteria and vary greatly in gross morphology. They can be isolated intact, as judged by spectroscopic criteria and electron microscopy. Phycobilisomes consist entirely of phycobiliproteins and proteins that function in the assembly of the particle. Quantitatively the major biliproteins are phycocyanin, allophycocyanin, and allophycocyanin B. Some cyanobacterial phycobilisomes also contain phycoerythrin. In phycobilisomes containing both phycoerythrin and phycocyanin, the periphery contains phycoerythrin, the area proximal to the core contains phycocyanin, and the core itself contains allophycocyanin. From studies of mutants and chromatic adaptation it is evident that the structure is assembled in a stepwise manner outward from the thylakoid membrane. The phycobilisome appears to be anchored to the thylakoid membrane through a segment of a high molecular weight biliprotein.

Phycobilisomes serve as antennas for the capture of light energy. The energy transfer pathway in phycobilisomes starts with phycoerythrin which transfers the energy to phyco-

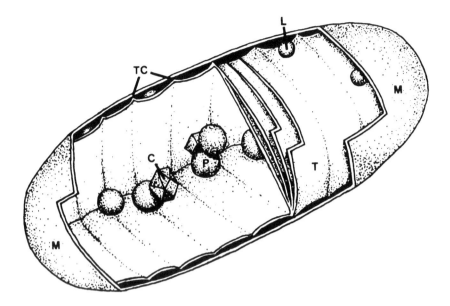

FIGURE 10. Artist's representation of the overall three dimensional architecture of *A. quadu-plicatum*. (C) Carboxysome, (L) lipid body, (P) polyphosphate body, (D) photosynthetic thylakoid membrane system, and (TC) contacts between thylakoids and cytoplasmic membrane. Thylakoids are depicted as solid sheets, each representing a pair of closely apposed unit membranes. The spacing between thylakoids-cytoplasmic membrane contact points is theoretical, as this was not determined precisely (illustration does not include cell wall, ribosomes, and nuclear material). (From Nierzwicki-Bauer, S. A. et al., *J. Cell. Biol.*, 97, 713, 1983. With permission.)

cyanin. Phycocyanin, in turn, transfers energy to allophycocyanin. Phycobilisomes donate energy primarily to PS II. Cyanobacteria do not contain chlorophyll b and as much as 50% of their light harvesting capacity resides in the phycobiliproteins.

Chlorophyll and the carotenoids are localized in the thylakoid lamellae, which can be isolated by differential centrifugation, and account for about 26% of the dry weight of the intact cells. Light harvested by chlorophyll is almost exclusively transferred to PS I and can only mediate cyclic photophosphorylation. PS II activity is almost entirely dependent upon energy transferred from the phycobilisomes.[10]

Eukaryotes

Despite the tremendous variation in form, organization, and size of eukaryotic algae, their cells contain essentially the same organelles as do other eukaryotic organisms. Nuclei, mitochondria, plastids, Golgi apparatus, endoplasmic reticulum, ribosomes, and vacuoles are universally present, and additional structures such as pyrenoids, contractile vacuoles, trichocysts, and haptonemata may also be present (Figures 11 and 12).[6,11]

Envelope

Some unicellular algae are surrounded only by a plasma membrane, which may be enclosed in mucilage, but the majority of algae are surrounded by cell walls. There is a wide range of form and composition of cell coverings. The mucilage of red algae is composed of sulfated polysaccharides. In members of the Prasinophyceae and the Haptopyceae families the cells are externally covered with scales of little-known chemical composition, often of more than one type. In addition to scales, some cells bear calcified bodies known as coccoliths. Some algae possess cell walls consisting largely of cellulose and hemicellulose, occasionally with oriented cellulose microfibrils. Most cell walls contain considerable amounts of protein. Some members of the Dinophyceae are covered by a wall composed of cellulose plates and

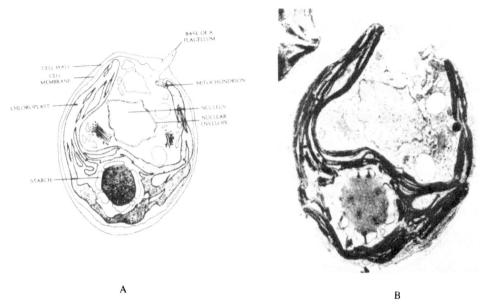

A   B

FIGURE 11.   Electron micrograph of *Chalamydomonas*, a photosynthetic eukaryotic cell.[11]

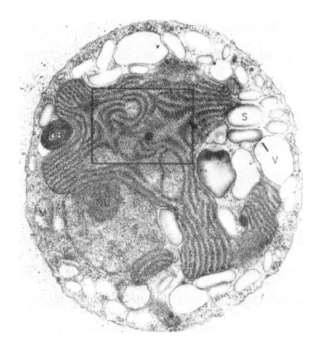

FIGURE 12.   Cell of the red alga *Porphyridium* with the membrane-bound nucleus, plastid, and mitochondrion. Rows of phycobilisomes are visible in the boxed area.[6] (S) Starch, (V) vacuoles, (Nu) nucleus, (M) mitochondrion, and (Cl) chloroplast. (From Trainor, F. R., *Introductory Phycology*, Wiley-Interscience, New York, 1978. With permission.)

many Euglenophyceae are surrounded by a pellicle consisting of helically arranged, flexible overlapping strips largely composed of protein. Bacillariophyceae (diatom) cells are surrounded by perforated frustules composed of silica and occasionally also some chitin. In the larger multicellular green, brown, and red algae, highly oriented polysaccharides such as cellulose, xylan, and mannan constitute the basic cell wall structure. In addition, sulfated polysaccharides and other compounds contribute to mechanical flexibility. Calcification of the walls also occurs in some members of the green, brown, and red algae. In many multicellular algae cytoplasmic continuity is provided by pores or pits in the walls between adjacent cells. In some red algae, apertures in the walls between neighboring cells contain a discrete, membrane-bound plug.

Nucleus

As in other eukaryotes, algal nuclei are enclosed in a nuclear envelope consisting of two unit membranes separated by a perinuclear space. The nuclear envelope is frequently continuous with the endoplasmic reticulum and it is characteristically perforated by pores 80 to 90 nm in diameter. Nuclear pores probably permit the passage of large molecules such as RNA and protein in and out of the nucleus. The nucleoplasm often contains one or two nucleolar regions rich in RNA.

Mitochondria

Mitochondria can be of various shapes and sizes. A few types of algae have only one mitochondrion per cell. The mitochondria occasionally branch profusely and can reach overall lengths exceeding the diameter of the cell. The basic structure includes a smooth outer membrane surrounding a highly infolded inner membrane that encloses a central lumen or matrix. The infoldings of the inner membrane (cristae) are occasionally swollen at the tips. The mitochondria contain circular DNA and ribosomes that are smaller than those of the cytoplasm (16 nm in diameter vs. 22 nm).

Golgi Apparatus

The Golgi apparatus is important in intracellular transport. The number of Golgi bodies per cell varies from one to several thousand. The stack of parallel, disc-shaped cisternae constituting a Golgi body are often slightly curved so as to give the organelle concave (secreting) and convex (forming) faces. Analysis of Golgi membranes shows that those on the forming face are similar to the membranes of the nucleus and endoplasmic reticulum and those on the secreting face resemble the plasma membrane, the central ones being intermediate in composition. In addition to serving as part of the internal transport system in the cell, the Golgi complex is involved in the formation and packaging of substances for extracellular transport. The substances packaged in this way include polysaccharides, glycoproteins, mucopolysaccharides, and lipoproteins. Often complex molecular structures are assembled in the Golgi apparatus. Examples include surface scales of Prasinophycean algae and the scales and coccoliths of some marine coccolithophorids. The activity of the Golgi complex as a dynamic membrane system involved in the synthesis and organized transport of soluble and insoluble molecules varies with metabolic conditions and with the stage of cell development.

Locomotory Structures

The majority of unicellular algae are motile due to the presence of one, two, four, or many flagella. Flagella consist of a cylinder of nine double, peripheral, longitudal fibers enclosing two single axial fibers, the whole being surrounded by a membrane that is continuous with the plasma membrane. Flagella may be smooth, i.e., devoid of hairs (whiplash) or bear hairs which in some algae are called mastigonemes (Flimmer). Some flagella bear appendages such as spines.

Haptophyceae possess a different type of filamentous appendage called a haptonema. A haptonema is anatomically and functionally different from a flagellum. The number and the arrangement of the fibers is also different. Haptonemata usually serve as the means of attachment to the substratum.

Photosynthetic Apparatus

The photosynthetic apparatus includes the entire complex of structures and components responsible for the photosynthetic processes of the organism. In algae, as in higher plants, this includes the pigment-containing structures, together with the components associated with the photochemical reactions and a set of components involved with carbon dioxide fixation. In eukaryotic algae the entire photosynthetic apparatus is located in the chloroplast, a cytoplasmic organelle enclosed in an envelope consisting of a pair of bilayer membranes. The matrix of the organelle, called the stroma, contains the photosynthetic lamellae (thylakoids). Electron-translucent DNA-containing areas (genophores) and 70S ribosomes (found also in mitochondria) are also seen in the chloroplast, and in certain algae, also eyespots, pyrenoids, and various crystalline inclusions.[12]

Thylakoids

The arrangement of thylakoids in various groups of algae can be classified on the basis of electron microscopic observations as follows:

• **Type I:** No association between individual thylakoids (thylakoids neither banded nor stacked); phycobilisomes located on the surfaces of the thylakoids; no endoplasmic reticulum in the chloroplast; peripheral thylakoids present or absent; when present those enclose the internal set of thylakoids; this type of arrangement is typical of the Rhodophyceae.
• **Type II:** Thylakoids associated into bands; no peripheral thylakoids; some nonaggregated phycobiliproteins within the intrathylakoid spaces; chloroplast endoplasmic reticulum is present; this type is found in the Cryptophyceae.
• **Type III:** Association of thylakoids into three bands; no grana; chloroplast endoplasmic reticulum present; this type of arrangement is found in the Dinophyceae, Haptophyceae, Eustigmatophyceae, Xanthophyceae, Chrysophyceae, and Phaeophyceae.
• **Type IV:** Bands of three fused thylakoids; grana present; found in Euglenophyceae.
• **Type V:** Bands of 2 to 6 fused thylakoids; grana present; present in Chlorophyceae, Prasinophyceae, and Charophyceae.

Although some variations are found within each type of thylakoid arrangement, this feature of algal ultrastructure is extremely useful in taxonomic and possibly in phylogenetic determinations; however, the indications of thylakoid structure must be used with caution since only a few representatives of each class of algae have been investigated to date, and the range and distribution of variations is not fully known.

The structure of algal thylakoids is very similar to that of higher plants. The thylakoid membrane of *Scenedesmus* seems to include spherical subunits approximately 10 nm in diameter. Freeze etching reveals that the fractured faces of algal thylakoids are similar to those described in higher plant chloroplasts.

Whenever thylakoids are closely appressed to form partitions, the association formed is usually strong and able to survive osmotic shock and such drastic treatments as sonication. The adhesion between thylakoids is probably due to electrostatic bridges formed by divalent ions, localized protein-protein interactions, or localized hydrophobic bonding. Apparently some component of PS II is involved in thylakoid stacking. As a result, stacked thylakoids are enriched in PS II while stroma thylakoids are relatively rich in PS I.

The assembly of the chloroplast thylakoid has been studied in some detail in *Chlamy-*

domonas reinhardtii. The chloroplast thylakoids neither develop from the chloroplast envelope nor the pyrenoidal membranes. New material is incorporated into preexisting membranes by a multistep process. The membrane is initially built up with a relatively small number of components and the residual constituents are incorporated at later stages. The new membrane becomes fully functional only after a certain lag period.

Pyrenoids

The pyrenoid is a differentiated region of the chloroplast occurring in many algal species. It can be within the chloroplast or project from its surface. In the Chlorophyceae and the Prasinophyceae the embedded pyrenoid is surrounded by a sheath of starch, while in certain diatoms and dinoflagellates it is wrapped in a membrane. Occasionally the pyrenoid is recognizable by the density of its matrix, which consists of granular or fibrous proteinaceous material. In some members of the Phaeophyceae, Euglenophyceae, Dinophyceae, and Haptophyceae, stalked pyrenoids were observed. These pyrenoids are bordered by the chloroplast envelope and outside the envelope there is a "pyrenoid cap", i.e., a swollen membrane sack. In some algae pyrenoids are present only during certain phases of the life cycle. In general, the presence of pyrenoids is associated with a high rate of metabolic activity and the accumulation of reserve products. The pyrenoid is thought to function in the conversion and translocation of photosynthetic products in chloroplasts. The matrix of the pyrenoid is probably the site of concentrated enzymes and substrates for these functions. In some algae the storage polysaccharide is closely associated with the pyrenoid, while in others the pyrenoid may function in a different manner. Whenever a pyrenoid is present starch grains form a sheath around it, whereas in its absence starch is randomly deposited inside or outside the chloroplast. In algae that accumulate reserve polysaccharides other than starch (i.e., cyanophycean starch, floridean starch in red algae, laminaran in brown algae, leucosin in chrysophytes and diatoms, paramylon in euglenoids, etc.), a spatial relationship between the polysaccharide and the pyrenoid occurs only when the pyrenoid is of the stalked type. Pyrenoids can form either *de novo* following cell division or by the division of a preexisting pyrenoid. Due to the difficulty of isolating the structure, its precise composition as well as the nature of its various activities remain obscure.

Eyespots

In some motile algae tightly packed, photosensitive, carotenoid-containing globules serve as a primitive photoreceptor also called "eyespot" or "stigma". Some eyespots are located within the chloroplast and of these a few appear to be associated with the flagella.

Chloroplast Division

Chloroplasts are always derived from preexisting chloroplasts. In most algae the chloroplasts divide by fission, producing two smaller daughter chloroplasts. In others, the chloroplasts develop from proplastids derived from mature plastids. The genophore of the chloroplast contains its genetic material and the number of genophores per chloroplast may vary from 1 to about 100. The DNA molecules within the genophore are attached to the thylakoid membranes. Since multiple copies of the chloroplast genome exist within each chloroplast, the complete genome can be transmitted to daughter chloroplasts even when division is unequal as proplastid formation.

Chloroplasts of different classes of algae containing similar pigments apparently also have similar thylakoid structure. It can probably be safely assumed that thylakoid structure and pigment composition in algae evolved interdependently.

Evolution of Chloroplasts

The endosymbiotic origin of several eukaryotic organelles, especially chloroplasts, is well supported by morphological and biochemical comparisons.[13] Several examples are known of unicellular eukaryotes with endosymbiotic cyanobacteria still recognizable as such by, for example, vestigial cell wall structures. *In situ* the endosymbiotic cyanobacteria are termed cyanelles. Endosymbiosis of cyanobacteria is considered a likely origin of chloroplasts in the Rhodophyta (red algae). The origin of chloroplasts in the Chlorophyta and higher plants is more problematical, because of their characteristic content of chlorophyll b, which is conspicuously absent from Cyanobacteria and Rhodophyta. A more probable candidate for the role of the endosymbiont having given rise to the chloroplasts of chlorophyll b containing plants is Prochloron, a prokaryotic organism containing chlorophyll b, lacking phycobilisomes, and displaying appression of thylakoids and formation of rudimentary grana stacks, all typical of higher plant chloroplasts.

The remaining orders of eukaryotic algae, the Chromophyta, are more diverse, including both chlorophyll c and d containing algae and the chlorophyll b containing Euglenophyta. There is reasonable support for the suggestion that these arose from several occurrences of endosymbiosis between two or even more eukaryotic organisms.[14]

PHOTOSYNTHETIC PIGMENTS

The first act of photosynthesis is light absorption. Of all the electromagnetic radiation falling on photosynthesizing plants only the visible light (wavelength 400 to 720 nm) is absorbed and used for photosynthesis. The pigments involved in the light harvesting process of a given algae can often be identified by the measurement of the absorption spectrum (absorbance as a function of wavelength) and the action spectrum (efficiency of oxygen evolution as a function of wavelength) of its photosynthesis.

Algae contain three major groups of pigments: (1) chlorophylls (Chl) that absorb blue and red light, e.g., Chl a, present in all algae and Chl b, present in green algae; (2) carotenoids that absorb blue and green light, e.g., β-carotene present in all algae and fucoxanthin, a characteristic pigment of brown algae; (3) phycobilins that absorb green, yellow, and orange light, e.g., R-phycoerythrin, a pigment of red algae and C-phycocyanin, found in cyanobacteria.

These pigments provide the algae with antennae to capture the light energy. It appears that most pigments are present in both photosystems but in different proportions. PS II of the green algae has a lower ratio of Chl a/b than PS I, and in phycobilin containing algae the majority of these pigments is found in association with PS II.

A major identifying characteristic of a pigment molecule is its absorption spectrum; however, absorption spectra of most pigments change upon extraction, and different spectra of the same pigment molecule will be obtained in different solvents and in different temperatures. Therefore, absorption spectra can be compared only when obtained under similar conditions.

Chlorophylls

A certain complication arises from the fact that in algae and higher plants chlorophyll is complexed with protein in vivo. The various chlorophyll-protein complexes have different absorption spectra depending on the protein as well as the chlorophyll species. Chlorophyll a occurs in all algae. Two forms (Chl a 670 and 680) can be directly observed in several algae at room temperature. Usual room temperature absorption spectra show only a broad band at 675 nm for chlorophyll a in the red. Cooling the algae to 77 K allows the resolution of the band into Chl a 670 and 680 and the observation of a new band at 705 to 710 nm.

Chlorophylls b, c, and d are limited to certain species of algae (Table 5).[15] For definite identification of the chlorophylls, these have to be first extracted using a suitable solvent (80 to 90% acetone, 90% methanol, or *N,N*-dimethyl formamide). Following the extraction the individual pigments must be isolated and purified by chromatography. The concentration

Table 5
DISTRIBUTION OF CHLOROPHYLLS AMONG
THE ALGAE

| | Chlorophyll | | | |
Algal group	Chl a	Chl b	Chl c	Chl d
Cyanophyceae	+	−	−	−
Rhodophyceae	+	−	−	+
Cryptophyceae	+	−	+	−
Dinophyceae	+	−	+	−
Rhaphidophyceae	+	−	+	−
Chrysophyceae	+	−	+	−
Haptophyceae	+	−	+	−
Bacillariophyceae	+	−	+	−
Xanthophyceae[a]	+	−	+	−
Phaeophyceae	+	−	+	−
Prasinophyceae	+	+	−	−
Euglenophyceae	+	+	−	−
Chlorophyceae[b]	+	+	−	−

Note: + Means the chlorophyll is present in at least some members, −
means the chlorophyll has not been recorded for any member.

[a] Includes Eustigmatophyceae.
[b] Includes Charophyceae.

From *Algal Physiology and Biochemistry*, Stewart, D. P., Ed., Blackwell
Scientific, London, 1974. With permission.

of chlorophylls is generally determined spectrophotometrically using the specific extinction
coefficient of the chlorophyll in the particular solvent. The specific extinction coefficient is
defined as A/dC, where A is the absorbance of the extract, d the inside pathlength of the
spectrophotometer cuvette in centimeters, and C is the concentration of the pigment in grams
per liter.

The biosynthetic pathway for chlorophyll formation in algae is similar to that of higher
plants and has been reviewed recently by Rebeiz and Lascelles;[16] however, algae in general
differ from higher plants in the last steps of chlorophyll synthesis.

Most higher plants grown in the dark accumulate protochlorophyllide (PChl) and its
conversion to chlorophyll a is light protochlorophyllide (PChl) and its conversion to chlorophyll a is light dependent. Most algae, on the other hand, synthesize chlorophyll while
growing heterotrophically in the dark and very little is known about this dark conversion of
PChl. Algae that requires light for chlorophyll biosynthesis include strains of *Euglena,
Ochromonas,* and *Cyanidium caldarium;* mutants of *Chlamydomonas, Chlorella,* and *Scenedesmus,* and glucose grown *Chlorella prototothecoides*. The photoreceptor for the conversion
in most algae and higher plants is apparently protein-bound protochlorophyllide (holochrome).

Primary control over the chlorophyll and heme biosynthetic pathway in algae is generally
assumed to operate at the level of 5-aminolevulinate (ALA) formation. In addition the
accumulation of the pigment appears to be intimately coordinated with thylakoid development
and photosynthetic activity. Chlorophyll biosynthesis is generally reported to be sensitive
to inhibitors of both cytoplasmic and organelle ribosomes (cycloheximide and chloramphenicol, respectively). The general conclusion of these studies is that chlorophyll synthesis,
though it may occur in plastids, is under nuclear control, and that later steps require cooperation between nuclear and plastid genomes for the synthesis of the structural components
of the photosynthetic apparatus.

Several nutrients have a marked effect on chlorophyll formation in algae. Deficiencies of iron, nitrogen, and magnesium inhibit chlorophyll synthesis and accumulation. An abundance of organic carbon in the medium and high light intensities also inhibit chlorophyll formation in some algae.[15]

Carotenoids

Carotenoids are tetraterpenes made up of eight branched, 5-carbon (isoprenoid) units. The first C-40 polyene formed biosynthetically is phytoene, which is stepwise desaturated to form lycopene. Lycopene is probably the precursor of all the carotenoids found in algae.[17]

β-Carotene is a typical carotene found in all algae and higher plants. The more complex carotenoids may be derived from certain carbon skeletons by a progression of chemical modifications, with accompanying changes in their oxidation states. The variety of carotenoids in algae is greater than in higher plants. Several groups of algae have common names that reflect their carotenoid content, e.g., the brown algae, Phaeophyceae, which contain several xanthophylls, most notably fucoxanthin. Most carotenoids are yellow or orange, but their color can be masked by the predominant chlorophyll.

Many types of xanthophylls exist in algae. The similarity between higher plants and Chlorophyceae is again attested by the high lutein content of these algae. The Rhodophyceae and Cryptophyceae also contain considerable amounts of this pigment. The other important xanthophylls are myxoxanthin and myxoxanthophyll, characteristic only of the blue-green algae; peridinin, the primary xanthophyll of the Dinophyceae; and fucoxanthin, present primarily in members of the Phaeophyceae and Bacillariophyceae.

Often it is difficult to determine the precise absorption band of any one carotenoid in algae because besides the usual presence of more than one type of carotenoid, their absorption bands also overlap with those of the chlorophylls. In general, the absorption bands for the carotenoids in vivo lie between 400 and 540 nm.

Phycobilins

The water-soluble pigments, the phycobilins, are found in abundance only in the blue-green and red algae, although trace amounts of chemically different types of phycobilins have been found in isolated species of other algal groups. Members of the Cyanophyceae mostly contain the blue pigment phycocyanin, although several species may contain the red phycoerythrin in addition to, or in place of, phycocyanin. The reverse is true of members of the Rhodophyceae. Both groups also contain small amounts of allophycocyanin. The existence of these pigments is very important for the light-harvesting capabilities of these algae as both groups contain, in addition to the phycobilins, only chlorophyll a (if chlorophyll d is considered to be an artefact of isolation) and carotenoids. The phycobilins fill in, or at least narrow, much of the light-energy gap left by chlorophyll a and the carotenoids, allowing the algae to use the solar radiation much more efficiently in photosynthesis in a manner like that of fucoxanthin in the brown algae. This is possible because of the absorption characteristics of phycobilins: phycocyanins usually have a broad absorption band around 620 nm, phycoerythrins around 545 nm, and allophycocyanin around 650 nm in vivo.

REFERENCES

1. **Keeton, W. T.,** *Biological Science,* W. W. Norton, New York, 1980.
2. **Nobel, P. S.,** *Introduction to Biophysical Plant Physiology,* W. H. Freeman, San Francisco, 1974.
3. **Lehninger, A. L.,** *Principles of Biochemistry,* Worth Publishers, New York, 1982.
4. **Edwards, G. and Walker, D. A.,** C_3, C_4: *Mechanisms, and Cellular and Environmental Regulation, of Photosynthesis,* Blackwell Scientific, Oxford, 1983.

5. **Stryer, L.,** *Biochemistry,* W. H. Freeman, San Francisco, 1981.

6. **Trainor, F. R.,** *Introductory Phycology,* John Wiley & Sons, New York, 1978.

7. **Nierzwicki-Bauer, S. A., Balkwill, D. L., and Stevens, S. E., Jr.,** Three dimensional ultrastructure of a unicellular cyanobacterium, *J. Cell Biol.,* 97, 713, 1983.

8. **Wolk, C. P.,** Physiology and cytological chemistry of blue-green algae, *Bacteriol. Rev.,* 37, 32, 1973.

9. **Stanier, R. Y. and Cohen-Bazire, G.,** Phototropic prokaryotes: the cyanobacteria, *Ann. Rev. Microbiol.,* 31, 225, 1977.

10. **Glazer, A. N.,** Comparative biochemistry of photosynthetic light-harvesting systems, *Ann. Rev. Biochem.,* 52, 125, 1983.

11. **Curtis, H.,** *Biology,* Worth Publishers, New York, 1983.

12. **Bisalputra, T.,** *Plastids,* in *Algal Physiology and Biochemistry,* Stewart, W. D. P., Ed., University of California Press, Berkeley, 1974, 124.

13. **Margulis, L.,** *Origin of Eukaryotic Cells,* Yale University Press, New Haven, Conn., 1970.

14. **Wraight, C. A.,** Current attitudes in photosynthesis research, in *Photosynthesis,* Vol. 1, Govindjee, Ed., Academic Press, New York, 1982, 1.

15. **Meeks, J. C.,** Chlorophylls, in *Algal Physiology and Biochemistry,* Stewart, W. D. P., Ed., University of California Press, Berkeley, 1974, 161.

16. **Rebeiz, C. A. and Lascelles, J.,** Biosynthesis of pigments in plants and bacteria, in *Photosynthesis,* Vol. 1, Govindjee, Ed., Academic Press, New York, 1974, 699.

17. **Goodwin, T. W.,** Carotenoids and biliproteins, in *Algal Physiology and Biochemistry,* Stewart, W. D. P., Ed., University of California Press, Berkeley, 1974, 176.

CELL RESPONSE TO ENVIRONMENTAL FACTORS

A. Richmond

The outdoor production of algal mass involves a continuous response of the cells to a changing environment, both diurnally and along the seasons. The nature of these environmental factors is elucidated in this chapter.

LIGHT

The major effect of light concerns photosynthesis, discussed by Lips and Avissar in "Photosynthesis and Ultrastructure in Microalgae". There are yet other effects of light which do not relate directly to the process of photosynthesis.

Photoinhibition and Photoinactivation

Impairment of photoautotrophic growth by supraoptimal intensities of light is a well-known phenomenon readily observed in the laboratory as well as in the field. The extent of such photoinhibition is known to depend on the incident irradiance as well as the spectral quality of the light. The time of exposure to a given quantum flux area density is also an important factor governing the onset of photoinhibition. Much attention in the study of photoinhibition was given to the quality of light, both UV and visible light having been implicated as potentially damaging.[1] Adaptations to different quantum flux area densities on the biochemical and physiological level vary greatly. Dinoflagellates investigated by Ryther[2] exhibited saturation of photosynthesis at relatively high intensities of incident radiation, whereas in a marine dinoflagellate *Amphidinium carterae*, maximal rates of photosynthesis were attained at as low an incident irradiance as 15 μE m^2/sec. Photoinhibition in this algae occurred when exposed to 80 or 150 μE m^2/sec (Figure 1).

The enzymatic mechanisms involved in photoinactivation relate to a fundamental sensitivity of the key enzyme of photoautotrophy, ribulose bisphosphate (RuBP) carboxylase, to inactivation by blue wavelength.[3] Of significance to the effects of high light intensities in outdoor cultures is the finding that RuBP carboxylase is sensitive to a number of reactive oxygen species, including singlet oxygen, hydrogen peroxide. and possibly superoxide[4] (see "Treatment of Photooxidation Outdoor Mass Cultures of Microalgae").

As summarized by Soeder and Stengel,[5] quoting many reports, high light intensities may also inhibit the respiration of actively photosynthesizing cells. In *Chlorella* such photoinhibition depended on respiratory activity prior to illumination and was greatest in blue light. The mechanism involved apparently relates to the finding that blue light leads to a destruction of cytochromes. Also, blue light is most active in the inhibition of DNA synthesis by high light during regreening of *Chlorella*.[6] In synchronous cultures of *Chlorella*, sensitivity to high light or to a sudden increase in light varies with age and the destruction of chlorophyll by intense light is favored by high O_2 which stimulates chlorophyllase activity.[7] An interesting aspect of photoinhibition in *Chlorella* was revealed by Lorenzen,[8] who reported that close to the upper threshold of temperature tolerance, *Chlorella* can be grown only in a dark-and-light cycle and not in continuous light.[5]

Light-Shade Adaptation

In unicellular algae, light-shade adaptation is characterized by changes in intracellular pigment content, which is often accompanied by changes in the photosynthetic response and the chemical composition. Soeder and Stengel[5] distinguished two types of adaptive reactions. The most usual is the "*Chlorella* type", which is characterized by an inverse relationship

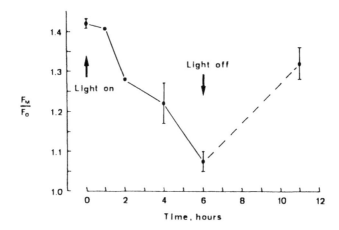

FIGURE 1. Time course for onset of photoinhibition for *A. carterae* grown at 15 μE m²/sec and treated at 350 μE m²/sec.[1] (From Samuelsson, C. and Richardson, K., *Mar. Biol.*, 70, 21, 1982. With permission.)

between the light intensity to which the algae are exposed and their chlorophyll-a content. Light adaptation of this type, then, is mainly accomplished by changes in pigment concentration. The other pattern is exhibited by algae belonging to the *"Cyclotella* type", which show an inverse relationship between light intensity and the activities or concentrations of photosynthetic enzymes.

In the *Chlorella* type chlorophyll metabolism is highly dynamic, i.e., changes in pigment content can occur within a relatively short time and may partially compensate for changes in light intensity by optimizing the ability of the cell to harvest the available light. By themselves, such changes do not confer an adaptive advantage unless the light harvested is transferred to photosynthetic reaction centers.

Falkowski and Owens[9] defined two strategies of light-shade adaptation in marine phytoplankton. Using chlorophyll/P700 ratio, which comprises one method for estimating the average size of the photosynthetic unit, the number and size of photosynthetic units in *Skeletonema costatum* (a diatom) and *Dunaliella tertiolecta* (a chlorophyte) was estimated as a function of light-shade adaptation. In the diatom, light-shade adaptation was characterized primarily by changes in the size and not the number of P700 units, whereas in the chlorophyte, overall changes in chlorophyll content were related to changes in the number and not the size of P700 units. Both strategies of light-shade adaptation effectively harvest and transfer light energy to reaction centers, but the *Skeletonema* strategy was found to be more effective at subsaturating intensities.

Wide variations are known in the time required to adapt to new light intensity — from a few hours to several days. Some species of algae do not have a significant capacity to respond to variations in light intensity and will die quickly when exposed to relatively small elevations in light intensity. Such species are obviously not suitable for mass culture, where the exposure to light may abruptly be greatly elevated after harvest.

Wide variations in photosynthetic characteristics and pigment contents were measured in *Oscillatoria redekei* under a range of photoperiods (6:18 light-dark to continuous light) and irradiances (13 to 260 μE/m²/sec) at 15°C. The light saturated rate of photosynthesis (P_{max}) per cell protein was found to be comparatively constant under different light regimes, but cells grown under low irradiances and/or short light-dark (LD) cycles showed marked increases in the efficiency with which low light was harvested. The increase in efficiency under low light doses corresponded to an increase in the phycocyanin and chlorophyll-a content of the cells, phycocyanin content increasing by a greater proportion than chloro-

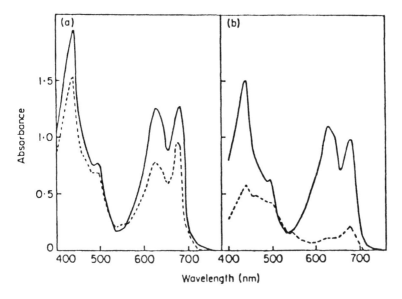

FIGURE 2. Absorption scan of 100 mg dry weight *Oscillatoria redekei* cells grown at 13 μE/m^2/sec (—) and 250 μE/m^2/sec (---) under a 6:18 LD cycle (a) and continuous light (b). (From Foy, R. H. and Gibson, C. E., *Br. Phycol. J.*, 17, 183, 1982. With permission.)

phylla. The increase in the ratio of phycocyanin to chlorophyll-a was highly correlated with increases in the efficiency of light harvesting.[10] In continuous light (LL) cultures, the cells were bleached under high irradiance and had a yellow-brown appearance. LD cells, on the other hand, retained a green or blue-green color under all irradiances. This is demonstrated in Figure 2 which shows in vivo spectra of 6:18 LD and LL cultures grown under the highest and lowest light intensities employed. The high light (LL) cells showed weak pigmentation throughout with a particularly large drop in the phycocyanin peak at 628 nm compared to the low light cells. The 6:18 LD low light cells were similar to the LL low light cells, but there was a much smaller reduction in absorption of the light cells under a 6:18 LD cycle than was observed for high light LL cells.

Changes in the composition of 6:18 LD and LL cells are plotted in Figure 3 and show that the pigmentation of 6:18 LD cells resembles that of low light LL cells. However, the pigment content of 6:18 LD 250 μE/m^2/sec cells was greater than LL cells receiving the same light dose spread over 24 hr. Under continuous light the carbohydrate content of the cells increased with increasing light intensity. By contrast, in 6:18 LD cells harvested at the end of the dark period, the levels of carbohydrates was relatively low at all light intensities. The range of compensation irradiances for continuous light cells was 4 to 82 μE/m^2/sec compared to 3 to 6 μE/m^2/sec for 6:18 LD cells. In conclusion, Foy and Gibson[10] suggested that the large difference between continuous light and LD cells was due in part to the high respiration rates of continuous light cells.

In addition to changes in light intensity and light regime, changes in the spectral quality of the incident light can have a profound effect on several metabolic processes. According to Owens and Esaias,[11] the major process operating in this context is the exit point of fixed carbon from the Calvin Cycle leading to a relative increase in the production of proteins under blue light and carbohydrates under red light.

Chromatic adaptation of some cyanobacteria is manifested by changes in phycobiliprotein and carotenoid synthesis. In cells of *Lyngbya-Plectonema Phormidium* (L.P.P.) sp., *Frenyella diplosiphon*, and *Tolypothrix tenuis* grown in red light, the total carotenoid content of L.P.P. sp. nearly doubled with a selective increase in the concentrations of β-carotene and

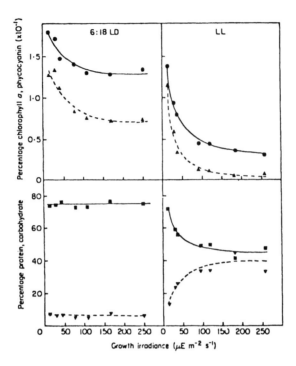

FIGURE 3.   Percentage composition of *Oscillatoria redekei* cultures grown under a 6:18 LD cycle and continuous light (LL) at different irradiances: ● Chlorophyll-a, ▲ phycocyanin, ■ protein, ▼ carbohydrate. (From Foy, R. H. and Gibson, C. E., *Br. Phycol. J.*, 17, 183, 1982. With permission.)

zeaxanthin compared with the effect of white light under the same conditions of illumination. In contrast, *F. diplosiphon* showed a 45% and *T. tenuis* a 21% reduction in total carotenoids with no significant selective effect on individual carotenoids.[12] The direct relationship between total carotenoid content and phycobilisome size which was found for L.P.P. sp. and *F. diplosiphon*, was suggested to be associated with the protective effect of carotenoids against photosensitized oxidation or photoxidation.[12]

Effect of Light on Algal Movement

Movement, active or passive, is widespread among the algae. Motile vegetative cells occur in all groups with the exception of brown algae, in which motility is restricted to gametes and zoospores. Movement can be observed in single cells as well as in colonies and filaments. Moreover, intracellular movements occur in many algae. Several different mechanisms affecting movement have been described. In some organisms visible organelles exist which are responsible for movement, while in others, no corresponding organelles or structures have been found. The speed of movement varies widely depending on the different mechanisms and on internal and external factors.[13]

The study of movement-response of algae is of great interest from the standpoint of mass cultures. This is because harvesting the algal mass represents an important segment of the cost of production, particularly in the small, unicellular algae. Movement of algae in response to stimuli may be exploited, in principle, to concentrate the biomass, facilitating thereby efficient removal of the algal cells from the growth medium.

Several types of movement are recognized: movements caused by swinging organelles, such as flagella and cilia; amoeboid movements; movements due to active contractions of contractile elements inside the cells and gliding movement. The speed of movement is extremely variable, ranging from a maximal rate of 0.5 μm/sec for *Anabena variabilis*, 1 or 2 μm/sec for *Porphyridium cruentum* and the chlorophyte *Micrasterias denticulata*, and up to approximately 200 μm/sec for *Chlamidomonas reinhardtii* and *Euglena gracilis*.[13]

Irrespective of its mechanism, movement of algae can be affected by several external factors, such as light, temperature, gravity, various chemicals, and electric and mechanical stimuli. Furthermore, movement can be influenced by the same stimulus in different ways, resulting in quite distinct types of responses. Two main types have been distinguished: kinesis (an effect on the speed of movement) and taxis (a reaction that leads to a distinct pattern in spatial arrangement or distribution of the organisms).[13]

Light is one of the most important ecological factors affecting the movement of almost all the motile algal species investigated so far and all the reaction types mentioned above have been observed. Responses may be obtained not only to visible light but also to UV and IR radiation.

Photo-Phobotactic Reactions

These are among the most common reactions of algae to light. They are variously called phobic responses, shock reactions, stop responses, or motor responses and are caused by sudden changes in light intensity. Photo-phobotaxis is widespread in algae, but the mode of the phobic responses, i.e., the behavior of the individual after stimulation, is different in the various groups and depends on the morphology of the organism and the mechanism of movement. In slowly moving organisms, such as diatoms and blue-green algae, it is simply a stop, sometimes preceded by a slow down, and normally followed by a return. In the fast moving *Euglena* the phobic response results in a sharp turn of the cell.[13]

Photo-Topotactic Reactions

These are also very common in algae. In photo-topotaxis the response is regarded as positive, when the organisms move toward the light source, and negative when they move away from it. Normally, positive reactions are observed at lower light intensities and negative reactions at higher light intensities. Between the positive and the negative range there is the so-called indifference zone, in which the cells display random orientation. As mentioned above, the photo-topotactic response can result either from a steering act or from positioning due to trial and error.[13]

A well-illustrated case, where photophobic response causes accumulation of cells in a lighted area which acts as a trap is described by Cruetz et al.[14] Visual qualitative observations of cells entering and leaving the trap was done with a microscope, as diagrammed in Figure 4, delineating the illumination occurring within the specimen plane. Cells leaving the illuminated area experience a rapid decrease in light intensity and respond by changing from their usual behavioral pattern of swimming in nearly straight lines. The cells change swimming direction one or more times.

In a typically weak response, cells change direction once, perhaps by 120° within 1 sec after experiencing the decrease in intensity. In a very strong response, cells undergo continuous changes in direction, spinning 360° a few times each second. Such a response gradually weakens over 10 or 15 sec into a behavior typical of weaker responses.

The observed result of this behavior is that cells originally swimming in the vicinity of the trap slowly accumulate within the trap. In light trap experiments with *Porphyridium cruentum*, a species which seems to have a promising economic potential, the density of photoaccumulations was strongly influenced by photokinesis and phototaxis. Both photosystem I (PS I) and photosystem II (PS II) pigments were the photoreceptors for the photophobic response in *Porphyridium*.[15]

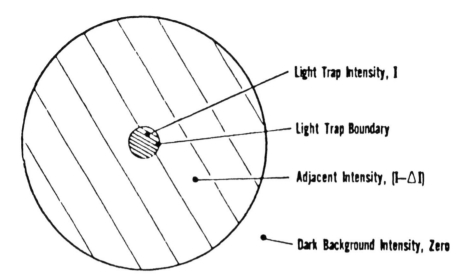

FIGURE 4.   Light trap in the specimen plane. Cells swimming in this plane experience light intensities as diagrammed. The punched-out hole acts as a light trap in which cells accumulate and contain light at a uniform intensity, I. The region adjacent to the trap has a lesser intensity, whose values depend upon the absorbance of the neutral density filter in which the hole was punched. The intensity in this region is (I-I), with a change of intensity. I, occurring at the boundary of the light trap, a circle of 1.0 cm in diameter. (From Creutz. C., et al., *Photochem. Photobiol.*, 27, 611, 1978. With permission.)

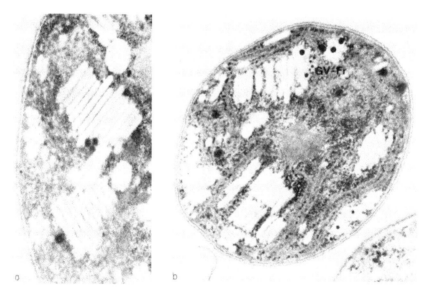

FIGURE 5.   Disappearance of gas vacuoles in *Mycrocystis*. (From Lehmann, H. and Weincke. C., *Plant Cell Environ.*, 3, 319, 1980. With permission.)

The Effect of Light and Nutrients on Buoyancy

Blue-green algae (cyanobacteria) can form dense surface and deepwater population maxima in lakes because their cells possess gas vacuoles. Gas vacuoles are aggregates of vesicles, with gas-permeable, proteinaceous walls. They are hollow cylindrical structures of low density, approximately 0.12 g/cm^3 in *Anabaena flosaquae*, and if they occur in sufficient numbers, they can make the algal cells buoyant[16] (Figure 5). In a stable water column the relative cellular volume occupied by these structures can determine whether an alga will rise to the surface or accumulate at some depth.

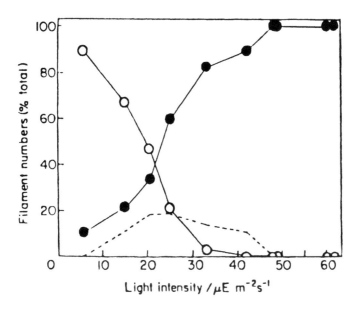

FIGURE 6. Percentage of total filaments present which were -o- positively, -●- negatively, and --- neutrally buoyant. The hemacytometer method was used in this study. (From Walsby, A. E. and Booker, M. J., *Br. Phycol. J.*, 15, 311, 1980. With permission.)

Buoyancy forms a prominent consideration in mass culture of algae. When buoyancy is negative or altogether absent, the cells have a strong tendency to sink. This phenomenon, typical for example to *Porphyridium cruentum*, relates to two major aspects of biomass production, i.e., the input of energy required for stirring and that required for harvesting. Obviously, significantly more energy is required to keep cells of negative buoyancy in suspension. On the other hand, such cells readily sink when stirring is stopped, facilitating an easy removal of the algal mass from the culture. For commercial production of *Spirulina*, buoyant types seem to be preferable. They may require less mixing energy and there is some indication that they may be more resistant to photooxidation.

Planktonic blue-green algae were found to regulate their buoyancy in response to light intensity; when algal colonies were exposed to high light intensities some of the gas vesicles collapsed and the algal colonies sank. In calm weather, subsurface population maxima of *Aphanizomenon* would sometimes occur in the early afternoon because of gas vesicle collapse. Collapse of the pressure-sensitive gas vesicles was accomplished by an increase in cell turgor pressure, driven by photosynthesis.[17] Paerl[18] pointed to the major effect of the extent of CO_2 fixation on the mechanism-promoting formation of surface scums. In *Microcystis aeruginosa*, addition of dissolved inorganic carbon decreased the buoyancy. In mixed blue-green populations, it was observed that not all species were equally responsive to the effect of light upon buoyancy and the amount of light required to cause a loss of buoyancy was not constant throughout the summer.

According to Walsby and Booker,[19] roughly equal numbers of floating and sinking filaments occurred in cultures left at light intensities of 13 to 22 μE/m²/sec. At higher intensities more of the filaments sank, and at lower intensities more floated (Figure 6). This buoyancy response was found to permit stratification of the alga in an experimental stabilized water column with a vertical light intensity gradient. The algal population formed a peak at a depth where the intensity was 1.0 klx, equivalent to 17 μE/m²/sec.

The most satisfactory method to determine the proportion of filaments in a culture of *Anabaena flos-aquae* which sank or floated was to leave an aliquot of the culture standing

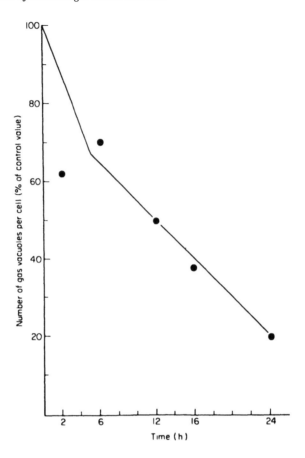

FIGURE 7.   Gas vacuole number per cell after transfer from nutrient medium into distilled water. (From Lehman, H. and Wiencke,
C., *Plant Cell Environ.*, 3, 319, 1980. With permission.)

on a hemocytometer slide and then to count the number of filaments which floated up in
contact with the coverslip and the number which sank on the platform of the slide. The
position of the filaments could be distinguished by the plane of focus. Only a short time
was needed for filaments to float or sink to the top or bottom of the hemocytometer chamber,
and this minimized any buoyancy change during the assay.[19]

The effect of nutrient availability on the formation or disappearance of gas vacuoles was
elucidated by Lehmann and Wiencke,[20] by transferring *Microcystis aeruginosa* to distilled
water (Figure 7). The number of gas vacuoles per cell decreased and reached a value of
20% of the control 24 hr after transfer. In senescent cells grown on a mineral agar for several
weeks, gas vacuoles also disappeared. The conclusion was that the disappearance of the gas
vacuoles may be a response to a nutrient deficiency in both cultures.

Among colonies of *M. aeruginosa* located at the air-water interface of nonmixed eutrophic
river water, internal cells were distinguished from peripheral cells. Internal cells showed an
increased share of photosynthate production when atmospheric $^{14}CO_2$ was applied. When
CO_2 was not added, peripheral cells in illuminated colonies accounted for at least 90% of
the $^{14}CO_2$ assimilation, internal cells remaining unlabeled. Pearl[18] concluded that inorganic
carbon (Ci) transport was restricted in large colonies below the water surface, forcing internal
cells to maintain a high degree of buoyancy, thereby promoting the formation of surface
scums. At the surface, Ci restrictions were alleviated (Figure 8). Accordingly, scum formation appears to have an ecological function, allowing cyanobacteria access to atmospheric
CO_2 when the Ci concentration is growth limiting in the water column.

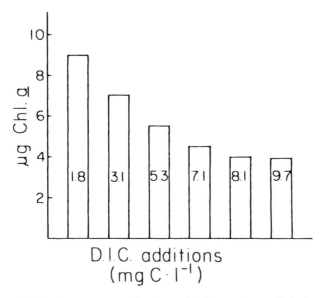

FIGURE 8. Buoyancy of freshly sampled *M. aeruginosa* colonies in pH 9.00 Tris-buffered Neuse River water in response to increasing dissolved inorganic carbon (DIC) (as HCO_3) additions. The natural river dissolved inorganic carbon concentration was 1.8 mg of C/ℓ. Decreased buoyancy, reported here as the decreased chlorophyll-a (Chl. *a*) accumulation in the upper 10 cm of graduated cylinders, coincided with increasing dissolved inorganic carbon concentrations. (From Paerl, H. W., *Appl. Environ. Microbiol.*, July, 252, 1983. With permission.)

Van Rijn and Shilo[21] studied buoyancy regulation in a natural population of *Oscillatoria* spp. in fish ponds. They concluded that the buoyancy of *Oscillatoria* is regulated by at least three different factors: the light regime, CO_2, and nutrient availability. The same conclusions were reached by Klemer et al.,[22] who found under chemostat conditions that carbon and nitrogen had opposite effects on the buoyancy of *O. rubescens*; CO_2 limitation increased buoyancy, whereas limited nitrogen decreased buoyancy.

When several of these factors are operative, the presence or absence of light is dominant over the effect of CO_2 availability. Thus, in darkness with added CO_2 the cyanobacteria retained their buoyancy. Furthermore, the observations indicated that CO_2 limitation overrode nitrogen limitation since the cells were highly buoyant in the pond as well as in the untreated enclosures in which both nitrogen and carbon were limiting. The effect of nitrogen could be demonstrated only after the cells had lost their buoyancy due to the addition of CO_2. A nitrogen addition at this stage resulted in regained buoyancy.

The wide distribution of surface blooms (composed of highly gas-vacuolate cyanobacteria) in most Israeli fishponds throughout summer and autumn was suggested to result from the unique combination of environmental conditions in these ponds.[21] These include the extremely high photosynthetic activity of the uppermost layers of the water, which leads to a drastic limitation of CO_2 and nitrogen. The entire water column is involved since in these shallow ponds (1.0 to 1.5 m) the daily afternoon winds cause total mixing of the water. The conclusion was that the high buoyancy of the cyanobacteria is the result of the reduction of CO_2 in the water and the effect of light limitation due to mutual shading.[21]

CELL RESPONSE TO TEMPERATURE

General Background

Environmental temperature is a factor to which the algal biomass responds continuously.

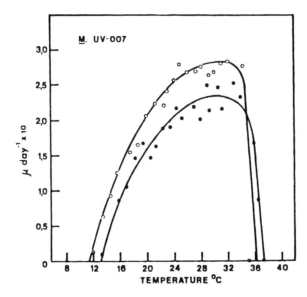

FIGURE 9.   Effect of incubation temperature on the specific growth
rate of *M.* UV-007 at two light intensities. (● 20 μE/m²/sec; o 33
μE/m⁻²/sec. Both light intensities are below saturation. (From
Kruger, G. H. J. and Elof, J. N., *J. Limnol. Soc. S. Afr.*, 4(1),
9, 1978. With permission.)

Cell temperature equals the temperature of the culture medium, in contrast with other
parameters of the medium, such as the pH. In addition to affecting the rates of cellular
reactions, temperature also affects the nature of metabolism, the nutritional requirements,
and the composition of biomass.

The effects of temperature on biomass growth and activity must be explained basically
in terms of two factors: one relates to the temperature dependence of the structure of cell
components (especially proteins and lipids), and the other concerns the temperature coefficients of reaction rates, which in turn depend on the activation energies of the reactions.
In response to these primary effects there will probably be many secondary effects on
metabolic regulatory mechanisms, specificity of enzyme reactions, cell permeability, and
cell composition. The secondary effects may, to some extent, be counteracted by special
environmental conditions.[23]

The temperature dependence of algal growth displays an exponential increase of yields
until the optimum temperature is reached. Important for large-scale production of microalgae
is the fact that the amplitude of temperature dependence decreases strongly with increasing
optical density of the suspension, i.e., with decreasing availability of irradiance to the cells
in the culture (see "Outdoor Mass Cultures of Microalgae"). Temperature requirements of
algae vary over a wide range, from cryophilic (cold loving), over mesophilic, which includes
all the candidates for commercial algaculture, to thermophilic (heat loving), which includes
algal species for which the optimal temperature for growth is over 40°C. Over most of the
temperature range below the optimum, the temperature coefficient of growth rate corresponds
to a Q_{10} of about 2, i.e., a twofold increase in growth rate per 10°C rise in temperature.
The growth rate approaches 0 at 10 to 25°C below the optimum temperature.[23]

The Effect of Temperature on the Growth Rate

The influence of temperature on the specific growth rate of *Microcystis* and *Synechococcus*
was studied in great detail by Kruger and Eloff[24] (Figure 9). A characteristic feature of the
response of many algal species to variations in temperature is depicted in Figure 9; i.e., the

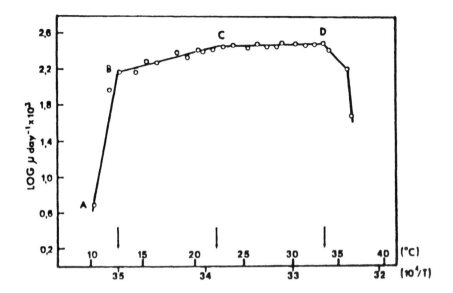

FIGURE 10. Arrhenius plot of the relationship between growth rate and temperature for *Microcystis incerta*, at light intensity of 20 μE/m²/sec.[24] (From Kruger, G. H. J. and Eloff, J. N., *J. Limnol. Soc. S. Afr.*, 4(1), 9, 1978. With permission.)

descending part of the curve being much steeper than the ascending part, indicating that a sudden decrease in growth rate occurred when the upper temperature limit is surpassed. This response is accentuated when the light intensity is elevated.

Payer et al.[31] investigated the response to temperature of 34 different green and blue-green species. The majority of the algae exhibited a temperature growth response curve with a wide plateau, culminating in a sudden decline from the optimum, thereby reflecting a pronounced inhibitory effect by temperatures which exceeded the optimum by only 2 to 3°C. In general, light saturating intensities in blue-green algae are reached earlier at low temperatures, whereas for temperatures above optimal, the higher the light intensity the sooner a decrease in growth was observed.[24,25]

When light intensity is held constant and no nutritional limitation interferes with potential growth, it is possible to describe the maximum growth rate solely as a function of temperature by applying the Arrhenius equation.[26]

The Arrhenius equation relates chemical reactions to temperature:

$$K = Ae - E/RT \qquad (1)$$

where K is the reaction rate, R is the gas constant, T the absolute temperature, A is a constant dependent on the frequency of formation of activated complexes of the reactants, and E is a constant known as the "activation energy" or "temperature characteristics".

The logarithmic version of Equation 1:

$$\log K = \log A - E/2.30 \, RT \qquad (2)$$

Hence a plot of logK against 1/T should be a straight line, with slope E/2.30 RT. The specific growth rate (u) may be substituted for the reaction rate (K) in Equation 2, yielding straight line relationships between log u and 1/T over a limited range.[23]

Figure 10 depicts Arrhenius plots, i.e., specific growth rates as a function of absolute temperature. Three distinct inflection points become evident, B, C, and D, indicating abrupt changes in the slopes of the curve. These points demonstrate that different Arrhenius rela-

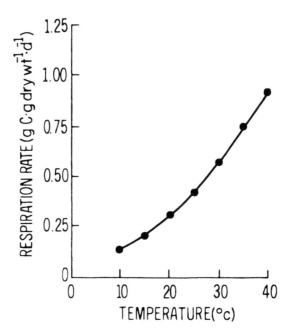

FIGURE 11.   Effect of temperature on the dark respiration rate
of *A. variabilis.* Cells were pregrown at 30°C and 500 μE/m²/
sec. (From Collins, C. D. and Boylen, C. W., *J. Phycol.,* 18,
206, 1982. With permission.)

tionships are applicable in different temperature ranges. Thus, the activation energy or
temperature characteristics for a particular organism is constant only over a short temperature
range, the changes in activation energy indicating differences in rate-controlling reactions
or in metabolic regulation.

The decrease in the growth rate at the upper extreme of the temperature range may reflect
either a disruption of metabolic regulation or death of the cells. If death occurs the growth
rate of the viable biomass (x) is given by

$$dx/dt = (u - k)x \qquad (3)$$

where u = specific growth rate and k = specific death rate. The death rate will become
dominant at high temperatures if the activation energy for death exceeds that for growth.
An increase in temperature will eventually cause the breakdown of protein structure so that
affinity for substrate and enzyme regulators will be affected.[23]

The Effect of Temperature on Dark Respiration

A most prominent effect of temperature on cell metabolism is reflected in the influence
of temperature on dark respiration. As a rule, the respiration rate, particularly in warm-
temperature algae, increases exponentially with temperature (Figure 11). This fact carries
far-reaching ramifications on the production of algal biomass, for when the temperature is
high, particularly at night, the loss in biomass due to intensive respiration may be most
damaging to the output (see "The Production of Biomass. A Challenge to Our Society").

An effect of the nutrients in the growth medium on cell response to temperature was
reported by McCombie.[27] Doubling the concentration of nutrient salts caused appreciable
changes in the form of the growth-temperature relation for *Chlamidomonas reinhardi,* one
effect being that the growth curve rose to a higher maximum than it did for the experiments

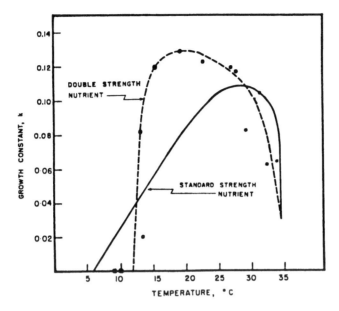

FIGURE 12.   Effect of doubling the concentration of all salts in the medium on the growth-temperature relation of *C. reinhardi*. The dots represent data for the cultures with double salt concentration and 200 fc illumination; the curve with the higher peak is fitted to these. (From McCombie, A. M., *J. Fish. Res. Board Canada*, 17(6), 871, 1960. With permission.)

with standard strength medium (Figure 12). A similar effect was revealed by Spoehr and Milner.[28] Cultures of *Chlorella* grown on one nutrient salt medium produced the best yield at 20°C, whereas on a different nutrient medium, the highest yield was obtained at 25°C (see "Productivity of Algae Under Natural Conditions").

High Temperature Strains

A singular aspect of the effect of temperature that concerns commercial production of algal biomass relates to the effect of relatively high temperature, such as may readily occur in hot regions. Sorokin[29] studied temperature optima for growth for both low and high temperature strains of *Chlorella* illuminated above the light saturating points. At 25°C, the rates of growth, respiration, and photosynthesis were close or slightly higher for the high temperature *Chlorella* than for the low temperature strain. However, if compared at a temperature optimal for their growth, the low temperature strain had about two doublings while the high temperature strain had more than nine doublings of cell material per 24-hr period. The rate of photosynthesis in the high temperature algae was four times higher at light saturation and three to six times higher at half-saturation than in low temperature algae.

The greater productivity of high temperature strains in comparison with those of low temperature is related to the ability of the former to utilize higher temperature and illuminance levels as indicated by their higher light saturating values for growth and photosynthesis.

In *Spirulina platensis* and *Chlorella sorokiniana*, both warm temperature algae, the highest O_2 production in an outdoor pond was recorded with the highest rates of solar irradiance and temperature, which effected the largest output in biomass. In contrast, with *Chlorella vulgaris*, the highest O_2 formation was noted at 18°C, a higher temperature resulting in lower O_2 concentrations.[30] This demonstrates well the importance of selecting and developing suitable algal species and strains for biomass production in accordance with the local environmental conditions.

While the advantages of high temperature strains for biomass production are rather obvious in the summer, with the advent of low winter temperatures such strains have a profound disadvantage. Warming the pond with geothermal water or growing the algae in closed systems that would be in effect well heated even in mid-winter may be the practical means by which to extend the growing season of thermophilic algae well into the winter and even throughout the year.

Ambient and Pond Temperature

The relationship between pond and ambient temperature is of practical significance. The temperature of the culture in the pond is determined by the temperature of the surrounding air, the extent and duration of solar irradiance (most of which is converted to heat while being absorbed in the algal mass), and the relative humidity of the air which governs the extent of evaporative cooling. In addition, the depth and surface of a culture, as well as the materials from which the pond was constructed, are important factors in stabilizing pond temperature. The relative humidity has a most profound effect on the difference between air and pond temperature. In tropical Bangkok, measurement of the daily temperature course indicated that culture temperatures may exceed air temperatures by as much as 13°C on sunny days during January or June. Only on cloudy days would the temperature of the air and the culture be the same.[31] In contrast, results from the author's laboratory, which is situated in the arid Negev desert in Israel, indicate that without exception, pond daytime temperature is lower by a few degrees than air temperature. Even in ponds covered with 0.2-mm thick transparent polyethylene sheets to increase the temperature in the winter, the temperature of the culture lags behind the air temperature by 6 to 8°C throughout the diurnal cycle.

The Effect of Temperature on the Chemical Composition

Temperature may have a significant influence on the chemical composition of algae. Sato and Murata[32] concluded that temperature was one of the most important environmental factors influencing the fatty acid composition. When *Anabaena variabilis* was shifted from a temperature of 38 to 22°C, lipid synthesis was markedly suppressed in the first 10 hr. During this period most of the palmitic acid of the diacylmonogalactosylglycerol was desaturated to palmitoleic acid. Thereafter lipid synthesis resumed, and in the following hours, the relative contents of palmitic and palmitoleic acids were almost restored to original levels. On the other hand, the oleic and linoleic acids were desaturated to α-linoleic acid in all the lipid classes.

Lipid synthesis was considerably stimulated in the first 5 hr after a growth-temperature shift from 22 to 38°C. During this period, the relative content of palmitic acid increased and that of palmitoleic acid decreased in the diacylmonogalactosylglycerol, becoming restored to the original levels after 20 hr. The oleic and linoleic acids increased with a concomitant decrease in α-linoleic acid in all the lipid classes. A decrease in unsaturation in the C16 and C18 acids was found and thought to be due to the stimulated synthesis of more saturated fatty acids.

Among the major molecular species of the lipids a particular change was seen in 1-oleoyl-2-palmitoylmonogalactosyl-*sn*-glycerol. This species rapidly decreased after a downward temperature shift and rapidly increased after the upward temperature shift. Sato and Murata[32] suggested that this compound is involved in regulating membrane fluidity during temperature acclimation.

Temperature had a strong influence on the chemical composition of several species of marine phytoplankton studied by Goldman.[33] In mass culture experiments, Goldman[33] found that temperature strongly influences cellular chemical composition, not only at the lower, but at the higher temperatures as well. Five marine phytoplankton species [*Skeletona costatum*

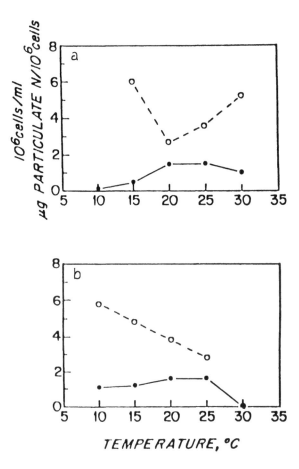

FIGURE 13.   Effect of temperature on steady state cell numbers
(●) and PN:CC ratios (O) in *Dunaliella tertiolecta* (above) and in
Skeletonema costatum (below) grown in continuous culture at a
dilution rate of 0.6/day. (From Goldman, J. C., *Limnol. Oceanogr.*, 22(5), 1977. With permission.)

(Skel); *Monochrysis lutheri* (Mono); *Phaeodactylum tricornutum* (TX-1); *Dunaliella tertiolecta* (Dun) and *Thalassiosira pseudonana* (3H)] were grown in continuous monocultures
at a constant dilution rate of 0.6/day with constant lighting (about 0.03 ly per minute visible
region) on enriched media consisting of 50% secondarily treated wastewater and 50% seawater. Steady-state measurements were made in the temperature range 10 to 30°C for cell
counts (CC) and particulate carbon (PC) as well as for particulate nitrogen (PN). Each species
responded to temperature variations somewhat differently; both *D. tertiolecta* and *S. costatum*
showed an increase in nitrogen and carbon content per cell (PN:CC ratio) at low temperature.
While this ratio steadily decreased in *S. costatum* with a rise in temperature, it increased
from 20°C on in *D. tertiolecta* (Figure 13). Payer et al.[31] reported that whereas no significant
influence on cell composition was observed in one strain of outdoor grown *Scenedesmus*,
the protein content of another strain showed a statistically significant decrease with increasing
temperature.

The Effect of Temperature on Species Dominance

Temperature had a very marked effect on species dominance. Studying natural populations
of marine phytoplankton grown in continuous cultures on wastewater-seawater mixtures at
a wide temperature range, Goldman[34] found that virtually all of the influent inorganic nitrogen

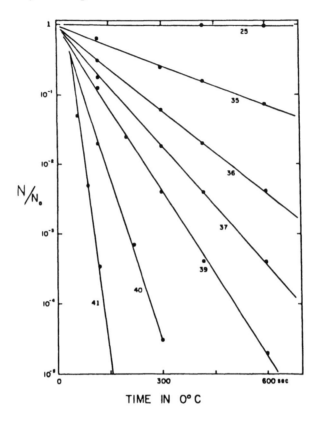

TIME IN 0° C

FIGURE 14. Survival vs. time in 0°C. The fraction of survivors (N/No) was estimated in terms of colony-forming units. Numbers of the curves describe the preceding temperature of culture. (From Rao, K. et al., *Plant Physiol.*, 59, 965, 1977. With permission.)

(14.0 mg/ℓ) was assimilated at every temperature tested. Temperature, however, affected a distinct change in species dominance; below 19.8°C *Phaeodactylum tricornutum* was dominant, at 27°C *Nitzschia* sp. was the main species, and as the temperature increased above 27°C, a blue-green alga, *Oscillatoria* sp., became increasingly dominant.

In commercially grown mass cultures, where the culture must be maintained as monoalgal, the decisive effect that temperature exerts on species competition must always be considered. In *Spirulina* mass cultures grown in the author's laboratory at Sede Boqer, it was observed that as soon as the summer heat receded and the average day temperature reached approximately 20°C, i.e., approximately 15°C below the optimum for *Spirulina*, *Chlorella* sp. quickly proliferated in the culture to become the dominant species with 2 or 3 weeks.

Chilling Temperature

During the winter, the temperature in outdoor cultures may descend close to the freezing point over the course of a clear night the adaptive response of algae to chilling temperatures is clearly of practical significance in mass cultures. Siva et al.[35] studied the phenomenon of cold shock on *Anacystis nidulans* whereby cells were exposed to 0°C within 22 sec. For cells grown at temperatures above 40°C loss of viability was maximal, becoming negligibly small for cells grown below 34°C prior to exposure to the cold shock (Figure 14). The conclusion was that the multiple effects which comprise the cold shock syndrome appear to be membrane-related phenomena. In higher plants, what has indeed been proposed but not yet fully supported is that chilling injury is a result of the phase change in lipids in the cellular membranes at low temperatures.[36] Ono and Murata[37] also concluded that the temperature critical for chilling susceptibility of *Anacystis nidulans* depended on the growth temperature. The midpoint values for the critical temperature regions were 4, 6, and 12°C in cells grown at 28, 33, and 38°C, respectively.

There may be several mechanisms responsible for chilling damages. Ono and Murata[38] proposed to interpret the chilling susceptibility of *A. nidulans* as follows: at chilling temperatures, the bilayer lipids of the cytoplasmic membrane are in the phase separation state and ions and solutes of low molecular weights leak from the cytoplasm to the outer medium. Decreases in the intracellular concentrations of ions and solutes affect, in turn, degradation of the physiological activities of the cells.

Thermotaxis

A special effect of temperature is on cell movement. Thermotaxis refers to the orientation of motile microorganisms in a given temperature gradient. As in other taxes, topic and phobic responses are recognized. Thermotaxis has been observed in several algae, some display positive reactions, e.g., *Navicula radiosa,* which moves from colder areas to 28 to 30°C, while others, e.g., *Haematococcus,* react negatively and move from warmer areas to 5 to 10°C.[13]

THE EFFECT OF THE HYDROGEN ION CONCENTRATION

The pH of the medium in which the cells grow is known to affect many biological processes. A very marked effect of the pH is observed in bodies of water that support high population densities such as eutrophic lagoons or fish ponds. In such habitats extreme diurnal fluctuations in pH may exist. These range from 6.5 at late night before sunrise due to accumulation of respiratory release of CO_2, all the way to pH 11.0 late in the day, when CO_2 and HCO_3 were depleted in the course of photosynthesis.

Plasma membranes are not freely permeable to hydrogen or hydroxyl ions so that intracellular and extracellular hydrogen ion concentrations do not necessarily equilibrate and a gradient of hydrogen ion concentration across the membrane can be expected. According to the chemiosmotic theory[39] this gradient of hydrogen concentration together with the membrane electrical potential determine the "proton motive force" which drives membrane reactions.[23]

Algae exhibit a clear dependency on the pH of the growth medium and different species vary greatly in their response to the pH. The relative distribution of the carbon species (i.e., CO_2, HCO_3^- and $CO_3^=$; see "Productivity of Algae Under Natural Conditions") determines the pH which thereby governs the growth and predominance of algal species both under natural conditions and in mass cultures. In many cyanobacteria, such as *Coccochloris peniocystis*, the rate of photosynthesis as measured by O_2 evolution exhibited a broad optimum over the pH range of 7.0 to 10.0 (Figure 15). Since the bicarbonate ion is the predominant species of dissolved inorganic carbon at pH values in this range, the capacity of *Coccochloris* as well as most cyanobacteria to grow best in this pH range provided the basis to suggest that they are capable of utilizing HCO_3 as a substrate for photosynthetic carbon fixation.[40]

Cook[41] reported that visible light inhibited cell division in *Euglena gracilis*, the process being pH-dependent, and was observed to be most severe in the pH range 3.5 to 5.0. Transfer of phototrophic cells from pH 6.8 to pH 4.2 lead to cell death at modest light intensities (500 fc) and inhibition of division at lower intensities (300 fc). The inhibition of cell division was preceded by a large influx of phosphate, most of which remained in the cold PCA pool. It was suggested that light may act to reduce control over phosphate entry into the *Euglena* cell at these intermediate pH levels; excess phosphate apparently leads to inhibition of cell division or death.[41]

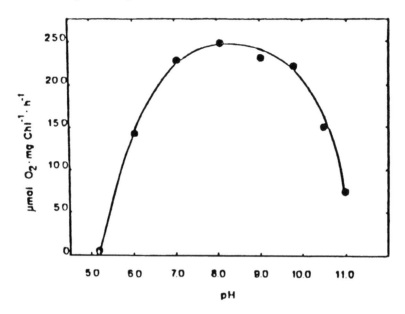

FIGURE 15.   Effect of external pH on photosynthetic O_2 evolution of *Coccochloris* at an external inorganic C concentration of 5 mM and 30°C. (From Coleman, J. R. and Coleman, B., *Photosynthesis IV*, Balaban International Science Services, Pa., 1981.)

The pH of the growth medium, by its effect on the dissociation of various salts and complexes, may influence their toxicity or inhibitory action. Additionally, the pH has a marked effect on the solubility of various metal compounds. A rise in the pH may cause a deficiency in some trace element which in turn may play havoc in the culture.

The pH may also affect the toxicity of algae, such as the blue-green *Microcystis aeruginosa*. Toxic strains of this algae have been responsible for numerous instances of livestock deaths in South Africa.[42] The pH influenced both the growth rate and the toxicity of *Microcystis;* cells became more toxic as the pH descended from the optimum. It appeared that the slower the cells grew, the more toxic they became (Figure 16). Significantly, maximum toxicity per biomass unit was not correlated with maximum biomass per unit volume of water.[43]

LIGHT AND TEMPERATURE INTERACTIONS

Standardization of the light saturation curve is often delineated in terms of only two parameters, the initial slope (alpha) and the chlorophyll-a-specific carbon production rate at saturating light (P_{max}), often referred to as the assimilation number.[44] Light saturation of photosynthesis is nevertheless influenced by several other factors such as nutrients and chemical composition, and particularly by the temperature.[45]

A good example for a strong interaction of temperature and light on the growth of *Chlorella* was provided in the work of Sorokin and Krauss.[46] As depicted in Figure 17, the number of doublings of cells per day was remarkably responsive to rather small variations in light and temperature. The light curve did not reach its full expression in the higher temperature, for as the temperature increased, a given radiation flux, which at lower temperature caused a decrease of growth, was not sufficient to affect light saturation.

The interrelationships between the effects of light and temperature are well illustrated in the study of Collins and Boylen[47] who investigated physiological responses of the blue-green *Anabaena variabilis* to instantaneous exposure to various combinations of light intensity and temperature. As observed for *Chlorella* and for other algae, there was an increase in the

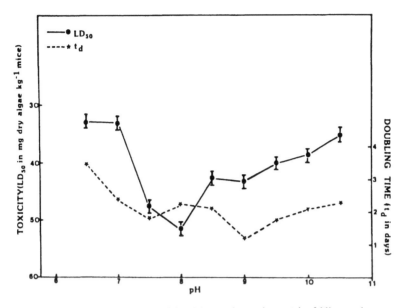

FIGURE 16. Effect of pH on toxicity (LD_{50}) and growth rate (td) of *Microcystis aeruginosa* (UV-006). (td = Doubling time of culture turbidity [Klett units] during logarithmic growth phase; I = 95% confidence interval.) (From Van der Westhuizen, A. J. and Eloff, J. N., *Z. Pflanzenphysiol.*, 110, 157, 1983. With permission.)

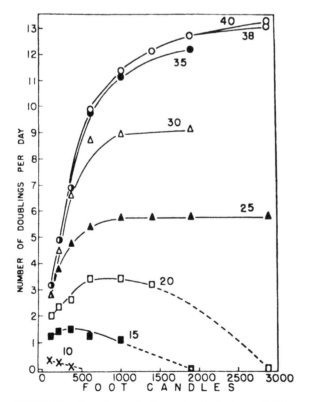

FIGURE 17. Growth rates for the synchronized cultures of *Chlorella pyrenoidosa* (strain 7-11-05), measured at various light intensities and temperatures from 10 to 40°C. Rates are given as the number of doublings of cell material per 24 hr period. Temperatures are indicated on the curves. (From Sorkin, C. and Krauss, R. W., *Plant Physiol.*, May, 37, 1961. With permission.)

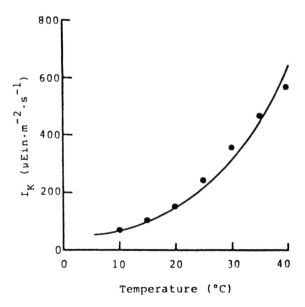

FIGURE 18.   Effect of temperature on the values for the light-
saturation parameter, Ik. (From Collins, C. D. and Boylen, C.
W., *J. Phycol.*, 18, 206, 1982. With permission.)

saturating-light intensity for photosynthesis (I SAT) with increasing temperature. Also, the
initial linear slope of each of the light curves decreased with increasing temperature; as a
result, the light saturation parameter I_K increased (Figure 18).

For each temperature there was a specific light intensity at which the maximum photosynthetic rate (P_{max}) was reached. At the lowest light intensity tested (42 $\mu E/m^2/sec$) P_{max}
was achieved at 15°C. At this low light level, higher temperatures drastically decreased the
photosynthetic rate. At higher light intensities the photosynthetic rate increased with an
increase in temperature. At 40°C, however, the cells were apparently unable to compensate
for the high temperature-dependent respiration rate which greatly reduced photosynthetic
rates above 35°C. An important aspect of the interaction of light and temperature was that
the optimum temperature for photosynthesis (Topt) increased with increasing light intensities.
At low light intensities (42 and 99 $\mu E/m^2/sec$) the photosynthetic rate increased with increasing temperature, whereas at high temperature (40°C) the photosynthetic rate increased
with increasing light intensities. The mid-range of combinations exhibit the characteristic
photosynthetic light response (Figure 19) (see also treatment of light-temperature interactions
in "Outdoor Mass Cultures of Microalgae").

Verity[45] investigated the interrelationships of temperature irradiance and daylength on the
marine diatom *Leptocylindrus danicus* Cleve. At each temperature, photosynthesis exhibited
a daylength-dependent curvilinear relationship with irradiance, the maximal rate of photosynthesis (P_{max}) increasing by an order of magnitude over 5 to 20°C in cells grown under
15:9 and 12:12 hr of light and darkness (LD). Cells cultures under 9:15 LD show a similar
trend, but P_{max} during this photoperiod at 15 and 20°C is considerably reduced relative to
longer daylengths (Figure 20).

Verity's[45] work illustrated that alpha, the initial slope of the light saturation curve for
photosynthesis (see Chapter "Elements of Pond Design and Construction"), showed a
temperature-dependent daylength effect. Thus, the initial slope of the light saturation curve
should not necessarily be considered as a physiological constant, but should rather be expected
to vary with environmental conditions.

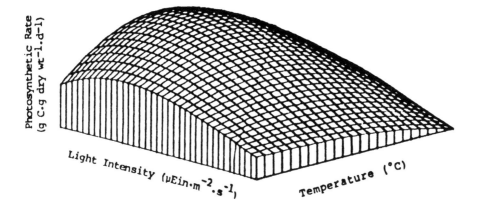

FIGURE 19. Trend surface for the light intensity and temperature interaction on the photosynthetic rate of *A. variabilis*. (From Collins, C. D. and Boylen, C. W., *J. Phycol.*, 18, 206, 1982. With permission.)

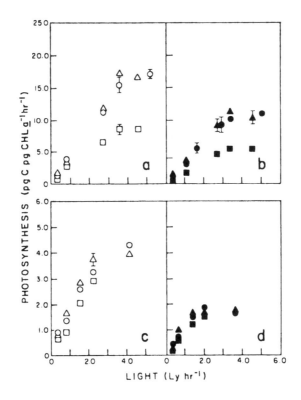

FIGURE 20. Photosynthesis-irradiance curves as a function of temperature and daylength: (a) 20°C; (b) 15°C; (c) 10°C; (d) 5°C; daylengths at each temperature are 15:9 ▲, 12:12 o, and 9:15 ■. (From Verity, *J. Exp. Mar. Biol. Ecol.*, 55, 29, 1981. With permission.)

SALINITY AND OSMOREGULATION

General

Plant cells are hydraulic systems that conduct chemical reactions in aqueous media. The cells are separated from their environment by a membrane which is largely permeable to water and largely impermeable to solutes. The adjustment to water stress which results from increased salinity in the growth medium imposes a demand for maintenance of cell hydration which is essential for both chemical and physical processes. Net movement of water occurs in response to physical forces, i.e., passive movement of water to regions of lower chemical potential, the driving force being the difference in chemical potential which develops across the cell membrane. Thus, in response to high salt concentration in its surroundings or, in the case of mass cultures, in response to continuous salinization of the medium due to continuous evaporation from the pond surface, the cell counterbalances the rise in outside solutes by enhanced synthesis of solutes inside the plasma membrane and/or enhanced uptake of solutes from the surroundings. Such osmoregulatory response to an environmental stress is a key factor in the survival of the cell and the species. Osmoregulation may be commercially exploited in certain species of microalgae to induce synthesis of high quantities of organic solutes with a defined market value. It therefore merits special attention.

The algae as a group exhibit an extremely wide range of tolerance to salts in their surroundings. Some species can tolerate only millimolar amounts of salt while others survive in saturated brine. What constitutes lethal saline stress for one group of algae is easily tolerated by others. With regard to the adaptation to salinity, algae may be roughly divided into halotolerant and halophilic, the latter requiring salt for optimum growth and the former having response mechanisms that permit their existence in saline medium.

The unicellular green alga *Dunaliella* is unique in its ability to survive extreme salt stress and therefore attracts the attention of many research groups. *Dunaliella* sp. may serve as useful models to comprehend the strategies of cell response to high salt concentration. (See ''Algae of Economic Potential.)

The survival under high salt stress requires extra energy, e.g., for the synthesis of organic osmotica, the maintenance of an appropriate ion balance between the medium and the cytoplasm (against the chemical concentration gradient) and for repair reactions of salt injuries. As pointed out by Gimmler et al.,[48] who studied the metabolic response of the halotolerant *Dunaliella parva* to hypertonic shocks, a unicellular photoautotrophic algae has difficulty in this respect, as compared to higher plants. This is because such an alga can liberate only limited amounts of phosphorylation energy and reduction equivalents from the breakdown of storage products, especially if the situation of young starch-depleted autospores at the end of the dark phase is considered. It is an absolute necessity for *D. parva* under high salt stress to improve photosynthesis and the energy budget of the cell, if growth is not to be severely inhibited. Some examples for this strategy are shown in Table 1. High salt algae have a higher chlorophyll content per cell and are able to absorb more light per cell. They grow better under high light intensities and exhibit a higher chlorophyll a/b ratio. Also, high salt algae perform a higher rate of photosynthesis and need more light and a higher bicarbonate concentration for half salturation of CO_2-assimilation. In some respects high salt *D. parva* resembles typical ''sun type'' plants, whereas other algae, especially marine algae, may be normally considered to belong to ''shade type'' plants. The salt-induced increase of photosynthesis is to be calculated on the basis of the protein content, instead of the chlorophyll content, since the amount of chlorophyll per cell responds per se to the salinity of the medium. High salt *Dunaliella* also follows the strategy of increasing the amount of some critical enzymes, e.g., carbonic anhydrase.[48]

Osmoregulation with Glycerol

The main organic osmoticum of *D. parva*, as with other *Dunaliella* species, is glycerol, and the higher the NaCl concentration of the medium, the higher the glycerol concentration in the cells (Figure 21). Glycerol concentrations up to 5 *M* have been observed (see Chapter 8), but the concentration ratio of endogenous glycerol to external NaCl never exceeds unity. Glycerol as osmoticum is economical with respect to energy and carbon requirements, and

Table 1
PIGMENTS AND
PHOTOSYNTHETIC
CHARACTERISTICS OF *D. PARVA*
CELLS ADAPTED TO DIFFERENT
SALINITIES[48]

	NaCl conc of the medium (M)	
	0.3	3.0
pg Chlorophyll per cell	0.91	1.15
Chlorophyll a/b	3.74	4.33
Chlorophyll/carotenoids	4.30	4.58
CO_2 assimilation (μmol) CO_2 x mg/chlxh	84	136
K_m of CO_2 fixation for HCO_3^--(mM)	2.0	4.7
Half saturation of CO_2 fixation with incandescent light (W $\times$ m^{-2})	9	22

From Gimmler, H., Wiedemann, C., and Moller, E. M., *Ber. Deutsch. Bot. Ges.*, 94, 613, 1981. With permission.

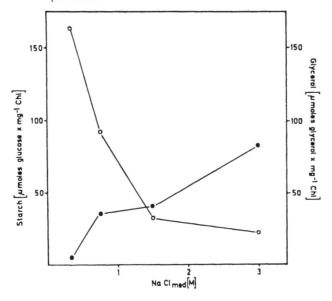

FIGURE 21.   The starch (O-O) and glycerol (●-●) content of *D. parva* cells adapted to different salinities. (From Gimmler, H. et al., *Ber. Deutsch. Bot. Ges.*, 94, 613, 1981. With permission.)

is a very compatible solute for enzymes and membranes with almost no toxic effects even at high concentrations.

The strategy of synthesizing glycerol has some drawbacks. Glycerol is osmotically less active than other physiological organic osmotica such as sucrose or sorbitol. Furthermore, equimolar solutions of NaCl and glycerol by no means give similar osmotic potentials. In fact, up to 45% of the actual external NaCl concentration was found inside the *D. parva* cell and osmotic compensation of the external medium is roughly obtained only by adding the osmotic potentials arising from both glycerol and NaCl inside the cells.

Response to Salinity by Ionic Regulation

Ehrenfeld and Cousin[49] investigated the ionic regulation of *Dunaliella tertiolecta* adapted to a large range of salinity. The concentrations of sodium and chloride increased linearly as the salinity of the incubation medium increased; the cells maintaining low cellular Na and Cl concentrations were about five times smaller than those of the external medium. On the other hand, the potassium (K) concentrations of *Dunaliella* cells were 6 to 13 times higher than the K concentrations of the external medium. A good correlation was found between gain of K and loss of Na, suggesting a stoichiometric exchange of these two ions. The magnitude of this apparent Na/K exchange increases as the salinity increases, the external K concentration necessary to mediate half-saturation of the Na/K exchange being a function of the NaCl concentration in the medium. This mechanism may be involved in the regulation of the ionic composition of *Dunaliella*, which is exposed to high salinity.

Another difficulty with glycerol, an uncharged compound of low molecular weight, is associated with its ready diffusion through the permeable plasmalemma at physiological temperatures. According to Gimmler et al.,[48] the higher the endogenous glycerol concentration, the higher the diffusion of glycerol into the medium. Thus, in order to maintain a given glycerol concentration in the cell, leakage of glycerol has to be compensated for by increased glycerol synthesis. This reflects a strategy of glycerol efflux tolerance rather than glycerol efflux avoidance.

Somewhat similar patterns of response to salinity were observed in the halophylic alga, *Asteromonas gracilis*, a green wall-less alga which grows on salt concentrations from 0.5 *M* NaCl (seawater) to saturation (4.5 *M* NaCl). The alga accumulated large amounts of intracellular glycerol in response to saline conditions and glycerol content of the cells varied in direct proportion to the extracellular salt concentration, being about 50 and 400 pg glycerol per cell in algae grown at 0.5 and 4.5 *M* NaCl, respectively. When the extracellular salt concentration was increased or decreased, the intracellular glycerol varied accordingly, reaching its new intracellular level after a few hours. In salt concentrations lower than 3.5 *M* and at growth temperatures below 40°C, there was no leakage of glycerol. Yet above 3.5 *M* NaCl, about 25% of the total glycerol leaked slowly from the cells to the medium.[50]

The leakage of glycerol has a great disadvantage from the standpoint of mass cultures. The process in itself is wasteful, but more important, it carries a potentially dangerous threat to the culture in that enrichment of the growth medium in organic solutes bestows an advantage on halobacteria which may proliferate much more rapidly than the algae. Indeed, experience from pilot-plant size *Dunaliella* ponds indicated that halophilic bacteria may bring about the deterioration of the culture.

Sucrose and Proline as Organic Osmotica

In contrast to *Dunaliella* and to *A. gracilis*, which have an obligate requirement for at least 100 m*M* NaCl or isosmotic concentrations of other solutes and always synthesize osmotic solutes for their continued growth, *Chlorella* grows best at high osmotic potentials, i.e., low-salt media. In species such as *Chlorella*, where high internal solute concentrations are required only when the cells are grown at low osmotic potentials, levels of organic solutes may well be regulated by mechanisms different from those found in osmophiles such as *Dunaliella*. Proline and sucrose occur at high concentration in salt-adapted cells of *Chlorella emersonii;* these solutes rapidly increased when cells grown at 1 m*M* NaCl were transferred to 50 and to 150 m*M* NaCl (Figure 22). Most of the proline accumulated at high NaCl was not derived from protein hydrolysis, since its formation ceased altogether both in

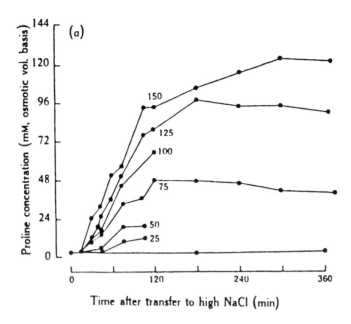

FIGURE 22.   Effect of NaCl on proline formation after transfer of cells from 1 mm NaCl to a range of external NaCl concentrations (m*M*), as shown on the curves. (From Greenway, H. and Setter, T. L., *Aust. J. Plant Physiol.*, 6, 69, 1979. With permission.)

the presence of DCMU (an inhibitor of photosynthesis) in the light, and when cells were placed in the dark. *De novo* synthesis of enzymes involved in proline formation was indicated by inhibition of net proline synthesis by cycloheximide (an inhibitor of protein synthesis), and in that, following addition of NaCl, a 15-min lag in proline accumulation was observed.

In contrast, regulation of sucrose synthesis by *C. emersonii* is presumably due either to activation and deactivation of enzymes, or to reduced growth with a consequent increase in substrate levels. This is indicated by the absence of a lag phase in the synthesis of sucrose when the alga is exposed to high salinity, and in that sucrose accumulated in the presence of cycloheximide at concentrations which inhibit proline synthesis. Also, large accumulation of sucrose took place when glucose was supplied to cells grown at 1 m*M* NaCl and rapid sucrose synthesis occurred while the cells were plasmolysed.[51]

An important conclusion from the work on the accumulation of proline and sucrose in *Chlorella emersonii* during the first hours in high NaCl was that turgor potential has a regulatory role in the synthesis of osmotic solutes. If cell turgor was not threatened, such as in the case of enhancing the osmoticum of the growth medium by rapidly permeating ethylene glycol (mol wt 62), algal growth was only slightly inhibited on the one hand and proline and sucrose did not accumulate on the other.[52]

Many changes in cell constituents take place in the course of osmotic response of the halophilic cyanobacterium Synechocystis DUN 52. Increased salinity affected a certain elevation in the cell content of K^+ and Na^+, and significant changes in the free amino acid pool composition took place with glycine and serine increasing with increasing salinity. Nevertheless, it was clear that the quantities of these amino acids as well as the elevation in K^+ and in Na^+ were insignificant for osmotic regulation. The concentration of free intracellular quaternary ammonium almost trebled on increasing the salinity threefold and the quantities detected were approximately 10 to 100 times higher than those of any intracellular solutes. Thus quaternary ammonium compounds, particularly glycinebetaine, were concluded to represent the major osmoticum in Synechocystis DUN 52. The increased intracellular concentrations of glycine and serine in high salt may be related to their role in the biosynthesis of glycinebetaine.[53]

Accumulation of β-Carotene

A singular response mechanism to high salinity, as well as to other factors which impede growth, exist in *Dunaliella bardawil*. It accumulates large amounts of β-carotene in addition to glycerol when cultivated under appropriate conditions. These include high light intensity, a high sodium chloride concentration, nitrate deficiency, and extreme temperatures. Under conditions of maximal carotene accumulation *D. bardawil* contains at least 8% of its dry weight as β-carotene while *D. salina* grown under similar conditions contains only about 0.3%. Electron micrographs of *D. bardawil* grown under conditions affecting high β-carotene accumulation show many β-carotene-containing globules located in the interthylakoid spaces of the chloroplast. The same alga grown under conditions in which β-carotene does not accumulate contain few to no β-carotene globules. When the carotene-rich globules were released from the algae into an aqueous medium they were shown by electron microscopy to be free of significant contamination and were composed of membrane-free osmiophilic droplets with an average diameter of 150 nm. β-Carotene accounted for essentially all the pigment in the purified globules.[54]

Clearly, many osmoregulatory systems exist in microalgae and no doubt many which may be of commercial potential await discovery. A unique system was recently revealed in blue-green algae isolated from its marine environment. These "marine" blue greens could be identified by their ability to synthesize and accumulate 2-O-alpha-D-glucopyranosylglycerol (glucosylglycerol) as a major osmoregulatory compound.[55]

Effect of Salinity on Photosynthesis and Respiration

It should be stressed that not all changes that take place when cells are exposed to salinity relate to osmoregulation. The response to increased salinity involves some succinct physiological changes in cells such as various degrees of loss in photosynthetic activity. One example may be seen in the work of Grodzinski and Colman,[56] who observed progressively low photosynthetic activity occurring when cells of *Anacystis nidulans* and *Coccochloris peniocystis* were incubated in increasing concentrations of osmotica from 0.2 to 0.7 M. Another observation was that pteridine was released from the cells proportionally to the loss of photosynthetic activity.

Vonshak and Richmond[57] made a distinction between the initial response to salinization and the steady-state response after an adaptive process took course; when *Anacystis nidulans* cultures growing photoautotrophically in a minimal medium were exposed to different concentrations of NaCl, a marked decrease in photosynthetic activity took place initially. This was in direct relation to the salt concentration and was not associated with a change in endogenous respiratory activity. After a period of exposure to salinity, a process of adaptation became apparent, being manifested in a partial reversal of the decline in photosynthesis and a marked increase in endogenous respiration.

In the course of adaptation to an elevation in the osmotic potential of the culture medium, a two-step process seems to have taken place. First, following the initial exposure to stress, a marked decrease in photosynthetic activity occurred which was not associated with a change in respiratory activity. Second, after a period of adaptation, a partial reversal of the decline in photosynthetic activity became apparent. This was accompanied by a marked increase in respiratory activity, the apparent aim of which was to supply the energy needed for correction of the Na^+-K^+ balance and/or for the synthesis of molecules needed for osmotic equilibration.[57]

Effect of Salinity on Cell Division and N_2-Fixation

High NaCl concentrations inhibit growth of *Chlorella emersonii* mainly via inhibitory

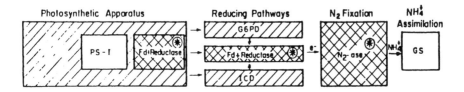

FIGURE 23.   Schematic presentation of the response to salt in N_2-fixing cyanobacteria. Symbols: ■, extremely sensitive; ▨, moderately sensitive; ☐ insensitive; ✱, adaptable to NaCl. (From Tel-Or, E., *Environ. Microbiol.*, October, 689, 1980. With permission.)

effects on cell division, delaying the time at which the subsequent generation of daughter cells was released from the mother cells.[58] Electron microscopy showed the delay in release of daughter cells to be due to a delay in cell division. At 18 hr, all cells grown in 1 m*M* NaCl had completed the third division, i.e., had formed 8 cells. In contrast, cells transferred from 200 m*M* NaCl had either completed only the first or second nuclear and cytoplasmic division, or had not divided at all. There was a lag before net DNA synthesis commenced and there were reductions in rates of net DNA synthesis in cells at 200 m*M* NaCl relative to 1 m*M* NaCl.

The effect of salt on photosynthetic activity and N_2-fixation was examined by Tel-Or[58] in two species of cyanobacteria, *Nostoc muscorum* and *Calothrix scopulorum*. Photosynthesis was found to be more resistant to a high salt concentration than was N_2 fixation, the salt resistance of both activities increasing after a period of exposure to salinity. Schematic presentation of the response to salt in these N_2-fixing cyanobacteria is presented in Figure 23.

The transfer of electrons via ferredoxin and ferredoxin nicotinamide adenine dinucleotide phosphate reductase is extremely sensitive to salt. In comparison, the transfer of reducing power by glucose-6-phosphate dehydrogenase, isocitric dehydrogenase, and PS I and glutamine synthetase exhibited higher tolerance to salt.[58]

An interesting feature of the adaptive response of algal cells to salt stress is the observation that such stress imparts resistance to other forms of stress which cause physical drought such as heat, subzero temperatures, and water stress. This phenomenon has been observed in higher plants. Also, the resistance to damages inflicted by an excess of heavy metals is more pronounced in cells adapted to salt stress. Gimmler et al.[48] suggested that the higher copper resistance exhibited by high-salt *D. parva* (which was shown not to be the result of different copper uptake) may be caused by copper complexation of glycerol and by binding of copper to the increased amounts of salt-induced SH-groups.[48]

ENVIRONMENTAL EFFECTS ON CELL MORPHOLOGY

Environmental and nutritional conditions exert very pronounced effects on the morphology of several algae. Some examples for morphological variability within the species are presented in Figure 24.[59] Morphological modifications may have significant bearing on the economics of algal mass cultures. This is illustrated in the case of modifications in the filament length of *Spirulina platensis*. The length of the filament effects the efficiency by which the filaments may be removed from the growth medium: the shorter the filament, the greater the difficulties in harvesting. In general, assuming an optimal nutrient status in mass cultures may be readily maintained. The most important parameters that may exert an effect on cell morphology are light and temperature.

Kullberg[60] made several observations on the effects of light and temperature on cell length of the cyanobacteria *Synechococcus lividus*. This species was found in steadily flowing water in springs at temperatures up to 74.5°C. The size of this hot spring alga was reported to be

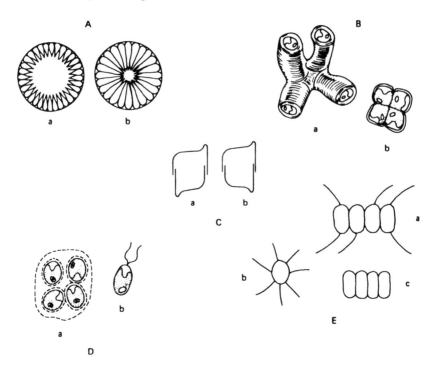

FIGURE 24. Examples of morphological variability. (A) One clonal culture can produce individuals of *Cyclotella* that resemble form a or form b. (B) A culture of *Chlorosarcinopsis* can produce packets that resemble typical members of the genus b, or Hormotila-like forms with abundant mucilage a. (C) These are diagrammatic cross-sections of a pennate diatom. The individual was observed dividing and some daughter cells had the keels opposite (a) and thus would be called Nitzschia, while others had keels on the same side (b) and thus resembled Hantzschia. (D) Chlamydomonas cells (b) have been produced as palmelloid colonies (a). (E) *Scenedesmus* culture 16 can form colonies with spines (a) or without spines (c), or unicells which resemble Franceia (b). (From Trainor, F. R., *Introductory Phycology*. Reproduced by permission from Wiley & Sons, New York, 1978.)

greatest at high temperatures and bright light, the mean change in length being 0.132 μm/ 1°C (Figure 25).

The environmental effect on cell length, however, was quite complex, for when light intensity was approximately 415 cal/cm²/day or less, there was no appreciable effect of temperature on cell length. When light intensity was 490 cal/cm²/day, the response of length to temperature was observed. In contrast, when the water flowed from an exposure of 256 cal/cm²/day at 70°C, to 507 cal/cm²/day at 63°C, the cells become longer at the lower temperature. In an artificially shaded thermal stream when the light was reduced from 650 to 159 cal/cm²/day, cells became shorter at the end of 6 weeks, changing from a length to temperature ratio of 0.218 to 0.050 μm/1°C. The temperature range where both temperature and light had the greatest effect on cell length was at the upper limit for the alga's growth (71 to 74.5°C).[61] Preliminary observations in the author's laboratory on factors affecting filament length in a small floating clone of *Spirulina platensis* indicated that the filaments of this strain, in which 4 to 8 helical turns occur most frequently, become shorter as temperature declines to the lower limit of the temperature response curve.

Many other interactions of environmental factors are known to affect morphological changes in algae. One example is the cyanophyte *Chlorogloea fritschii*, in which the filament was reported to grow in length and width when exposed to high light and low nitrogen. When nitrogen was added, cell diameter and filament length further increased. Under low light, the cells became considerably narrower and the filaments were greatly shortened.[61]

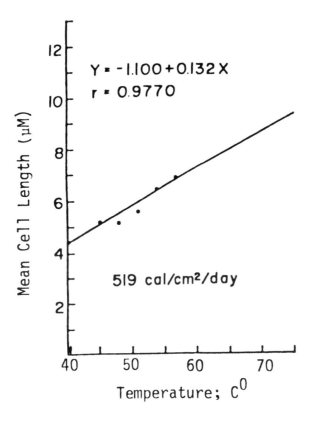

FIGURE 25. Length of *Synenchococcus lividus* as a function of temperature. (From Kullbergh, R. G., *Trans. Am. Microsci. Soc.*, 100(2), 150, 1981. With permission.)

Another example for extreme morphological changes was observed in the cyanobacteria *Nostoc muscorum*.[62] This species appears in two basic forms: coccoid form, aseriate stage and a filamentous form, seriate stage. These cell forms were dependent on the conditions of growth. When grown under conditions which supported rapid growth, the seriate stage prevailed, whereas the aseriate stage was prevalent under conditions in which growth was not rapid. The quality of light had a marked effect on morphological changes in this alga: red light induced the dark aseriate to seriate conversion and at the same time enhanced the dark growth after illumination ceased. Green light suppressed both the conversion of form and the dark growth stimulated by red light.[62]

REFERENCES

1. **Samuelsson, G. and Richardson, K.,** Photoinhibition and low quantum flux densities in a marine dino-flagellate *(Amphinidium carterae)*, *Mar. Biol.*, 70, 21, 1982.
2. **Ryther, J. H.,** Photosynthesis in the ocean as a function of light intensity, *Limnol. Oceanogr.*, 1, 61, 1956.
3. **Codd, G. A. and Stewart, R.,** Photoinactivation of ribulose biphosphate carboxylase from green algae and cyanobacteria, *FEMS Microbiol. Lett.*, 8, 237, 1980.
4. **Whitelam, G. C. and Codd, G. A.,** Photoinactivation of *Microcystis aeruginosa* ribulose 1,5-biphosphate carboxylase: effects of endogenous and added sensitizers and the role of oxygen, *FEMS Microbiol. Lett.*, 16, 269, 1983.

5. **Soeder, C. and Stengel, E.**, Physico-chemical factors affecting metabolism and growth rate, in *Algal Physiology and Biochemistry*, Stewart, W. D. P., Ed., University of California Press, Berkeley, 1974.

6. **Sokawa, Y. and Hase, E.**, Suppressive effect of light on the formation of DNA and on the increase of deoxythymidine monophosphate kinase in *Chorella protothecoides, Plant Cell Physiol.*, 9, 461, 1968.

7. **Ziegler, R. and Schanderl, S. H.**, Chlorophyll degradation and the kinetics of dephytylated derivatives in a mutant of *Chlorella, Photosynthetica (Praha)*, 3, 45, 1969.

8. **Lorenzen, H.**, Temperatureinflusse auf *Chlorella pyrenoidosa* unter besonderer Berucksichtigung der Zellentwicklung, *Flora (Jena)*, 153, 554, 1963.

9. **Falkowski, P. G. and Owens, T. G.**, Light-shade adaptation. Two strategies in marine phytoplankton, *Plant Physiol.*, 66, 592, 1980.

10. **Foy, R. H. and Gibson, C. E.**, Photosynthetic characteristics of planktonic blue green algae: changes in photosynthetic capacity and pigmentation of *Oscillatoria redekei* van Goor under high and low light, *Br. Phycol. J.*, 17, 183, 1982.

11. **Owens, O. H. and Esaias, W. E.**, Physiological responses of phytoplankton to major environmental factors, *Ann. Rev. Plant Physiol.*, 27, 461, 1976.

12. **Fiksdahl, A. E., Foss, P., Liaaen-Jensen, S., and Siegelman, H. W.**, Carotenoids of blue green algae-II. Carotenoids of chromatically-adapted cyanobacteria, *Comp. Biochem. Physiol.*, 76B, 599, 1983.

13. **Nultsch, W.**, Movements, in *Algal Physiology and Biochemistry*, Stewart, W. D. P., Ed., University of California Press, Berkeley, 1974.

14. **Creutz, C., Colombetti, G., and Diehn, B.**, Photophobic behavioral responses of *Euglena* in a light intensity gradient, and the kinetics of photoreceptor pigment interconversions, *Photochem. Photobiol.*, 27, 611, 1978.

15. **Schuchart, H.**, Photomovement of the red alga *Porphyridium cruentum* (Ag.) naegeli. III. Action spectrum of the photophobic response, *Arch. Microbiol.*, 128, 105, 1980.

16. **Walsby, A. E. and Armstrong, R. E.**, Average thickness of the gas vesicle wall in *Anabaena flos-aquae*, *J. Molec. Biol.*, 129, 279, 1979.

17. **Konopka, A., Brock, T. D., and Walsby, A. E.**, Buoyancy regulation by planktonic blue green algae in Lake Mendota, Wisconsin, *Arch. Hydrobiol.*, 83, 524, 1978.

18. **Paerl, H. W.**, Partitioning oc CO_2 fixation in the colonial cyanobacterium *Mycrocystis aeruginosa:* mechanism promoting formation of surface scums, *Appl. Environ. Microbiol.*, July, 252, 1983.

19. **Walsby, A. E. and Booker, M. J.**, Changes in buoyancy of a planktonic blue green alga in response to light intensity, *Br. Phycol. J.*, 15, 311, 1980.

20. **Lehmann, H. and Wiencke, C.**, Disappearance of gas vacuoles in the blue green alga *Mycrocystis aeruginosa, Plant, Cell Environ.*, 3, 319, 1980.

21. **Van Rijn, J. and Shilo, M.**, Buoyancy regulation in a natural population of *Oscillatoria* sp. in fishponds, *Limnol. Oceanogr.*, 28(No. 5), 1034, 1983.

22. **Klemer, A. R., Feuillade, J., and Feuillade, M.**, Cyanobacterial blooms: carbon and nitrogen limitation have opposite effects on the buoyancy of *Oscillatoria, Reprint series, Am. Assoc. Adv. Sci.*, Vol. 215, 1629, 1982.

23. **Pirt, S. J.**, *Principles of Microbe and Cell Cultivation*, Blackwell Scientific, Oxford, 1975.

24. **Kruger, G. H. J. and Eloff, J. N.**, The effect of temperature on specific growth rate and activation energy of microcystis and synechoccocus isolates relevant to the onset of natural blooms, *J. Limnol. Soc. S. Afr.*, 4(1), 9, 1978.

25. **Sorokin, C. and Krauss, R. W.**, The dependence of cell division in Chlorella on temperature and light intensity, *Am. J. Bot.*, 52(4), 331, 1965.

26. **Goldman, J. C. and Carpenter, E. J.**, A kinetic approach to the effect of temperature on algal growth, *Limnol. Oceanogr.*, 19, 756, 1974.

27. **McCombie, A. M.**, Actions and interactions of temperature, light intensity and nutrient concentration on the growth of the green alga, *Chlamydomonas reinhardi* Dangeard, *J. Fish. Res. Board Canada*, 17(6), 871, 1960.

28. **Spoehr, H. A. and Milner, H. W.**, The chemical composition of *chlorella;* effect of environmental conditions, *Plant Physiol.*, 84, 120, 1948.

29. **Sorokin, C.**, Tabular comparative data for the low- and high-temperature strains of *Chlorella, Nature (London)*, 184, 613, 1959.

30. **Vonshak, A. et al.**, Production of Spirulina biomass. Effects of environmental factors and population density, *Biomass*, 2, 175, 1982.

31. **Payer, H. D., Chiemvichak, Y., Hosakul, K., Kongpanichkul, C., Kraidej, L., Nguitragul, M., Reungmanipytoon, S., and Buri, P.**, Temperature as an important climatic factor during mass production of microscopic algae, in *Algae Biomass*, Shelef, G. and Soeder, C. J., Eds., Elsevier North Holland, Amsterdam, 1980.

32. **Sato, N. and Murata, N.,** Temperature shift-induced responses in lipids in the blue green alga, *Anabaena variabilis.* The central role of diacylmonoalactosylglycerol in thermo adaptation, *Biochim. Biophys. Acta,* 619, 353, 1980.

33. **Goldman, J. C.,** Biomass production in mass cultures of marine phytoplankton at varying temperatures, *J. Exp. Mar. Biol. Ecol.,* 27, 161, 1977.

34. **Goldman, J. C.,** Temperature effects on phytoplankton growth in continuous culture, *Limnol. Oceanogr.,* 22(5), 932, 1977.

35. **Siva, K., Rao, K., Brand, J. J., and Myers, J.,** Cold shock syndrome in *Anacystis nidulans, Plant Physiol.,* 59, 965, 1977.

36. **Lyons, J. M.,** Chilling injury in plants, *Ann. Rev. Plant Physiol.,* 24, 445, 1973.

37. **Ono, T. A. and Murata, N.,** Chilling susceptibility of the blue green alga *Anacystis nidulans.* I. Effect of growth temperature, *Plant Physiol.,* 67, 176, 1981.

38. **Ono, T. A. and Murata, N.,** Chilling susceptibility of the blue green alga *Anacystis nidulans.* II. Stimulation of the passive permeability of cytoplasmic membrane at chilling temperatures, *Plant Physiol.,* 67, 182, 1981.

39. **Mitchell, P.,** in *J. Bioenerg.,* 4, 63, 1973.

40. **Coleman, J. R. and Colman, B.,** The effect of pH on photosynthesis and inorganic carbon accumulation in a blue-green alga, in *Photosynthesis IV. Regulation of Carbon Metabolism,* Akoyunoglou, G., Ed., Balaban International Science Services, Philadelphia, Pa., 1981.

41. **Cook, J. R.,** Phosphate incorporation by *Euglena gracilis* during pH-dependent photo-inhibition, *J. Protozool.,* 28(2), 157, 1981.

42. **Toerien, D. F., Scott, W. E., and Pitout, M. J.,** Microcystis toxins: isolation, identification, implications, *Water S. Afr.,* 2, 160, 1976.

43. **Van der Westhuizen, A. J. and Eloff, J. N.,** Effect of culture age and pH of culture medium on the growth and toxicity of the blue green alga *Mycrocystis aeruginosa, Z. Pflanzenphysiol.,* 110, 157, 1983.

44. **Jassby, A. and Platt, T.,** Mathematical formulation of the relationship between photosynthesis and light for phytoplankton, *Limnol. Oceanogr.,* 21, 540, 1976.

45. **Verity, P. G.,** Effects of temperature, irradiance, and daylength on the marine diatom *Leptocylindrus danicus* Cleve. I. Photosynthesis and cellular composition, *J. Exp. Mar. Biol. Ecol.,* 55, 79, 1981.

46. **Sorokin, C. and Krauss, R. W.,** Effects of Temperature and Illuminance on Chlorella Growth Uncoupled from Cell Division, *Plant Physiol.,* May, 37, 1961.

47. **Collins, C. D. and Boylen, C. W.,** Physiological responses of *Anabaena variabilis* (Cyanophyceae) to instantaneous exposure to various combinations of light intensity and temperature, *J. Phycol.,* 18, 206, 1982.

48. **Gimmler, H., Wiedemann, C., and Moller, E. M.,** The metabolic response of the halotolerant green alga *Dunaliella parva* to hypertonic shocks, *Ber. Deutsch. Bot. Ges.,* 94, 613, 1981.

49. **Ehrenfeld, J. and Cousin, J. L.,** Ionic regulation of the unicellular green alga *Dunaliella tertiolecta, J. Membrane Biol.,* 70, 47, 1982.

50. **Ben-Amotz, A. and Grunwald, T.,** Osmoregulation in the halotolerant alga *Asteromonas gracilis, Plant Physiol.,* 67, 613, 1981.

51. **Greenway, H. and Setter, T. L.,** Accumulation of proline and sucrose during the first hours after transfer of *Chlorella emersonii* to high NaCl, *Aust. J. Plant Physiol.,* 6, 69, 1979.

52. **Setter, T. L. and Greenway, H.,** Growth and osmoregulation of *Chlorella emersonii* in NaCl and neutral osmotica, *Aust. J. Plant Physiol.,* 6, 47, 1979.

53. **Mohammad, F. A. A., Reed, R. H., and Stewart, W. D. P.,** The halophilic cyanobacterium *Synechocystis* DUN52 and its osmotic responses, *FEMS Microbiol. Lett.,* 16, 287, 1983.

54. **Ben Amotz, A., Katz, A., and Avron, M.,** Accumulation of β-carotene in halotolerant algae: purification and characterization of β-carotene-rich globules from *Dunaliella bardawil* (Chlorophyceae), *J. Phycol.,* 18, 529, 1982.

55. **Mackay, M. A., Norton, R. S., and Borowitzka, L. J.,** Marine blue green algae have a unique osmoregulatory system, *Mar. Biol.,* 73, 301, 1983.

56. **Grodzinski, B. and Colman, B.,** Loss of photosynthetic activity in two blue green algae as a result of osmotic stress, *J. Bacteriol.,* July, 456, 1973.

57. **Vonshak, A. and Richmond, A.,** Photosynthetic and respiratory activity in *Anacystis nidulans* adapted to osmotic stress, *Plant Physiol.,* 68, 504, 1981.

58. **Tel-Or, E.,** Response of N₂-fixing cyanobacteria to salt, *Appl. Environ. Microbiol.,* October, 689, 1980.

59. **Trainor, F. R.,** *Introductory Phycology,* John Wiley & Sons, New York, 1978.

60. **Kullberg, R. G.,** Effects of light and temperature on cell length of *Synechoccocus lividus* (Cyanophyta), *Trans. Am. Microsc. Soc.,* 100(2), 150, 1981.

61. **Peat, A. and Whitton, B. A.,** Environmental effects on the structure of the blue green alga, *Chlorogloea fritschii, Archiv. Mikrobiol.,* 57, 155, 1967.

62. **Isono, T. and Fujita, Y.,** Studies on morphological changes of the blue green alga *Nostoc muscorum* A with special reference to the role of light, *Plant Cell Physiol.,* 22(2), 185, 1981.

PRODUCTIVITY OF ALGAE UNDER NATURAL CONDITIONS

Zvy Dubinsky

ALGAL MASS CULTURE AND WATER BLOOMS IN NATURE

The phenomena loosely christened "water blooms" have always attracted the attention of the practitioners of algal mass culture. The green, blue-green, reddish, and orange waters described from virtually any conceivable aquatic biotope, freshwater, marine, natural, or man-made, seemed to be the philosopher's stone of applied algology. If only deciphered, the massive concentrations of microscopic algae occurring during blooms would be obtained by emulating the conditions which had led to the natural algal bloom, thereby creating a continuous algal crop to be harvested by man at will.

This optimistic line of reasoning is in most cases unwarranted. The visible, occasionally noisome algal biomass represents in many instances a concentration in space and time of algal cells only slowly accumulated in a given locality or transported to it from vast expanses of water.[1-3] In both such cases the true primary productivity per unit area and time may be low.

Opposite extreme situations are also well known. Very high primary productivities of phytoplankton do not necessarily result in any noticeable buildup of algal populations either because of sinking of the algal cells out of the euphotic zone or because of extensive grazing by zooplankton, which consumes the increment of algal biomass.[2,4] Another rather wide-spread phenomenon in tropical seas is the symbiotic association of dinoflagellates with invertebrate hosts such as corals.[5] These dense populations of endozoic "captive phyto-plankton" exhibit very high rates of photosynthesis; nevertheless, virtually all (>95%) of their photosynthate is translocated to the host leading to extremely low doubling rates of the algae.[6]

Let us examine a few additional illustrative situations. In extremely hypertrophic aquatic ecosystems such as heavily polluted estuaries or oxidation lagoons with heavy organic loading, the water may become brilliant emerald-green due to extremely shallow, dense populations of flagellate algae such as Euglena and Chlamydomonas. These algae may form a film at the surface, only a few millimeters thick, because of the prevailing anaerobic conditions below the air-water interface and limited penetration of light. The depth integral of photosynthesis is inevitably low due to the extremely restricted euphotic zone. In cases where certain algal populations become buoyant because of formation of gas vacuoles, like in so many blue-green algal blooms,[7-9] or in the case of the red-orange *Botryococcus braunii* blooms, due to lipid accumulation, we witness dense populations converging near the sur-face.[10] These frequently moribund algae may create scums or almost solid "crusts" on the heavily polluted waters,[1] as in Lake Erie or many fishponds[11] on which water fowl may safely "walk on the water". Again, in most such cases these blooms represent terminal phases of an algal population, rather than the expression of an energetically growing one (Figure 1). Primary productivities are low, losses due to excretion and lysis are high, and subsequent bloom cycles are initiated by resting cells resuspended from the sediment or residual populations that were not a part of the conspicuous bloom.[12]

Extensive streaks, patches, and "cells" of dense algal populations are formed by various physical flow patterns in their respective water bodies, the most common of these being Langmuir cells formed by combinations of wind and water movements (Figure 2).[3] Between the dense, narrow streaks where algae converge at the surface, rather wide water cells with sparse phytoplankton prevail. The average cell densities and areal primary productivity rates may again be rather limited.

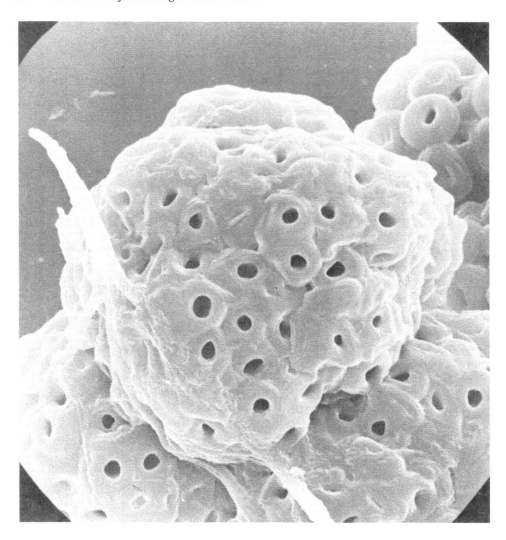

FIGURE 1. Scanning electron micrograph of orange-red colonies from a surface-scum of the "oil alga" *Botryococcus braunii* in Lake Kinneret, Israel. Such "empty" colonies do not initiate subsequent blooms. (Photo courtesy of Dina Rubin.)

The well-studied and extensively documented annual spring bloom of the dinoflagellate *Peridinium cinctum* (Figure 3) in Lake Kinneret (Israel)[3,37,47] is an illustration of a combination of these factors. Here we witness a development of high standing crops of this alga, exceeding 500 mg chlorophyll a/m² (Figure 4). Moreover, these swimming algae optimize nutrient gradients and irradiance levels in the water column, resulting in very dense horizontal strata.[16] Circulation patterns, internal seiches, and wind events superimpose horizontal inhomogeneities on this pattern.[17] In general, very high cell densities accumulate over a few months, in spite of low photosynthetic activities and division rates, mainly due to very limited grazing.

In the English Channel and North Sea, Arctic, and Antarctic waters, many workers have reported very high rates of primary productivities during the spring. Very heavy grazing pressure by zooplankton and salps prevent any buildup of phytoplankton numbers. Daily food intake by the herbivores was reported to be between 40 to 390% of their body weight. Such intensive grazing allows very fast turnover of nutrients that sustain the high primary productivity rates.[2]

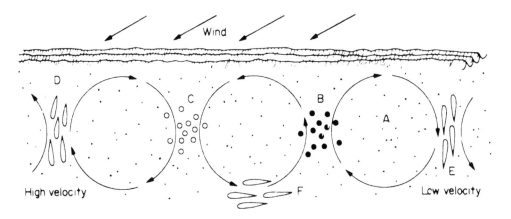

FIGURE 2. The effect of Langmuir cells on plankton distribution. (A) Neutrally buoyant plankton; (B) heavier-than-water cells concentrate in upwellings; (C) buoyant particles gather in downwellings; (D,E) actively swimming organisms; (F) organisms aggregated in areas of low current velocity, between upwellings and downwellings. (Parsons and Takahashi, 1973.[30])

FIGURE 3. The dinoflagellate *Peridinium cinctum*, which dominates the annual spring bloom in Lake Kinneret, Israel. (Photo courtesy of Tamar Berner.)

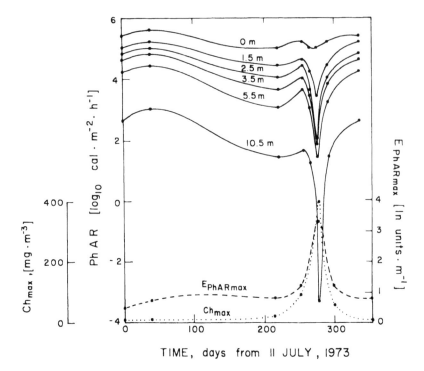

FIGURE 4.   Seasonal changes in downwelling irradiance (PhAR, photosynthetically available radiation), maximal light attenuation coefficients ($E_{PhAR\ max}$) and maximal chlorophyll concentrations (Chl_{max}). In 1973 the *Peridinium cinctum* bloom peaked on April 15, chlorophyll-a concentration in the euphotic zone reached 580 mg/m³. (Dubinsky Z. and Berman, T., *Limnol. Oceanogr.*, 2, 66, 1979. With permission.)

The examples in the previous paragraph demonstrated situations in various ecosystems where there was poor coupling between rates of primary productivity and phytoplankton densities.

CONDITIONS AND PROCESSES LEADING TO ALGAL BLOOMS

For an increase in phytoplankton biomass, a number of conditions have to be met. Gross photosynthetic rates have to exceed the combined losses caused by excretion of dissolved organic compounds, respiration of the algae, sinking, parasitism, grazing, and death (Equation 1).

$$dB/dt = B(P_G\text{-}E\text{-}R\text{-}S\text{-}P\text{-}G\text{-}D),\qquad(1)$$

where dB/dt is the rate of change in biomass; P_G is gross photosynthesis; E is excretion; R is respiration; S is sinking; P is parasitism; D is death; P_G, E, R, S, P, G, and D are rates, proportional to the standing crop B.

The dependence of photosynthesis on light is one of the most thoroughly studied topics in algal physiology.[18-24] In our context it is worth remembering that at low irradiance levels, photosynthetic rates are limited by the ambient photon flux densities, by the efficiency by which this flux is harvested by the photosynthetic pigment array, and the efficiency by which light, once trapped, is converted into photochemical products. The initial slope of the photosynthesis vs. irradiance (P vs. I) curve, is proportional to the product of light harvesting and processing efficiencies of the cells (Figure 5). As irradiance levels approach I_k, the average intervals between arrivals of photons to the photosynthetic reaction centers falls

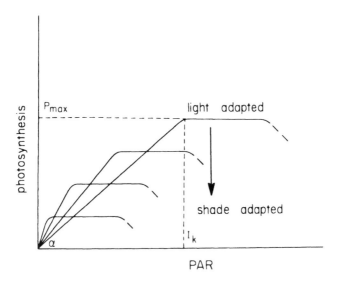

FIGURE 5. Changes in P_{max}, I_k, and α with light and shade adaptation. For explanation, see text.

below the turnover time of the photosynthetic electron flow. These "excess" photons cannot be utilized and their energy is "wasted" as fluorescence and heat. Above this irradiance level (I_k), the cells are light saturated, and they operate at their maximal photosynthetic rate, P_{max}. These rates, when not limited by supply of such nutrients as CO_2 or unfavorable temperatures, depend on the number of photosynthetic units in the cell and their turnover time.[22-24] P_{max} is species-dependent and may increase as phytoplankton adapts to high light levels and decreases in the course of shade adaptation (Figure 5). In many studies at high irradiance levels algae were found to be photoinhibited, showing a marked decrease in photosynthetic rates near the surface. This effect becomes more pronounced in poorly mixed water columns or in long incubation times in enclosed samples incubated with [14]C (Figure 6).

As we descend in a water column, light is attenuated, its intensity is reduced, and the spectral distribution becomes progressively restricted.[25] Photosynthetic rates in the light-limited part of the water column are roughly proportional to irradiance levels, which decrease exponentially with depth. At the compensation depth photosynthesis equals respiration (Equations 2 to 3):

$$P_G = R \tag{2}$$

therefore

$$P_N = P_G - R = 0 \tag{3}$$

P_N is net photosynthesis. Below the compensation depth, also known as the euphotic depth z_{eu}, any biomass increment due to photosynthesis during the day is consumed by respiratory losses over 24 hr. The euphotic depth at which Equation 3 is reached is usually assumed to be where the light level reaches 1% of its subsurface value (Equation 4).

$$I_{z_{eu}} = 0.01 \, I_o \tag{4}$$

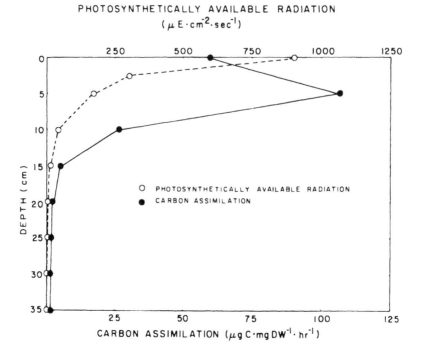

FIGURE 6.   Photosynthetic profile in a shallow hypertrophic pond. Upper 5 cm show light inhibition.

I_{zeu} is the irradiance at the euphotic depth and I_o is the subsurface irradiance. From the Beer-Lamberth law

$$I_z = I_o e^{-\eta z} \tag{5}$$

I_z is the irradiance at depth z, η light attenuation coefficient, in units, per meter; e is the basis of natural logarithms; z is depth, in meters.

From Equations 4 and 5 follows

$$0.01\ I_o = I_o e^{-\eta z_{eu}} \tag{6}$$

and

$$z_{eu} = -\ln 0.01/\eta \tag{7}$$

If, as during most blooms, light is attenuated mainly by algal pigments, then

$$\eta \simeq k_c{:}chl\ a \tag{8}$$

k_c is the in vivo specific absorbance of phytoplankton per milligram chlorophyll-a $(mg^{-1}/chl\ a/{\cdot}m^2)$[26] and chlorophyll-a is the concentration of this pigment (mg/m^{-3}). The values of k_c usually fall between 0.005 and 0.02. For this range, Equations 7 and 8 predict that areal chlorophyll-a concentrations between 230 ($k_c = 0.02$) and 921 ($k_c = 0.005$) mg/m^2 will absorb 99% of the incident light (Figures 6 and 7).[18,27-29]

From these considerations it follows that whenever algal populations become dense enough to reduce (in a vertically mixed water body, the average irradiance to which the cells are

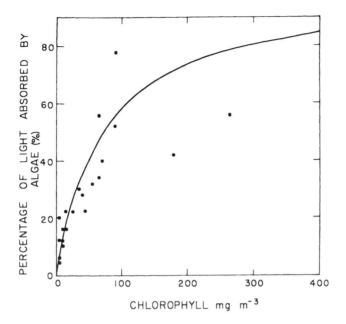

FIGURE 7.   Fractional light absorbance by algae of total light absorbed by water, as function of chlorophyll concentration. (Dubinsky, Z. and Berman, T., *Limnol. Oceanogr.*, 26, 665, 1981. With permission.)

exposed to I_k or below), P_G falls below P_{max} and the specific growth rate of the population declines below its maximal potential.

At algal concentrations above these, average light levels are further reduced until, for such a mixed layer, the depth integral of photosynthesis falls below that of respiration, dB/dt becomes negative, and the algal standing crop will be decreasing. This may be stated in another way. The critical depth in a water body is that at which the depth integrals of P_G and R for a 24-hr period are equal.[2,30] Therefore, if the mixing depth exceeds the critical depth, the bloom will decline.

It must be noted that in reality Equation 8 is not necessarily true but Equation 9,[19,21]

$$\eta = k_w + k_c:\text{chl a} \qquad (9)$$

where k_w is the light attenuation due to all substances other than living phytoplankton. If k_w is a large fraction of η, light available for photosynthesis is reduced accordingly, and so are the euphotic depth, the critical depth and maximal phytoplankton crops.[29]

Under conditions where light is not limiting, various inorganic nutrients will limit algal growth. Algal cells, like any other living matter, have rather rigid boundaries of the ratios of various elements in their biomass, usually close to the "Redfield ratio" which for P:N:C is 1:16:106.[31] Ambient concentrations of these and other elements must provide the influxes necessary for multiplication. Under steady-state, light-saturated conditions, growth rates will be set by the concentration of the limiting nutrient following the Monod equation (Equation 10),[32]

$$\mu = \mu_m S/(K_s + S) \qquad (10)$$

μ being the specific growth rate, μ_m the maximum growth rate, S the resource concentration, K_s a constant concentration.

It seems that many green and blue-green algal blooms in nutrient-rich waters do follow the Monod relationship fairly closely.

This does not hold true for many other blooms. Some algae succeed in uncoupling their growth rates from prevailing nutrient concentrations in the surrounding water.[33] Such algae are capable of nutrient uptake in excess (luxury uptake) of their instantaneous growth rates. This out-of-phase uptake results in large intracellular reserve pools of a given nutrient, allowing many subsequent divisions, even long after no detectable levels of the limiting nutrient can be found in the water. The storage may be accomplished either during transient high-nutrient pulses caused by storms, vertical mixing, zooplankton excretion, or by specialized phases of the algal life cycle. *Peridinium cinctum* in Lake Kinneret (Israel) continues to divide when virtually no phosphorus is available in the lake. Following the collapse of the bloom in late spring, surviving resting spores sink to the sediment, accumulating high levels of phosphorus, and when resuspended initiate the following annual bloom.[12-15] In this and similar cases relating the growth rates of various algae to different nutrients the Monod model will not apply, because of the uncoupling of cellular division rates from the environment.[34,35] Under such nonequilibrium conditions, growth depends on the intracellular concentrations of limiting nutrients, or on the cell quota (Equation 11).

$$\mu/\mu_{max} = 1 - k_Q/Q \qquad (11)$$

Q being the cell quota, or the weight of the intracellular pool of the nutrient per unit biomass; k_Q is the "subsistence quota", the minimal quota for life.

Once light conditions and nutrient concentrations favor algal growth, its rate has to exceed that of all the combined losses (Equation 1) to allow the buildup of a bloom.

Extracellular excretion by healthy algae usually involves only small losses (<5%) of photosynthate. Under conditions of nutrient imbalance, if photosynthesis may proceed, because of sufficient light and CO_2, but cell division is limited by nutrient shortages, the cells may lose much of the assimilated carbon as dissolved organic compounds,[2,36] or even shedding of carbohydrate structures such as thecae.[37]

Similarly, respiration in healthy algae consumes somewhere between 10 to 15% of gross photosynthesis,[2,30,38] and up to twice that fraction on a 24-hr basis. This value increases with temperature, stress, senescence, and above-optimal irradiance levels. Additional losses will be discussed in the context of their effects on species succession and competition in natural blooms.

ALGAL DEVELOPMENT IN THE PROCESS OF EUTROPHICATION

Water bodies containing low levels of nutrients essential for algal growth are termed oligotrophic, while nutrient rich waters are called eutrophic. In the marine environment regions of upwelling, where deep, nutrient rich water masses reach the surface high primary productivity results leading through the food web to high fish stocks.[2,4,39] In high latitudes, the marine water column is well mixed and nutrients in the euphotic zone are constantly replenished, supporting high primary productivity. Tropical seas, because of high surface temperatures are permanently stratified, a condition resulting in very low nutrient concentrations of nutrients in the euphotic zone. The "blue deserts" sustain only extremely low primary productivities, which are limited by the rates of recycling of nutrients within the upper layers of the sea, rather than influx from the deep waters below the thermocline. For example, in the productive Long Island Sound, production on an annual basis averages 389 g carbon/m²/year as compared to 78 in the oligotrophic Sargasso Sea. The Eastern Mediterranean is another example of a very oligotrophic marine region.[2]

Lakes, in general, proceed from being oligotrophic to increasingly euphotic.[1,40] This transition is accompanied by corresponding increases in algal productivity, and the development of massive seasonal, or persistent water blooms. While natural eutrophication proceeds slowly by accumulation of nutrients from the watershed, it may be accelerated to a great extent by human activities such as fertilization of agricultural lands around the lake, river, or estuary and the discharge of sewage or other nutrient-rich waters. The paleolimnological record of the sediments from several lakes, of which Lakes Biwa (Japan), Zurich (Switzerland), Erie and Washington (U.S.) are among the best studied, shows slow changes in algal populations as natural eutrophication proceeds. As a result of growth of adjacent population centers and of human impact on the waters and surrounding land, the pace of eutrophication was highly accelerated and these lakes became dominated by dense algal bloom of a few hardy species, replacing much sparser, diverse assemblages.[1,41] These processes have adversely affected the transparency, color, taste, and smell of the water of many lakes in all parts of our planet.[1,41,42] The importance of the problem of water quality has prompted worldwide research efforts, publications, and symposia,[41] resulting in scientific advice and public action. The well-documented spectacular reversal of this trend in Lake Washington since the diversion of municipal sewage in 1973, as well as in other cases such as the Thames in England, show the potential of such efforts.[1] The sewage supplies algae with all nutrients, especially phosphorus, which is the most important single limiting nutrient in fresh waters. The phosphorus is mainly derived from domestic detergents. In the 1970s the detergent industry has responded to the challenge of eutrophication by the development of phosphate-free detergents which have since become the norm in all developed countries.

In many cases the addition of high concentrations of phosphorus lowers the N:P ratio in the water, thereby giving nitrogen-fixing blue-green algae a competitive edge over other groups of phytoplankton.[8,43,44] The resulting massive blooms of cyanobacteria float toward the surface where high irradiance and oxygen concentrations lead to their photodynamic death and decomposition.[8,9,45] The latter process may lead to bacterial utilization of virtually all the dissolved oxygen, and to the development of anoxic conditions, accompanied by black sulfide deposits and the foul smells of ammonia, hydrogen-sulfide, and methane. Such situations cause fish kills and limit the flora and fauna to very few species.

SUCCESSION AND COMPETITION

A remarkable feature of many aquatic, freshwater, and marine ecosystems is the succession of dominant species proceeding in a recurring annual cycle.[1,2,13,46] Some of the many examples are the seasonal cycles occurring in temperate lakes, where a spring burst of diatom species will replace each other in quick succession. This is followed by a minor summer population of minute flagellates and blue-greens, and then a second major peak of blue-greens, diatoms, and dinoflagellates (Figure 8). Increased insolation in spring increases photosynthesis, while causing stratification and reducing the mixed layer to above the critical depth. The sequence between the diatom species is set by their uptake characteristics for silicate and phosphorus. The nutrient-rich epilimnion is being depleted by algal uptake, a process resulting in senescence of the diatom population. Zooplankton grazing and fungal parasites terminate the diatom blooms. The low populations of flagellates and blue-greens enjoy organic substances released into the water by the diatoms[47] and multiply on recycled nutrients within the epilimnion. In autumn, most lakes mix again as surface cooling eliminates stratification. This overturn makes nutrients available in the euphotic zone, but at this time zooplankton grazing is relatively heavy, increasing the ratio of large forms like dinoflagellates, avoided by the herbivores.[1,2,48]

Blue-green algae-cyanobacteria manage in many cases to displace other groups by a variety of strategies.[8,9,49-51] Many species have the capacity to combine molecular nitrogen, thereby overcoming local or seasonal low concentrations of this element. Such is the case with the blooms of some *Trichodesmium* species, a filamentous alga forming extensive blooms in tropical oligotrophic waters.[43] Other species optimize nutrient and light gradients by alternate

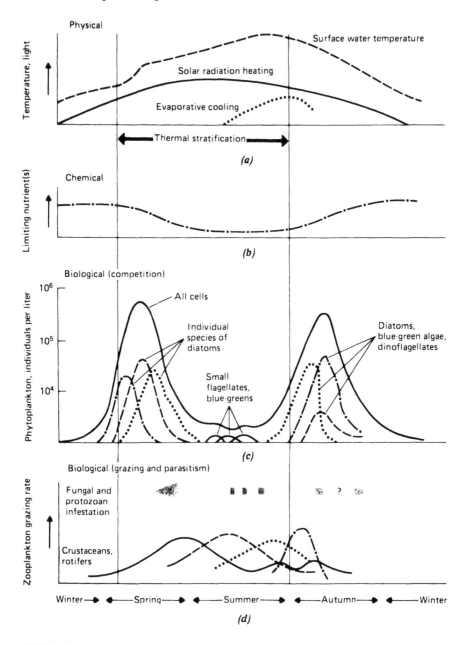

FIGURE 8.   Main abiotic a,b and biological c,d factors controlling succession of phytoplankton blooms in temperate lakes. (Goldman, C. R. and Horne, A. J., *Eds.*, *Limnology*, McGraw-Hill, New York, 205. With permission.)

sinking and flotation, accomplished by the formation and collapse of gas vacuoles.[7] This flotation may also allow such algae to use CO_2 directly from the water-air interface in poorly buffered lakes. The surface scum formed by the blue-green finds itself in optimal conditions for photosynthesis and nitrogen fixation, while at the same time successively competing with other algae, left in virtual darkness below the surface bloom.[50] Cyanobacteria have also been shown to excrete substances detrimental to other, competing species. However, under conditions of high irradiances and oxygen concentrations, cyanobacterial blooms developing in fish ponds may become trapped near the surface subject to the danger of photodynamic death. It may be that healthy algae may counter this danger by an increase of cellular carotenoid levels which screen out much of the harmful UV radiation.[51]

These and other sequences of algal blooms have been studied *in situ,* revealing in many cases correlations between changing environmental allogenic factors such as hydromechanics, temperature, irradiance, and nutrient composition.[58,59] In many cases, these correlations could be proven causative in subsequent laboratory studies. However, in many other cases, biotic factors were shown to be the major forces driving species dominance and succession. Such factors include physiological adaptations to special environmental conditions, life histories, predation, parasitism, and symbiosis. Whenever one species becomes dominant in a given environment, displacing any other, this outcome is considered proof of the principle of competitive exclusion,[52] derived from the mathematical treatment of competing species by Volterra[53] and Gause.[54]

When a nearly unialgal population dominates a water body or forms a bloom, it is most commonly because this species has the lowest requirement for a limiting nutrient or resource for which a number of species compete. Since the water bloom itself alters the nutrient ratios in the water as well as other environmental key parameters, such as the underwater light field, a different resource may in turn become limiting, causing dominance to be shifted from the species with the lowest requirement for the first resource to another species having the lowest equilibrium need for the subsequent limiting resource. For instance, under nitrogen-limited conditions, cyanobacterian species capable of N_2 fixation may become dominant, since they will exhibit the lowest need for this limiting nutrient. However, if limnological, seasonal or biological events now cause phosphorus to become limiting in the same lake, nitrogen-fixing blue-greens may be replaced by other algae. Such was indeed the case in Lakes Washington and Trummen.[55] Similar outcomes were found with low Si:P ratios which affect the ability of silicon-requiring diatoms to compete with other algal groups.[56]

In many natural systems, competitive exclusion may be modified by spatial or temporal heterogeneity in the distribution of key nutrients and such environmental parameters as salinity, temperature, and light.[3] Such structuring within the system allows coexistence of a number of species in apparent violation of the exclusion principle. In a series of recent, elegant studies, Tilman[55,56] has shown how any number of species may coexist indefinitely, provided that they have different sequences of threshold sensitivities for minimal values for critical limiting resources or biological factors. Thus, a given species may be excluded under one P:N ratio, dominate the system under another, and coexist over some specific intermediate range of such ratios. Similarly, temperature may alter the equilibrium requirements of various species for the same resource, thereby shifting the dominance among the competing species. Such an outcome between diatom species competing for silica under a wide (4 to 24°C) temperature range was found by Tilman et al.[56] Attempts to control or manipulate the dominant algal species in open algal mass culture installations were met with only limited success,[57,58] except under special circumstances, where the rather exotic requirements of the algae grown excluded most competitors. Such is the case with the brine alga, *Dunaliella salina,* as well as with *Spirulina platensis* grown on alkaline, high bicarbonate media.

MAXIMAL PRIMARY PRODUCTIVITY UNDER NATURAL CONDITIONS

Because of seasonal fluctuations, the levels of the various environmental conditions affecting algal photosynthesis, and division rates, maximal levels of primary productivity are rarely sustained over long periods under natural conditions. Moreover, algal growth itself is cybernetically self-regulating by many negative feedback loops. Such are the reduction of dissolved nutrients as they are incorporated into the increasing algal standing crop, while being removed from the water body itself. A similar effect is commonly observed in respect to the underwater light field (Figure 4). The growing algal population harvests an increasing

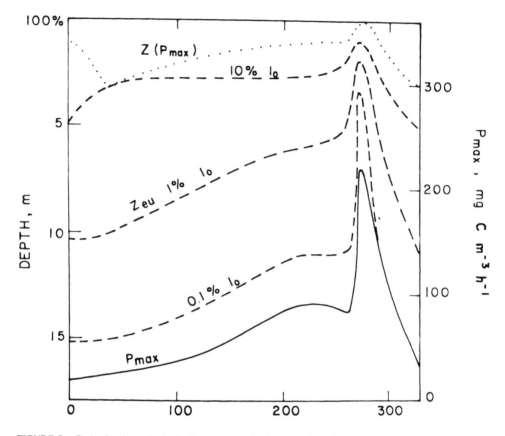

FIGURE 9. Reduction in euphotic depth, z_{eu} and vertical compression of photosynthetic profile during dino-flagellate bloom in Lake Kinneret. Israel. During the bloom, maximal values of photosynthesis, P_{max} increase, and the depth at which these values occur, $z (P_{max})$ is reduced. (Dubinsky, Z. and Berman, T., *Limnol. Oceanogr.*, 26, 665, 1981. With permission.)

fraction of the light, reducing its intensity, and narrowing its spectral range to the wavelengths which are the least useful to the algae.[25,29] The euphotic depth becomes more and more restricted (Figure 9) and more algae are found at any time below the compensation depth. The nutrient-starved or light-limited dense algal populations may become very susceptible to the adverse effects of such additional environmental factors as high temperature, wind-induced shear, parasites, and grazers.

During periods of maximal primary productivity, the rate of the process is limited only by efficiency of harvesting and utilization of light by the algae. Since at high irradiances near the water surface much light is available for photosynthesis (but because of the saturation of all photosynthetic units by the high photon-flux), much of the energy is wasted. The result is very low quantum yield for the conversion of the absorbed light energy in assimilation of CO_2 into photosynthate. Conversely, toward the lower part of the euphotic zone, the little light available is used very efficiently.[39,59] These opposed trends have been analyzed by Ryther who concluded that if primary productivity is not limited by any nutrient, gross photosynthesis may reach about 38 g of dry algal weight per square meter per day. Of this value, about 10 g/m²/day are used up by the respiration of the algae themselves. The value of 27 g/m²/day is very close to observed average values in intensive algal mass cultures.[60] In some isolated cases, mostly in the East African soda lakes, such as Lake Aranguadi, even slightly higher values have been reported.[28,61-63] It may be worth noting that some of these lakes support continuous high primary productivity rates. These may result from a combination of optimal temperature and light regimes and a shallow, usually mixed water column. Nutrient supply is provided by the mixing and grazing of zooplankton and water fowl. In all cases, the high pH and alkalinities prevent nighttime losses of respiratory CO_2 and always maintain considerable reserves of this nutrient that might otherwise limit photosynthetic rates, a well-known problem in algal mass cultures.

REFERENCES

1. **Goldman, C. R. and Horne, A. J.**, *Limnology*, McGraw-Hill, New York, 1983.
2. **Raymont, J. E. G.**, *Plankton and Productivity in the Oceans. I. Phytoplankton*, Pergamon Press, Oxford, 1980, Chap. 8.
3. **Steele, J. H.**, *Spatial Patterns in Plankton Communities*, Plenum Press, New York, 1978.
4. **Walsh, J. J.**, Death in the sea: enigmatic phytoplankton losses, *Prog. Oceanogr.*, 12, 1, 1983.
5. **Droop, M. R.**, Algae and invertebrates in symbiosis, in *Symbiotic Associations*, Nutman, P. S. and Mosse, B., Eds., Cambridge University Press, London, 1963, 171.
6. **Muscatine, L., Falkowski, P. G., Porter, J. W., and Dubinsky, Z.**, Fate of photosynthetically fixed carbon in light and shade-adapted corals, *Proc. R. Soc. London*, B222, 181, 1984.
7. **Okada, M. and Aiba, S.**, Simulation of water-bloom in a eutrophic lake. III. Modelling the vertical migration and growth of *Microcystis aeruginosa*, *Water Res.*, 17, 883, 1983.
8. **Shilo, M.**, Photosynthetic microbial communities in aquatic ecosystems, *Phil. Trans. R. Soc. London Ser. B:*, 297, 565, 1982.
9. **Walsby, A. E. and Klemer, A. R.**, The role of gas vacuoles in the microstratification of a population of *Oscillatoria agardhii* var. *isothrix* in Denning Lake, Minnesota, *Arch. Hydrobiol.*, 74, 375, 1974.
10. **Aaronson, S., Berner, T., Gold, K., Kushner, L., Patni, N. J., Repak, A., and Rubin, D.**, Some observations on the green planktonic alga *Botryococcus braunii* and its bloom form. *J. Plankton Res.*, 5, 693, 1983.
11. **Shilo, M.**, Study on the isolation and control of blue-green algae from fish ponds, *Bamidgeh*, 17, 83, 1965.
12. **Pollingher, U. and Serruya, C.**, Phased division of *Peridinium cinctum* fa. *westii* and the development of the blooms in Lake Kinneret, *J. Phycol.*, 11, 155, 1976.
13. **Berman, T. and Pollingher, U.**, Annual and seasonal variations of phytoplankton, chlorophyll, and photosynthesis in Lake Kinneret, *Limnol. Oceanogr.*, 19, 31, 1974.
14. **Pollingher, U. and Berman, T.**, Quantitative and qualitative changes in the phytoplankton of Lake Kinneret, Israel, *Oikos*, 29, 419, 1978.
15. **Serruya, C., Ed.**, *Lake Kinneret*, W. Junk, The Hague, 1978.
16. **Berman, T. and Rodhe, W.**, Distribution and migration of *Peridinium* in Lake Kinneret, *Mitt. Int. Ver. Theor. Angew. Limnol.*, 18, 588, 1971.
17. **Serruya, S.**, Wind, water temperature and motions in Lake Kinneret: general pattern, *Verh. Int. Verein Limnol.*, 19, 73, 1975.
18. **Talling, J. F.**, Generalized and specialized features of phytoplankton as a form of photosynthetic cover, in *Prediction and Measurement of Photosynthetic Productivity*, Centre for Agriculture Publication Document, Wageningen, 1970, 431.
19. **Bannister, T. T.**, Production equations in terms of chlorophyll concentration, quantum yield and upper limit to production, *Limnol. Oceanogr.*, 19, 1, 1974.
20. **Jassby, A. D. and Platt, T.**, Mathematical formulation of the relationship between photosynthesis and light for phytoplankton, *Limnol. Oceanogr.*, 21, 540, 1976.
21. **Dubinsky, Z.**, Light utilization efficiency in natural phytoplankton communities, in *Primary Productivity in the Sea*, Falkowski, P. G., Ed., Plenum Press, New York, 1980, 83.
22. **Falkowski, P. G.**, Light-shade adaptation in marine phytoplankton, in *Primary Productivity in the Sea*, Falkowski, P. G., Ed., Plenum Press, New York, 1980, 99.
23. **Prezelin, B. B.**, Light reactions in photosynthesis, in *Physiological Basis of Phytoplankton Ecology*, Platt, T., Ed., Can. Bull. Fish Aquat. Sci, 210, Ottawa, 1981, 251.
24. **Richardson, K., Beardall, J., and Raven, J. A.**, Adaptation of unicellular algae to irradiance: an analysis of strategies, *New Phytol.*, 93, 157, 1983.
25. **Dubinsky, Z. and Berman, T.**, Seasonal changes in the spectral composition of downwelling irradiance in Lake Kinneret (Israel), *Limnol. Oceanogr.*, 24, 652, 1979.
26. **Atlas, D. and Bannister, T. T.**, Dependence of mean spectral extinction coefficient of phytoplankton on depth, water color, and species, *Limnol. Oceanogr.*, 25, 157, 1980.

27. **Dubinsky, Z. and Berman, T.,** Light utilization by phytoplankton in Lake Kinneret (Israel), *Limnol. Oceanogr.,* 26, 660, 1981.
28. **Talling, J. F., Wood, R. B., Prosser, M. V., and Baxter, R. M.,** The upper limit of photosynthetic productivity by phytoplankton: evidence from Ethiopian soda lakes, *Freshwater Biol.,* 3, 53, 1973.
29. **Tilzer, M. M.,** The importance of fractional light absorption by photosynthetic pigments for phytoplankton productivity in Lake Constance, *Limnol. Oceanogr.,* 28, 833, 1983.
30. **Parsons, T. R. and Takahashi, M.,** *Biological Oceanographic Processes,* Pergamon Press, Oxford, 1973.
31. **Redfield, A. C.,** The biological control of chemical factors in the environment, *Am. Sci.,* 46, 205, 1958.
32. **Monod, J.,** *Recherches sur la Croissance des Cultures Bacteriennes,* Herman, Paris, 1942.
33. **Droop, M. R.,** 25 years of algal growth kinetics, a personal view, *Bot. Mar.,* 26, 99, 1983.
34. **Ketchum, B. M.,** The absorption of phosphate and nitrate by illuminated cultures of *Nitzschia closterium, Am. J. Bot.,* 26, 399, 1939.
35. **Eppley, R. W. and Strickland, J. D. H.,** Kinetics of phytoplankton growth, in *Advances in Microbiology of the Sea,* Vol. 1, Academic Press, London, 1980, 23.
36. **Fogg, G. E.,** Dissolved organic matter in oceans and lakes, *New Biol.,* 29, 30, 1959.
37. **Criscuolo, C. M., Dubinsky, Z., and Aaronson, S.,** Skeleton shedding in *Peridinium cinctum* from Lake Kinneret — a unique phytoplankton response to nutrient imbalance, in *Developments in Arid Zone Ecology and Environmental Quality,* Shuval, H., Ed., Balaban ISS, Philadelphia, 1981, 169.
38. **Sharp, J. H.,** Excretion of organic matter by marine phytoplankton: do healthy cells do it? *Limnol. Oceanogr.,* 22, 381, 1977.
39. **Ryther, J. H.,** Potential productivity of the sea, *Science,* 130, 602, 1959.
40. **Edmondson, W. T.,** The sedimentary record of the eutrophication of Lake Washington, *Proc. Natl. Acad. Sci. U.S.A.,* 71, 5093, 1974.
41. **Likens, G. E.,** Nutrients and eutrophication: the limiting-nutrient controversy, *Am. Soc. Limnol. Oceanogr.,* Allen Press, Lawrence, Kan., 1972.
42. **Schwimmer, M. and Schwimmer, D.,** *The Role of Algae and Plankton in Medicine,* Grune & Stratton, New York, 1955.
43. **Saino, T. and Hattori, A.,** Nitrogen fixation by *Trichodesmium* and its significance in nitrogen cycling in the Kuroshio area and adjacent waters, in *The Kuroshio, IV,* Takenouti, Y., Ed., Saikon Publishing, Tokyo, 1980, 697.
44. **Smith, V. H.,** Low nitrogen to phosphorus ratios favor dominance by blue-green algae in lake phytoplankton, *Science,* 221, 669, 1983.
45. **Eloff, J. N., Steinitz, Y., and Shilo, M.,** Photooxidation of cyanobacteria in natural conditions, *Appl. Environ. Microbiol.,* 31, 119, 1976.
46. **Keating, K. I.,** Allelopathic influence on blue-green bloom sequence in a eutrophic lake, *Science,* 196, 885, 1977.
47. **Keating, K. I.,** Blue-green algal inhibition of diatomic growth: transition from mesotrophic to eutrophic community structure, *Science,* 199, 971, 1978.
48. **Ryther, J. H. and Sanders, J. G.,** Experimental evidence of zooplankton control of the species, composition and size distribution of marine phytoplankton, *Mar. Ecol. Prog. Ser.,* 3, 279, 1980.
49. **Ganf, G. G. and Oliver, R. L.,** Vertical separation of light and available nutrients as a factor causing replacement of green algae by blue-green algae in the plankton of a stratified lake, *J. Ecol.,* 70, 829, 1982.
50. **Paerl, H. W. and Ustach, J. F.,** Blue-green algal scums: an explanation for their occurrence during freshwater blooms, *Limnol. Oceanogr.,* 27, 212, 1982.
51. **Paerl, H. W., Tucker, J., and Bland, P. T.,** Carotenoid enhancement and its role in maintaining blue-green algal (*Microcystis aeruginosa*) surface blooms, *Limnol. Oceanogr.,* 28, 847, 1983.
52. **Hardin, G.,** The competitive exclusion principle, *Science,* 131, 1292, 1960.
53. **Volterra, V.,** Fluctuations in the abundance of a species considered mathematically, *Nature (London),* 118, 558, 1926.
54. **Gause, G. F.,** *The Struggle for Existence,* Hafner Press, New York, 1934.
55. **Tilman, D.,** *Resource Competition and Community Structure,* Princeton University Press, Princeton, N.J., 1982.
56. **Tilman, D., Kilham, S. S., and Kilham, P.,** Phytoplankton community ecology: the role of limiting nutrients, *Ann. Rev. Ecol. Syst.,* 13, 349, 1982.
57. **Goldman, J. C. and Ryther, J. H.,** Temperature-influenced species competition in mass cultures of marine phytoplankton, *Biotechnol. Bioeng.,* 18, 1125, 1976.
58. **Azov, Y., Shelef, G., Moraine, R., and Levi, A.,** Controlling algal genera in high-rate wastewater oxidation ponds, in *Algal Biomass, Production and Use,* Shelef, G. and Soeder, C. J., Eds., Elsevier, Amsterdam, 1980, 523.
59. **Dubinsky, Z., Berman, T., and Schanz, F.,** *J. Plankton Res.,* 6, 339, 1984.
60. **Goldman, J. C.,** Outdoor algal mass cultures. II. Photosynthetic yield limitations, *Water Res.,* 13, 119, 1979.

61. **Melack, J. M. and Kilham, P.,** Photosynthetic rates of phytoplankton in East African alkaline, saline lakes, *Limnol. Oceanogr.*, 19, 743, 1974.

62. **Tuite, C. H.,** Standing crop densities and distribution of *Spirulina* and benthic diatoms in East African alkaline saline lakes, *Freshwater Biol.*, 11, 345, 1981.

63. **Young, T. C. and King, D. L.,** Interacting limits to algal growth: light, phosphorus, and carbon dioxide availability, *Water Res.*, 14, 409, 1980.

LABORATORY TECHNIQUES FOR THE CULTIVATION OF MICROALGAE

A. Vonshak

INTRODUCTION

The manner in which algae are cultivated varies widely, depending not only on the specific organism, but also on the use to be made of the culture. The aim of this chapter is to describe the basic approaches in measuring algal growth and some of the most common techniques used, as well as to indicate the problems associated with them. It is not intended to provide a detailed manual on the different procedures used. More detailed information may be found in the literature referred to in the text.

ALGAL GROWTH IN LIQUID CULTURE

Cultivation of microalgae in liquid culture may be considered as a process in which the bulk concentration of a given population increases. To obtain an actively growing culture requires (1) a viable inoculum of minimal size, (2) a supply of the needed nutrients and microelements, (3) suitable physicochemical conditions (temperature, pH, etc.), and (4) light as a source of energy. When measuring growth in liquid culture, some precautions must be taken: that the culture is homogeneous and well mixed, that growth on the walls of the container is prevented, and that concentration gradients of nutrients and gases are avoided. Assessing growth also requires a precise definition of what should be measured. The most common is based on the idea of growth as the ability of cells to undergo a complete cell cycle. This definition implies that the most direct way of monitoring growth will be by counting cells, i.e., monitoring the increase in cell number by microscopic or electronic methods. Nevertheless, growth can be less directly monitored, by measurement of the content of protein, ribonucleic acid, deoxyribonucleic acid, etc.

The most appropriate parameter for monitoring growth must be chosen. In an asynchronous culture, this problem can be solved by causing a culture to be in balanced growth, i.e., so that all its extensive properties increase exponentially at the same rate. In such a culture, the increase in any extensive property (dx) such as the amount of protein, pigment, DNA or biomass, will be proportional to the initial amount present (x) and to the time interval (dt)

$$dx = \mu dt \tag{1}$$

where dx/dt represents the actual population growth rate and μ, having the dimensions of the reciprocal of time, is termed the specific growth rate and represents the growth per unit time. Equation 1 can be integrated ($x = x_o$ at $t = 0$) to:

$$x = x_o e^{\mu t} \tag{2}$$

Growth fits this relation during its exponential or logarithmic phase. Equation 2 can be further solved to:

$$\ln x/x_o = \mu t \tag{3}$$

When $x = 2x_o$

$$\ln 2 = \mu t_2 \tag{4}$$

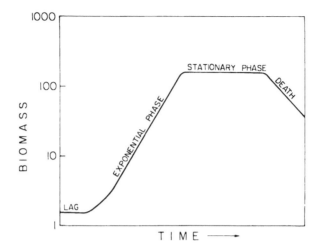

FIGURE 1.   Idealized normal growth cycle for a microalgae population in a batch culture.

and

$$t_2 = \ln 2/\mu = 0.693/\mu \tag{5}$$

where t_2 is the doubling time of the biomass and Equation 5 relates the specific growth rate to the doubling time.

Equations 1 to 5 predict the growth of microalgae in simple systems where the factors influencing growth are constant, but do not allow prediction of the deviation from exponential growth when the culture enters the stationary-state phase, which is part of the growth cycle of a batch system.

Batch Cultures (Closed Systems)

In a batch system, nothing is added to or removed from the liquid phase after inoculating an appropriate medium with a viable inoculum. It follows, therefore, that a batch system can support cell multiplication for only a limited time and with progressive changes in the composition of the medium and the light intensity within the culture.

Figure 1 shows an idealized normal growth curve for a simple homogeneous batch culture of algae. Growth proceeds through a lag phase during which the cell number does not increase and then into a growth phase characterized by an exponential increase in biomass. Ultimately, changes in the physicochemical environment result in a phase of no net increase in cell number, the stationary phase. Cells in the stationary phase still require an energy source to maintain viability. The availability of an energy source in a concentrated batch culture is necessarily limited and hence a death phase follows, which is often characterized by an exponential decrease in the number of living cells.

Synchronized Batch Cultures

These are used in cell cycle studies when growth patterns representative of individual cells are to be followed. In an ideal synchronized culture, growth is unbalanced, since all the cells divide nearly simultaneously. Thus, a plot of log cell number vs. cultivation time will increase in steps rather than as a straight line. The pattern of growth and the changes in some of the growth parameters are shown in Figure 2. Synchronized cultures can be obtained by (1) induction of synchrony by repeated shifts in environmental or nutritional conditions or (2) physical separation of cells from a random population and subsequent

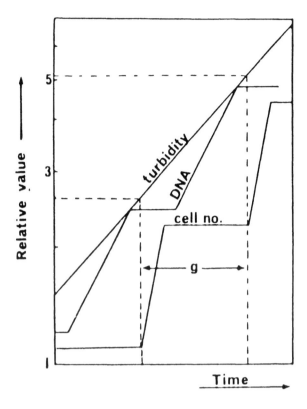

FIGURE 2. Changes in some growth parameters in a synchronized culture.

reculturing of those that are at the same stage of the cell cycle. Table 1 summarizes the most common techniques and algal species used in synchrony studies.

Continuous Culture (Open System)

The continuous culture technique is widely used for bacterial as well as for algal cultivation.[15,16] It differs from batch culture in that a fresh supply of nutrients is continuously added at the same rate as medium is withdrawn from the culture so that the culture volume remains constant with time. This theoretically permits continuous exponential growth of the culture. Steady-state (exponential phase) growth is possible when all factors promoting cell growth are balanced by those factors contributing to loss of cells, so that the cell concentration (biomass) is maintained as constant. This occurs when the flow rate of fresh medium into the culture (f = dv/dt) equals the flow rate of culture throughout the overflow. If V is the culture volume, then

$$(dv/dt) \, (1/v) = \mu$$

If we define D as the dilution rate, which also equals $(dv/dt)/V$, then $D = \mu$.

There are basically three different ways to cultivate microalgae in a continuous system, as outlined below.

Chemostat

Chemostat cultivation is based on the assumption that the algal growth rate is limited by the slowest step of nutrient metabolism under the particular conditions and thus, by the concentration of a particular nutrient. The chemostat system consists of a reservoir containing

Table 1
PROCEDURES USED FOR OBTAINING
SYNCHRONIZED CULTURES OF
MICROALGAE

Organism	Procedure	Ref.
Chlorella	DL & temp. shifts	1
Chlorella	Light/dark changes	2
Chlorella	Separation of cells by size fol-lowed by light/dark cycle	3
Chlorella	Centrifugation in Ficoll (selection)	4
Eudorina elegans	Light/dark cycles	5
Olisthodiscus luteus	Light/dark cycles	6
Chlorella and *Scenedesmus*	Light/dark cycles under photo-synthetical controlled dilution	7
Chlamydomonas	Light/dark shifts	8
Scenedesmus	Light/dark shifts	9
Anacystis nidulands	Light/dark shifts and CO_2 starvation	10
	Light/dark cycles	11
	Light intensity and dark changes	12
	Temp. shifts	13
	Temp. and light shifts	14

the culture medium, a constant-flow pump, and a culture vessel with constant volume (maintained by an overflow). Culture liquid leaves the vessel at the same rate as new medium is fed into it. The culture has to be homogeneously mixed, which may be done with the aeration stream, a mechanical impeller, or in small culture vessels, magnetic stirring.

Semicontinuous Culture

Semicontinuous cultures are a type of batch culture which are diluted at frequent intervals. The concentration of biomass is monitored to estimate the proper frequency of dilution and the dilution ratio.

Turbidostat Culture

This type of culture depends on continuous monitoring of the biomass by an optical device that measures the turbidity and controls the dilution rate, maintaining the culture at a preset optical density. Nutrients are always nonlimiting. Light, on the other hand, may be a limiting factor unless the culture is very dilute.

PROCEDURES FOR DETERMINATION OF BIOMASS CONCENTRATION AND CELL GROWTH

Cell Counts

Direct Microscopic Counting

This procedure is commonly used to determine algal growth. For counting natural dilute populations, an inverted microscope and settling chambers are recommended. The major difficulty in microscopic enumeration is reproducibility; thus, adequate attention should be given to sampling, diluting, and filling the chamber, as well as choosing the right counting chamber, microscope magnification, and range of cell concentration. Recommended counting chambers are listed in Table 2 and depend on the cell size and concentration. The exact steps of counting depend on the counting device used and are described in detail elsewhere.[17]

Table 2
CELL COUNTING CHAMBERS AND THEIR PROPERTIES

Commercial name of chamber	Chamber vol (ml)	Depth (mm)	Objective used for magnification	Cell size (μm)	Cell conc counted easily
Redgwick Rafter	1.0	1.0	2.5—10	50—100	30—10^4
Palmer Malony	0.1	0.4	10—45	5—150	10^2—10^5
Speirs Levy hematocytometer	$4/10^3$	0.2	10—20	5—75	10^4—10^7
Improved Neaubouer	$2/10^4$	0.1	20—40 (phase)	2—30	10^5—10^7
Petroff Houser	$2/10^5$	0.02	40—100	0.5—5	10^5—10^8

In general, counting chambers need to be cleaned and dried before use. The algal cells should settle for 4 to 6 min before counting, especially if small cells are counted in a deep chamber. The use of two counting chambers is recommended, so that the cells can settle in one while the other is being counted. After the chamber is filled, it should be examined at low magnification to assure satisfactory distribution and concentration of cells.

Colony Counts

These are commonly done in bacteriology, but are rarely used in phycology to determine biomass. Nevertheless, this procedure allows counting colonies to arise from single viable cells, which may be important in the study of stationary phase cultures. The exact procedure has to be adapted to the strain used. The most common involve (1) spread plates, in which a small volume (0.1 ml) of a culture, after dilution if needed, is spread on top of solidified agar in a Petri dish, or (2) layered plates, in which 2 to 3 ml of diluted culture are added to a tube containing 0.6 to 0.8% soft agar is then mixed and poured on top of solidified agar in a Petri dish. The Petri dishes are then kept inverted for at least 3 to 4 days under a humid environment before colonies can be counted.

Electronic Counting

The Coulter® Counter and similar instruments are very useful for counting nonfilamentous round algal cells, but very difficult to apply to rod-shaped or other irregular cells. The principle of electronic counting is as follows. A diluted algal suspension is pumped through a very small orifice between two fluid-filled compartments. Electrodes in each measure the electrical resistance of the system. The orifice is small enough that its electrical resistance is high. When a particle passes through the orifice, the resistance increases further since the conductivity of the algal cell is less than that of the medium. The change in resistance is converted into an electrical pulse which is counted. Two main problems, failure of daughter cells to separate after cell division and clogging of the orifice, exist in the outline use of this device. Thus, highly pure reagents and careful cleaning procedures are recommended. Useful information and detailed operating procedures are usually provided with the instruction manuals for the instruments.

Light Scattering (Turbidity)

Light-scattering methods are the most general techniques used to follow the growth of pure cultures. They can be very powerful but can lead to erroneous results. The major advantages are that they can be performed quickly and nondestructively. They mainly give information about increase in biomass (dry weight) and not directly about the number of cells. The common practice is to use any available colorimeter or spectrophotometer to measure turbidity. At low turbidities there is a simple geometric relationship between the number of cells in the light path and the measured light intensity, since the intensity of the unscattered light decreases exponentially as the number of cells increases. Of the many instruments available, the Klett-Summersen colorimeter is the most popular, most likely due to its stability and ease of use.

Dry Weight

Measuring the increase in dry weight is one of the most direct ways to estimate biomass production and involves the following steps.

Sampling

Taking representative aliquots from the algal culture is one of the most crucial steps. Special attention should be paid to achieving a well-stirred culture, rapid pipeting of samples to avoid settling, and a large enough sample.

Separation

After sampling, cells are separated from the medium by membrane filtration or by centrifugation. The cells have to be washed to remove salts and other contaminants, usually with diluted medium or buffer. Marine algae must not be washed with distilled water to avoid plasmolysis and bursting of the cells.

Drying

A wide range of temperatures (70 to 110°C) is mentioned in the literature. It is best to optimize the drying temperature to the particular organism according to some basic rules: (1) avoidance of excessive heat, (2) good reproducibility, and (3) identical weights obtained for the same sample taken at intervals of 1 hr.

Weighing

After the centrifuge tube or the dish containing the filter membrane is cooled in a desiccator (15 to 30 min) to room temperature, it is weighed quickly to avoid moisture absorption by the samples. (Centrifuge tubes or filter membranes should be dried by the same procedure before use.) The results are expressed as dry weight per volume or per illuminated surface area. The most common factors affecting the accuracy are: (1) poor sampling, not representative of the culture (duplicates are recommended); (2) loss of cells during separation due to gas vaculated cells, not sedimenting in centrifugation, or to lysis of cells by washing with distilled water; and (3) improper handling of samples, e.g., insufficient or excessive drying, inadequate cooling before weighing.

Chlorophyll Determination

One of the most rapid chemical methods for estimating the amount of living plant material is the determination of chlorophyll. The procedures described here are the most commonly used. However, in some strains, special extraction treatments may be required.

Separation of Cells

Cells may be removed from concentrated laboratory cultures by centrifugation. With large volumes, cells are collected on nitrocellulose or other filters (GFC or, preferably, HA).

Extraction

Pigments can be extracted with a variety of solvents: acetone,[18] methanol, and ether.[19] In some cases, brief heating is required for complete extraction. This stage is most important, since complete pigment extraction is required.

Determination of Absorbance

After extraction, the cell debris are removed by filtration or centrifugation, and the extracted pigments are transferred to a clean tube for spectroscopic determination. It is most important that the appropriate wavelength be used, since the absorption spectra of the various chlorophyll species differ and are affected by the solvents used.

For example:

$$chl_a = 15.6\ A_{663} - 2.0\ A_{645} - 0.8\ A_{630}$$

$$chl_b = 25.4\ A_{645} - 4.4\ A_{665} - 10.3\ A_{630}$$

$$chl_c = 109\ A_{630} - 12.5\ A_{665} - 28.7\ A_{645}$$

when the procedure of Richard and Thompson[18] is used.

The UNESCO procedure[20] called for the following:

$$chl_a = 11.64\ A_{663} - 2.16\ A_{645} - 0.10\ A_{630}$$

$$chl_b = 20.97\ A_{645} - 3.94\ A_{663} - 3.66\ A_{630}$$

$$chl_c = 54.22\ A_{630} - 14.81\ A_{645} - 5.53\ A_{663}$$

ISOLATION AND PURIFICATION OF MICROALGAE

Before going into a detailed description of these procedures, it should be mentioned that, due to the differences between algal species, all procedures must be adapted to meet the specific requirements of each algal strain to be isolated.

Collection

The major source of algal strains for most phycological studies are culture collections and yet, the initial supply of algal material is often from a natural population. Algae can be collected and concentrated efficiently with plankton nets or by filtration through nitrocellulose filters. Attached forms can be collected by scraping the surface. When algal specimens are collected directly, it is recommended to make observations and isolations within a very short time, to increase the chances for obtaining a uniform sample.

Isolation

Unialgal cultures are started from clones propagated from a single cell or filament. Many techniques can be used to obtain such a culture, depending on the cell size, equipment available, etc.

Washing by Micropipette

This is one of the simplest techniques for isolating microscopic forms in the use of micropipettes to carry individual cells through a series of sterile washes. A micropipette made from soft glass tubing or from a Pasteur pipette is used for picking up single cells or filaments under a dissecting microscope. A microhook can be used to pick up filamentous algae. This procedure requires some training and practice before satisfactory results are obtained.

Atomizer Technique

This fairly simple technique can be used to both isolate and purify collections of algae. Algae, 10 mℓ, are washed by centrifugation and after the final wash all but 2 mℓ of the supernatant is discarded. The pellet is suspended in the remaining supernatant. A 15-cm long microtube is inserted down to the bottom of the centrifuge tube and held in place with a cotton stopper (Figure 3). Compressed air is directed through a small opening so that the

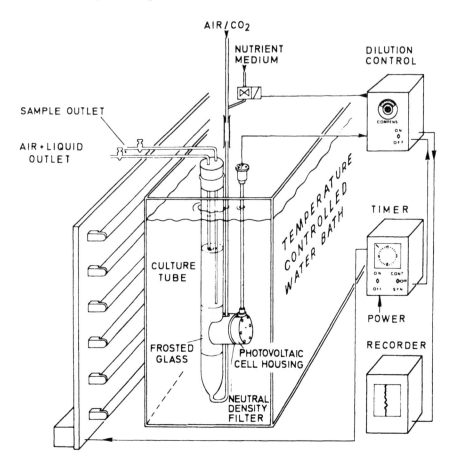

FIGURE 3. A schematic illustration of an atomizer apparatus.

air stream crosses the top of the microtube extending out of the centrifuge tube. The algal suspension is drawn up the microtube, finely atomized, and sprayed onto a sterile plate of agar medium. The plate is then recovered and placed under lights. After several days single cells or colonies free of bacteria and fungi can be picked up from the plate with a sterile toothpick and transferred to a sterile liquid medium or another agar plate.

Several additional techniques may be used which exploit the special physiological requirements or features of various algal strains.

Taxis

Motile algae that react positively or negatively to light, electrical current or other stimuli can be preferentially recovered.

Osmotic Changes

Some algae are more resistant to rapid changes in osmotic pressure; thus, transferring the culture from distilled water to a salt solution and back to distilled water will facilitate removal of some protozoa and some other sensitive organisms.

Agar Plates

A small volume (0.1 to 0.5 mℓ) of a culture is mixed with soft agar and poured on top of solidified agar plates. After a few days colonies can be picked up with a piece of agar and placed into fresh sterile medium.

Table 3
SOME COMMON ANTIBIOTICS
INHIBITING GROWTH OF ALGAE

Antibiotic	Conc (mg/ℓ)	Alga	Ref.
Erythromycin	700	*Chlorella*	21
Neomycin	5	*Clamydomonas*	22
		Nostoc	
		Scenedesmus	
Penicillin	0.1	*Anabaena Variabilis*	23
	2	*Microcystis*	24
Streptomycin	0.1	*Anabaena Variabilis*	23
	10	*Clamydomonas*	22
	2—4	*Nitzschia*	

Purification (Axenic Cultures)

The simplest way to obtain an axenic agar culture is by separating algal cells from bacteria by centrifugation. It is recommended to test this procedure before trying others.

UV Irradiation

Most of the eukaryotic algae are more resistant to UV light than are bacterial cells. However, it is best to use short exposure times since UV irradiation may cause mutations in the surviving algae.

Filtration

Membrane filters can be used to separate filamentous algae from bacteria. Filaments are usually broken to short lengths (3 to 5 cells) by sonication and diluted samples are vacuum filtered through a sterile membrane filter.

Antibiotics

Various antibiotics have been effectively used in purification of algae from bacteria. It is important to use the minimum effective concentration of the drugs because chloroplasts and blue-green algae are sensitive to most of the antibiotics that kill bacteria. The effects of some antibiotics on algae are summarized in Table 3.

GROWTH, MAINTENANCE, AND PRESERVATION

Once algal cultures have been obtained by direct isolation from nature or from a culture collection, it is recommended to establish a locally maintained culture.

Sterilization

This is mandatory even for algae that are not bacteria-free so as to avoid contamination from other sources. All glassware, pipettes, and media should be sterilized before use by autoclaving for 15 min at 15 pounds pressure. In certain instances, such as with seawater where autoclaving causes precipitation of salts from the medium, filter sterilization should be done.

Illumination

After a small amount of algal cells is transferred to fresh medium, it is placed under cool-white fluorescent tubes (200 to 400 fc). In liquid cultures for rapid multiplication, 500 fc are used. After good growth is obtained (1 to 2 weeks), the cultures are transferred to lower illumination (50 to 100 fc) for slower growth and storage.

Algae stocks on agar media are illuminated after transfer with an intensity of 250 fc for 6 or 7 days until good growth has been obtained. Such stocks are then moved to areas with an illumination level of 50 to 75 fc.

The use of cool-white fluorescent tubes for culturing algae is recommended. Using incandescent bulbs or direct sunlight causes problems with heat. Cultures grown in direct sunlight will often reach a temperature as much as 10°C above that of the room. Most cultures do best when given a dark period each day. The two time sequences most commonly used are 12 hr light alternated with 12 hr dark or 16 hr light and 8 hr dark.

Temperature

For preservation, most algae do well at room temperature: 15 to 20°C. Only few may prefer higher temperatures for survival or have 20 to 25°C as a lethal temperature.

Frequency of Transfer

Cultures are transferred at different intervals depending on the maintenance conditions and species. Unicellular and filamentous nonmotile species may be transferred once every 6 to 12 months. Flagellated species may require more frequent transfers. With some species long preservation under liquid nitrogen has been successful.

Culture Media

There are many recipes for media for the cultivation of microalgae under laboratory conditions. Most of them are modifications of previously published formulas and some are derived from analysis of the water in the native habitat and ecological considerations. Only few are the result of a detailed study on the nutrient requirements of the organisms.[25] The main considerations in developing nutrient recipe for algal cultivation are

1. The total salt concentration, mostly dependent on the ecological origin of the organism.
2. The composition and concentration of major ionic components such as potassium, magnesium, sodium, calcium, sulfate, and phosphate.
3. Nitrogen sources: nitrate, ammonia, and urea are widely used as the nitrogen sources, mainly dependent on the species performance and the pH optimum. Growth is highly dependent on nitrogen availability. Most microalgae contain 7 to 9% nitrogen per dry weight. Thus, for the production of 1 g cell in 1 ℓ of culture, a minimum of 500 to 600 mg/ℓ of KNO_3 will be required.
4. Carbon source: inorganic carbon is usually supplied as CO_2 gas in a 1 to 5% mixture with air. Another mode of supplying carbon is as bicarbonate. The preference is highly dependent on the pH optimum for growth.
5. pH: usually, acidic pH values are used to prevent precipitation of calcium, magnesium, and some of the trace elements.
6. Trace elements: these are usually supplied in a mixture in microgram per liter concentrations previously found to be effective. Nevertheless, the necessity of such components for growth has not always been demonstrated. For stability of the mixture of trace elements, chelating agents such as citrate and EDTA are used.
7. Vitamins: many algae require vitamins such as thiamin and coblamin for growth.

Recipes for Some Algal Growth Media
Modified Allen's Medium (Widely Used for Blue-Green Algae)
For each liter of medium, add the following to 999 mℓ of glass-distilled water:

$NaNO_3$	1.500 g
K_2HPO_4	0.039 g
$MgSO_4 \cdot 7H_2O$	0.075 g

Na_2CO_3	0.020 g
$Ca(NO_3)_2 \cdot 4H_2O$	0.020 g
$Na_2SiO_3 \cdot 9H_2O$	0.058 g
EDTA	0.001 g
Citric acid	0.006 g
$FeCl_3$	0.002 g
Microelements*	1 mℓ

* Microelements: to 1000 mℓ of glass-distilled water, add the following:

H_3BO_4	2.86 g
$MnCl_2 \cdot 4H_2O$	1.81 g
$ZnSO_4 \cdot 7H_2O$	0.222 g
$Na_2MoO_4 \cdot 2H_2O$	0.391 g
$CuSO_4 \cdot 5H_2O$	0.079 g
$Co(NO_3)_2 \cdot 6H_2O$	0.0494 g
Adjust the pH to 7.8	
and autoclave.	

This medium can be solidified with 1.5% agar, which should be autoclaved separately and mixed when cooled to 47°C.

BG-11 Medium

$NaNO_3$*	1.500 g
$K_2HPO_4 \cdot 3H_2O$	0.040 g
$MgSO_4 \cdot 7H_2O$	0.075 g
$CaCl_2 \cdot 2H_2O$	0.036 g
Citric acid	0.006 g
Ferric ammonium citrate	0.006 g
Ethylenediaminetetraacetic acid (EDTA), disodium magnesium salt	0.001 g
Na_2CO_3	0.020 g
Trace metal mix A5 (see below)	1.0 mℓ
Deionized water	1000 mℓ

After autoclaving and cooling, the pH of the medium should be 7.4.

Trace metal mix A5:

H_3BO_3	2.860 mg/mℓ
$MnCl_2 \cdot 4H_2O$	1.810 mg/mℓ
$ZnSO_4 \cdot 7H_2O$	0.222 mg/mℓ
$Na_2MoO_4 \cdot 2H_2O$	0.390 mg/mℓ
$CuSO_4 \cdot 5H_2O$	0.079 mg/mℓ
$Co(NO_3)_2 \cdot 6H_2O$	0.0494 mg/mℓ

*Omitting the nitrogen sources makes the medium suitable for cultivating N_2-fixing cyanobacteria.

Medium for Chlorella

Many formulas are used for the cultivation of this genus. The following was used by Sorokin and Krauss:[26]

	g/ℓ
KNO_3	1.250
KH_2PO_4	1.250
$MgSO_4 \cdot 7H_2O$	1.000
$CaCl_2$	0.084
H_3BO_3	0.114
$FeSO_4 \cdot 7H_2O$	0.050
$ZnSO_4 \cdot 7H_2O$	0.088
$MnCl_2 \cdot 4H_2O$	0.014
MoO_3	0.007
$CuSO_4 \cdot 5H_2O$	0.016
$Co(NO_3)_2 \cdot 6H_2O$	0.005
EDTA	0.5

The pH of the medium is 6.8.

Medium for Spirulina

	g/ℓ
NaCl	1.00
$MgSO_4 \cdot 7H_2O$	0.20
$CaCl_2$	0.040
$FeSO_4 \cdot 7H_2O$	0.010
EDTA	0.080
K_2HPO_4	0.500
$NaNO_3$	2.500
K_2SO_4	1.000
$NaHCO_3$	16.800

and 1 mℓ/ℓ of A_5 and B_6 as below:

A_5	g/ℓ
H_3BO_3	2.860
$MnCl_2 \cdot 4H_2O$	1.810
$ZnSO_4 \cdot 7H_2O$	0.222
$CuSO_4 \cdot 5H_2O$	0.074
MoO_3	0.015

B_6	g/ℓ
NH_4NO_3	229.6×10^{-4}
$K_2Cr_2(SO_4)_4 \cdot 24H_2O$	960×10^{-4}
$NiSO_4 \cdot 7H_2O$	478.5×10^{-4}
$Na_2SO_4 \cdot 2H_2O$	179.4×10^{-4}
$Ti(SO_4)_3$	400×10^{-4}
$Co(NO_3)_2 \cdot 6H_2O$	439.8×10^{-4}

Modified Chu 10

	amt./ℓ (mg)
$Ca(NO_3)_2 \cdot 4H_2O$	20.0
KH_2PO_4	6.2
$MgSO_4 \cdot 7H_2O$	25.0
Na_2CO_3	20.0
Na_2SiO_3	25.0

Trace elements final conc in 1 ℓ

$HCl(1\ N)$	0.25 mℓ
Na_2 EDTA	2.0 mg
$FeCl_3$	1.0 mg
H_3BO_3	2.48 mg
$MnCl_2/4H_2O$	1.39 mg
$(NH_4)_6Mo_7O_{24} \cdot 4H_2O$	1.00 mg
Vitamin B_{12}	0.01 mg
Vitamin B_1	0.001 mg
Biotin	0.001 mg

N-8 used for *Chlorella* and other green algae

	mg/ℓ
$Na_2HPO_4 \cdot 2H_2O$	260
KH_2PO_4	740
$CaCl_2$	10
Fe EDTA	10
$MgSO_4 \cdot 7H_2O$	50
KNO_3	1000
Trace elements	1 mℓ

Trace element stock

	g/ℓ
$Al_2(SO_4)_3 \cdot 18H_2O$	3.58
$MnCl_2 \cdot 4H_2O$	12.98
$CuSO_4 \cdot 5H_2O$	1.83
$ZnSO_4 \cdot 7H_2O$	3.2

Bold's Basal Medium

This is a useful medium for many algae. It may be supplemented with soil extract for growing algae isolated from soils. Six salt solutions, 400 mℓ in volume, are employed with each containing one of the following salts in the amount listed:

1.	$NaNO_3$	10 g
2.	$MgSO_4 \cdot 7H_2O$	3 g
3.	K_2HPO_4	3 g
4.	KH_2PO_4	7 g
5.	$CaCl_2 \cdot 2H_2O$	1 g
6.	$NaCl$	1 g

For trace element solution:

7.	EDTA	50 g/ℓ
	KOH	31 g/ℓ
8.	$FeSO_4 \cdot 7H_2O$	4.98 g/ℓ

(Water acidified with 1 mℓ of H_2SO_4)

9.	H_3BO_3	11.42 g/ℓ
10.	$ZnSO_4 \cdot 7H_2O$	8.82 g/ℓ
	$MnCl_2 \cdot 4H_2O$	1.44 g/ℓ
	MoO_3	0.71 g/ℓ
	$CuSO_4 \cdot 5H_2O$	1.57 g/ℓ
	$Co(NO_3)_2$	0.49 g/ℓ

10 mℓ of each stock solution (1 to 6) and 1 mℓ of each solution 7 to 10 are added to a final volume of 1 ℓ. The medium may be solidified by 15 g/ℓ of agar.

Euglena gracilis Medium

	g/100 mℓ
Sodium acetate (hydrated)	0.1
Beef extract	0.1
Yeast extract	0.2
Bactotryptone	0.2
$CaCl_2$	0.001

May be solidified adding 15 g agar.

Ochromonas Medium

Glass distilled water	960 mℓ
Glucose	1.0 g
Tryptone	1.0 g
Yeast extract	1.0 g
Liver extract (infusion)	40 mℓ

Chlamydomonas reinhardii Medium

	Minimal salt medium	High salt minimal medium
$NH_4Cl(g)$	0.05	0.50
$MgSO_4 \cdot 7H_2O(g)$	0.02	0.02
$CaCl_2 \cdot 2H_2O(g)$	0.01	0.01
$K_2HPO_4(g)$	0.72	1.44
$KH_2PO_4(g)$	0.36	0.72
Hutner's trace elements (mℓ)	1	1
Distilled water (ℓ)	1	1

**Hutner's trace elements
solution** (g)

EDTA	50.0
$ZnSO_4 \cdot 7H_2O$	22.0
H_3BO_3	11.4
$MnCl_2 \cdot 4H_2O$	5.1
$FeSO_4 \cdot 7H_2O$	5.0
$CoCl_2 \cdot 6H_2O$	1.6
$CuSO_4 \cdot 5H_2O$	1.6
$(NH_4)_6Mo_7O_{24} \cdot 4H_2O$	1.1
Distilled water (mℓ)	750

Boil, cool slightly, and bring to pH 6.5 to 6.8 with KOH (do not use NaOH). The clear solution is diluted to 1000 mℓ with distilled water and should have a green color which changes to purple on standing. It is stable for at least 1 year.

Phaeodactylum/Nitzschia Medium

	g/ℓ
NaCl	5.0
$MgSO_4 \cdot 7H_2O$	1.2
$NaNO_3$	1.0
KCl	0.60
$CaCl_2$	0.30
K_2HPO_4	0.10
Tris (Hydroxymethyl) amino methane	1.0
Micro elements	10 mℓ

Micro elements stock

	g/ℓ
Na_2 EDTA	3.0
H_3BO_3	0.60
$FeSO_4 \cdot 7H_2O$	0.20
$MnCl_2$	0.14
$ZnSO_4 \cdot H_2O$	0.033
$CO(NO_3)_2 \cdot 6H_2O$	0.0007
$CuSO_4 \cdot 5H_2O$	0.0002

Recipes for Seawater and Halotolerant Algae
Dunaliella parva Medium

	Final conc
NaCl	1.5 M
$MgSO_4 \cdot 7H_2O$	24 mM
$MgCO_2 \cdot 6H_2O$	20 mM
$CaCl_2 \cdot 2H_2O$	10 mM
$NaNO_3$	4 mM
KNO_3	1 mM

K₂HPO₄	0.1 mM
FeCl₃·6H₂O	1.5 μM
Na₂ EDTA	30 μM
H₃BO₃	7 μM
MnCl₂·4H₂O	0.8 μM
ZnCl₂	0.02 μM
CaCl₂·2H₂O	0.02 μM
Tris·HCl	20 μM

pH — 7.4

Artificial Seawater Used for Porphyridium

	g/ℓ
NaCl	27
MgSO₄·7H₂O	6.6
MgCl₂·6H₂O	5.6
CaCl₂·2H₂O	1.5
KNO₃	1.0
KH₂PO₄	0.07
NaHCO₃	0.04
Tris·HCl (1 μm, pH 7.6)	20 mℓ
Micro elements	1 mℓ

Micro element stock

	mg/ℓ
ZnCl₂	4.0
H₃BO₃	60.0
CaCl₂·6H₂O	
CaCl₂·2H₂O	4 mg
MnCl₂·4H₂	40 mg
(NH₄)₆O₂₄·4H₂O	37 mg
+ 1 mℓ of Fe EDTA solution	
Fe EDTA solution	
100 mℓ of 0.05 M Na₂ EDTA pH 7.6	
240 mg FeCl₃/4H₂O	

Enriched Seawater

	Stock	For 1 ℓ of filter sterilized seawater
KNO₃	1 M (10 g/ℓ)	2 mℓ
KH₂PO₄	0.1 M (13.61 g/ℓ)	2 mℓ

Na$_2$	30 mM	
EDTA·2H$_2$O	25 mM	1.5 mℓ
+ FeCl$_3$·6H$_2$O		
Tris·HCl	1 M (pH 7.6)	20 mℓ
NaHCO$_3$	1 M (84 g/ℓ)	5 mℓ
Na$_2$SiO$_3$*	0.1 M (30 g/ℓ)	5 mℓ
Trace elements		10 mℓ

Trace elements stock

	g/ℓ
CaSO$_4$·5H$_2$O	0.98
ZnSO$_4$·7H$_2$O	2.2
CaCl$_2$·6H$_2$O	1.0
MnCl$_2$·4H$_2$O	18.0
Na$_2$MoO$_4$·2H$_2$O	0.63

* Used for cultivation of diatoms.

Solidified Media Used for Maintenance

Cyanobacteria Agar

	g/ℓ
KNO$_3$	5.0
K$_2$HPO$_4$	0.1
MgSO$_4$·7H$_2$O	0.05

Fe ammonium citrate 1 mℓ of 1% solution solidified with 15 g of agar.

Porphyridium Agar (For 1 ℓ of Medium)

Glass-distilled water	400 mℓ
Filter-sterilized seawater	500 mℓ
Soil water supernatant	100 mℓ
Yeast extract	1.0 g
Tryptone	1.0 g
Agar	15g

Diatom Seawater-Agar

Bold's Basal Medium	500 mℓ
Filter-sterilized seawater	500 mℓ

Soil supernatant	50 mℓ
Agar	15 g

Allen and Arnon Medium for Blue-Green Algae*

	Stock conc M	Final conc mM	mℓ/ℓ medium
$MgSO_4$	0.1	1	1
$CaCl_2$	0.05	0.5	10
NaCl	4.0	0.004	1
K_2HPO_4**	2.0	0.002	1
$NaNO_3$	2.5	2.500	1
KNO_3	2.8	2.500	1
Micro elements			1

Micro element stock

	g/ℓ
Fe EDTA	4.00
$MnSO_4 \cdot 4H_2O$	0.50
MoO_3	0.10
$ZnSO_4 \cdot 4H_2O$	0.05
$CuSO_4 \cdot 5H_2O$	0.02
H_3BO_3	0.50
NH_4NO_3	0.01
$CO(NO_3)_2 \cdot 6H_2O$	0.01
$NiSO_4 \cdot 6H_2O$	0.01
$Cr_2(SO_4)_3K_2SO_4 \cdot 24H_2O$	0.01
$Na_2NO_4 \cdot 2H_2O$	0.01
$TiOSO_4$	0.01

* For liquid cultures a 1:8 dilution may be used,
supplied with 1 to 5% CO_2 in air.
** For a solidified medium 10 g/ℓ of purified agar
is used.

Soil Water Medium

This medium is used especially for isolation procedures. The success in using this medium is highly dependent on the selection of a suitable soil. A culture vessel is filled with about $^1/_2$ in. of good garden soil. The soil should be of about medium humus content and should not contain commercial fertilizer. Any good potting soil can be used successfully. The vessel is then filled with glass distilled water, and plugged or capped. It is then steamed (not autoclaved) for 2 hr on 2 consecutive days. This medium is used for nonsterile conditions where "normal" morphology is studied. For phototrophic algae from alkaline habitats a pinch of $CaCO_3$ is placed in the bottom of the vessel before the soil and water are added.

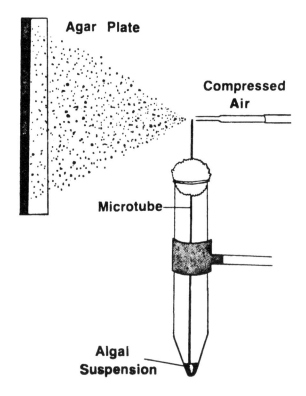

Agar Plate

Compressed Air

Microtube

Algal Suspension

FIGURE 4. Schematic diagram of the arrangement of the culture tube and the photovoltaic cell in a light-thermostate, with the automatic dilution control device, timer, and recorder.[7] (From Pfau, J., et al., *Arch. Microbiol.*, 75, 338, 1971. With permission.)

SPECIAL DEVICES

Since microalgae are easy to handle as an experimental material, several special devices have been developed for different research needs. The aim of this section is to mention some of those devices (references to full descriptions are given).

Mass Cultivation of Microalgae Under Controlled Conditions

For studies where mass quantities of algal cells are needed, few devices that can meet this supply have been developed. A full description of one is given by Juttner[27] for the cultivation of *Anacystis nidulans*, where cells are cultured in glass tubes arranged in parallel, through which they are pumped. The device is fully controlled with respect to CO_2 supply and temperature. The device can easily be used for other algae and photosynthetic bacteria.

Automatic Synchronization

A special tube for the cultivation of microalgae is used. An automatic dilution device regulated by a photocell and a timer for dark/light cycles is attached to the system. This experimental set-up is used for obtaining synchronized cultures of *Scenedesmus* and *Chlorella*. Figure 4 shows a schematic diagram of the device.

A Thermal Gradient Device

A temperature gradient device was developed to study the effect of temperature and its interaction with other variables on the growth, reproduction, and physiology of many aquatic

Thermal gradient plate

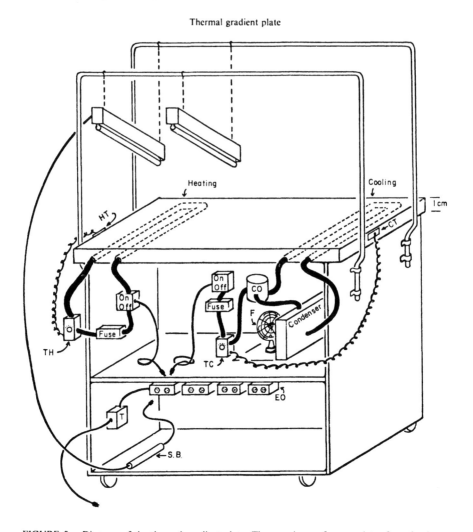

FIGURE 5.   Diagram of the thermal gradient plate. The growing surface consists of an aluminum plate that has one end cooled by a standard refrigeration evaporator and the opposite end heated using a heating element. The result is the formation of a linear temperature gradient down the length of the plate. Fluorescent lights are suspended above the growing surface and the entire apparatus is on rollers for ease in relocation. Key: (TH) heating thermostat; (HT) heating thermocouple; (TC) cooling thermostat; (CT) cooling thermocouple; (CO) compressor; (T) interval timer; (SB) starter ballast; (F) fan; (EO) electrical outlet. (From Siver, P. A., *Br. Phycol. J.*, 18, 159, 1983. With permission.)

organisms. A stable temperature gradient was produced along the length of a 1 m × 1 m × 1.25 cm aluminum plate by heating one end of the plate with a heating element and cooling the opposite end with a standard refrigeration evaporator. Figure 5 illustrates one of these devices. The device can be adapted to various requirements. One modification is described by Yanish et al.,[29] where a light intensity gradient is combined with the temperature gradient plate so that the interrelation of temperature and light can be studied easily.

A Device for Simulation of the Variability in Sunlight Intensity

A device that simulates variations in sunlight intensity by controlling the voltage applied to a tungsten lamp according to a predetermined program is described. The technique allows considerable latitude in color correction and enables including another variable in experiments. The device has been used in research on marine algae and is claimed to be easy to

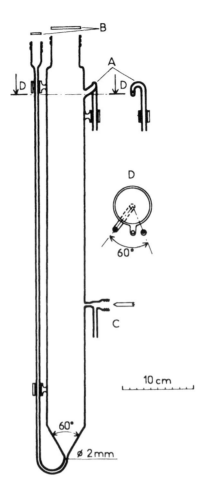

FIGURE 6. A modified growth tube. (A) Intermitted siphon; (B) silicone seals; (C) glass valves with Teflon® needle; (D) section view. (From Feuillade, J., and Feuillade, M., *Limnol. Oceanogr.*, 24, 562, 1979. With permission.)

use. Other devices producing diurnal variation of light intensity were constructed with sequence timers[30] or motor-driven rotating shutters.[31]

A Chemostat for Planktonic Filamentous Algae

Continuous culture of planktonic filamentous alga may cause problems due to poor mixing, growth on the walls, etc. Several devices have been described that overcome such problems. Figure 6 demonstrates such a tube successfully used for the cultivation of *Oscilatoria*.[32]

Turbidostat and Chemostat

A variety of devices have been described for this purpose. The main problems concern maintaining an axenic culture for prolonged periods in a chemostat or turbidostat where dilutions are frequent or continuous. Groeneweg and Soeder[33] constructed a culture tube (Figure 7) for the cultivation of axenic microalgae in a chemostat or a turbidostat (Figure 8).

Light Measurement in Algal Suspension

The precise measurement of illumination in algal suspensions is fairly complicated due

SCREW CAP
WITH BORE
Ø 8MM

SEALING RING

SCREW THREAD
Ø 15 MM

SCREW THREAD
Ø 42MM

SCREW
CONNECTOR

SEALING RING

CULTURE TUBE
Ø 36 MM

FIGURE 7.   Dismantled culture tube to show the various parts. (From Groeneweg, J. and Soeder, C. J., *Br. Phycol.*, 13, 337, 1978. With permission.)

to the geometry of the culture vessels and shapes of the sensors. Van Liere et al.[34] used a photodiode that can be fixed at any place inside a culture vessel so that useful information on light intensity within the culture can be obtained. A somewhat different device is described by Griffith et al.,[35] where precise measurement of incident illumination at the center of an algal suspension within the reaction chamber of an oxygen electrode allows reliable comparisons to be made of the photosynthetic characteristics of cultures of differing population densities and provides estimations of the degree of shade adaptation in *Chlorella*. A schematic view of such a device is given in Figure 9.

Oxygen Electrode

Algae, like all photosynthetic organisms, evolve oxygen. This activity may be used as a parameter for measuring metabolic activity under various environmental conditions and as an indication of the well-being of algal cultures outdoors. The oxygen electrode is an electrochemical cell where a current proportional to the oxygen concentration is generated. The cell consists of a platinum cathode and a silver anode with a KCl solution as the electrolyte. When the two electrodes are connected through a battery, a current proportional to the oxygen concentration will flow. To avoid poisoning of the electrodes, they are shielded by a thin Teflon® membrane. The response of the electrode and the permeability of the membrane to oxygen are temperature-sensitive, so the temperature should be constant during

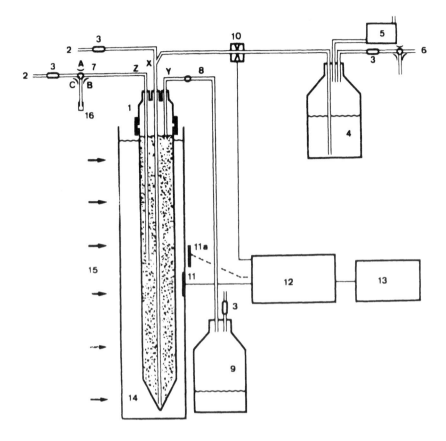

FIGURE 8. Diagram of algal tubidostat (or chemostat). (1) Culture tube with screw connector, (2) air-CO$_2$ inlet, (3) sterile glass-wool filter, (4) nutrient reservoir, (5) sterile filter for nutrient solution, (6) air inlet for supplying constant pressure in the nutrient reservoir, (7) sampling device, (8) overflow tube, (9) overflow reservoir, (10) solenoid valve (tubidostat) or perisaltic pump (chemostat), (11) photocell, (11a) reference photocell, (12) dilution control unit, (13) recorder, (14) thermostatic waterbath, (15) illumination, (16) rubber cap filled with alcohol. (From Groeneweg, J. and Soeder, C. J., *Br. Phycol. J.*, 13, 337, 1978. With permission.)

measurement. Detailed instructions for the operation of oxygen electrodes can be found in manufacturers manuals and in Jassbey.[36]

SPECIAL PROCEDURES

Most of the techniques developed for physiological studies in microorganisms are easily adapted to algae cultures. Some procedures that needed special modification or were specially developed to be used with algae are reported here.

Microscopical Methods
Fluorescence Microscopy
The applicability of fluorescence microscopy to routine algal studies was initially discussed by Wood,[37] as a procedure to easily distinguish between photosynthetic (chlorophyll-bearing) and nonpigmented flagellates. The standard Zeiss filter sets excitation wavelengths of 380 to 490 and 520 to 560 nm were the most useful for examinations of algae. The brightest fluorescence of blue-green algae *(Cyanophyta)*, red algae *(Rhodophyta)*, and cryptomonads *(Cryptophyta)* always occurred with the green excitation wavelengths (540 to 560 nm), whereas all the other algae examined fluoresced much brighter when excited with the blue-

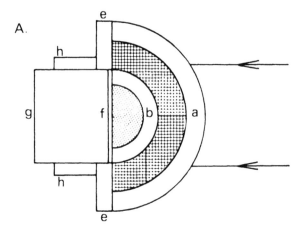

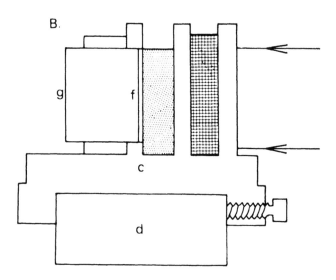

FIGURE 9.   Perspex half-cell viewed from above (A) and in section from side (B) showing arrangement for insertion of quantum sensor (g) of light meter in a position corresponding to center of algal suspension (stippled) in oxygen electrode chamber. Arrows show path of beam of incident light. (a,b) Two tubular pieces of Perspex cut longitudinally, (c) Perspex base, (d) stirrer plate of electrode, (e) Perspex sheet, (f) glass window, (h) Perspex cylinder. (From Griffiths, D. et al., *Limnol. Oceanogr.*, 23, 368, 1978. With permission.)

violet wavelengths (380 to 490 nm). The fluorescence emission of diatoms *(Bacillariophyta),* green algae *(Chlorophyta),* yellow-green algae *(Chrysophyta),* and euglenoids *(Euglenophyta)* was enhanced by removing the LP-445 filter from the set, since the peak excitation wavelengths for these algae are below 455 nm. The substantially brighter fluorescence of blue-green algae, red algae, and cryptomonads at excitation wavelengths between 520 and 560 nm compared to their fluorescence at excitation wavelengths of 380 to 500 nm was apparently due to light absorption by biliproteins. Fluorescence microscopy offers few advantages: (1) easy distinction between algae, protozoa, and bacteria in water samples; (2) differentiation of biliprotein-containing algae from other groups; (3) provision of additional criteria to estimate the percent of viable cells.

Visualization of Nuclei by Lomofungungin[38]

Lomofungin, an antibiotic, stains living algal nuclei red when it is present in the culture medium for 15 to 20 min at a concentration of 40 μg/mℓ. In *Spirogyra* and *Maugeotia* the nucleus is stained orange and the nucleolus is deep red. Higher concentrations or longer incubation times affect cell viability.

Using Fluorochromes for Determination of Nuclear DNA[39]

Fluorochromes have been used in conjunction with epi (incident) UV illumination for sensitive and selective determination of nuclear DNA in ten species from six algal genera: *Mougeotia, Oedogonium, Sirogonium, Spirogyra,* and *Zygnema* among the green algae, and the marine red alga *Polysiphonia boldii.* The cytofluorometric procedure is simple and sensitive. Following staining with 4′,6-diamidino-2-phenylindole (DAPI), nuclei fluoresce blue-white. Algae stained with 2,5-bis[4′aminophenyl 1(1′)]-1,3,4-oxadiazole (BEO) also show brilliant blue-white nuclear fluorescence. Although BAO staining requires freshly prepared dye and sulfite water and careful control of hydrolysis, the fluorescence of the stained specimens does not fade under UV irradiation as rapidly as with some other fluorochromes. A more useful fluorochrome is the fungal antibiotic mithramycin. The staining protocol is simple and the bright orange-yellow fluorescence of the nuclei is associated with an exceptional degree of sensitivity and specificity for DNA.

An Improved Procedure for Feulgen DNA Staining[40]

The old procedure for Feulgen staining of DNA is improved by treatment of cells with cupraammonium to remove polysaccharide wall material, followed by neutralization with propionocarmine, which enables thinner squashes and better chromosome spreads without loss of differential staining. Fe-propionocarmine applied as a gradient to the slide provides cells stained with the Feulgen stain alone or with the Feulgen/Fe-propionocarmine stain, thereby facilitating comparison. Where dilute, the Fe-propionocarmine enhances nuclear staining without staining other organelles; where more concentrated it also stains the nuclear, spindle, spindle polar bodies, pyrenoid body, and protoplasts. The procedure was used for a variety of algae such as diatoms, green flagellates, chrysophytes, and dinoflagellates.

Fluorescence Microscopy for Estimation of Filamentous Blue-Green Alga[41]

Algal samples were stained with primuline and filtered on black membrane filters. The algae cells are visualized under wet mount at high magnification. For filamentous blue-green algae of identified lengths, their volume can be estimated by counting the times that filaments cross a grid placed in the ocular. Identification of algae on the filters is relatively easy and the method enables rapid counting of dilute samples.

Separation and Isolation of Algae

Obtaining Axenic Clones of Blue-Green Algae[42]

This procedure involves a combination of steps that provide a 1000-fold reduction in the bacteria-algae ratio and permit bacteria-free filaments or cells to be isolated and grown from agar pour plates. In the first step, phenol is added to a dark-treated culture to selectively reduce the numbers of actively growing bacteria while leaving the resting algal cells viable. Next the treated algal suspension is washed on a Millipore filter pad or membrane and then plated on washed agar containing buffered mineral medium plus vitamins and soil extract. The final steps consist of incubating the agar pour plates, removing cores of agar containing bacteria-free filaments or cells, culturing the cores in a buffered mineral medium, and rigorously testing the resulting cultures for bacterial contamination. Algae grow on 50 to 90% of the cores. The method, with appropriate adaptations, should be suitable for obtaining axenic clones of freshwater and marine algae.

This procedure can be modified omitting the phenol treatment and adding a step that should increase the effectiveness of antibiotics such as penicillin (2 mg/ℓ) that affect cell wall biosynthesis. This is achieved by osmotic shock and sonication.[43]

Surface Plating of Blue-Green Algae[44,45]

The growth of single cells to colonies on solidified agar is almost essential in the isolation of mutants, determination of percent viable cells in a culture, or isolation of strains and axenic cultures. While green algae are easily grown on agar plates, blue-green and filamentous alga may cause some problems. The important steps for obtaining high plating efficiency are

1.   Avoid drying of plates during incubation. This can be achieved by maintaining high humidity in the incubation area.
2.   Include an appropriate carbon source in the solidified agar.
3.   Remove any inhibitors. This can be done by autoclaving the agar and nutrients separately and combining them only after cooling.

Another possibility is purifying agar by washing it with water and acetone. In this case, only 1% agar is used, and the agar and nutrients may be autoclaved together. Filamentous blue-green algae usually must be fragmented by sonication.

Separation of Microalgae by Density Gradients

Different microalgae are separated from mixed cultures by using density gradient centrifugation in the silica sol Percoll,[46] achieving separation on the basis of differences in buoyant density. The microalgae form discrete bands at particular positions within the gradient. If the banding positions for several algae are sufficiently different, they can be readily separated by fractionation of the gradient. Photosynthetic activity and subsequent growth of microalgae are unaffected by centrifugation in Percoll.

Estimation of Biomass and Growth

The most commonly used procedures have been mentioned previously. A detailed comparison of eight methods for estimating algal biomass was published,[47] wherein the colonial diatom *Asterionella formosa* was used as a test organism. They were based on (1) cell counts by light microscopy and electronic means; (2) in vivo determination of the optical properties, scattering, attenuation, and fluorescence; and (3) chemical estimations on filtered cell aliquots of the reducing capacity (C-equivalent) and the amount of solvent-extracted chlorophyll-a. The main criteria for evaluating the methods were the precision, sensitivity, limits of detection, time required, and quantity of sample needed. Most methods had an acceptable precision over wide ranges of algal concentration, although for visual counting and nephelometry the coefficient of variation typically exceeded 10%. The in vivo estimations were the most rapid, but were unsuitable at very low biomass concentrations. The chemical methods and (on diluted samples) electronic counting generally required larger quantities of sample, with the notable exception of fluorometry of extracts. The chemical methods were relatively slow, but allowed several samples to be processed at once and gave more generalized measures of algal biomass (e.g., C-equivalent, chlorophyll-a). Visual cell counts, although relatively low and fatiguing, were unsurpassed for low limit of detection, economy of sample, and assessment of cell condition.

ADDRESSES OF CULTURE COLLECTIONS

1. Sammlung von Algenkulturen
   Pflanzenphysiologysches Institute
   Universitat Gottigen
   18 Nikolausbergerweg
   Gottigen D 3400
   Federal Republic of Germany

2. The Culture Collection of Algae and
   Microorganisms
   Institute of Applied Microalgae
   University of Tokyo
   Tokyo, Japan

3. Culture Center of Algae and Protozoa
   (CCAP)
   Institute of Terrestrial Ecology
   Natural Environmental Research
   Council
   36 Storeys Way
   Cambridge, CB3 ODT
   United Kingdom

4. American Type Culture Collection
   (ATCC)
   12301 Parklawn Drive
   Rockville, Maryland 20852

5. Carolina Biological Supply Company
   2700 York Road
   Burlington, North Carolina 27215

6. University of Texas Culture Collection
   (UTEX)
   Department of Botany
   University of Texas
   Austin, Texas 78712

7. Prof. G. S. Venkataraman
   Indian Agricultural Research Institute
   Culture Collection of Microalgae
   New Delhi, India

8. Czechoslovak Academy of Sciences
   Botanical Institute
   Laboratory of Hydrobotanics
   Trebon, Czechoslovakia

REFERENCES

1. **Hase, E., Morimura, Y., and Tamiya, H.**, Some data on the growth physiology of *Chlorella* studied by the technique of synchronous cultures, *Arch. Biochem. Biophys.*, 69, 149, 1957.
2. **Schmidt, R. R. and King, K. W.**, Metabolic shifts during synchronous growth of *Chlorella pyrenoidosa*, *Biochim. Biophys. Acta*, 47, 391, 1961.
3. **Tamiya, H., Morimura, Y., Yokota, M., and Kunieda, R.**, Mode of nuclear division of synchronous cultures of *Chlorella:* comparison of various methods of synchronization, *Plant Cell Physiol.*, 2, 383, 1961.
4. **Sitz, T. O., Kent, A. B., Hopkins, H. A., and Schmidt, R. R.**, Equilibrium density-gradient procedure for selection of synchronous cells from asynchronous cultures, *Science*, 168, 1231, 1970.
5. **Kemp, C. L. and Lee, K. A.**, Synchronous growth in colonial *Eudorina elegans*, *J. Phycol.*, 12, 105, 1975.
6. **Cattolico, R. A., Boothroyd, J. C., and Gibbs, S. P.**, Synchronous growth and plastid replication in the naturally wall less alga *Olisthodiscus luheus*, *Plant Physiol.*, 57, 497, 1976.
7. **Pfau, J., Werthmiller, K., and Senger, H.**, Permanent automatic synchronization of microalgae achieved by photoelectrically controlled dillution, *Arch. Microbiol.*, 75, 338, 1971.
8. **Bernstein, E.**, Synchronous division in *Chlamydomonas noewusii*, *Science*, 131, 1528, 1960.
9. **Singer, H. and Bishop, N. I.**, Emerson enhancement in synchronous *Scenedesmus* cultures, *Nature (London)*, 221, 975, 1969.
10. **Herdman, M., Faulkner, B. M., and Carr, N. G.**, Synchronous growth and genome replication in the blue-green alga *Anacystis nidulans*, *Arch. Mikrobiol.*, 73, 238, 1970.
11. **Csatorday, K. and Horvath, G.**, Synchronization of *Anacystis nidulands*. Oxygen evolution during the cell cycle, *Arch. Microbiol.*, 111(3), 245, 1976.
12. **Lorenzen, H. and Kaushik, B. D.**, Experiments with synchronous *Anacystis nidulands*, *Ber. Dstch. Ges. Bd.*, 89, 491, 1976.
13. **Venkataraman, G. S. and Lorenzen, H.**, Biochemical studies on *Anacystis nidulands* during its synchronous growth, *Arch. Microbiol.*, 69, 34, 1969.
14. **Lorenzen, H. and Venkataraman, G. S.**, Synchronous cell division in *Anacystis nidulands Richter*, *Arch. Microbiol.*, 67, 251, 1969.
15. **Myres, J. and Clark, L. B.**, Culture conditions and the development of the photosynthetic mechanism. II. An apparatus for the continuous culture of *Chlorella*, *J. Gen. Physiol.*, 98, 103, 1944.
16. **Calcott, H. P.**, *Continuous Cultures of Cells*, CRC Press, Boca Raton, Fla., 1981.
17. **Guillard, R. R. L.**, Division rates, in *Handbook of Phycological Methods: Culture Methods and Growth Measurements*, Stein, J. R., Ed., Cambridge University Press, London, 1973.
18. **Richard, F. A. and Thompson, T. G.**, The estimation and characterization of plankton population by pigment analysis, *J. Mar. Res.*, 2, 156, 1952.
19. **Vollenweider, R. A.**, A manual on methods for measuring primary production in aquatic environments, in *International Biological Program Handbook No. 12*, Blackwell Scientific, Oxford, 1969.
20. **Strickland, J. D. H. and Parson, T. R.**, *A Practical Handbook of Seawater Analysis*, Fish. Res. Board Can. Bulletin 1968, 167.
21. **Tomisek, A., Reid, R. M., Short, W. A., and Skipper, H. E.**, Studies on the photosynthesis reaction. III. The effect of various inhibitors upon growth and carbonate fixation in *chlorella pyrenoidosa*, *Plant Physiol.*, 32, 1, 1957.
22. **Foter, M. J., Palmer, C. M., and Maloney, T. E.**, Antialgal properties of various antibiotics, *Antibiot. Chemother.*, 3, 305, 1953.
23. **Galloway, R. A. and Krauss, R. W.**, The differential action of chemical agents on certain alga, bacteria and fungi, *Ann. J. Bot.*, 46, 40, 1959.
24. **Palmer, C. and Malloney, T. E.**, Preliminary screening for potential algicides, *Ohio, J. Sci.*, 55, 1, 1955.
25. **Provasoli, L. and Printner, I. J.**, Ecological implication of in vivo nutritional requirements of algal flagellate, *Ann. N.Y. Acad. Sci.*, 56, 839, 1953.
26. **Sorokin, C. and Krauss, R. W.**, The effect of light intensity on the growth rates of green algae, *Plant Physiol.*, 33, 109, 1958.
27. **Juttner, F.**, Mass cultivation of *Anacystis nidulands*, in *The Biology of Blue-Green Algae*, Carr, N. G. and Whitton, B. A., Eds., Blackwell Scientific, Oxford, 1973.
28. **Siver, P. A.**, A new thermal gradient device for culturing algae, *Br. Phycol. J.*, 18, 159, 1983.
29. **Yanish, C. H., Lee, K. W., and Edwards, P.**, An improved apparatus for the culture of algae under varying regimes of temperature and light intensity, *Bot. Mar.*, 22, 395, 1979.
30. **Woording, D.**, A Laboratory Model of Diurnal Rhythms in Phytoplankton, Ph.D. thesis, Rutgers University, New Brunswick, N.J., 1968.
31. **Quraishi, F. Q. and Spencer, C. P.**, Studies on the responses of marine phytoplankton to light fields of varying intensities, in *4th Mar. Biol. Symp.*, Crisp, D. Y., Ed., Cambridge University Press, London, 1971.

32. **Feuillade, J. and Feuillade, M.,** A chemostat device adapted to planktonic *Oscillatoria* cultivation, *Limnol. Oceanogr.,* 24, 562, 1979.
33. **Groeneweg, J. and Soeder, C. J.,** An improved tube for axenic cultures of microalgae, *Br. Phycol. J.,* 13, 337, 1978.
34. **Van Liere, L., Loogman, J. G., and Mur, L. R.,** Measuring light irradiance in cultures of phototrophic microorganisms, *FEMS Microbiol. Lett.,* 3, 161, 1978.
35. **Griffiths, J., Dilwyn, L., Van Thinh, L., and Florian, R.,** A modified chamber for use with an oxygen electrode to allow measurement of inciden illumination within the reaction suspension, *Limnol. Oceanogr.,* 23, 368, 1978.
36. **Jassby, D. A.,** Polarographic measurements of photosynthesis and respiration, in *Handbook of Phycological Methods, Physiological and Biochemical Methods,* Hellebust, J. A. and Craigie, J. S., Eds., Cambridge University Press, London, 1978.
37. **Wood, E. J. F.,** A method for phytoplankton study, *Limnol. Oceanogr.,* 7, 32, 1962.
38. **Kopecka, M. and Gabriel, M.,** Staining the nuclei in cells and protoplasts of living yeasts, moulds and green algae with the antibiotic lomofunging, *Arch. Microbiol.,* 119, 305, 1978.
39. **Hull, H. M., Hoshaw, R. W., and Wang, J.-C.,** Cytofluorometric determination of nuclear DNA in living and preserved algae, *Stain Technol.,* 57(no. 5), 273, 1982.
40. **Hanic, L. A.,** Feulgen, iron-propionocarmine and cupra-ammonium in preparing algal chromosomes for light microscopy, *Stain Technol.,* 54(no. 3), 129, 1979.
41. **Brock, T. D.,** Use of fluorescence microscopy for quantifying phytoplankton, especially filamentous blue-green algae, *Limnol. Oceanogr.,* 23, 158, 1978.
42. **Carmichael, W. W. and Gorham, P. R.,** An improved method for obtaining axenic clones of planktonic blue-green algae, *J. Phycol.,* 10, 238, 1974.
43. **Brown, L. M.,** Production of axenic cultures of algae by an osmotic method, *Phycologia,* 21(3), 408, 1982.
44. **Van Baalen, C.,** Quantitative surface plating of coccoid blue-green algae, *J. Phycol.,* 1, 19, 1965.
45. **Wolk, C. P. and Wojciuch, E.,** Simple methods for plating single vegetative cells of, and for replica-plating, filamentous blue-green algae, *Arch. Mikrobiol.,* 91, 91, 1973.
46. **Whitelam, G. C., Lanaras, T., and Codd, G. A.,** Rapid separation of microalgae by density gradient centrifugation in percoll, *Br. Phycol. J.,* 18, 23, 1983.
47. **Butterwick, C., Heaney, S. I., and Talling, J. F.,** A comparison of eight methods for estimating the biomass and growth of planktonic algae, *Br. Phycol. J.,* 17, 69, 1982.

ALGAL NUTRITION

D. Kaplan, A. E. Richmond, Z. Dubinsky, and S. Aaronson

NUTRITIONAL MODES

Attempts have been made to classify algae according to their nutritional modes, but much ambiguity remains. The complexity and confusion in the trophy-terminology of algae originates in the real overlapping and intergrading of the various modes, in the ability of many algae to change their nutritional modes in response to environmental circumstances, and to a great extent in the different purposes of the nutritional classification systems (e.g., biochemical, taxonomic, or ecological).

Algae have two major forms of nutrition: lithotrophy (from the Greek lithos — stone and trophein — nourishment) or autotrophy, and heterotrophy (from the Greek heteros — other). Each of these nutritional forms can in turn be subdivided into variants on the same general theme. Nutritional intergradients between the major nutritional forms also occur. These will be described below and their relationships may be seen in Figure 1.

Lithotrophy (Autotrophy)

Autotrophs (or lithogrophs) are organisms that obtain all the elements they need from inorganic compounds and the energy for their metabolism from light or the oxidation of inorganic compounds or ions. Lwoff[1,2] subdivided lithotrophs according to their energy source; those using light energy are *phototrophs* and those using energy obtained from the oxidation of inorganic nitrogen, sulfur, iron, or manganese are *chemotrophs*.

Heterotrophy

Heterotrophs[3] are those organisms whose material and energy needs come from organic compounds synthesized by other organisms. For organisms whose energy is derived from the oxidation of organic compounds that also serve as a carbon source, the term *organotrophy* or *chemoheterotrophy* is used.[4,5]

Auxotrophy

This is a form of heterotrophy[6] in which organisms require only very small quantities of essential organic compounds (e.g., vitamins, etc.). The term *auxotroph* is also used in microbial genetics for mutants that can no longer synthesize essential nutrients that the parental strain or *prototroph* could synthesize. The nutrient must now be supplied exogenously.

Mixotrophy[6]

This is equivalent to *photolithotrophic heterotrophy*[5] and designates a nutritional mode in which photosynthesis is the main source of energy, although both organic compounds and CO_2 are essential. Mixotrophy may also apply to the *chemolithotrophic heterotrophy* of Nelson and Lewin[5] in which energy is supplied by chemical oxidation rather than from light, but other conditions are as described for photolithotrophic heterotrophy. *Amphitrophy* was proposed by Pringsheim[6] for organisms able to live either auto- or heterotrophically.

Photoheterotrophy

This term (photo-organotrophy = photoassimilation = photometabolism) refers to a nutritional mode in which light is required to use organic compounds as carbon sources. Stanier[4] defines this as the ability to grow in light in the presence of DCMU (3-(3,4-dichlorophenyl)-1-dimethyl urea), an inhibitor of photosynthesis.

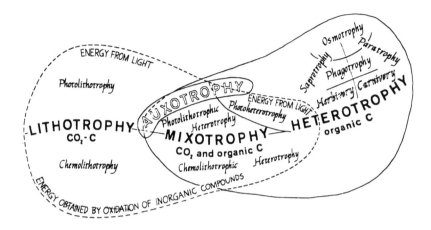

FIGURE 1.   Relationships among terms used to describe the main nutritional modes found in nature, in particular in the algae.

Paratrophy

This is a term used by Lwoff[1] for parasitic nutrition where energy is obtained from the host cell.

Phagotrophy

This is a term applied to the uptake of solid food particles and their subsequent digestion in food vacuoles.

Saprotrophy

This is the nutritional mode in which nonliving organic matter is used, in contrast to paratrophy. Other terms sometimes found in the literature are *oxytrophy,* the use of acetate; *halotrophy,* the use of sugar by colorless flagellates: and *methylotrophy,* the use of methane and its derivatives as either carbon or energy sources.

In addition to the above forms of nutrition, algae also form symbiotic associations with bacteria on a community scale, e.g., in sewage oxidation ponds, where bacteria decompose the organic matter in the raw sewage providing the necessary nutrients for algal growth, the algae supply photosynthetically produced oxygen for bacterial metabolism.[7-9] Bacteria have been shown to be essential or at least beneficial for algae in a number of studies.[10,11]

Several algae have been reported to enlarge, in effect, their nutritional potential by secretion of digestive enzymes into their environment. Miller[12] reported that *Monodus subterraneous* secreted a glutaminase, and Pringsheim[13] showed that *Ochromonas variabilis* secreted amylase, invertase, and what appeared to be proteases and lipases. Aaronson[14] reported the secretion of phosphatases, RNAse, DNAse, cathepsin arylsulfatase, and several glycosidases by *Ochromonas danica.*

Most algae are *phototrophs.* They obtain energy from the absorption of light and electrons for the reduction of carbon dioxide from the oxidation of an inorganic substrate, usually water, with the evolution of oxygen. They cannot grow significantly in the dark. If this is their sole form of nutrition, they are *obligate phototrophs.*

Photolithotrophy seems to predominate among brown and red algae and *Cyanobacteria,* while heterotrophy is more common among green algae, diatoms, and flagellated algae.

THE OPTIMAL RANGE OF NUTRIENT CONCENTRATIONS

Much research has been conducted to determine the optimum mineral composition of

Table 1
A GENERAL LIST OF NUTRIENTS NEEDED BY ALGAL CELLS

Element	Compounds	Conc range/ℓ media
C	CO_2, HCO_3^-, CO_3^{2-} organic molecules	g
O	O_2, H_2O, organic molecules	g
H	H_2O, organic molecules, H_2S	g
N	N_2, NH_4^+, NO_3^-, NO_2^-, amino acids, purines, pyrmidines, urea, etc.	g
Na	Several inorganic salts, i.e., NaCl, Na_2SO_4, Na_3PO_4	g
K	Several inorganic salts, i.e., KCl, K_2SO_4, K_3PO_4	g
Ca	Several inorganic salts, i.e., $CaCO_3$, Ca^{2+} (as chloride)	g
P	Several inorganic salts, Na or K phosphates, Na_2 glycerophosphate · $5H_2O$	g
S	Several inorganic salts, $MgSO_4 \cdot 7H_2O$, amino acids	g
Mg	Several inorganic salts, Co_3^{2+}, SO_4^{2-}, or Cl^- salts	g
Cl	As Na^+, K^+, Ca^{2+}, or NH_4^+ salts	g
Fe	$FeCl_3$, $Fe(NH_4)_2SO_4$, ferric citrate	mg
Zn	SO_4^{2-} or Cl^- salts	mg
Mn	SO_4^{2-} or Cl^- salts	mg
Br	As Na^+, K^+, Ca^{2+}, or NH_4^+ salts	mg
Si	$Na_3SiO_3 \cdot 9H_2O$	mg
B	H_3BO_3	mg
Mo	Na^+ or NH_4^+ molybdate salts	µg
V	$Na_3VO_4 \cdot 16H_2O$	µg
Sr	SO_4^{2-} or Cl^- salts	µg
Al	SO_4^{2-} or Cl^- salts	µg
Rb	SO_4^{2-} or Cl^- salts	µg
Li	SO_4^{2-} or Cl^- salts	µg
Cu	SO_4^{2-} or Cl^- salts	µg
Co	Vitamin B_{12}, SO_4^{2-}, or Cl^- salts	µg
I	As Na^+, K^+, Ca^{2+}, or NH_4^+ salts	µg
Se	Na_2SeO_3	ng

From Aaronson, S., *Experimental Microbiology*, Ecology Academic Press, New York, 1970. With permission.

growth media for various algal species. A general list of the nutrients needed by algal cells, together with the orders of magnitude of the concentration used in the growth media is provided in Table 1. Clearly, most formulas for culture media differ greatly from what the algae would have in their natural environments, for the concentrations of all the required nutrient elements usually far exceed their natural levels. Such high concentrations of nutrients are well tolerated by most species. Thus, particularly in laboratory batch cultures where the population density increases severalfold without a change of medium, high nutrient concentrations limit the possibility for the development of a deficiency in some mineral element. In systems designed for the commercial production of algae, the concentrations of minerals needed in the growth media must be more carefully determined. This is because of the risk that some elements may precipitate due to changes in the temperature and other factors. Also, the capital invested in growth media represents a significant portion of the operating costs and if a culture must be terminated and the growth medium discarded, the loss would obviously be greater if the concentrations of nutrients were unnecessarily high.

It must be borne in mind, however, that determining an exact and narrow concentration range for each nutrient in the medium so as to maintain its level at a minimal value without imposing restrictions on growth may not be possible. This is because the optimum nutrient concentration for a given algal strain may vary considerably, since it depends on many factors such as the population density, light, temperature, and pH. A very important factor is the potential growth rate under the environmental conditions prevailing outdoors. When the conditions permit a fast growth rate, the optimum concentration range for a given mineral

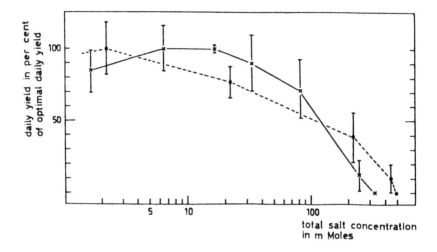

FIGURE 2.   Optimal salt concentration and salt resistance in *Chlorella fusca* as dependent on other culture conditions. Solid line: cultivation at 30°C, 1% CO_2 in air, 10 klux, during the light period of the 16:8 hr light/dark cycle, complete synchrony. Broken line: cultivation at 25°C, 4 klux during the light period of the 12:12 hr light/dark cycle, gassing with atmospheric air, synchrony in groups. Indication of standard deviation of daily yields which were measured in terms of daily increments of cell number per milliliter. (From Sorder, C. J., et al., *Arch Hydrobiol.*, 33(Suppl.), 127, 1969. With permission.)

nutrient shifts toward a substantially higher concentration. Soeder et al.[16] described a set of experiments that demonstrated the influence of growth conditions on the optimum nutrient levels; *Chlorella fusca* was first grown under suboptimal conditions, varying the macro-element concentrations over a wide range while keeping their proportions and the absolute microelement concentrations constant. The experiments were then repeated under improved conditions for the growth of *Chlorella* and a growth-vs.-concentration curve significantly different from the first one was obtained. The combined improvements in temperature, CO_2 supply, and light substantially reduced the salt tolerance. At 25°C (4 klux) the concentration of salts inhibitory to growth was about 0.44 *M*, whereas for cultures at 30°C (10 klux) it was 0.24 *M*. All cultures growing at higher salt concentrations passed through a slow stepwise adaptation to the respective conditions (Figure 2).

The elementary composition of many algal species based on chemical analyses is given in Table 2.

These data represent rough approximations because of the great variation between different species and populations of the same alga grown under different conditions. According to Healey,[17] most of the variation in C, H, and O is between species, and variations in chemical composition that stem from extreme nutrient deficiencies are small relative to the variations between species. The highest values of Na, K, Ca, and Mg tend to occur in marine algae, whereas the highest concentrations of Si are found in diatoms. In general, increases in the concentrations of a variety of elements in the growth medium are accompanied by an increase in their levels within the cells.

IDENTIFYING NUTRITIONAL LIMITATIONS

It is essential in mass algal culture to identify and quantify the factors that limit cell growth and development. Plants have two common responses to changes in concentration of limiting factors:[18] a change in the final yield (type I response; Figure 3) or a change in the growth rate (type II response). Type II is typical of phytoplankton algae in nature, yet some experiments have failed to show growth rate changes because of inappropriate design.

Table 2
ELEMENTARY COMPOSITION OF
ALGAE, INCLUDING THOSE
ELEMENTS KNOWN TO BE REQUIRED
BY AT LEAST SOME ALGAE

	μg/mg Dry weight		Relative no. of atoms
Element	Average	Range	
H	65	29—100	8,140,000
C	430	175—650	4,460,000
O	275	205—330	2,120,000
N	55	10—140	487,000
Si	54	0—230	237,000
K	17.3	1—75	55,000
P	11	0.5—33	43,800
Na	6.1	0.4—47	32,500
Mg	5.6	0.5—75	28,700
Ca	8.7	0.0—80	27,500
S	5.9	1.5—16	23,800
Fe	5.9	0.2—34	13,800
Zn	0.28	0.005—1.0	540
B	0.03	0.001—0.25	350
Cu	0.10	0.006—0.3	200
Mn	0.06	0.02—0.24	138
Co	0.06	0.0001—0.2	125
Mo	0.0008	0.0002—0.001	1

Note: Cl, I, and V are omitted because of insufficient information.

From Healy, F. P., *CRC Critical Reviews in Microalgae,* CRC Press, Boca Raton, Fla. 1973. With permission.

O'Brien[18] explained that in many of these experiments, one rate-limiting factor was in such low concentration that it obscured the effect that a change in the concentration or intensity of some other limiting factor may have had on the algal growth rate. The intensity of light in laboratory experiments can be such a factor, as could the concentration or form of carbon in the culture medium, or perhaps a lack of vitamins or other organic growth factors. A distinct interaction between light intensity and nutrient concentration in affecting the growth rate of *Carteria* sp. was demonstrated by Jones.[19] At low light intensities changes in the concentrations of nitrogen and phosphorous had no effect on the rate of growth, whereas at medium light intensities the growth rate was greater with higher concentrations of nitrogen and phosphorous. Such situations may be common in laboratory studies, where natural light intensities are often not available. This alone may account for the fact that many laboratory studies yield data that fit growth patterns of type I rather than of type II.

Another source of difficulty in observing a change in the rate of growth may be the experimental technique used. Laboratory bioassays involving a batch culture technique, in which a large amount of water to be tested is incubated with a relatively small initial inoculum of a test organism, rarely show a type II pattern. With this experimental design a diminished growth rate caused by the low concentration of some limiting factor appears only as the population approaches maximum yield. This change in rate, however, is almost impossible to observe, because while the population is growing exponentially the nutrient concentration is decreasing exponentially and the population grows most of the time in a relatively high, nonrate-limiting concentration of the limiting nutrient.[18]

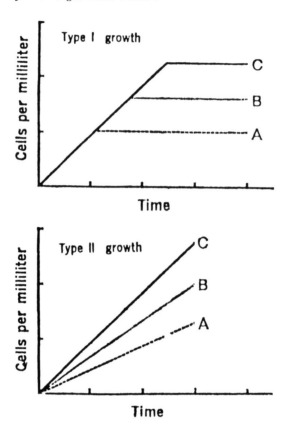

FIGURE 3. Type I and II growth patterns, The ordinate is a logarithmic scale. Increasing concentrations of the limiting factor are represented by A, B, and C. (From O'Brien, W. J., *Science*, 128, 616, 1972. With permission.)

Various kinetic models may be used to identify a growth-limiting nutrient. The most widely used is the hyperbolic function developed by Monod,[20] which may be formulated as follows:

$$u = \hat{u}[S/(K_s + S)] \tag{1}$$

where u = specific growth rate per day; $\hat{u}$ = maximum specific growth rate, 1/day; S = nutrient concentration, mg/ℓ; and K_s = half-saturation coefficient (nutrient concentration at u = $\hat{u}$/2), mg/ℓ.

At low values of S, the Monod equation approximates a first-order equation in which the specific growth rate is linearly related to the concentration of the limiting nutrient:

$$u = \hat{u}[S/K_s] \tag{2}$$

When S>>K_s, the zero-order relationship

$$u = \hat{u} \tag{3}$$

and the specific growth rate is at its maximum and no longer depends on the concentration of the varied nutrient, but rather on the fixed environmental conditions such as light and temperature (Figure 4).

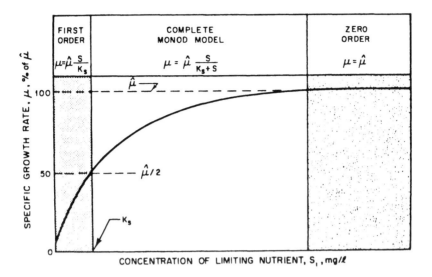

FIGURE 4. Relationship between limiting nutrient and specific growth rate for the Monod model. (From Goldman, J. E. et al., *J. Water Pollut. Cont. Fed.*, 46(3), 553, 1974. With permission.)

In aquatic biotopes, the concentration of a particular mineral often determines the distribution and growth rate of algal species. The production of algal mass in carefully controlled reactors, however, facilitates ready access by the algal cells to all the minerals required for their maximal growth and development. Under such conditions it is nevertheless imperative to determine, for each species, the optimum concentration range for each mineral.

Optimum concentrations or essential requirements for mineral nutrients are usually described in terms of amounts per unit volume of the culture medium. When dealing with saturating concentrations where the essential needs of an alga are met or even exceeded, an indication of the concentrations in the medium is certainly sufficient. If, however, a mineral nutrient is the factor limiting algal production, the proportion of nutrient to biomass will be more meaningful. Under natural or laboratory conditions at the lower end of the concentration range, the nutrient supply is frequently expressed as micromoles per cell or per milligram of protein. An example of this is provided by Davis,[22] who described growth limitation of *Dunaliella tertiolecta* by iron concentration in terms of iron per 10^6 cells.

As soon as the net uptake of a nutrient is zero, growth should be completely inhibited. However, many nutrients are stored in excess of their actual, essential requirements. This means that normal growth may continue for some time after the concentration of nutrient in the medium has reached its lowest possible value. Growth at the expense of stored nutrients is especially well known in the case of phosphate[23,24] but also occurs with iron.[16] The cessation of plasmatic growth and reproduction caused by acute nutrient deficiency usually does not exclude some continuation of photosynthetic and biosynthetic activity.[25] In laboratory studies of the symptoms of nutrient deficiency, a broad range of partial limitation in which the growth rate is to a greater or lesser extent below its maximum value has been described.

It should be noted that since the Monod function considers only the steady-state situation in continuously growing cultures, it cannot always be applied directly to natural populations of algae. At least in the development of blooms, the phytoplankton community resembles a batch system more closely, in that there is a steady decline in the nutrient supply per algal cell. Within certain limits, then, the half-life of this decline depends linearly on the population density, and the concentration of the limiting nutrient should be related to the algal biomass and not to the volume of the body of water.[26]

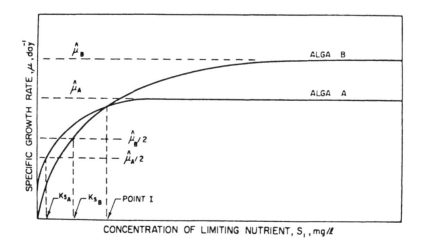

FIGURE 5.   Comparison of Monod growth curves for two algae having different maximum growth rates and half-saturation coefficients. (From Goldman, J. E. et al., *J. Water Pollut. Cont. Fed.*, 46(3), 553, 1974. With permission.)

The use of the Monod model in identifying limiting nutrients in natural waters is based on measuring the magnitude of the K_s values. The K_s value marks approximately the upper nutrient concentration at which the growth rate ceases to be proportional to that concentration. Thus, for a nutrient to be limiting, its concentration must be approximately equal to or less than the K_s value. By comparing the K_s value for a particular nutrient and algal species with the amount of that nutrient remaining in solution during an algal bloom in a natural body of water, it is possible to gain some insight into the role of nutrients in controlling algal growth in nature.[21]

In artificial systems for algal production, the likelihood of mineral deficiencies is small, because the concentration of the nutrients in the growth medium is routinely monitored. The determination of the K_s value is potentially valuable in formulating the levels of some elements in the growth medium, with the idea of affecting species competition, a major problem in the maintenance of monoalgal cultures. Thus, the concentration of a mineral for which a contaminating algal species has a K_s value significantly larger than the equivalent K_s for the cultivated species should be kept as low as possible. By continuous, slow addition of this element to the growth medium, its concentration should be kept as low as possible on the threshold of becoming growth-limiting for the desired alga. A hypothetical example is presented in Figure 5. Species A, which is cultivated for its biomass, has a slower growth rate than species B, which contaminates the culture and may become, with time, the dominant species in the pond. Species B has a significantly higher K_s for nutrient S. To check the growth of species B and prevent it from overcoming the cultivated species A, the concentration of S should be carefully maintained slightly above $K_s(A)$ and as much as possible below $K_s(B)$.

In practice, determining the optimum nutrient concentrations for the growth of an algal species may be a formidable task, because growth and development in mass algal cultures are often determined by a variety of factors and because most mineral nutrients are usually in excess.

It is important to realize the limitations of the Monod model of substrate-limited microbial growth which is mentioned above. As pointed out by Droop[27] the Monod model is essentially a single compartment model that has the following limitations: it can be applied only to systems in equilibrium with their surroundings and it assumes constant cell composition with respect to the rate-controlling substrate at all growth rates. These limitations may not be strictly applicable to outdoor mass algal cultures.

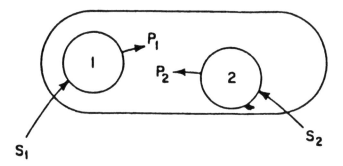

FIGURE 6. Conceptual representation of the noninteractive model. (From Bader, F. G., *Biotechnol. Bioengineer.*, 20, 183, 1978. With permission.)

The success of any attempt to understand the mineral metabolism of microorganisms will depend on the choice of the proper culture system. As a method for studying nutritional requirements, continuous culture (see "Laboratory Techniques for the Cultivation of Microalgae") is more useful than batch systems. Continuous culture techniques are the only means by which the relations between growth rates and nutrient concentrations can be determined accurately. Moreover, the adaptation of algae to changes in the nutrient supply can be analyzed better in continuous cultures. On the other hand, the qualitative detection of nutrient requirements will certainly continue to rely on batch culture.

Synchronous cultures of algae have to a certain extent been used to study changes in the uptake of minerals during the course of the cellular life cycle.[28,29] This type of culture exhibits, at intervals, very high growth rates which are useful in elucidating trace element requirements, e.g., as in the study of Soeder and Thiele[30] of calcium effects in *Chlorella*.

MODELS FOR DOUBLE-SUBSTRATE LIMITED GROWTH

It has become apparent in recent years that the growth rate of organisms may be limited simultaneously by two or more substrates. Bader[31] pointed out that when an organism is growing in an environment in which two of its required substrates are at less than saturating levels, double-substrate limitation may occur. Two different concepts, interactive and noninteractive double-substrate limited growth models, have been developed.

A noninteractive model basically implies that the growth rate of the organism can be limited by only one substrate at a time. Therefore, the growth rate of the organism will be equal to the lowest growth rate that would be predicted from the separate single-substrate models. A conceptual representation of the noninteractive model is shown in Figure 6, in which systems 1 (S_1) and 2 (S_2) operate independently of each other.

An interactive model is based upon the assumption that if two substrates are at less than saturating concentrations, then both must affect the overall growth rate of the organism. The simplest type of interactive model may be constructed by simply multiplying two single-substrate limited models together.

Bader[31] described two cases that provide a rational basis for the interactive model (Figures 7a and b). In the first case (Figure 7a) a cell has a certain number of enzymes present which, in the presence of a cofactor (S_2), converts a substrate (S_1) to a product (P_1) which is required for the growth of the cell. If the external substrate (S_1) and cofactor (S_2) concentrations are both at half velocity levels, then only one half of the total enzyme would be active, and the active enzyme would be producing product at one half its maximum rate. The overall rate of P_1 production and the growth rate of the cell would be one fourth the maximum possible rate.

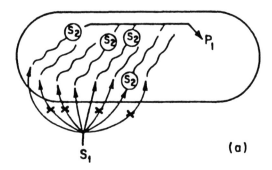

(a)

$$S_1 \xrightarrow{\ E_1\ } S_1' \xrightarrow{\ E_2\ } S_1'' \xrightarrow{\ E_3\ } S_1'''$$

$$E_4 \searrow$$
$$\longrightarrow P_1$$
$$S_2 \nearrow$$

(b)

FIGURE 7. Conceptual representation of the interactive model. (a) S_1 is converted to P_1 by an enzyme which requires S_2 as a cofactor. (b) Substrates S_1 and S_2 from two parallel pathways are combined by an enzyme E_T to produce a product P_1 which is required for growth. (From Bader, F. G., *Biotechnol. Bioengineer.*, 20, 183, 1978. With permission.)

In the second case (Figure 7b) two substrates (S_1 and S_2) or their derivatives are required to produce a single product (P_1), which is required for the growth of the cell. This is probably the most common type of interaction between substrates which are required for the anabolic functions of the organism.

A clear example for a direct, multiplicative interaction of light, phosphorous, and carbon dioxide availability in affecting the growth rate of *Anacystis nidulans* was provided by Young and King.[32] In Figure 8 specific growth rates of algal cultures grown at four combinations of light intensity and initial phosphorous concentration are given as a function of the carbon dioxide concentration measured at intervals during batch growth. As these data illustrate, algae incubated at high light with ample phosphorous (915 lux, 580 μg Pi/ℓ) grew faster, over a wider range of carbon dioxide concentrations, and to lower concentrations of carbon dioxide than algae incubated at low light with limited phosphorous (280 lux, 53 μg Pi/ℓ). The data in Figure 8 show further that intermediate growth rates and carbon dioxide minima were obtained with cultures incubated at high light with limited phosphorous, or at low light with ample phosphorous. Thus, the growth rate of *A. nidulans* as a function of carbon dioxide concentration shows a simultaneous, direct dependence on the availability of light and phosphorous.[32]

MACRONUTRIENT ELEMENTS

The elements that are essential constituents of algal biomass are usually divided into major and minor elements, according to the quantities required to obtain optimal growth, hence the terms "macro" and "micro" elements or nutrients.

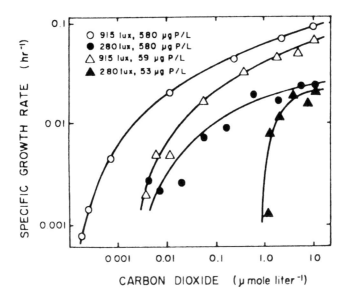

FIGURE 8. Specific growth rate of *A. nidulans* as a function of carbon dioxide for individual and simultaneous increases in illumination and initial phosphorus. (From Young, T. C. and King, D. L., *Water Res.*, 14, 409, 1980. With permission.)

Major elements, or macronutrients, are those elements that contribute to the molecules which make up the structure of algae and are therefore required in rather large amounts. All algae require carbon, nitrogen, oxygen, hydrogen, and phosphorous and also calcium, magnesium, sulfur, and potassium, in ''macro'' quantities.

Minor nutrients are those elements required in milligrams per liter or lower concentrations that are involved as components of essential molecules such as growth factors or enzymes or are required for the activation of certain enzymes. A list of the common concentration ranges of these elements in culture media for algae is shown in Table 1.

Oxygen

Although not usually considered a nutrient, oxygen is required by all algae for structural and metabolic purposes. This element is a constituent of almost all organic compounds in the cell and is usually the final electron acceptor in biological oxidations. Molecular oxygen is obtained by algae directly from the atmosphere or from oxygen dissolved in water. Oxygen evolved through the photosynthetic photolysis of water can also be used directly. Some algae can fix CO_2 in the absence of O_2 using H_2 as a reductant, surviving in oxygen-poor or even anaerobic environments. In such algae, the oxygen requirements are probably met from organic compounds.

The O_2 requirements of algae must be viewed on the background of the clear sensitivity of many algae to elevated O_2 concentrations. Inhibition of photosynthesis in algae by concentrations of O_2 above that in the air, especially when the CO_2 concentration is low, has been well documented (see ''Outdoor Mass Cultures of Microalgae''). In some cyanobacteria, photosynthesis and N_2 fixation appear even more sensitive to O_2, inhibition being found even below atmospheric O_2 concentrations.[33]

Hydrogen (Electrons)

Instead of the widespread utilization of H_2O as the source of electrons for the photoreduction of CO_2, some algae have the ability to use other substances. The use of molecular hydrogen required hydrogenase, which catalyzes the following reaction:

$$2H_2 + CO_2 \rightarrow (CH_2O) + H_2O \tag{4}$$

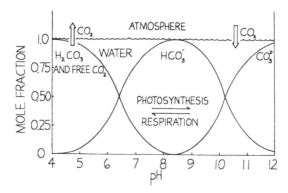

FIGURE 9.   Effects of photosynthesis and respiration on pH,
the ionic forms of inorganic carbon in the water, and CO_2
exchange at the air-water interface.

So far, about 50% of the algal species tested exhibit hydrogenase activity. The species
that contain this enzyme belong to the *Cyanobacteria, Euglenophyta, Chlorophyta, Rhodophyta,* and *Phaeophyta*[34] (with the noteworthy exclusion, so far, of the *Bacillariophyta*
and *Pyrrophyta*). This process, when it occurs in the light, is termed photoreduction.

Some algal species can use hydrogen to reduce substrates such as FAD, FMN,[35] NAD,
NADP, pyruvate,[36] and oxygen.[37]

Carbon Requirement and Metabolism
Inorganic Carbon

All algae growing chemo- or photolithotrophically use dissolved CO_2 or one of its hydrated
forms for the synthesis of organic compounds. In water, CO_2 may appear as H_2CO_3, HCO_3^-
or CO_3^{-2}, depending on the pH (Figure 9).

In most natural freshwaters the major pH buffer is the CO_2-H_2CO_3-HCO_3^--CO_3^{2-} system,
which is also a very useful buffer system for mass algal culture maintained at an alkaline
pH. The total available concentration of dissolved inorganic carbon is as follows:

$$C_r = CO_2(aq) + H_2CO_3 + HCO_3^- + CO_3^= \qquad (5)$$

where C_r = total dissolved inorganic carbon concentration (mg/ℓ or mol/ℓ), $CO_2(aq)$ =
aqueous carbon dioxide concentration (mol/ℓ), H_2CO_3 = carbonic acid concentration (mol/
ℓ), HCO_3^- = bicarbonate concentration (mol/ℓ), and CO_3^{2-} = carbonate concentration
(mol/ℓ).

The relative concentrations of the inorganic carbon species determine the pH and in turn
are determined by the pH (Figure 9). At equilibrium the $CO_2(aq)$ concentration is much
greater than the H_2CO_3 concentration, and the $CO_2(aq)$ concentration may be considered to
represent the sum of the dehydrated and hydrated forms of CO_2. Thus, the relationship

$$C_r = H_2CO_3^* + HCO_3^- + CO_3^{2-} \qquad (6)$$

where

$$H_2CO_3^* \rightleftharpoons CO_2(aq) + H_2CO_3 \qquad (7)$$

is considered valid.

The carbonate system can provide CO_2 through the following interconversions:

$$2HCO_3^- \rightleftharpoons CO_3^{2-} + H_2O + CO_2$$

$$HCO_3^- \rightleftharpoons CO_2 + OH^- \tag{8}$$

$$CO_3^{2-} + H_2O \rightleftharpoons CO_2 + 2OH^- \tag{9}$$

Typically the pH of natural waters at or near equilibrium with atmospheric CO_2 is about 8 to 8.5, and HCO_3^- is the major carbon species. Since CO_2 is the form taken up by growing algae in this pH range, additional CO_2 is provided and OH^- is produced by the reactions of Equations 7 and 8. The reaction of Equation 7 predominates at pH values less than 8, and the reaction of Equation 8 is more significant at pH values greater than 10. Between pH values of 8 and 10, both reactions are important.[21]

As the pH increases CO_3^{2-} becomes the major inorganic carbon species. It can be converted to CO_2 by hydration (Equation 9), which also increases the pH. Because of OH^- production by these reactions, it is not uncommon to observe pH values as high as 10 or 11 in active algal systems such as sewage treatment ponds, fish ponds, or *Spirulina* production ponds. In mass algal cultures, this rise in pH usually demonstrates that the CO_2 supplied from aqueous CO_2 is sufficient to meet the demands of the growing algae.

In mass algal cultures, the pH of the culture medium must be maintained to keep it in the optimum range for the cultivated species and to prevent depletion of carbon. This is done by introducing into the medium either CO_2 or $NaHCO_3$. As algal growth proceeds in natural waters, the concentration of total carbon decreases, the pH rises, and the entire carbonate buffer system is disturbed. It is worth noting, however, that only a portion of the total inorganic carbon can be extracted during intense algal activity in water buffered mainly by the $H_2CO_3^*$-HCO_3^--CO_3^{2-} system. If all the HCO_3^* and CO_3^{2-} were converted to CO_2 and OH^-, the pH would approach a value of 14. Inhibition of algal growth usually occurs at pH values of 10 to 11; thus, the inorganic carbon remaining at these pH values must be considered unavailable for algal growth. For most natural waters approximately one half of the total inorganic carbon can be utilized as CO_2 before the pH rises to about 11.[21]

One should note that the rates of the principal chemical reactions that involve inorganic carbon in water are very different. Whereas the ionization reactions

$$CO_3^{2-} + H^+ \rightleftharpoons HCO_3^- \tag{10}$$

and

$$HCO_3^- + H^+ \rightleftharpoons H_2CO_3 \tag{11}$$

are essentially instantaneous (reaction half-lives on the order of 10^{-15}/sec), the dehydration of H_2CO_3 to CO_2 is a relatively slow step.

$$H_2CO_3 \xrightarrow{kH_2CO_3} H_2O + CO_2(aq) \tag{12}$$

and

$$k(H_2CO_3) = 26.6/sec \text{ at } 25°C$$

Reaction 7, the principal source of $CO_2(aq)$ at pH values less than 8, is the summation of the very rapid reactions 10 and 11. The direct conversion of HCO_3^- to $CO_2(aq)$ and OH^-, predominant at pH values greater than 10 (Equation 8), is a much slower reaction (rate constant equal to 2×10^{-4}/sec at 25°C) than the dehydration of H_2CO_3.[21]

The carbon form that apparently enters the algal cell most readily is CO_2, and many investigators think that CO_2 is the only molecule that can be used directly in the condensation of CO_2 and ribulose biphosphate to yield two phosphoglyceric acid molecules in the reaction catalyzed by ribulose biphosphate carboxylase (RuBPcase). There is no agreement at present on the use of HCO_3^-; it may enter the alga by active transport or it may have to be dissociated by carbonic anhydrase located on the surface or exogenously, which reversibly catalyzes the dissociation of HCO_3^- to CO_2 and water. These possibilities are discussed below.

Miller and Colman[38,39] report that in the unicellular Cyanobacterium *Coccochloris peniocystis*, which photosynthesizes optimally at alkaline pH, most of the carbon fixation is based upon exogenous HCO_3^- rather than CO_2. This conclusion rests largely on the rate of CO_2 fixation at alkaline pH, which was as much as 50 times the maximum rate of CO_2 production by the spontaneous dehydration of HCO_3^- in the external medium. It seemed obvious that a large HCO_3^- influx (at least 100 meq/mg of chlorophyll per hour) must take place across the cell membrane in exchange for the OH^- produced within the cell.

Convincing evidence for the intracellular accumulation of HCO_3^- was reported for *Anabaena variabilis* by Badger et al.[40] After cells had been rapidly separated from the incubation solution by centrifugation through a layer of silicone oil, the internal inorganic carbon concentration was as much as 1000 times the external concentration. This accumulation was drastically reduced by various inhibitors of energy metabolism. The inorganic carbon within the intracellular pool appeared to serve as an intermediate in photosynthesis.[40]

Miller and Colman[38,39] measured the active transport and accumulation of bicarbonate by *Coccochloris peniocystis*. A substantial pool of inorganic carbon was accumulated within the cells, presumably as HCO_3^-, before the onset of the maximum rate of photosynthesis. A typical time course of incorporation of [^{14}C] bicarbonate into various fractions of illuminated cells is shown in Figure 10, which shows that for about 30 sec the amount of carbon taken up by the cells (but remaining unfixed) exceeds the abount fixed into acid-stable products. The relationship between these two carbon pools indicates a precursor-product relationship, with acid-labile carbon eventually becoming fixed into acid-stable forms by photosynthesis.

The unfixed inorganic carbon (mainly HCO_3^-) in the cells reached an estimated internal concentration of 2.9 mM, which was 200 times the external inorganic carbon (also mostly HCO_3^-) concentration. Even larger accumulation ratios were observed, greater than 1000 times the external HCO_3^- concentration. Accumulation did not occur in the dark and was greatly suppressed by the inhibitors of photosynthesis 3-(3,4-dichlorophenyl)-1,1-dimethyl urea (DCMU) and 3-chlorocarbonylcyanide phenylhydrazone (CCCP).[40]

In contrast to these conclusions as to the mode of HCO_3^- utilization by algal cells, Tsuzuki[41] showed that while low-CO_2 cells of *Chlamidomonas reinhardtii* utilized HCO_3^- and CO_2 for photosynthesis at pH 7.4, their affinity for HCO_3^- was lower than that for CO_2 (Figure 11). Also, this affinity was further decreased as the pH was elevated. Tsuzuki[41] proposed that these results are most simply explained if the cells actually absorb CO_2, which is obtained from HCO_3^- by the action of carbonic anhydrase located outside the plasmalemma. Carbonic anhydrase (EC 4.2.1.1), which appears to control the entry of CO_2 into the cell and to concentrate it at the carboxylation site, is most active when the external CO_2 concentration is low. Indeed, Ingle and Colman[42] could not detect carbonic anhydrase activity in four species of Cyanobacteria grown with high CO_2 (5%).

Regardless of the pathway of HCO_3^- assimilation, it seems generally true that algae able to use HCO_3^- as a carbon source have a marked advantage over species lacking this ability at a pH above 8.3.

CO_2 may also be supplied by algal respiration, if it is reassimilated in the light before it has a chance to leave the cell.

Organic Carbon

The modes or organic carbon nutrition vary greatly in algae, even within a single species.

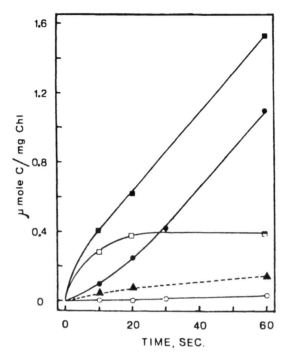

FIGURE 10. Inorganic carbon accumulation by illuminated *C. peniocystis*. The initial inorganic carbon concentration was $42\mu M$, pH 8.0. Symbols :(■), total carbon taken up; (●), carbon photoassimilated into acid-stable products; (□) inorganic, unassimilated carbon within the cells; (○), ether-extractable carbon (lipids and keto acid phenylhydrazones); and (▲), calculated uptake supportable solely by CO_2 transport. (From Miller, A. G. and Colman, B., *J. Bacteriol.*, September, 1253, 1980. With permission.)

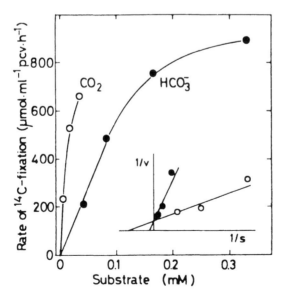

FIGURE 11. Rates of $^{14}CO_2$ (○) and $H^{14}CO_3^-$ fixation (●) vs. concentrations of respective substrates in low-CO_2 cells of *Chlamydomonas reinhardtii*. (From Tsuzuki, M., *Z. Pflanzenphysiol.* 110, 29, 1983. With permission.)

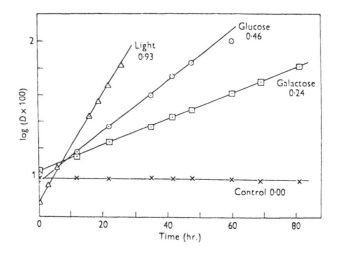

FIGURE 12.   Growth curves in terms of optical density (D). *Chlorella pyrenoidosa* at 25°C in basal medium aerated with 5% (v/v) CO_2 in air. in light, in darkness (control), and in darkness with addition of 1% of the substrate indicated. The figures underneath the curves represent values of the specific growth rates. (From Samejima, H. and Myers, J., *J. Gen. Microbiol.*, 18, 107, 1958. With permission.)

A good example is *Euglena gracilis* grown under different environmental conditions. *E. gracilis* strain L incorporates [^{14}C] acetate efficiently in the light but not in the dark;[43] however, other *E. gracilis* strains can use a variety of organic compounds for growth in the dark, e.g., var. bacillaris, which uses sucrose, glucose, fructose, galactose, pyruvate, succinate, fumarate, malate, ethanol, acetate, butyrate, alanine, aspartate, and glutamate. Another strain, Vischer, uses only acetate and butyrate.[44]

Organic carbon nutrition in a number of species of the genera *Chlorella* and *Scenedesmus*, which are of interest with respect to biomass production, has been extensively studied. These species lie near the midpoint of the spectrum between obligate autotrophy and obligate heterotrophy in the sense that they can shift rapidly and reversibly between growth in darkness on organic substrates and growth in light on carbon dioxide.[45] In *Chlorella pyrenoidosa*, heterotrophic cellular synthesis proceeds with high efficiency albeit from a very limited number of substrates. Growth of *C. pyrenoidosa* in darkness with the addition of various sugars is shown in Figure 12. The type of sugar and the light intensity had marked effects on the growth rates of *C. pyrenoidosa* and *C. ellipsoidea* (Table 3).

From the studies of Samejima and Myers,[46] a lack of versatility in the use of organic substrates appears to be a result of permeability restrictions, e.g., sucrose, does not serve as a carbon source for growth but is found within the cell.[47] Likewise, many of the other organic compounds that fail to support growth have been identified as intermediary metabolites.

The maximum rate of cell synthesis provided by photosynthesis under saturating light is not increased by adding glucose,[45] but at low light intensities the rates supported by photosynthesis and glucose assimilation become approximately additive. However, the rate supported by glucose in darkness is not augmented by the addition of the other utilizable substrates galactose and acetate. Apparently, there is a common reaction in cell synthesis from all three substrates which is rate-saturated during glucose assimilation; in photosynthesis this reaction is bypassed by an alternative pathway or is not required.

Comparison of the efficiencies of heterotrophic and autotrophic metabolism in *Chlorella pyrenoidosa* is instructive. The efficiency observed for total cell synthesis from glucose was 55%, while that for total cell synthesis from carbon dioxide in light was only 20 ± 2%.[46]

Table 3
GROWTH RATES OF *CHLORELLA*
PYRENOIDOSA AND *C. ELLIPSOIDEA*
ON VARIOUS SUGARS AT
DIFFERENT LIGHT INTENSITITES

	Light intensity		
	None	Weak	Medium
Chlorella pyrenoidosa			
None	0.00	0.01	0.17
Glucose	0.46	0.54	0.64
Galactose	0.24	0.25	0.32
Fructose	0.00	0.05	0.17
Maltose	0.00	0.03	0.17
Sucrose	0.00	0.03	0.16
Chlorella ellipsoidea			
None	0.00	0.02	0.19
Glucose	0.47	0.54	0.64
Galactose	0.19	0.14	0.41
Fructose	0.00	0.03	0.20
Maltose	0.00	0.00	0.17
Sucrose	0.00	0.03	0.18

Note: Growth at 25°C in basal medium + sugar 1% (w/
v). Specific growth values are expressed in $\log_{10}$
unit/day.

From Samejima, H. and Myers, J., *J. Gen. Microbiol.*,
18, 107, 1958. With permission.

An interesting variation on heterotrophic nutrition is found in algae that grow under anaerobic conditions, which are rather rare. Among 217 strains of unicellular algae, Nakayama et al.[48] were able to grow only two strains of *Chlorella* sp. anaerobically. Organic broth medium, in an L-shaped test tube filled with oxygen and carbon dioxide absorbers and closed by a rubber stopper, was inoculated with an algal strain, the head space of the tube was filled with pure nitrogen, and the tube was shaken under light for 21 days. Vitamins, glucose or acetate, and peptone or yeast extract were also needed in addition to salts. *p*-Chlorophenyl-1,1-dimethylurea (CMU) at 10^{-5} M inhibited the anaerobic growth of the two strains.

When oxygen and carbon dioxide absorbers were not used, many strains grew in the closed system with an organic medium and a nitrogen atmosphere under light, with a high efficiency of conversion. *Chlorella pyrenoidosa* strain C-28 yielded 71.8, 66.8, and 47.9 mg of biomass from 100 mg of glucose in a closed culture under light, an open culture under light, and an open culture with light, respectively.[48]

Nitrogen Requirements and Metabolism
General Aspects

After carbon (ignoring hydrogen and oxygen, which can be obtained from water), nitrogen is quantitatively the most important element contributing to the dry matter of algal cells. The proportion of nitrogen as the percent of dry weight can vary from 1 to 10%. It is low in diatoms where the silica in the cell wall makes a substantial contribution to dry matter and in nitrogen-deficient organisms that have accumulated large amounts of carbon compounds, such as oils or polysaccharides.[49] However, in exponentially growing cells of

Table 4
COMPARISON OF THE GROWTH RATES OF SOME ALGAE ON
VARIOUS N SOURCES

Organism	Specific rate contants (k) (log₁₀ units/day)			Conc of N (nmol/ℓ)		
	NH_4^+	NO_3^-	Urea	NH_4^+	NO_3^-	Urea
Cyanophyta						
Agmenellum quadruplicatum	3.3	3.0	1.7	0.6	Variable	4.3
Anabaena variabilis	0.69	0.69	0.81	3.0	10.0	1.0
Nostoc muscorum	0.48	0.50	0.54	3.0	10.0	1.0
Chlorophyta						
Chlorella ellipsoidea	—	0.50	0.40	—	12.0	66.0
Chlorella pyrenoidosa	0.5	0.45	0.47	13.0	12.0	66.0
Chrysophyta						
Amphiphora alata	1.14	1.30	1.30	0.1	0.1	0.1
Chaetoceros simplex	1.54	1.58	1.35			
Chaetoceros sp.	0.96	1.30	1.03			
Chrysochromulina sp.	1.12	1.14	1.35			
Cyclotella cryptica	0.57	0.57	0.49		0.9	0.2
Skeletonema sp.	1.41	1.66	0.99		0.1	0.1
Stephanopyxis costata	1.23	1.35	1.58			

From Syrett, P. J., *Canadian Bulletin of Fisheries* and *Aquatic Sciences*, 1981, p. 182. With permission.

nondiatomaceous microalgae, nitrogen accounts for about 7 to 10% of the dry matter and carbon about 50%.[50]

A variety of nitrogen compounds, both inorganic and organic, can serve as sole nitrogen sources for the growth of various microalgae, which include both prokaryotes and eukaryotes. The ability to fix gaseous nitrogen is confined to prokaryotes, and thus among algae only some of the Cyanobacteria can do so. The ability to use nitrate (NO_3^-), nitrite (NO_2^-), or ammonia (NH_4^+) appears to be general among algae, maximum growth rates being similar with either NO_3^- or NH_4^+ as N source (Table 4).

When ammonium is used as the sole N source the pH of the medium may fall sharply, causing undesirable side effects. Some algae are sensitive to high ammonium concentrations and their growth may be inhibited by 1 mM ammonium. The reason for the inhibition by ammonium is unknown but it may be correlated with an increase in internal pH due to the penetration of undissociated ammonium hydroxide molecules.[52]

Nitrite can serve as a nitrogen source for many species only at low concentrations, approximately 1 mM. At higher concentrations nitrite inhibits growth.[52]

Algae from various taxonomic groups can use a variety of organic N compounds as the sole nitrogen source. In general, the amides, urea, glutamine, and asparagine are good sources of nitrogen,[53-55] and growth rates with urea are generally similar to those with NO_3^- or NH_4^+ (Table 4). Amino acids, particularly glycine, serine, alamine, glutamic acid, and aspartic acid, also serve as nitrogen sources.[54] Purines and purine derivatives may also serve as the sole nitrogen source for certain species. The ability of various algae to utilize various organic N compounds for growth is shown in Table 5.[55] Utilization of compounds such as xanthine, hypoxanthine, and uric acid by algae has been reported.[56-59]

Table 5
DISTRIBUTION OF THE ABILITY OF ALGAE TO UTILIZE ORGANIC NITROGEN FOR GROWTH

Nitrogen source	Chlamydomonas	Chlorella	Chlorella: 16	Scenedesmus	Selenastrum	Raphidonema	Cosmarium	Platymonas	Bumilleriopsis	Tribonema	Monodus	Ochromonas	Pavlova	Cyclotella	Phaeodactylum	Nitzschia	Euglena	Ectocarpus	Porphyridium	Cyanidium	Synechococcus	Synechocystis	Pseudanabaena: 6903	Pseudanabaena: B2	LPP 6402	LPP 7310	Anabaena
Glycine	+	+	+	+	+	+	+	+	+	+	−	−	+	+	−	+	+	+	−	+	−	−	−	−	−	−	−
Glycylglycine	−	−	+	+	−	−	−	−	−	−	−	−	−	−	−	−	−	+	−	−	−	−	−	−	−	−	−
Glutamate	+	−	+	−	−	−	+	+	−	+	−	+	−	+	−	+	−	+	−	+	−	−	−	+	−	−	−
Glutamine	+	+	+	+	+	+	+	+	+	+	+	+	−	+	−	+	+	+	−	+	+	+	+	+	−	+	+
Aspartate	−	−	+	+	+	−	−	+	−	+	−	+	+	+	−	+	−	+	−	+	−	−	−	−	−	−	−
Asparagine	−	−	+	+	+	−	−	+	−	+	−	+	−	+	−	+	−	+	−	+	−	−	−	+	−	+	−
Histidine	−	−	+	+	−	−	+	−	−	−	−	−	−	+	−	−	−	−	−	+	−	−	−	−	−	−	−
Methionine	−	−	+	+	−	−	+	+	+	−	−	+	−	−	+	+	−	+	−	+	−	−	−	−	−	−	−
Leucine	−	−	+	+	+	−	−	+	−	+	−	+	+	−	−	−	+	+	−	+	−	−	−	−	−	−	−
Alanine	−	−	+	+	+	−	+	+	+	+	−	+	−	−	−	+	+	+	−	+	−	−	−	−	−	−	−
Serine	−	−	+	+	−	−	+	+	−	+	−	−	−	−	−	+	+	−	−	+	−	−	−	−	−	−	−
Proline	−	−	+	+	−	−	+	+	+	−	−	−	−	−	−	−	−	−	−	+	−	−	−	−	−	−	−
Arginine	+	+	+	+	+	−	+	+	+	+	−	+	+	+	+	+	−	+	−	+	−	−	+	−	−	+	+
Ornithine	+	+	+	+	+	−	−	+	−	−	−	+	−	+	−	−	+	−	−	+	−	−	−	−	−	+	+
Betaine	−	−	+	−	−	−	−	−	−	−	−	−	−	−	−	−	−	−	−	−	−	−	−	−	−	−	−
Acetamide	+	+	+	+	−	+	−	+	+	+	−	−	−	+	−	+	−	+	−	+	−	−	−	−	−	−	−
Putrescine	−	−	+	+	−	−	−	−	−	−	+	−	−	−	−	−	−	−	−	+	−	−	−	−	−	−	−
Adenosine	−	−	+	+	−	−	−	+	−	+	−	−	−	−	+	+	−	−	−	−	−	−	−	−	−	−	−
Anosine	+	−	+	+	−	−	−	+	−	+	−	−	+	+	+	+	−	−	−	+	−	−	−	−	−	−	−
Urate	+	+	+	+	−	+	+	+	+	+	−	−	−	+	+	+	−	+	−	−	−	−	−	−	−	+	−
Uridine	−	−	−	−	−	−	−	−	−	−	−	−	−	−	−	−	+	+	−	−	−	−	−	−	−	−	−
Urea	+	+	+	+	+	+	+	+	+	+	+	+	+	+	−	+	+	−	+	+	+	+	+	+	+	+	+
Ammonium	+	+	+	+	+	+	+	+	+	+	+	+	+	+	+	+	+	−	+	+	+	−	+	+	+	+	+
Nitrate	+	+	+	+	+	+	+	+	+	+	+	+	−	+	+	+	+	+	+	−	+	+	+	−	+	+	+

From Neilson, A. H. and Larson, T., *Physiol. Plant*, 48, 542, 1980. With permission.

Table 6
NITROGEN-FIXING *CYANOBACTERIA*

Group	Genus	Total	Aerobic	Anaerobic microaerobic	Assay conditions[a]
Chroococcacean	*Alphanothece*	1	1	1	T.N.
	Gloeothece[b]	5	5	5	C_2H_2
	Synechococcus	27	0	3	C_2H_2
Pleurocapsalean	*Dermocarpa*	6	0	2	C_2H_2
	Xenococcus	3	0	1	C_2H_2
	Myxosarcina	2	0	1	C_2H_2
	Chroococcidiopsis	8	0	8	C_2H_2
	Pleurocapsa	12	0	7	C_2H_2
Nonheterocystous[c]	*Oscillatoria*	9	0	5	C_2H_2
filamentous	*Pseudoanabaena*	8	0	4	C_2H_2
forms	*Lyngbya-plectonema*				
	Phormidium	25	0	16	C_2H_2, $^{15}N_2$, T.N.
Heterocystous	*Anabaena*	15	15	15	C_2H_2, $^{15}N_2$, T.N.
filamentous	*Anabaenopsis*	2	2	2	C_2H_2, $^{15}N_2$, T.N.
forms	*Aulosira*	1	1	1	T.N.
	Calothrix	4	4	4	C_2H_2, $^{15}N_2$, T.N.
	Cylindrospermum	5	5	5	C_2H_2, T.N.
	Fischerella	2	2	2	T.N.
	Hepalosiphon	1	1	1	T.N.
	Mastigocladus	1	1	1	T.N.
	Nostoc	13	13	13	C_2H_2, $^{15}N_2$, T.N.
	Scytonema	3	3	3	T.N.
	Stigonema	1	1	1	T.N.
	Tolypothrix	2	2	2	T.N.
	Westiella	1	1	1	T.N.
	Westiellopsis	1	1	1	$^{15}N_2$, T.N.

[a] Certain, or all, of the cyanobacteria have been tested by these methods. T.N. = total nitrogen.
[b] Includes strains previously designated as N_2-fixing gloeocapsa strains.
[c] The data given here are those of Rippka and Waterbury, but the exact numbers of strains tested and shown to have N_2-ase may be larger since various earlier workers had examined and obtained positive results with strains which may or may not correspond to those tested by Rippka and Waterbury.

From Stewart, W. D. P., et al., *Nitrogen and Rice*, I.R.R.I., Los Banos, Laguna, Philippines, 1979. With permission.

Assimilation of Inorganic Nitrogen Compounds
Nitrogen Fixation

As mentioned above, some of the Cyanobacteria are capable of fixing atmospheric nitrogen. Nitrogen fixation takes place by reduction of N_2 to NH_4^+, a reaction catalyzed by the enzyme nitrogenase. Stewart et al.[60] listed the Cyanobacteria that have been tested for the ability to fix N_2 (Table 6).

The general conclusions that can be drawn from this compilation are (1) not all the Cyanobacteria can fix N_2; (2) cyanobacterial N_2 fixation is carried out by unicellular organisms, filamentous heterocystous forms, and by filamentous strains lacking heterocysts; (3) unicellular and filamentous but nonheteroxystous Cyanobacteria can fix N_2 only at low external O_2 levels, except for *Gloeothece* and *Aphanothece*, unicellular Cyanobacteria that fix N_2 under both aerobic and anaerobic conditions (see also Bothe[61]); and (4) heterocystous algae can fix N_2 in the presence of atmospheric levels of O_2. In the heterocystous Cyanobacteria there is overwhelming evidence that the site of aerobic N_2 fixation is the heterocyst, which is a specialized cell in the filament.[62,63]

The best known examples among the N_2-fixing Cyanobacteria are the heterocystous forms *Anabaena* and *Nostoc* sp. Vegetative cells usually differentiate to heterocysts only when they are grown in the absence of combined nitrogen. Their nitrogenase is active under aerobic and unaerobic conditions.

Nitrogenase is a complex enzyme with properties that are remarkably similar in the many organisms in which it has been studied. The enzyme is composed of a molybdenum-iron protein and a smaller iron protein, which in combination fix N_2. Neither protein can fix nitrogen by itself. It catalyzes the reaction:

$$N_2 + 6H^+ + 6 \text{ electrons} + 12 \text{ MgATP} \rightarrow 2NH_3 + 12 \text{ MgADP} + 12 \text{ P}_i \quad (13)$$

Besides N_2, the enzyme catalyzes the reduction of a variety of other substrates, i.e., N_3^-, N_2O, HCN, CH_3NC, CH_3CN, C_2H_2, H^+, and cyclopropane.[61,64] The reduction of H^+ to H_2 is referred to as ATP-dependent H_2 evolution. The enzyme activity depends on the presence of ATP, Mg^{2+}, and a reductant, and on low O_2, because oxygen inactivates the enzyme and prevents its synthesis. Glutamine synthetase (GS), the enzyme that assimilates newly fixed NH_4^+, also appears to be important in the synthesis and regulation of nitrogenase.[65] The reductant for nitrogenase comes from carbon compounds produced in photosynthesis,[66] and ATP may be generated by various metabolic mechanisms. (For more details on cyanobacterial nitrogenase, structure, function, and regulation, see Stewart.[65])

Nitrate and Nitrite Assimilation

When N is taken up in an oxidized form as nitrate (NO_3^-) or nitrite (NO_2^-), it must be reduced before it can be incorporated into organic molecules. The oxidation/reduction state of the N-atom in nitrate is $+5$ and in ammonia -3, so eight electrons are needed for the reduction. In the past it was assumed that three intermediates occur between nitrate and ammonia, associated with four reductive steps, each adding a pair of electrons:

$$
\begin{array}{ccccccccc}
NO_3^- & \rightarrow & NO_2^- & \rightarrow & N_2O_2 & \rightarrow & NH_2OH & \rightarrow & NH_4^+ \\
+5 & & +3 & & +1 & & -1 & & -3
\end{array} \quad (14)
$$

More recent studies with higher plants[67,68] and algae[69-72] suggest that only two enzymes catalyze the entire reduction of nitrate to ammonium:

$$NO_3^- \xrightarrow{\; 2e^- \;} NO_2^- \xrightarrow{\; 6e^- \;} NH_4^+ \quad (15)$$
$$\quad\quad\;\; (1) \quad\quad\;\; (2)$$

(1) nitrate reductase:NAD(P)H:nitrate oxidoreductase (NR) catalyzes the reduction of nitrate to nitrite (two electron reaction); and (2) nitrite reductase:NAD(P)H:nitrite oxidoreductase (NiR) catalyzes the reduction of nitrite directly to ammonium (six electron reaction). There are two types of nitrate reductase known in algae. One is found in eukaryotic algae and resembles higher plant nitrate reductases in many respects. It is found in the soluble portion of cell-free extracts and catalyzes the reaction:

$$NO_3^- + NAD(P)H + H^+ \xrightarrow{\;\; NR \;\;} NO_2^- + NAD(P)^+ + H_2O \quad (16)$$

The enzyme has been purified to some extent from various algae. It contains molybdenum, heme, and flavin adenine dinucleotide,[78-76] probably two molecules of each per molecule of the enzyme. A model of *Chlorella* NR as proposed by Solomonson[75] is shown in Figure 13.

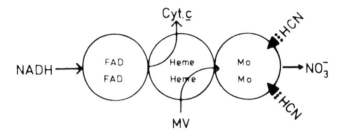

FIGURE 13.   Structure of *Chlorella* nitrate reductase according to Solomonson. (From Solomonson, L. P., *Nitrogen Assimilation of Plants*, Academic Press, New York, 1979. With permission.)

Like the NR of higher plants, the enzyme has three activities that can be measured and distinguished experimentally: (1) cytochrome *c* reductase activity — reduction of cytochrome *c* with NAD(P)H; (2) reduction of NO_3^- with electron donors such as reduced methyl viologen (MV) or flavin mononucleotide (FMN); and (3) reduction of NO_3^- with NAD(P)H. The first two activities are partial reactions, while the last requires the fully functional enzyme. The pyridine nucleotide specificity of NR differs among various algae.[51]

The second type of NR is found in Cyanobacteria and is apparently associated with chlorophyll-containing particles. Its molecular weight is about 7500 and it contains molybdenum but not flavin or cytochrome.[77] This type of enzyme uses reduced ferredoxin as an electron donor and not pyridine nucleotide. It catalyzes the reaction:

$$NO_3^- + 2Fd_{red} + 2H^+ \rightarrow NO_2^- + 2Fd_{ox} + H_2O \qquad (17)$$

Nitrite reductase (NiR) reduces nitrite to ammonia without the release of any free intermediates. The reaction is catalyzed by ferredoxin:

$$NO_2^- + 6Fd_{red} + 8H^+ \xrightarrow{\text{NiR}} NH_4^+ + 6Fd_{ox} + 2H_2O \qquad (18)$$

The enzyme has now been studied in algae of various taxonomic groups.[78-81] Its molecular weight is 60,000 to 70,000, and it contains a siroheme and an iron-sulfur center. In general, it appears to be very much like the NiR from higher plants.[68]

Ammonia Assimilation

Ammonium can be utilized as a nitrogen source by most algae (Table 4). When algae are supplied with both ammonium N and nitrate N, nitrate is often not utilized until all the ammonium has disappeared.[82-87] The effect of ammonium addition on cells of *Anacystis nidulans* utilizing nitrate is shown in Figure 14. As soon as ammonium is added, the uptake of nitrate is inhibited. Nitrate utilization resumes immediately after the exhaustion of ammonium.[87]

The effect of ammonium is related to the control of nitrate assimilation. Ammonium is the end product of nitrate reduction and it causes feedback inhibition and repression of the system responsible for nitrate uptake and reduction.

Supply of ammonium N to the growth medium of N_2-fixing Cyanobacteria inhibits nitrogenase activity and is also accompanied by a decrease in the differentiation of heterocysts. Removal of ammonium ions from the growth medium stimulates heterocysts differentiation and nitrogenase activity up to the original level.[88] The effects of several inorganic nitrogen sources on growth, nitrogenase activity, and heterocysts formation in *Anabaena* sp. strain 7120 and *Anabaena cylindrica*[89] are shown in Table 7 and Figure 15.

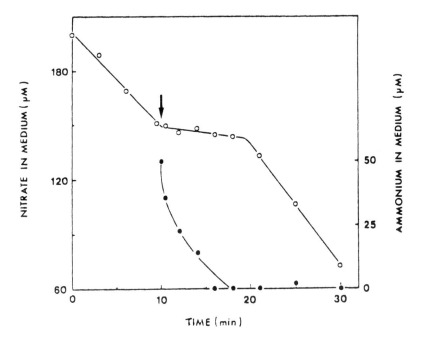

FIGURE 14. Time-course of nitrate utilization and of its inhibition by ammonium in *A. nidulans*. At the time indicated by the arrow NH₄Cl was added to give a concentration of 50 μ*M*. Nitrate (o——o); ammonium (●——●). (From Flores, E., et al., *Arch. Microbiol.*, 128, 137, 1980. With permission.)

Nitrogen-deficient algal cultures are an excellent tool for the study of the uptake and assimilation of ammonium N. Ammonium-N is assimilated rapidly by nitrogen-starved algae, sometimes four to five times more rapidly than by normal cells.[90] Assimilation takes place at the expense of endogenous carbohydrate reserves, and the NH_4^+ is converted into organic N compounds.[90,91]

Various ammonium-assimilating enzymes — glutamine dehydrogenase (GDH), glutamine synthetase (GS), glutamate synthase (GOGAT), alanine dehydrogenase, and carbamoyl phosphate synthatase — occur in algae.

The enzyme glutamine dehydrogenase catalyzes the formation of glutamic acid from NH_4^+ and α-glutaric acid:

$$
\begin{array}{c}
COOH \\
| \\
CH_2 \\
| \\
CH_2 \\
| \\
CO \\
| \\
COOH
\end{array}
+ NH_3 + NAD(P)H + H^+
\xrightarrow{\text{GDH}}
\begin{array}{c}
COOH \\
| \\
CH_2 \\
| \\
CH_2 \\
| \\
CHNH_2 \\
| \\
COOH
\end{array}
+ NAD(P)^+ + H_2O
$$

α-Glutaric acid   Glutamic acid   (19)

For a long time this reaction was accepted as the mechanism responsible for ammonium assimilation, with glutamic acid being the key compound in the conversion of inorganic N to organic N compounds. Thus, other amino acids, including alanine, were regarded as

Table 7
GROWTH RATES AND HETEROCYST
FREQUENCIES OF *ANABAENA* SP. STRAIN
7120 AND *A. CYLINDRICA* CULTURED WITH
DINITROGEN, NITRATE, AND AMMONIUM[a]

Species	Nitrogen source	Doubling time		Heterocystous frequency	
		hr	SE	%	SE
Anabaena sp.	N_2	21.5	1.0	8.4	0.3
strain 7102	NO_3^-	21.1	1.7	1.3	0.2
	NH_4^+	18.8	0.5	0	
A. cylindrica	N_2	18.2	1.0	6.3	0.5
	NO_3^-	15.0	1.6	4.3	0.3
	NH_4^+	14.3	0.3	0.1	0.1

[a] Growth was measured by changes in light scattering at 750 nm or by Chl-a content. In the experiments with combined nitrogen, the atmosphere was air with or without CO_2 or argon-oxygen-carbon dioxide, 80:19:1 (v/v/v), and at NO_3^- and NH_4^+ concentrations of 5 and 2.5 mM, respectively. There were no significant differences between cultures grown with combined nitrogen under air or argon-oxygen-carbon dioxide. In the case of N_2, the atmosphere was air with or without 1% CO_2. Buffer in all cases was 5 mM TES, pH 7.5. Each value is the mean standard error of the mean of 4 to 6 separate experiments.

From Meeks, J. C., et al., *Appl. Environ. Microbiol.*, 45(4), 1351, 1983. With permission.

being formed from glutamine by the addition of NH_4^+ to the γ-carboxyl group of glutamic acid in a reaction catalyzed by glutamine synthetase (GS) and requiring ATP.

$$
\begin{array}{c}
\text{COOH} \\
| \\
\text{CH}_2 \\
| \\
\text{CH}_2 + \text{NH}_3 + \text{ATP} \\
| \\
\text{CHNH}_2 \\
| \\
\text{COOH}
\end{array}
\xrightarrow{\text{GS}}
\begin{array}{c}
\text{CONH}_2 \\
| \\
\text{CH}_2 \\
| \\
\text{CH}_2 + \text{ADP} + \text{Pi} \\
| \\
\text{CHNH}_2 \\
| \\
\text{COOH}
\end{array}
$$

Glutamic acid   Glutamine   (20)

But work done during the last decade indicated that an alternative pathway of ammonium assimilation, which was first found in bacteria,[92,93] also operates in higher plants and in Cyanobacteria.[94-96] In this pathway glutamine is the first product of ammonium assimilation. The NH_4^+ incorporated into the amide group of glutamine and then transferred to α-oxoglutaric acid, a reaction catalyzed by glutamine oxoglutarate aminotransferase or glutamate synthase (GOGAT).

$$\text{glutamine} + \text{α-glutaric acid} + [2H] \xrightarrow{\text{GOGAT}} 2 \text{ glutamic acid} \qquad (21)$$

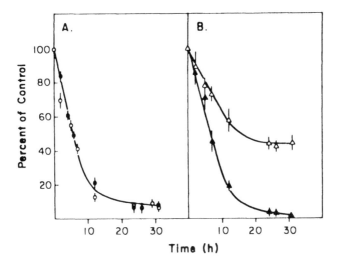

FIGURE 15.   Effect of added nitrate and ammonium to dinitrogen-grow-
ing cultures of *Anabaena* sp. strain 7120 (A) and *A. cylindrica* (B) on the
activity of nitrogenase (acetylene reduction) as a function of time after the
additions. Experiments were conducted on semicontinuous cultures converted to batch culture conditions at the time of addition of 5 m*M* nitrate
or 2.5 m*M* ammonium. Open symbols refer to nitrate additions and closed
symbols to ammonium additions. Each point is the mean standard error
of the mean of 3 to 5 experiments. Control rates of acetylene reduction
were 9.95, 0.7, 20.20, and 1.5 nmol/mg of protein per min for *Anabaena*
sp. strain 7120 and *A. cylindrica*, respectively. (From Meeks, J. E., et
al., *Appl. Environ. Microbiol.*, 45, 1351, 1983. With permission.)

The reaction results in the formation of two molecules of glutamic acid and requires a
reductant, which can be either NADH or reduced ferredoxin in higher plant and algae. The
use of L-methionine DL-sulfoximine (MSO or MSX), a powerful inhibitor for GS, and
azaserine, which inhibits GOGAT, demonstrated that the GS/GOGAT reaction[97-100] is the
primary pathway for NH_4^+ assimilation in various algae. A survey of the presence of the
key enzymes GDH, GS, and GOGAT in algae indicated that all three enzymes are present
in eukaryotic algae, but the levels of activity depend upon growth conditions. In most of
the Cyanobacteria the GS/GOGAT pathway is the main route of primary NH_4^+ assimilation.[65]

Alanine dehydrogenase catalyzes the direct formation of alanine from NH_4^+ and pyruvic
acid. The reaction requires NAD^+. This enzyme is significant in some Cyanobacteria that
lack GDH,[101] but seems to be the main route of alanine formation in the Cyanobacterium
Cylyndrospemun licheniform.[98] The enzyme is also present in *Chlamydomonas*,[102] but in
general it is considered to be less important than GS in overall ammonium assimilation.

Another pathway of NH_4^+ assimilation is that involving carbamoyl phosphate synthetase.
This enzyme is found in various Cyanobacteria and is essential for citrulline and arginine
production. The formation of carbamyl phosphate catalyzed by this enzyme is the initial
step for pyrimidine formation, but pyrimidine can also be formed by utilizing the amide
nitrogen of glutamine and not only by utilization of ammonium.

The main routes of NH_4^+ assimilation into organic N compounds are summarized in
Figure 16.

Assimilation of Organic-N Compounds
Urea

Urea is a good potential nitrogen source for almost all the algal species studied (Table
4). Urea is usually hydrolyzed before its N is incorporated into algal cells.[103-106] Two enzymes

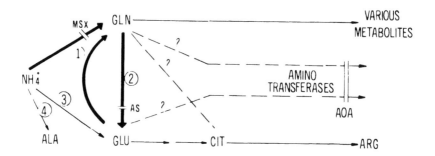

FIGURE 16.   The main metabolic routes of ammonium assimilation into organic-N compounds: (1) GS, glutamine synthetase; (2) GOGAT, glutamate synthase; (3) GDH, glutamine dehydrogenase; (4) ALDH, alanine dehdyrogenase; MSX, L methionine-DL-sulfoximine; AS, azaserine; and AOA, aminooxyacetate.

that metabolize urea are known in algae, urease and urea amidolyase (UALase). Urease catalyzes the simple hydrolytic reaction:

$$H_2O + CO(NH_2)_2 \rightarrow CO_2 + 2NH_3 \tag{22}$$

while urea amidolyase catalyzes the overall reaction

$$CO(NH_2)_2 + ATP + H_2O \rightarrow CO_2 + 2NH_3 + ADP + P_i \tag{23}$$

which is comprised of two reactions catalyzed by two separate enzymes. Urea carboxylase catalyzes the formation of allophanate:

$$CO(NH_2)_2 + HCO_3^- + ATP \rightarrow O=C \begin{smallmatrix} NH_2 \\ \\ NH \\ \\ COO^- \end{smallmatrix} + ADP + Pi + H_2O \tag{24}$$

and the allophanate lyase catalyzes the hydrolysis of allophanate:

$$O=C \begin{smallmatrix} NH_2 \\ \\ NH \\ \\ COO^- \end{smallmatrix} + H_2O \rightarrow CO_2 + HCO_3^- + 2NH_3 \tag{25}$$

Algae that metabolize urea have urease or UALase but not both.[107,108] It seems that Chlorophycea contain UALase and other algae contain urease.[109]

Amino Acids

Amino acids are taken up by some algae as was demonstrated by studies with ^{14}C-labeled amino acids.[17] In *Nitzchia ovalis* at least three amino acid uptake systems specific for transporting acidic, polybasic, and neutral amino acids were found.[110] At least three amino acid transport sites were found also in the diatom *Melosira nummuloides*.[111] Kirk and Kirk[112] showed the existence of specific amino acid carriers in various Chlorophyceae and such systems exist also in other groups of algae. Once taken up, all amino acid do not undergo the same conversions inside the cells.[113]

Other Organic N Compounds

The uptake of other organic N compounds by algae has not been studied much; however, purine and its derivatives seem to be taken up by *Chlorella pyrenoidosa* and metabolized[114] (see also the "Nutritional Modes" section in this chapter).

Effects of Nitrogen Deficiency

Deprivation of nitrogen is a serious threat to microalgae. Nitrogen-deficient algal cells can be prepared in the laboratory in two different ways: (1) cultures can be grown with a limited amount of available nitrogen so they become nitrogen-deficient when it is exhausted; or (2) nitrogen-sufficient cells can be transferred to a nitrogen-free medium and allowed to photosynthesize. The two methods do not necessarily produce cultures with identical properties.[115]

In general, microorganisms respond to nitrogen deprivation by preferential degradation of one or more nitrogen-containing macromolecules,[116] resulting in a remarkable decrease in the cells' nitrogen content[115] and an accumulation of carbon reserve compounds such as polysaccharides and fats.[90] Under conditions for nitrogen deficiency the content of photosynthetic pigments decreases and the rate of photosynthesis is reduced.[117] The ability to take up and assimilate combined nitrogen compounds is increased under these conditions.[17]

In Cyanobacteria two major endogenous nitrogen storage compounds seem to be utilized by nitrogen-deficient cells, cyanophycin granule polypeptide (CGP) and phycocyanin (PC). Cyanophycine granules are copolymers of aspartic acid and arginine. The polypeptide was characterized in *Anabaena cylindrica*[118-123] and *Aphanocapsa* 6308.[124,125] Cyanophycin accumulates during the stationary phase of nitrogen-sufficient cultures and is depleted during nitrogen deficiency.[126] The accessory pigment phycocyanin is rapidly and specifically degraded in nitrogen-limited cells, and reappears rapidly when nitrogen becomes available.[127-129] Proteases that degrade phycocyanin and appear to be activated or preferentially synthetized during nitrogen starvation have been described in *Anabaena cylindrica*[130,131] and *Spirulina platensis*.[132] The effects of nitrogen starvation on various growth parameters and on the phycocyanin content of *Spirulina platensis* are shown in Figure 17 and on protease activity in Figure 18. In various taxonomic groups the chlorophyll content was also found to decrease remarkably during nitrogen starvation.[117,127,133]

Changes in enzymatic activities were also observed during nitrogen starvation. For example, nitrate reductase activity was observed in ammonium grown cells of various algal species after a short period of nitrogen starvation.[134-136] The activity of other N-assimilating enzymes also depends on the nitrogen status of the algal cultures. As shown in a detailed study of nitrogen-starved cultures of *Ankistrodesmus braunii*,[137] the activities of nitrite reductase, glutamic dehydrogenase, glutamine synthetase, and urea amidolyase were derepressed during the development of nitrogen deficiency. The increase in activity of N-assimilating enzymes is accompanied by a decrease in the rate of photosynthesis, as measured by $^{14}CO_2$ incorporation in *Thalassiosira*[138] and by $^{14}CO_2$ incorporation and ribulosebiphosphate carboxylase activity in *Ankistrodesmus*.[137] The activities of uptake mechanisms for the various nitrogen compounds also change during nitrogen depletion. This was shown for nitrate, ammonia, urea, and amino acids.[17,52] In *Phaeodactylum tricornutum*, ammonium-grown cells that cannot take up nitrate develop this ability after a short time of nitrogen deprivation.[139] The kinetics of the development of the nitrate uptake system in washed ammonium-grown cells and the effect of ammonium ions on its development are shown in Figure 19.

Heterocyst differentiation and nitrogenase activity increase in N_2-fixing Cyanobacteria when the availability of combined nitrogen decreases (Table 7 and Figure 15). Ultrastructural changes were also reported in various filamentous heterocystous or nonheterocystous and unicellular Cyanobacteria as a result of deprivation of combined nitrogen.[140-143] These changes

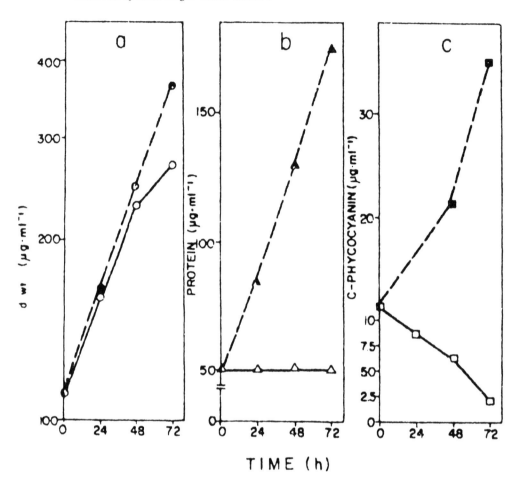

FIGURE 17. Effect of nitrogen starvation on (a) dry weight, (b) protein content, and (c) c-phycocyanin content. Filled symbols, control treatment; open symbols, nitrogen starvation treatment. (From Bussiba, S. and Richmond, A. E., *Arch. Microbiol.*, 125, 143, 1980. With permission.)

include phycobilisome depletion, polysaccharide accumulation, thylakoid rearrangement, and changes in the number of types of inclusion bodies.

Interaction of Nitrogen and Carbon Metabolism

It is well known that light stimulates inorganic nitrogen uptake and assimilation by algae[90] and higher plants,[68] probably in several ways. One may be by direct photoreduction of the N compound, with a reductant such as NAD(P)H or reduced ferredoxin produced by photosynthetic electron transport. Thus, a close relation between light and reactions that require reduced ferredoxin or NAD(P)H is expected. Work done by Kessler,[144] Kessler and Zumft,[145] and Thomas et al.[146] indicates that light affects mainly the nitrite reducing step, which requires ferredoxin. The dependence of nitrite reduction upon a continuous supply of reductant originating from photosynthetic electron flow is well established (Figures 20 and 21).[146] In nitrogen-limited cultures of *Chlorella* (30 μg NO_3^- N/L), light stimulates nitrite reduction, and DCMU, which inhibits photosynthetic electron flow, prevents the light effect (Figure 21). Under the experimental conditions used, about one half of the electrons used for nitrite reduction in the light were generated by photochemical reactions.[146]

Another possible way for light to stimulate the uptake and assimilation of nitrogen is by the supply of ATP via photophosphorylation. This possibility is supported by the finding

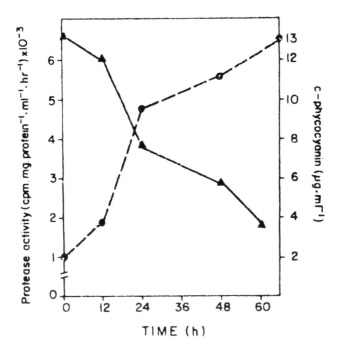

FIGURE 18. Effect of nitrogen starvation on protease activity and c-phycocyanin content. (●———●) Protease activity; (▲———▲) c-phycocyanin content. (From Bussiba, S. and Richmond, A. E., *Arch. Microbiol.*, 125, 143, 1980. With permission.)

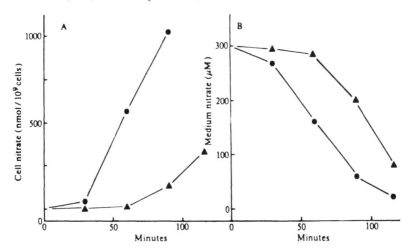

FIGURE 19. Effect of presence of ammonium ions on development of nitrate uptake system. Ammonium-grown cells, washed and suspended in N-free medium, were illuminated and aerated without (●———●) and with (▲———▲) added NH_4Cl (1 mm). After 3 hr cells in each suspension were collected by centrifuging, washed, and resuspended in N-free medium, KNO_3 was added and (A) nitrate appearance in the cells, and (B) nitrate disappearance from the medium followed. (From Cresswell, R. C. and Syrett, P. J., *J. Exp. Bot.*, 32, 19, 1981. With permission.)

that uncouplers of photophosphorylation inhibit both nitrate and nitrite uptake and reduction. ATP seems to be required for nitrate uptake rather than reduction,[147,148] suggesting that nitrate enters the cells by an active mechanism. Some ATP can also be supplied through the respiratory system, but apparently not enough to carry on N uptake at maximum rates.

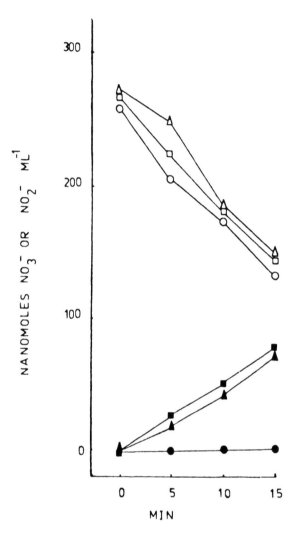

FIGURE 20.   The effect of DCMU on nitrate uptake. Cells from the 30 μg NO_3-N/mℓ input culture were incubated with 0.4 mM KNO$_3$ and nitrate disappearance from the medium together with any appearance of nitrite was followed over a 15 min period in light, (○———○), light + DCMU (△———△), and darkness (□———□). Open symbols designate nitrate disappearance, filled symbols nitrite appearance. (From Thomas, R. J., et al., *Planta*, 133, 9, 1976. With permission.)

A third way by which light may stimulate inorganic nitrogen uptake is more indirect through the photosynthetic production of carbon skeletons to accept the reduced nitrogen. This possibility is illustrated by the requirement for both light and CO_2 to obtain the highest rates of nitrate, nitrite, and ammonia uptake.[86,149-152] The requirement for CO_2 may be replaced by an organic carbon source such as glucose in *Chlorella*[152] or acetate in *Clamydomonas*.[86] Nitrogen-starved cultures that accumulate polysaccharides assimilate either NO_3^- or NH_4^+ in darkness at rates faster than do nitrogen-sufficient cells under ordinary photosynthetic conditions.[86] This indicates that carbon reserves in the nitrogen-starved cells serve as carbon skeletons for assimilated nitrogen upon addition of nitrogen compounds.

ATP and reduced ferredoxin are also required for NH_4^+ assimilation into organic N compounds by the GS/GOGAT pathway and for N_2 fixation by nitrogenase in N_2-fixing

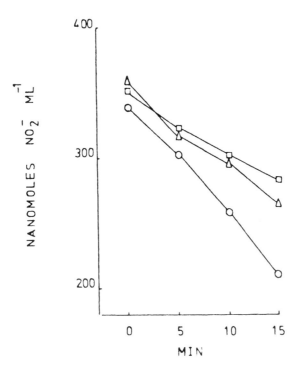

FIGURE 21. The effect of DCMU on nitrite uptake. Cells from the 30 μg NO₃-N/mℓ input culture were incubated with 0.4 mM NaNO₂ and nitrite disappearance from the medium was followed over a 15 min period in light (○——○), light + DCMU (△——△), and darkness (□——□). (From Thomas, R. J., et al., *Planta*, 133, 13, 1976. With permission.)

Cyanobacteria. Photosynthesis in the N_2-fixing algae is the ultimate source of electrons and a direct source of ATP for nitrogenase activity. N_2 fixation is stimulated by light, and when it occurs in darkness its rate rapidly declines, probably due to a lack of reductants.[153]

The overall interaction between nitrogen and carbon metabolism is rather complex, and the results obtained so far with algae and higher plants suggest that all three mechanisms of light-stimulated inorganic nitrogen uptake and assimilation are active in algae.

Phosphorus Requirement and Metabolism
General Aspects

Phosphorus is one of the major nutrient elements required for normal growth of algae. It plays a major role in most cellular processes, particularly those involved in energy transfer and in nucleic acid synthesis. The role of phosphorus in algal metabolism has also attracted the attention of ecologists since this element is frequently limited for algal growth in nature.[154,155]

Although the concentrations of organic phosphates in natural waters often exceed that of inorganic phosphate,[156] the major form in which microalgal cells acquire phosphorus is as inorganic phosphate ($H_2PO_4^-$ + HPO_4^{2-}) or, collectively, P_i.[21] For organic phosphate compounds to serve as the primary source of phosphorus, they must be hydrolyzed by extracellular enzymes such as phosphoesterases[157] or phosphatases,[158,159] and the resulting P_i is taken up.[160] In some Cyanobacteria, however, sugar phosphates can be taken up intact.[161]

The phosphorus requirements for optimum growth differ considerably from species to species, even if no other external factor is limiting. Rodhe[162] differentiated between three main groups of freshwater algae according to their ability to tolerate phosphate within the range below, around, or above 20 μg/ℓ (see also Soeder et al.[26]). Most of the algae fall into the groups with low or medium phosphorus tolerance.

Phosphorus Uptake and Assimilation

Phosphorus uptake and metabolism was studied with algae growing under phosphorus-deficient and phosphorus-sufficient conditions. The uptake of phosphorus from the surrounding medium by algae is generally stimulated by light. It is an energy-dependent reaction as shown by its sensitivity to uncouplers.[163-165] The uptake rate is also influenced by the phosphate concentration in the medium, the pH, and, in several alga species, on the availability of Na^+, K^+, or Mg^{2+}.[160,166]

The phosphate in the cells is channeled into various inorganic and organic phosphorus compounds. Algae and other green plants use three major processes to incorporate orthophosphate into organic "high energy" compounds: (1) substrate-level phosphorylation, (2) oxidative phosphorylation, and (3) photophosphorylation. The general reaction is

$$ADP + P_i \xrightarrow{\text{Energy}} ATP \qquad (26)$$

In the first two processes, energy is derived either directly from the oxidation of respiratory substrates or from the electron transport system of mitochondria. In the third process, which is typical of plants, light energy is converted into the "energy rich" phosphate bonds of ATP. This compound has a central role in the metabolism of living organisms, since its hydrolysis provides the energy needed for most energy requiring reactions such as CO_2 fixation, ion uptake and transport, formation of nucleic acids, and many other reactions.

Inorganic phosphate usually occurs in algae in the form of polyphosphates, but in some cases metaphosphates have been found.[167,168] The structure of these forms is shown below:

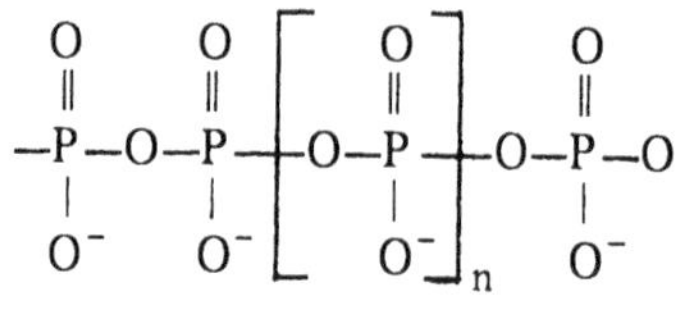

$$\begin{array}{ccccc}
O & O & \left[\begin{array}{c} O \\ \| \end{array}\right. & O & \\
\| & \| & \| & \| & \\
-P-O-P & O-P & O-P-O \\
| & | & | & | \\
O^- & O^- & \left. O^- \right]_n & O^-
\end{array}$$

Polyphosphate $\qquad\qquad\qquad (27)$

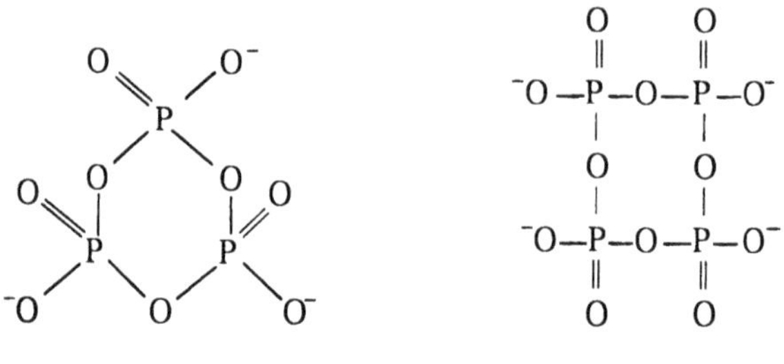

(a) Trimetaphosphate $\qquad\qquad$ (b) Tetrametaphosphate

Metaphosphates $\qquad\qquad\qquad (28)$

Polyphosphates are acid-labile substances that are completely split to orthophosphate units in 1 N HCl at 100°C. Most naturally occurring polyphosphates exist in either acid soluble or acid insoluble forms. They appear to be the principal form of phosphate storage in most algae, accumulating in distinct polyphosphate granules, which appear in normal cells under phosphate sufficient growth conditions and disappear under phosphate deficient conditions. It seems that in algae the main metabolic role of polyphosphates is to supply the cells with phosphorus for special reactions such as the synthesis of nucleic acids or reactions closely associated with cell division.[160]

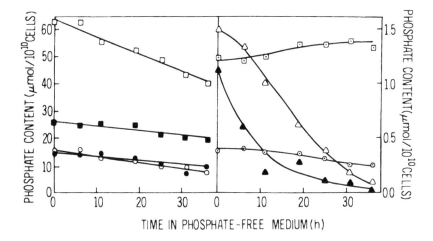

FIGURE 22.   Changes in phosphate fractions after transfer of *Chlorella* cells to phosphate-free medium. *Chlorella* suspension, after 5 days of growth in [³³P]orthophosphate (3.23 m*M*) was contrifuged, washed, and resuspended in fresh medium without phosphate. Total phosphate, □——□; phosphate in residue after alkaline extraction, ■——■; total trichloroacetic acid-soluble phosphate, ○——○; total KOH-soluble phosphate, ●——●; acid-soluble polyphosphate, ▲——▲; acid-insoluble polyphosphate, △——△; lipid phosphate, ◙ — ◙ ; orthophosphate, ☉—☉. (From Aitchinson, P. A. and Butt, V. S., *J. Exp. Bot.*, 24, 497, 1973. With permission.)

Phosphorus Deficiency Symptoms

Some of the symptoms of phosphorus deficiency are similar to those observed in nitrogen-deficient cultures. The contents of protein, chlorophyll-a, RNA, and DNA tend to decrease while the carbohydrate content increases in eukaryotic as well as prokaryotic algal cells.[169] A decrease in the cellular ATP content was found in many algal species of various taxonomic groups.[170] As phosphorus deficiency develops, the total phosphate content of algal cells decreases. In a study using *Chlorella vulgaris* Aitchison and Butt[171] found that during the period of phosphate starvation, the levels of most phosphate fractions declined, especially those of inorganic polyphosphates (Figure 22). When the cells were returned to a phosphate-containing medium, phosphate was taken up much faster than before starvation. The phosphate content of most cell fractions increased gradually until the prestarvation levels had been reached. A remarkable increase in the polyphosphate fractions was observed upon provision of phosphate to phosphorus-starved cells (Figure 23).

Maximal accumulation of acid-soluble polyphosphate was observed 5 to 8 hr after the addition of phosphate and its level at that time was much higher than in unstarved cells. The use of ³²P-labeled phosphate showed that virtually all the polyphosphate synthesized upon phosphate addition to deficient cells was derived from the phosphate present in the external medium.[171] Similar results were observed in many algae of various taxonomic groups.[163,172,173]

Induction or activation of the cell surface enzyme alkaline phosphatase occurred in phosphate-deficient cultures of several diatoms, Crysophytes, and Cyanobacteria.[17]

Morphological changes such as cells or trichoms size and shape were observed in Cyanobacteria, subjected to phosphorus starvation.[169] Ultrastructural studies indicated the loss of polyphosphate bodies in phosphate-deficient cells of eukaryotic and prokaryotic algae.[174,175] Accumulation of cyanophycin granules was observed in phosphate-deficient cells of some Cyanobacteria.[176,177] Phosphorus deficiency appears to induce akinate formation and to suppress heterocyst formation in some Cyanobacteria.[169] Nitrogenase activity in N₂-fixing Cyanobacteria is closely related to the availability of phosphate in cultures[173] and natural habitats.[178,179]

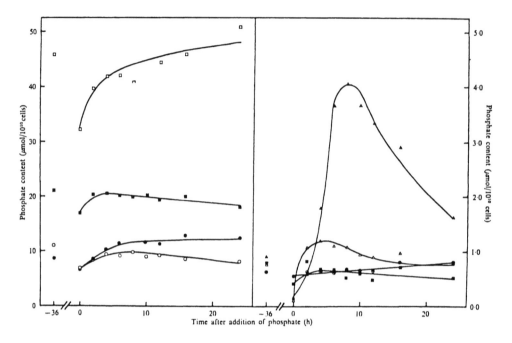

FIGURE 23.   Changes in phosphate fractions on addition of phosphate to phosphate-starved cells.[179] Experiments as in Figure 22. After 36 hr in phosphate-free medium, cells were centrifuged, washed, and resuspended at same cell-density in original medium containing 3.23 mM phosphate (containing 60 μCi/ℓ[^{32}P]orthophosphate). Symbols as Figure 22. (From Aitchinson, P. A. and Butt, V. S., *J. Exp. Bot.*, 24, 497, 1973. With permission.)

Relationship Between Phosphate and Other Factors

As already mentioned, phosphate uptake by algae is an energy-dependent reaction, as is the synthesis of polyphosphates. The energy needed for phosphate uptake can be provided by photosynthesis or respiration.[163] The inorganic polyphosphates are synthesized in the light at the expense of ATP generated during photophosphorylation.[160] The formation of polyphosphates in the light is influenced by several other conditions such as the presence of O_2 or CO_2 and the pH of the medium.[180-182] Phosphate and nitrogen metabolism are sometimes closely related and changes in the N/P ratio in lakes might cause changes in the phytoplanktonic population.[169]

Sulfur

Sulfur is essential to all organisms in the form of sulfur-containing amino acids (methionine, cystine, cysteine), biotin, pantothenic acid, thiamin, lipoic acid, some growth factors, sulfolipids, etc. Most algae obtain their sulfur from inorganic sulfate. A few are able to obtain their sulfur from organic sulfur compounds, e.g., *Chlorella vulgaris* was reported to utilize methionine,[183] and *Chlorella pyrenoidosa* grew on cysteine but lost its ability to grow photoautotrophically.[184] Of several *Chlorella* species and strains tested, four used D- or L-methionine and one used L-methionine, and streptomycin-bleached *Euglena* grown methionine, cysteine, or homocysteine.[185]

Like nitrate, the major part of the sulfate taken up by algae must be reduced before it can be incorporated into cellular material. Nevertheless, short-term sulfate uptake by *Chlorella pyrenoidosa* appeared to be independent of its subsequent reduction, and mutants incapable of growing on or reducing sulfate took up sulfate for over 3 hr at the same rate as wild-type cells capable of reducing it. In the absence of a mechanism for storing large amounts of sulfate, this must be only a short-term phenomenon in nonvacuolated algae.[17] Uptake of sulfate by *C. australis* is sensitive to uncouplers in both light and darkness. This, together with the temperature sensitivity of the reaction and the large concentration gradient between the vacuole and the medium, show the uptake to be active, presumably driven by phosphorylation. Uptake of sulfate was temporarily stimulated upon changing from white light to darkness or to far-red light or upon adding DCMU, indicating that this uptake is not strictly dependent on photosynthetic processes.[17]

Unlike in *C. australis*, uptake of sulfate by both *Chlorella pyrenoidosa*[184] and *Scenedesmus* sp.[186] is stimulated by light. As with N assimilation, light could be acting by providing reductants, energy from photophosphorylation, or C skeletons. The sensitivity of sulfate uptake by both these algae to uncouplers shows an energy requirement for uptake and/or reduction. With *Scenedesmus*, the stimulation by light is greater in the presence than the absence of CO_2, showing that here the provision of C skeletons is part of the explanation. Finally, the ability of isolated chloroplast membranes to reduce sulfate to sulfite and the latter to sulfide, if they are provided in the light with a soluble chloroplast extract and various cofactors, suggests a photoreductant may be involved in sulfate reduction.[187]

Under anaerobic reducing conditions, in environments rich in sulfide, many algae seem to be able at least to survive.[188-190] Some green algae and Cyanobacteria have been shown to grow either heterotrophically in the dark or in the light using the sulfide ion as the electron donor[190-192] and releasing elementary sulfur rather than oxygen. *Oscillatoria limnetica*, which was isolated from a hypersaline thermal pond near the Gulf of Eilat,[193] can shift, according to the availability of the Na_2S, from using H_2O as the electron donor in oxygenic photosynthesis to Na_2S.[188]

Calcium

Calcium requirements for maximum growth have been demonstrated for green algae, Cyanobacteria, and diatoms.[194-197] While the function of Ca remains largely unknown, it is involved in the deposition of calcareous scales and skeletons by several, mostly marine, algae.[198] In some green algae, calcium may be specifically required for the release of motile cells and sometimes it has a role in the formation of cell walls by zoospores. The activity of an enzyme mediating the release of zoospores by *Chlamydomonas reinhardii* is dependent on Ca and Mg. Some role in cell wall formation has been suggested for calcium. In general, of the elements required in relatively large amounts by at least some algae, Ca appears to be the least understood functionally.[17]

The calcium requirement could not be satisfied by beryllium, magnesium, barium, cobalt, iron, zinc, manganese, copper, molybdenum, nickel, thallium, gallium, aluminum, vanadium, germanium, titanium, zirconium, arsenic, bismuth, tin, chromium, mercury, sodium, potassium, boron, silver, or lead, when these ions were added singly to a medium of glucose-urea-EDTA-salts lacking calcium. Strontium did, however, promote growth in such a calcium-deficient medium, the addition of 1 to 3 μg strontium to 5 mℓ of medium resulting in a larger cell yield than the addition of an equimolar quantity of calcium (Figure 24).[199] It might appear that the real requirement of *Chlorella* might be for strontium rather than calcium. However, since EDTA binds calcium more strongly than strontium, a quantitative comparison of results obtained in a complex medium containing EDTA is quite difficult.[199]

Sodium, Potassium, and Chlorine

Sodium is required by some algae and not by others. The Na requirements of marine Cyanobacteria greatly exceed those of freshwater species.[200] The Na requirements of *Anabaena cylindrica* cannot be satisfied by K, Li, Rb, or Cs. Sodium stimulates the autotrophic but not the heterotrophic growth of *Chlorella pyrenoidosa*, is required by the halophilic *Ctenocladus circinnalis*, and is probably necessary for all marine and halophilic algae, perhaps for osmotic reasons. Na is an activator of several enzymes and algae that are deficient in K tend to replace intracellular K by Na.[17]

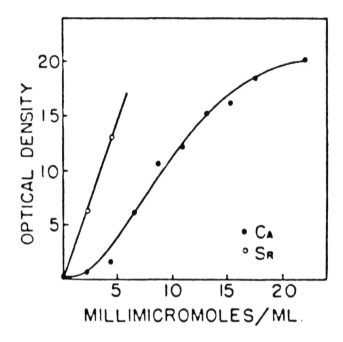

FIGURE 24.   Cell yield vs. calcium or strontium concentration for *Chlorella* grown in a glucose-urea-EDTA-salt medium deficient only in calcium. Since it will be shown later that calcium and strontium are approximately equally effective for growth in the absence of EDTA, these curves illustrate the usefulness of EDTA in demonstrating micronutrient requirements. In the presence of EDTA the concentration of the more strongly chelated calcium ion must be higher to achieve a given cell yield than the concentration of the less strongly complexed strontium ion. (From Walker, J. B., *Arch. Biochem. Biophys.*, 46, 1, 1953. With permission.)

Potassium is a cofactor for a variety of enzymes and is probably universally required by algae. In bacteria, K is involved in ribosome structure, protein synthesis, and osmotic regulation and it seems safe to assume that K also serves these functions in algae. Wherever a K requirement has been sought among algae, it has been found, but in several green algae K could be at least partially replaced by Rb.[17]

The regulation of K and Na uptake has been studied extensively. Chloride uptake appears to be closely associated with the uptake of K and Na. MacRobbie[201] proposed a model for the uptake of K, Na, and Cl, according to which chloride is taken up actively by the cells, and this uptake promotes passive carbon cation influx. The regulation of this uptake is effected by a K-Na exchange. The overall uptake mechanism would accordingly be as follows: a high intracellular K/Na ratio is maintained by an energy-dependent Na-K pump apparently linked preferentially to cyclic phosphorylation.

Magnesium

Magnesium has many crucial functions in cellular metabolism. Probably the major portion of Mg in most cells is involved in the aggregation of ribosomes into functional units. The function of Mg in reactions involving the transfer of "high energy" phosphate groups and as a component of chlorophyll is well established.[17] *Euglena gracilis* cells possess a magnesium-dependent ATPase and also a calcium-dependent ATPase that can be partly activated by magnesium.[202] Magnesium has an allosteric effect upon the ribulose 1,5 diphosphate carboxylase of *Chlorella ellipsoidea*.[203]

MICRONUTRIENTS

The elements belonging to this category are required in relatively very small amounts, in the concentration range of micro- to milligram per 1000 mℓ of culture medium. Thus, it requires very careful experimentation to establish their requirements by algal species. Holm-Hanson et al.[204] listed criteria for determining the essentiality of microelements:

1. The element must have a positive effect on total growth, i.e., it must permit completion of the normal life cycle.
2. It must exert a direct physiological effect on the algae, i.e., it must not influence growth indirectly through an effect on nutrient balance, the pH of the solution, etc.
3. It should not be replaceable by another element.
4. The deficiency should be "reversible". That is, upon addition of the element to cultures in incipient stages of deficiency, normal growth should resume.
5. The response to the element should be noted in a representative number of species.

Iron

Iron is involved in nitrogen assimilation, since ferredoxin is required as the electron donor for both nitrate and nitrite reductase.[205,206] Iron is also important in photosynthesis, affecting the synthesis of the photosynthetic pigments chlorophyll-a (Chl-a) and c-phycocyanin (CPC).[207] Fe has a functional role in ferredoxin and the iron-sulfur proteins on the acceptor side of photosystem I (PS I), making it a vital component of photosynthesis. It also affects cytochrome synthesis since it is incorporated into the heme ring.

The effects of iron starvation on the growth and physiology of the unicellular cyanobacterium *Agmenellum quadruplicatum* were studied by Hardie et al.[209] Most of the iron was taken up at two different times: when the cells were initially inoculated into the medium and after the cultures had become quite dense and had stopped growing. Iron became limiting for growth 16 hr after transfer to an iron-deficient medium, but the cultures retained full viability for at least 212 hr. Once Fe became limiting, c-phycocyanin and chlorophyll-a were degraded concurrently. This was followed by an accumulation of intracellular glucose in place of the c-phycocyanin. Nitrate and nitrite reductase activities were elevated for 50 hr, after which they steadily decreased. Once iron was restored to the culture medium, growth resumed, the intracellular pigment levels increased rapidly, and the amount of glucose decreased.

Hardie et al.[210] also studied the effects of iron starvation on the ultrastructure of *Agmenellum quadruplicatum* by thin sectioning and transmission electron microscopy (TEM). Intracellular polysaccharides began to accumulate at the onset of iron limitation. This was followed by degradation of ribosomes and then degradation of the thylakoid membranes, both of which were virtually absent by 200 hr. The thylakoids underwent structural modifications and rearrangement before they actually began to break down. All the changes were reversed when iron was re-added to cultures starved for 200 hr. The sequence of ultrastructural changes observed during iron starvation clearly differed from those reported to occur during nitrogen, phosphorus, or carbon limitation, pointing out the usefulness of electron microscopy in elucidating the effects of nutrient deficiencies.

According to Meisch et al.,[211] who studied ultrastructural changes in *Chlorella fusca* during iron deficiency, the presence of vanadate prevented iron deficiency symptoms.

Boron

Boron is required by several Cyanobacteria and diatoms, but is apparently not required by green algae. According to the review by Healey,[17] the boron requirement of marine diatoms appears to be greater than that of freshwater diatoms. This may be related to the

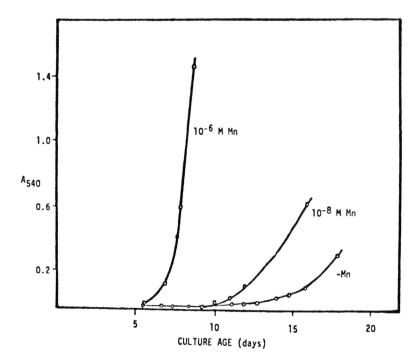

FIGURE 25.   Absorbance at 540 nm of *Scenedesmus obliquus* at 3 concentrations of Mn as a function of days after inoculation.[214] Inocula were grown at 10^{-8} *M* Mn and were diluted 1:200 in the inoculation of each culture. In the "-Mn" culture, no Mn was added to the medium, but some Mn, probably at a concentration of 10^{-9} *M*, was judged to be present in the media, originally from the cell inoculum or from trace contaminants in the chemicals which constituted the growth medium. (From Jahnke and Soulen, *Z. Pflanzenphysiol.*, 88, 83, 1978. With permission.)

higher B concentration in the oceans than in freshwater. The B requirement of *Cylindrotheca fusiformis* increases as does the silicon concentration, suggesting some relationship with Si metabolism.[212] This possibility if further supported by the stimulation by borosilicate of the growth of two diatom species;[213] however, the role of B in diatoms may be the same as in nonsiliceous algae, with the requirement exaggerated by a competitive effect of Si.

Manganese, Copper, and Zinc

Manganese and copper are essential components of the photosynthetic electron transport system and they may be assumed to be required by all algae growing photosynthetically. These elements also have other functions in the cell, such as components or cofactors of several enzymes.

Jahnke and Soulen[214] studied manganese deficiency in autotrophically grown cultures of the green alga *Scenedesmus obliquus* (Figure 25). *Scenedesmus* grew at an exponential but reduced rate in low Mn (10^{-8} *M* Mn). Cells grown with saturating levels of Mn appeared to store excess Mn, which delayed the formation of Mn deficiency when the cultures were transferred to Mn-deficient media. Mn was saturating above 10^{-8} *M*, and limiting between 10^{-8} and 10^{-7} *M* Mn. As little as 1 min incubation in the dark with saturating concentrations of Mn ($>10^{-6}$ *M*) allowed deficient cells to regain full photosynthesis, as measured by oxygen evolution rates in the light.

The complexity of the problems relating to heavy metal nutrition and toxicity is indicated by the work of Bates et al.[215] who studied the growth of *Chlamydomonas variabilis* Dangeard in the absence of presence of added zinc. The rate constants for Zn transport were smaller for cells grown with added zinc and decreased slightly with increasing culture age. The

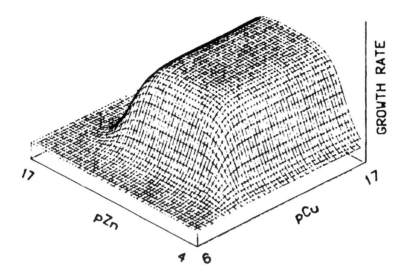

FIGURE 26. Graphical presentation of a model proposed to describe the growth of *Scenedesmus quadricauda* as a function of copper and zinc ion concentrations. (Peterson, R., *Environ. Sci. Technol.*, 16, 443, 1982. With permission.)

number of binding sites per unit cell surface decreased markedly with culture age, as did the amount of Zn extractable with EDTA. This points out that variations in metal absorption and transport as a function of the physiological state of the cells should be considered when trace metal uptake is modeled and when phytoplankton are used to assess metal toxicity.

Sandmann and Boger[216] reported on another aspect relating to the complexity of metal ion effects on alga, i.e., the tremendous capacity of algal cells to accumulate ions. The microalga *Scenedesmus acutus* can accumulate large amounts of copper ions. Depending on the concentration in the medium, an enrichment of up to 1000-fold within the cells is possible. For optimum growth a copper supply of 0.1 to 1 μM is necessary. Growth is reduced by 50% and chlorophyll becomes bleached when the medium is depleted of copper. Copper concentrations exceeding 10 μM are toxic, but the toxicity is only temporary and the algae can adapt to up to 50 μM copper ions within 24 hr. During this transient period, chlorophyll is degraded and lipids are oxidized.

Another source of complexity for interpreting the effects of heavy metals is that several may affect an alga additively. Chemical specification of a single metal, however, may be important in pinpointing nutritional or toxic effects. This is well illustrated in a study by Peterson[217] who demonstrated, in *Scenesdesmus quadricada* grown in a defined medium with a wide range of copper and zinc concentration, that the growth rate of the alga was influenced by and could be predicted from the concentrations of both Cu^{2+} and Zn^{2+}. Excess copper and zinc were toxic. The free metal ion is the chemical form that is significant in both nutrition and toxicity. On the basis of the experimental data, Petersen[217] proposed the model shown in Figure 26. According to the model, the growth rate should decrease at low zinc concentrations because of trace-metal deficiency and at high copper or zinc concentrations because of metal toxicity. A useful aspect of the model is that it permits a consideration of joint action of two metals as well as the chemical speciation of the metals and can be expanded to account for additional metals present in toxic concentrations.[217]

Molybdenum, Vanadium, Cobalt, and Nickel

Molybdenum functions in nitrate reductase and N_2 fixation.[218] It is thus required by all algae that obtain nitrogen by either of these processes. *Chlorella pyrenoidosa*, *Scenedesmus obliquus*, and *Anabena cylindrica* show no detectable Mo requirements when grown on

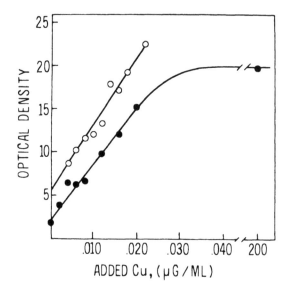

FIGURE 27. Cell yield vs. copper concentration for *Chlorella* grown in glucose-urea-EDTA-salts and glucose-nitrate-EDTA-salts media deficient only in copper. (From Walker, J. B., *Arch. Biochem. Biophys.*, 46, 1, 1983. With permission.)

reduced nitrogen.[17] A clear requirement for Mo in *Chlorella pyrenoidosa* in a glucose-nitrate-EDTA-salts medium was shown by Walker.[199] When urea was used as the sole nitrogen source or was present in the medium together with nitrate, however, no molybdenum requirement could be detected (Figure 27).[199]

Vanadium was clearly shown by Arnon and Wessel[219] to be essential for the growth of *Scenedesmus obliquus*. Figure 28 shows an experiment in which four parallel cultures were grown in a medium lacking vanadium for 3 days until a deficiency was apparent. Then vanadium was added to two of the cultures and the change in their growth was followed. Of 16 other elements (titanium, chromium, tungsten, aluminum, arsenic, cadmium, strontium, mercury, lead, lithium, rubidium, bromine, iodine, fluorine, selenium, and beryllium) added to the vanadium-deficient cultures, none could substitute for vanadium.

It is worth noting that while vanadium can replace molybdenum as a catalyst in nitrogen fixation by *Azotobacter* and *Clostridium*, vanadium also seems to have specific effects distinct from those of molybdenum. The minimum molybdenum requirement of *Scenedesmus* was low: approximately 1200 atm of molybdenum for a single cell. Growth was not improved at concentrations greater than 0.1 μg molybdenum per liter of nutrient solution. With vanadium, increasing initial rates of growth were observed with concentrations up to 100 μg vanadium per liter. Under conditions of molybdenum deficiency, *Scenedesmus* showed a marked decline in chlorophyll concentration, whereas with vanadium-deficient algae the decrease in chlorophyll was much less pronounced.[219]

Vanadium is also essential for marine algae. Fries[220] reported that 10 μg V/ℓ enhanced the fresh weight of *Fucus spiralis* and *Enteromorpha compressa* by 400 and 90%, respectively.

Cobalt or the Co-containing vitamin B_{12} is required by several algae. Co was shown to be a required element in the Cyanobacteria *Nostoc muscorum*, *Calothrix parietina*, *Coccochloris peniocystis*, and *Diplocystis aeruginosa*, in that its omission from the growth medium resulted in a 20 to 75% reduction in growth.[204] A role of Co other than in vitamin B_{12} which functions in the transfer of methyl groups has not been demonstrated.

Low concentrations of Ni are toxic to several algal species. The effect of nickel on the growth of the freshwater diatom *Navicula Pelliculosa* was studied by Fezy et al.,[221] who

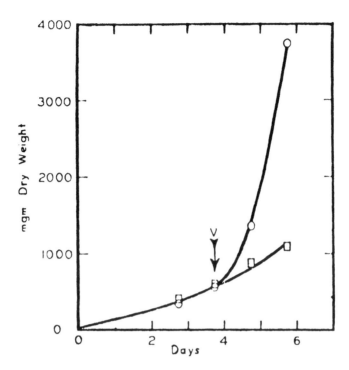

FIGURE 28. Effect of adding vanadium on the growth of vanadium-deficient *Scenedesmus obliquus* cultures.[219] At the time indicated by arrow, 20 μg vanadium (as ammonium vanadate) was added per liter of nutrient solution. Growth is expressed in milligrams dry weight per liter of nutrient solution. 1 ppm iron was supplied as ferric chloride. (Reprinted by permission from *Nature (London)*, 4388, 1039, copyright(c) 1953, Macmillan Journals Limited.)

observed that Ni concentrations as low as 100 μg/ℓ reduced growth by some 50%. Yet Van Baalen and O'Donell[222] isolated an *Oscillatoria* sp. that had an absolute requirement for Ni. Rees and Bekheet[223] found that Ni was required for urease synthesis by *Phaeodactylum tricornutum* and *Tetraselmis subcordiformis* as well as for growth on urea by *Phaeodactylum*. Neither copper nor cobalt could substitute for Ni, but cobalt partially restored urease activity in *Phaeodactylum*. The addition of Ni to nickel-deficient cultures of *Phaeodactylum* or *Tetraselmis* resulted in a rapid increase of urease activity (up to 30 times the normal level), which was not inhibited by cycloheximide.

Silicon

As summarized by Healey,[17] Si is present in the cell walls of members of several divisions of algae, but its metabolism has been extensively studied only in diatoms, where it is the major component of the cell wall. The only form of Si that can be taken up by diatoms appears to be orthosilicis acid (H_4SiO_4). Other Si compounds tested are utilizable only to the extent that they are hydrolyzed to soluble silicate. Germanium competitively inhibits the use of Si by diatoms, a molar ratio of Ge/Si of 0.1 or more inhibiting the growth of *Navicula pelliculosa* and of 0.2 or more that of *Cylindrotheca fusiformis*.[224]

Silicon uptake by both *Navicula pelliculosa* and *Nitzschia alba* is particularly sensitive to washing of the cells. In the former, high rates of Si uptake can be restored by adding reduced S compounds, and the uptake is sensitive to Cd, implicating a role for sulfhydryl groups in Si uptake.[225] In the latter case, thiol compounds are not effective in restoring Si uptake, but glutamate, glutamine, and aspartate are effective.[226]

The uptake of Si is energy dependent, as shown by its sensitivity to dinitrophenol and other respiratory inhibitors[227] and by the transients in nucleotide triphosphates and phosphate uptake induced by the addition of silicate to Si-deficient *N. pelliculosa*.[228,229] Silicon uptake appears to be largely confined to that part of the cell division cycle immediately preceding cell separation, during which the cell wall is laid down. The abruptness with which Si uptake begins and ends during this period suggests that the transport process is closely linked with wall deposition.[228,229] Darley[224] reported a specific requirement for silicon in the net synthesis of DNA in *Cylindrotheca fusiformis*.

Selenium

Selenium plays a role which is not yet clear in algal nutrition and toxicity. Moede et al.[230] reported on the inhibition by selenium of the growth of *Scenedesmus dimorphus* and *Anabaena cylindrica*. An interesting aspect of the role of selenium was reported by Patrick et al.[231] who found that changes in the selenium concentration in experimental streams resulted in altered abundances of the indigenous algal populations. Addition of selenate caused an increased abundance of Cyanobacteria and decreased amounts of diatoms. In a eutrophic lake the biomass of Cyanobacteria was correlated with the concentration of total dissolved selenium while that of other groups (green algae, diatoms, and flagellates) did not show a similar relationship.[232]

Pintner and Provasoli[233] reported that $1.25 \cdot 10^{-7} M$ SeO_3^{2-} stimulated the growth of some marine chrysomonads; another study demonstrated that selenium is absolutely necessary for *Peridinium cinctum*, a fresh water dinoflagellate, and also for some diatoms.[234]

General Responses

From the viewpoint of commercial algal production, it is obviously important that even a minor nutrient deficiency shall not prevail in the growth medium. The generalizations suggested by Healey,[17] who recognized three general patterns of response to nutrient deficiency, are instructive in this respect. One general response of algae to nutrient deficiency is a decrease in the content of photosynthetic pigments. As reported by Healey,[17] a decreased chlorophyll content has been reported in a variety of algae as a result of deficiencies in N,[235-237] P,[238,239] S,[239,240] Si,[241,242] Mg,[239,243] Fe,[207,239] K,[23,239] and Mo. In Cyanobacteria and red algae, the photosynthetically active biliproteins are also lost in response to various nutrient deficiencies.[7,207,235,244,245] Fucoxanthin, a major photosynthetic pigment of the diatom *Navicula pelliculosa*, is as sensitive to Si deficiency as is chlorophyll-a.[246]

The second rather general response to nutrient deficiency which is characteristic of many algal species is the accumulation of storage compounds such as starch, lipids, and protein. Accumulation of either or both carbohydrate and lipid serves as a sink for excess photosynthate formed when the synthesis of structural components or cell division is prevented by any of several nutrient deficiencies.

Finally, a general response to nutrient deficiency is a decrease in protein and nucleic acid synthesis, which results in a decrease in their quantities in the cell. The decrease in the relative protein and nucleic acid content is partly apparent, due to the accumulation of carbohydrate or lipid, but is nevertheless real, in view of the decreased rates of their synthesis.[17]

VITAMINS AND GROWTH REGULATORS

Vitamins

The nutritional requirements of algae are similar to those of higher plants, except that many algae require vitamins for full expression of their growth potential. Provasoli and Carlucci[247] proposed that the delay in recognition of auxotrophy in algae was mostly due to the success in the 1890s of Beijerinck, Molisch, and Miguel in culturing a few freshwater green algae and diatoms in inorganic media. It became clear that these two algal groups are outstanding in comprising a large, perhaps a preponderant, number of species that do not require vitamins. The belief that algae are autotrophic was furthered by the use of mineral media in isolating algae, thus selecting against vitamin-requiring species; however, many algae cannot be grown on mineral media. Pringsheim,[248] who pioneered the field, introduced the addition of soil water and soil extracts to cultures for growing the more demanding free-living algae. This addition, which contains nitrates, ammonia, vitamins, an array of bacterial metabolites, and trace metals chelated by humic acids, proved very successful.

Today, numerous algal species are known to be auxotrophic. Nevertheless, the auxotrophic algae are unusual among microorganisms, for they respond to only three vitamins. Vitamin B_{12} and thiamin are required alone or in combination by the majority of the auxotrophic algae, and B_{12} seems required more often than thiamin. Biotin has so far been shown to be necessary for a few chrysomonads and dinoflagellates and one euglenoid.[247]

Incidence of auxotrophy in algal groups reveals differences in their requirements for vitamins. Members of the Cyanophyceae, Chlorophyceae, Xantophyceae, and Phaeophyceae have the least number of species requiring vitamins. The predominance of autotrophs in these algal groups seems undisputed, even taking into consideration that the information available may not be completely reliable. To avoid bias in future studies, isolation of species from natural habitats should be done with media containing vitamins. The clones obtained in pure cultures should then be tested for autotrophy and the need for single vitamins, performing at least three serial transfers in the same type of medium to exhaust reserves and carry-over.[247]

Plant Hormones

Many observations suggest that plant hormones or other growth regulators may operate in algae. Nevertheless, few tests of the physiological roles of hormones belonging to the three major groups of plant growth regulators, auxins, cytokinins, and gibberellins, have been done. Two approaches have been used to detect growth substances in algae: (1) extraction, characterization by chromatography, and plant bioassay, and (2) testing the physiological response of algae to the application of growth substances. From numerous reports, only a general concept may be formulated, i.e., some favorable effects on cell division, cell size, and growth may be obtained in certain species of microalgae and under certain growth conditions, but no obligatory requirements for plant growth regulators have been recorded.

Nevertheless, from the viewpoint of mass cultivation, it may be economically feasible to add a variety of natural growth stimulators. A good example is 1-triacontanol (TRIA), a naturally occurring wax component of many plant species. TRIA is a saturated straight-chain primary alcohol with 30 carbon atoms[249] which increases the growth and yield of several field crops.[250]

Haugstad et al.[251] studied the effects of triacontanol on growth, photosynthesis, and photorespiration in *Chlamydomonas reinhardtii*, which exhibits C_3 properties and *Anacystis nidulans*, which exhibits C_4 properties. *Chlamydomonas* cultures increased significantly in cell number after 4 days and in chlorophyll content after 3 days of treatment with 2.3×10^{-8} M TRIA in chloroform/Tween®-20. In cultures of *Anacystis* the chlorophyll content became significantly higher 3 days after treatment with 2.3×10^{-9} M TRIA and the cell number was noticeably higher than in the controls.

CO_2 uptake by TRIA-treated *Chlamydomonas* cultures was about the same in both 2 and 21% O_2, and the inhibition by O_2 was significantly less than in the controls. Photosynthesis in *Anacystis* was O_2-insensitive with TRIA, and no changes were observed in the rate of O_2 uptake or in the O_2 sensitivity of CO_2 uptake. The authors concluded that TRIA affects some

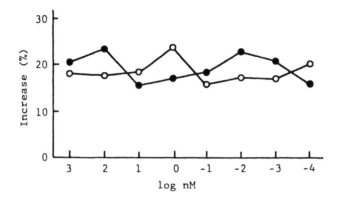

FIGURE 29.   Growth promoting activity of adenosine (●) and 2'-deox-
yadenosine (○).[252] The numbers of the increase indicate the percentage
increase of dry cell weight compared with that of the control. (From
Komoda, Y., et al., *Chem. Pharm. Bull.*, 31(10), 3771, 1983. With
permission.)

process that regulates the balance between photosynthesis and photorespiration, and that
other processes that result in increased growth are probably also affected.

Another example of a growth-promoting effect that has the potential to be of practical
significance is the effect of adenosine and 2'-deoxyadenosine, both of which promote growth
in *P. tricornutum* even at extremely low concentrations of 10^{-4} nM (Figure 29).[252]

REFERENCES

1. **Lwoff, A.,** L'Evolution Physiologique, *Etude des Pertes de Fonctions ches les Microorganismes,* Hermann.
   Paris, 1943.
2. **Lwoff, A.,** Introduction to biochemistry of protozoa, in *Biochemistry and Physiology of Protozoa,* Vol. 1.
   Lwoff, A., Ed., Academic Press, New York, 1951, 1.
3. **Pfeffer, W.,** *Handbuch der Pflanzenphysiologie,* Leipzig.
4. **Stanier, R. Y.,** Autotrophy and heterotrophy in unicellular blue-green algae, in *The Biology of Blue-Green
   algae,* Carr, N. G. and Whitton, B. A., Eds., University of California Press, Berkeley, 1973, 501.
5. **Nelson, A. H. and Lewin, R. H.,** The uptake and utilization of organic carbon by algae: an essay in
   comparative biochemistry, *Phycologia,* 13, 227, 1974.
6. **Pringsheim, E. G.,** Heterotrophic bei Algen und Flagellaten, in *Encyclopedia of Plant Physiology, XI.*
   Ruhland, W., Ed., Springer-Verlag, Berlin, 1959, 303.
7. **Oswald, W. J., and Gotaas, H. B.,** Photosynthesis in sewage treatment, J. Sanit. Eng. Div., *81, 1, 1955.*
8. **Hendricks, D. W. and Pote, W. D.,** Thermodynamic analysis of a primary oxidation pond. *J. Water
   Pollut. Cont. Fed.,* 46, 333, 1974.
9. **Hummenik, F. J. and Hanna, G. P., Jr.,** Algal-bacterial symbiosis for removal and conservation of
   wastewater nutrients, *J. Water Pollut. Cont. Fed.,* 43, 580, 1971.
10. **Hamburger, B.,** Bakterin Symbioses bei *volvox aurens* Ehrenberg. *Arch. Microbiol.,* 29, 291, 1958.
11. **Lang, W.,** Effect of carbohydrate on the symbiotic growth of planktonic blue-green algae with bacteria.
    Nature (London), 215, 1277, 1967.
12. **Miller, J. D. A.,** An extra-cellular enzyme produced by Monodus, *Br. Phycol. Bull.,* 7, 22, 1959.
13. **Pringsheim, E. G.,** On the nutrition of *Ochromonas, Q. J. Microsc. Sci.,* 93, 71, 1952.
14. **Aaronson, S.,** Digestion in phytoflagellates, in *Lysosomes in Biology and Pathology,* Vol. 3, Dingle. J.
    T., Ed., North-Holland, Amsterdam, 1973, 18.
15. **Aaronson, S.,** *Experimental Microbiological Ecology,* Academic Press, New York, 1970.
16. **Soeder, C. J., Schulze, G., and Thiele, D.,** Einfluss verschiedener kulturbedingungen auf das wachstrum
    in synchronkulturen von *Chlorella fusca* SH et Kr. *Arch. Hydrobiol.,* 33(Suppl.). 127, 1967.
17. **Healey, F. P.,** Inorganic nutrient uptake and deficiency in algae, *CRC Crit. Rev. Microbiol.,* 3, 69, 1973.

18. **O'Brien, W. J.,** Limiting factors in phytoplankton algae: their meaning and measurement, *Science,* 128, 616, 1972.
19. **Jones, R. F.,** in *Conference on Marine Biology: Proceedings of the Second International Interdisciplinary Conference,* Oppenheimer, C. H., Ed., New York Academy of Sciences, New York, 1966.
20. **Monod, J.,** *Recherches sur la Croissance des Cultures Bacterlenns,* Herman, Paris, 1942.
21. **Goldman, J. C., Oswald, W. J., and Jenkins, D.,** The kinetics of inorganic carbon limited algal growth, *J. Water Pollut. Cont. Fed.,* 46(3), 553, 1974.
22. **Davis, A. G.,** Iron chelation and the growth of marine phytoplankton. I. Growth kinetics and chlorophyll production in cultures of the euryhaline flagellate *Dunaliella tertiolecta* under iron limiting conditions, *J. Mar. Biol. Assoc. U.K.,* 50, 65, 1970.
23. **Grim, J.,** Der phosphor und die pflanzlich produktion in bodensee das gasu, *Wasserfach,* 108, 1261, 1967.
24. **Kuhl, A.,** Phosphate metabolism of green algae, in *Algae, Man and the Environment,* Jackson, D. F., Ed., Syracuse, N.Y., 1968, 37.
25. **Pirson, A. and Badour, S. S. A.,** Kennzeichnung von Mineralsalzman gelzustanden bei Grunalgen mit analytisch-chemischer Methodik. I. Kohlinhydratspiegel, organischer Stickstoff und Chlorophyll bei Kalimangel im Vergleich mit Magnesium-und Manganmangel, *Flora,* 150, 243, 1961.
26. **Soeder, C. J., Muller, H., Payer, H. D., and Schulle, H.,** Mineral nutrition of planktonic algae: some considerations, some experiments, *Mitt. Int. Verein Limnol.,* 19, 39, 1971.
27. **Droop, M. R.,** The Nutrient Control of Algal Growth, Scottish Marine Biological Association, Oban, Scotland, 1974.
28. **Pirson, A. and Lorenzen, H.,** Synchronized dividing algae, *Ann. Rev. Plant Physiol.,* 14, 439, 1966.
29. **Tamiya, H.,** Synchronous cultures of algae, *Ann. Rev. Plant Physiol.,* 17, 1, 1966.
30. **Soeder, C. J. and Thiele, D.,** Wirkungen des calcium mangels auf *Chlorella fusca* Sh. et Kr., *Z. Pflanzenphysiol.,* 57, 339, 1967.
31. **Bader, F. G.,** Analysis of double-substrate limited growth, *Biotechnol. Bioeng.,* 20, 183, 1978.
32. **Young, T. C. and King, D. L.,** Interacting limits to algal growth: light, phosphorus, and carbon dioxide availability, *Water Res.,* 14, 409, 1980.
33. **Stewart, W. D. P. and Pearson, H. W.,** Effects of aerobic and anaerobic conditions on growth and metabolism of blue-green algae, *Proc. R. Soc. London Ser. B:,* 175, 293, 1970.
34. **Kessler, E.,** Hydrogenase, photoreduction and anaerobic growth, in *Algal Physiology and Biochemistry,* Botanical Monographs, Vol. 10, Stewart, W. D. P., Ed., University of California Press, Berkeley, 1974, 456.
35. **Lee, J. K. H. and Stiller, M.,** Hydrogenase activity in cell-free preparation of Chlorella, *Biochim. Biophys. Acta,* 132, 503, 1967.
36. **Kessler, E. and Maifarth, M.,** Vorkommen und Leistugnsfahigkeit von Hydrogense bei einigen Grunalgen, *Arch. Mikrobiol.,* 37, 215, 1960.
37. **Gaffron, H.,** Reduction of carbon dioxide coupled with the oxyhydrogen reaction in algae, *J. Gen. Physiol.,* 26, 241, 1972.
38. **Miller, A. G. and Colman, B.,** Active transport and accumulation of bicarbonate by a unicellular cyanobacterium, *J. Bacteriol.,* September, 1253, 1980.
39. **Miller, A. G. and Colman, B.,** Evidence for HCO_3^- transport by the blue-green alga (cyanobacterium) *Coccochloris peniocystis, Plant Physiol.,* 65, 397, 1980.
40. **Badger, M. R., Kaplan, A., and Berry, J. A.,** A mechanism for concentrating CO_2 in *Chlamydomonas reinhardti* and *Anabaena variabilis* and its role in photosynthetic CO_2 fixation, *Carnegie Inst. Yearb.,* 78, 251, 1978.
41. **Tsuzuki, M.,** Mode of HCO_3^- utilization by the cells of *Chlamydomonas reinhardtii* grown under ordinary air, *Z. Pflanzenphysiol.,* 110(1), 29, 1983.
42. **Ingle, R. K. and Colman, B.,** Carbonic anhydrase levels in blue-green algae, *Can. J. Bot.,* 53, 2385, 1975.
43. **Cook, J. R.,** Photoassimilation of acetate by an obligately phototrophic strain of *Euglena gracilis, J. Protozool.,* 14, 382, 1967.
44. **Cook, J. R.,** The cultivation and growth of *Euglena,* in *The Biology of Euglena,* Vol. 1, Buetow, D. E., Ed., Academic Press, New York, 1968, 244.
45. **Killan, A. and Myers, J.,** A special effect on the growth of *Chlorella vulgaris, Am. J. Bot.,* 43, 569, 1956.
46. **Samejima, H. and Myers, J.,** On the Heterotrophic growth of *Chlorella pyrenoidosa, J. Gen. Microbiol.,* 18, 107, 1958.
47. **Milner, H. W.,** The fatty acids of *Chlorella, J. Biol. Chem.,* 176, 813, 1948.
48. **Nakayama, O., Ueno, T., and Tsuchiya, F.,** Heterotrophic culture of algae in a closed system, *J. Ferment. Technol.,* 52(1), 225, 1974.
49. **Fogg, G. E. and Collyer, D. M.,** The accumulation of lipids by algae, in *Algal Culture: From Laboratory to Pilot Plant,* Publ. no. 600, Burlew, J. S., Ed., Carnegie Institution, Washington, D.C., 1953, 177.

50. **Vaccaro, R. F.,** Inorganic nitrogen in sea-water, in *Chemical Oceanography,* Vol. 1, Riley, J. P. and Skirrow, G., Eds., Academic Press, New York, 1965, 365.

51. **Syrett, P. J.,** Nitrogen metabolism of microalgae, *Can. Bull. Fish. Aquat. Sci.,* 210, 182, 1981.

52. **Morris, I.,** Nitrogen assimilation and protein synthesis, in *Algal Physiology and Biochemistry,* Stewart, W. D. P., Ed., Blackwell Scientific, Oxford, 1974, 583.

53. **Thomas, W. H.,** Nutrient requirements and utilizations; algae, in *Metabolism,* Biological Handbooks, Altman, P. T. and Dittmer, D. S., Eds., Federation of American Societies for Experimental Biology, Bethesda, Md., 1968.

54. **Wheeler, P. A., North, B. B., and Stephens, G. C.,** Amino acid uptake by marine phytoplankters, *Limnol. Oceanogr.,* 19, 249, 1974.

55. **Neilson, A. H. and Larson, T.,** The utilization of organic nitrogen for growth of algae: physiological aspects, *Physiol. Plant.,* 48, 542, 1980.

56. **Van Baalen, C.,** Studies on the marine blue-green algae, *Bot. Mar.,* 4, 129, 1962.

57. **Van Baalen, C. and Marler, J. E.,** Characteristics of marine blue-green algae with uric acid as nitrogen source, *J. Gen. Microbiol.,* 32, 457, 1963.

58. **Antia, N. J., Berland, B. R., Bonin, D. J., and Meastrini, S. Y.,** Allantoin as nitrogen source for growth of marine benthic microalgae, *Phycologia,* 19, 103, 1980.

59. **Prasad, P. V. D.,** Hypoxanthine and allantoin as nitrogen sources for the growth of some freshwater green algae, *New Phytol.,* 93, 575, 1983.

60. **Stewart, W. D. P., Rowell, P., Ladha, J. K., and Sampaio, M. J. A. M.,** Blue-green algae (cyanobacteria) — some aspects related to their role as sources of fixed nitrogen in paddy soils, in *Nitrogen and Rice,* International Rice Research Institute, Los Banos, Laguna, Philippines, 1979.

61. **Bothe, H.,** Nitrogen fixation, in *The Biology of Cyanobacteria,* Carr, N. G. and Whitton, B. A., Eds., Blackwell Scientific, Oxford, 1982, 87.

62. **Haselkorn, R.,** Heterocysts, *Ann. Rev. Plant Physiol.,* 29, 319, 1978.

63. **Wolk, C. P.,** Heterocysts, in *The Biology of Cyanobacteria,* Carr, N. G. and Whitton, B. A., Eds., Blackwell Scientific, Oxford, 1982, 359.

64. **Bothe, H., Yates, M. G., and Cannon, F. C.,** Nitrogen fixation, in *Encyclopedia of Plant Physiology,* New Series, Lauchli, A. and Bieleski, R., Eds., Springer-Verlag, Berlin, 1982.

65. **Stewart, W. D. P.,** Some aspects of structure and function in N_2-fixing cyanobacteria, *Ann. Rev. Microbiol.,* 34, 497, 1980.

66. **Walk, C. P.,** Movement of carbon from vegetative cells to heterocysts in *Anabaena cylindrica, J. Bacteriol.,* 96, 2138, 1968.

67. **Beevers, L. and Hageman, R. H.,** Nitrate reduction in higher plants, *Ann. Rev. Plant Physiol.,* 20, 495, 1969.

68. **Hewitt, E. J.,** Assimilatory nitrate-nitrite reduction, *Ann. Rev. Plant Physiol.,* 26, 73, 1975.

69. **Hattori, A. and Myers, J.,** Reduction of nitrate and nitrite by subcellular preparations of *Anabaena cylindrica.* I. Reduction of nitrite to ammonia, *Plant Physiol. (Lancaster),* 41, 1031, 1966.

70. **Hattori, A. and Myers, J.,** Reduction of nitrate and nitrite by subcellular preparations of *Anabaena cylindrica.* II. Reduction of nitrate to nitrite, *Plant Cell Physiol. (Tokyo),* 8, 327, 1967.

71. **Zumft, W. G., Paneque, A., Aparichio, P. J., and Losada, M.,** Mechanism of nitrate reduction in *chlorella, Biochem. Biophys. Res. Commun.,* 36, 980, 1969.

72. **Aparichio, P. J., Cardenas, J., Zumft, W. G., Vega, J. M., Herrera, J., Paneque, A., and Losada, M.,** Molybdenum and iron as constituents of the enzymes of the nitrate reducing system from *chlorella, Phytochemistry,* 10, 1487, 1971.

73. **de la Rosa, M. A., Diez, J., Vega, J. M., and Losada, M.,** Purification and properties of assmililatory nitrate reductase [NAD(P)H] from *Ankistrodesmus braunii, Eur. J. Biochem.,* 106, 249, 1980.

74. **Nicholas, G. L., Shehata, S. A. M., and Syrett, P. J.,** Nitrate reductase deficient mutants of *Chlamydomonas reinhardii,* biochemical characteristics, *J. Gen. Microbiol.,* 108, 79, 1978.

75. **Solomonson, L. P.,** Structure of *chlorella* nitrate reductase, in *Nitrogen Assimilation of Plants,* Hewitt, E. J. and Curring, C. U., Eds., Academic Press, New York, 1979, 199.

76. **Giri, L. and Ramadoss, C. S.,** Physical studies on assimilatory nitrate reductase from *Chlorella vulgaris, J. Biol. Chem.,* 254, 11703, 1979.

77. **Losada, M. and Gurrero, M. G.,** The photosynthetic reduction of nitrate and its regulation, in *Photosynthesis in Relation to Model Systems,* Barber, J., Ed., Elsevier/North Holland, Amsterdam, 1979, 366.

78. **Hattori, A. and Uesugi, I.,** Purification and properties of nitrite reductase from the blue-green alga *Anabaena cylindrica, Plant Cell. Physiol.,* 9, 689, 1968.

79. **Grant, B. R.,** Nitrite reductase in *Dunaliella tertiolecta,* isolation and properties, *Plant Cell Physiol.,* 11, 55, 1970.

80. **Eppley, R. W. and Rogers, J. N.,** Inorganic nitrogen assimilation of *ditylum brightwellii, J. Phycol.,* 6, 344, 1970.

81. **Zumft, W. G.,** Ferredoxin: nitrite oxidoreductase from *chlorella*. Purification and properties, *Biochim. Biophys. Acta,* 276, 363, 1972.
82. **Syrett, P. J. and Morris, I.,** The inhibition of nitrate assimilation by ammonium in *chlorella, Biochim. Biophys. Acta,* 67, 566, 1963.
83. **Losada, M., Paneque, A., Aparicio, P. J., Vega, J. M., Cardenas, J., and Herrera, J.,** Inactivation and repression by ammonium of the nitrate reducing system in *chlorella, Biochem. Biophys. Res. Commun.,* 38, 1009, 1970.
84. **Pistorius, E. K., Funkhouser, E. A., and Voss, H.,** Effect of ammonium and ferricyanide on nitrate utilization by *chlorella vulgaris, Planta,* 141, 279, 1978.
85. **Eppley, R. W., Coatsworth, J. L., and Solorzano, L.,** Studies on nitrate reductase in marine phytoplankton, *Limnol. Oceanogr.,* 14, 194, 1969.
86. **Thacker, A. and Syrett, P. J.,** The assimilation of nitrate and ammonium by *Chlamydomonas reinhardii, New Phytol.,* 71, 435, 1972.
87. **Flores, E., Guerrero, M. G., and Losada, M.,** Short term ammonium inhibition of nitrate utilization by *Anacystis nidulans* and other cyanobacteria, *Arch. Microbiol.,* 128, 137, 1980.
88. **Stewart, W. D. P., Haystead, A., and Dharmawardene, M. W. N.,** Nitrogen assimilation and metabolism in blue-green algae, in *Nitrogen Fixation by Free-Living Microorganisms,* Stewart, W. D. P., Ed., Cambridge University Press, New York, 1975.
89. **Meeks, J. C., Wycoff, K. L., Chapman, J. S., and Enderlin, C. S.,** Regulation of expression of nitrate and dinitrogen assimilation by *Anabaena* species, *Appl. Environ. Microbiol.,* 45(4), 1351, 1983.
90. **Syrett, P. J.,** Nitrogen assimilation, in *Physiology and Biochemistry of Algae,* Lewin, R. A., Ed., Academic Press, New York, 1962, 171.
91. **Edge, P. A. and Rickets, T. R.,** The effect of nitrogen refeeding on the carbohydrate content into nitrogen starved cells of *Platymonas striata* Butcher, *Planta,* 136, 159, 1977.
92. **Brown, C. M., MacDonald-Brown, D. S., and Meers, J. L.,** Physiological aspects of microbial inorganic nitrogen metabolism, *Adv. Microb. Physiol.,* 11, 1, 1974.
93. **Tempest, D. W., Meers, J. L., and Brown, C. M.,** Synthesis of glutamate in *Acrobacter aerogenes* by a hitherto unknown route, *Biochem. J.,* 117, 405, 1970.
94. **Miflin, B. J. and Lea, P. J.,** The pathway of nitrogen assimilation in plants, *Phytochemistry,* 15, 873, 1976.
95. **Dharmawardene, M. W. N., Haystead, A., and Stewart, W. P. D.,** Glutamine synthetase of the nitrogen fixing alga *Anabaena cylindrica, Arch. Microbiol.,* 90, 281, 1973.
96. **Ramos, J. L., Flores, E., and Guerrero, M. G.,** Glutamine synthetase glutamate synthase: the pathway of ammonium in *Anacystis nidulans, Conference Proceedings II,* Congress of Federation of European Societies of Plant Physiology, Santiago, Chile, 1980, 579.
97. **Ramos, J. L., Guerrero, M. G., and Losada, M.,** Photosynthetic production of ammonia by blue-green algae, *Conference Proceedings II,* Congress of Federation of European Societies of Plant Physiology, Santiago, Chile, 1980, 581.
98. **Meeks, J. C., Wolk, C. P., Lockau, W., Schilling, N., Shaffer, P. W., and Chien, W. S.,** Pathways of assimilation of $(_{13}N)N^2$ and $_{13}NH^4_+$ by cyanobacteria with and without heterocysts, *J. Bacteriol.,* 134, 125, 1978.
99. **Rigano, V. D. M., Vona, V., Fuggi, A., and Rigano, C.,** Effect of L-methionine-DL-sulphoximine, a specific inhibitor of glutamine synthetase, on ammonium and nitrate metabolism in the unicellular alga *Cyanidium caldarium, Physiol. Plant,* 54, 47, 1982.
100. **Dohler, G. and Roblenbroich, H. J.,** Photosynthetic assimilation of $_{15}$N-ammonia and $_{15}$N-nitrate in the marine diatoms *Bellerochea yucatanensis* (von Stosch) and *Skeletonema costatum, Z. Naturforsch.,* 36c, 834, 1981.
101. **Neilson, A. H. and Dondroff, M.,** Ammonia assimilation in blue-green algae, *Arch. Mikrobiol.,* 89, 15, 1973.
102. **Kates, J. R. and Jones, R. F.,** Variation in alanine dehydrogenase and glutamate dehydrogenase during the synchronous development of *Chlamidomonas, Biochem. Biophys. Acta,* 86, 438, 1964.
103. **Allison, R. K., Skipper, H. E., Ried, M. R., Short, W. A., and Hogan, G. L.,** Studies on the photosynthetic reaction. II. Sodium formate and urea feeding experiments with *Nostoc muscorum, Plant Physiol.,* 29, 164, 1954.
104. **Hattori, A.,** Studies on the metabolism of urea and other nitrogenous compounds in *Chlorella ellipsoidea.* III. Assimilation of urea, *Plant Cell Physiol.,* 1, 107, 1960.
105. **Hodson, R. C. and Thompson, J. F.,** Metabolism of urea by *Chlorella vulgaris, Plant Physiol.,* 44, 691, 1969.
106. **Little, L. W. and Mah, R. A.,** Ammonia production in urea grown cultures of *Chlorella ellipsoidea, J. Phycol.,* 6, 277, 1970.
107. **Leftley, J. W. and Syrett, P. J.,** Ureas and ATP: urea amydolase activity in unicellular algae, *J. Gen. Microbiol.,* 77, 109, 1973.

108. **Bekheit, I. A. and Syrett, P. J.,** Urea degrading enzymes in algae, *Br. Phycol. J.,* 12, 137, 1977.
109. **Berns, D. S., Holohan, R., and Scott, E.,** Urease activity in blue green algae, *Science,* 152, 1077, 1966.
110. **North, B. B. and Stephens, G. C.,** Amino acid transport in *Nitzchia ovalis,* Arnott. *J. Phycol.,* 8, 64, 1972.
111. **Hellebust, J. A.,** The uptake and utilization of organic substances by marine phytoplankters, in Symp. Organic Matter in Natural Waters, Publ. No. 1, Hood, D. W., Ed., Institute of Marine Science, University of Alaska, Occass, 1970.
112. **Kirk, D. L. and Kirk, M. M.,** Amino acid and urea uptake in ten species of Chlorophyta, *J. Phycol.,* 14, 198, 1978.
113. **Wheeler, P. A. and Stephens, G. C.,** Metabolic segregation of intracellular free amino acids in *Platymonas* (Chlorophyta), *J. Phycol.,* 13, 193, 1977.
114. **Ammann, E. C. B. and Lynch, V. H.,** Purine metabolism by unicellular algae. II. Adenine, hypoxanthine, and xanthine degradation by *Chlorella pyrenoidosa, Biochem. Biophys. Acta,* 87, 370, 1964.
115. **Fogg, G. E.,** Nitrogen nutrition and metabolic patterns in algae, *Symp. Soc. Exp. Biol.,* 13, 106, 1959.
116. **Dawes, E. A.,** Endogenous metabolism and the survival of starved prokaryotes, *Symp. Gen. Microbiol.,* 26, 19, 1976.
117. **Fogg, G. E.,** *Algal Culture and Phytoplankton Ecology,* University of Wisconsin Press, Madison, 1966.
118. **Simon, R. D.,** Cyanophycin granules from the blue-green alga *Anabaena cylindrica:* a reserve material consisting of copolymers of aspartic and arginine, *Proc. Natl. Acad. Sci. U.S.A.,* 68, 265, 1971.
119. **Simon, R. D.,** The effect of chloramphenicol on the production of cyanophycin granule polypeptide in the blue-green alga *Anabaena Cylindrica, Arch. Microbiol.,* 92, 115, 1973.
120. **Simon, R. D.,** Measurement of the cyanophycin granule peptide contained in the blue-green alga *Anabaena cylindrica, J. Bacteriol.,* 114, 1213, 1973.
121. **Simon, R. D.,** The biosynthesis of multi-L-arginyl poly (L-aspartic acid) in the filamentous cyanobacterium *Anabaena cylindrica, Biochem. Biophys. Acta,* 422, 407, 1976.
122. **Lang, N. J., Simon, R. D., and Wolk, C. P.,** Correspondence of cyanophycin granules with structured granules in *Anabaena cylindrica, Arch. Mikrobiol.,* 83, 313, 1972.
123. **Simon, R. D. and Weathers, P. J.,** Determination of the structure of the novel polypeptide containing aspartic acid and arginine which is found in cyanobacteria, *Biochem. Biophys. Acta,* 420, 165, 1976.
124. **Allen, M. M. and Weathers, P. J.,** Structure and composition of cyanophycin granules in the cyanobacterium *Aphanocapsa,* 6308, *J. Bacteriol.,* 114, 959, 1980.
125. **Allen, M. M., Hutchinson, F., and Weathers, P. J.,** Cyanophycin granule polypeptide (CGP) formation and degradation in the cyanobacterium, *Aphanocapsa,* 6308, *J. Bacteriol.,* 114, 687, 1980.
126. **Allen, M. M. and Hutchinson, F.,** Nitrogen limitation and recovery in the cyanobacterium *Aphanocapsa* 6308, *Arch. Microbiol.,* 128, 1, 1980.
127. **Allen, M. M. and Smith, A. J.,** Nitrogen chlorosis in blue-green algae, *Arch. Microbiol.,* 69, 114, 1969.
128. **Lau, R. H., Mackenzie, M. M., and Coolittle, W. F.,** Phycocyanin synthesis and degradation in the blue-green bacterium Anacystis nidulans, Lemm. *Arch. Microbiol.,* 107, 15, 1976.
129. **Stewart, W. D. P. and Lex, M.,** Nitrogenase activity in the blue-green alga *Plectonema Borynum* strain 594, *Arch. Mikrobiol.,* 73, 250, 1970.
130. **Foulds, I. and Carr, N. G.,** A proteolic enzyme degrading phycocyanin in the cyanobacterium *Anabaena cylindrica, FEMS Microbiol. Lett.,* 2, 117, 1977.
131. **Wood, N. B. and Haselkorn, R.,** Protein degradation during heterocyst development in *Anabaena,* in *Proc. 2nd Int. Symp. Photosynthetic Prokaryotes,* Codd, G. and Stewart, W. D., Eds., Dundee, Scotland, 1976, 125.
132. **Bussiba, S. and Richmond, A. E.,** C-Phycocyanin as a storage protein in the blue-green alga *Spirulina platensis, Arch. Microbiol.,* 125, 143, 1980.
133. **Thomas, W. H. and Dodson, A. N.,** On nitrogen deficiency in tropical Pacific Oceanic phytoplankton. II. Photosynthetic and cellular characteristics of a chemostat-grown diatom, *Limnol. Oceanogr.,* 17, 575, 1972.
134. **Morris, I. and Syrett, P. J.,** The effect of nitrogen starvation on the activity of nitrate reductase and other enzymes in *Chlorella, J. Gen. Microbiol.,* 38, 21, 1965.
135. **Syrett, P. J. and Hipkin, C. R.,** The appearance of nitrate reductase activity in nitrogen-starved cells of Ankistrodesmus braunii, *Planta,* 3, 57, 1973.
136. **Rigano, C. and Violante, U.,** Effect of nitrate, ammonia and nitrogen starvation on the regulation of nitrate reductase in *Cyanidium caldarium, Arch. Mikrobiol.,* 90, 27, 1973.
137. **Hipkin, C. R. and Syrett, P. J.,** Some effects of nitrogen-starvation on nitrogen and carbohydrate metabolism in *Ankistrodesmus braunii, Planta,* 133, 209, 1977.
138. **Eppley, R. W. and Renger, E. H.,** Nitrogen assimilation of an oceanic diatom in nitrogen-limited continuous culture, *J. Phycol.,* 10, 15, 1974.
139. **Cresswell, R. C. and Syrett, P. J.,** Uptake of nitrate by the diatom *Phaeodactylum tricornutum, J. Exp. Bot.,* 32, 19, 1981.

140. **de Vasconcelos, L. and Fay, P.,** Nitrogen metabolism and ultrastructure in *Anabaena cylindrica.* I. The effect of nitrogen starvation, *Arch. Microbiol.,* 96, 271, 1974.
141. **Giesy, R. M.,** A light and electron microsocope study of interlamellar polyglucozide bodies in *Oscillatoria chalybia, Am. J. Bot.,* 51, 388, 1964.
142. **Peat, A. Whitton, B. A.,** Environmental effects on the structure of the blue-green alga *Chloroglea fritschii, Arch. Mikrobiol.,* 57, 155, 1967.
143. **Stevens, S. E., Jr., Balkwill, D. L., and Paone, D. A. M.,** The effects of nitrogen limitation on the ultrastructure of the cyanobacterium Agmenellum quadruplicatum, *Arch. Mikrobiol.,* 130, 204, 1981.
144. **Kessler, E.,** Reduction of nitrate by green algae, *Symp. Soc. Exp. Biol.,* 13, 87, 1959.
145. **Kessler, E. and Zumft, W. G.,** Effect of nitrite and nitrate on chlorophyll fluorescence in green algae, *Planta,* 3, 41, 1973.
146. **Thomas, R. J., Hipkin, C. R., and Syrett, P. J.,** The interaction of nitrogen assimilation with photosynthesis in nitrogen deficient cells of *Chlorella, Planta,* 133, 9, 1976.
147. **Ahmed, J. and Morris, I.,** The effects of 2,4-dinitrophenol and other uncoupling agents on the assimilation of nitrate and nitrite by *Chlorella, Biochem. Biophys. Acta,* 162, 32, 1968.
148. **Tischner, R. and Lorenzen, H.,** Nitrate uptake and nitrate reduction in synchronous *Chlorella, Plant,* 146, 287, 1979.
149. **Grant, B. R.,** The action of light on nitrate and nitrite assimilation by the marine chlorophyte *Dunaliella tertialecta, J. Gen. Microbiol.,* 48, 379, 1967.
150. **Grant, B. R.,** Effect of carbon dioxide concentration and buffer system on nitrate and nitrite assimilation in *Dunaliella tertiolecta, J. Gen. Microbiol.,* 54, 327, 1968.
151. **Grant, B. R. and Turner, I. M.,** Light simulated nitrate and nitrite assimilation in several species of algae, *Comp. Biochem. Physiol.,* 29, 995, 1969.
152. **Syrett, P. J. and Morris, I.,** The inhibition of nitrate assimilation by ammonium in *Chlorella, Biochem. Biophys. Acta,* 67, 566, 1963.
153. **Stewart, W. D. P., Rowell, P., and Apte, S. K.,** Cellular physiology and the ecology of N^2 fixing blue-green algae, in *Recent Developments in Nitrogen Fixation,* Newton, W. E., Postgate, J. R., and Rodriguez-Barrueco, C., Eds., Academic Press, New York, 1977, 287.
154. **Talling, J. F.,** Freshwater algae, in *Physiology and Biochemistry of Algae,* Lewin, A., Ed., Academic Press, New York, 1962, 743.
155. **Healey, F. P. and Hendzel, L. L.,** Physiological indicators of nutrient deficiency in lake phytoplankton, *Can. J. Fish. Aquat. Sci.,* 37, 422, 1980.
156. **Hutchinson, G. E.,** *A Treatise on Limnology,* Vol. 1, John Wiley & Sons, New York, 1957.
157. **Reichardt, W., Overbeck, J., and Steubing, L.,** Free dissolved enzymes in lake waters, *Nature (London),* 216, 1345, 1968.
158. **Overbeck, J. and Babenzien, H. D.,** Uber den Nachweis von freien enzymen in gewasser, *Arch. Hydrobiol.,* 60(1), 107, 1964.
159. **Strickland, J. D. H. and Solorzano, L.,** Determination of monoesterase hydrolysable phosphate and phosphorus esterase activity in seawater, in *Some Contemporary Studies in Marine Science,* Barnes, H., Ed., Georg Allen & Unwin, London, 1966, 665.
160. **Kuhl, A.,** Phosphorus, in *Algal Physiology and Biochemistry,* Stewart, W. D. P., Ed., Blackwell Scientific, Oxford, 1974, 636.
161. **Rubin, P., Zetooney, E., and McGrown, R. E.,** Uptake and utilization of sugar phosphates by Anabaena flos aquae, *Plant Physiol.,* 60, 407, 1977.
162. **Rodhe, W.,** Environmental requirements of freshwater plankton algae, *Symbol. Bot. Ups.,* 10, 1, 1948.
163. **Healey, F. P.,** Characteristics of phosphorus deficiency in Anabaena, *J. Phycol.,* 9, 383, 1973.
164. **Smith, F. A.,** Active phosphate uptake by *Nitella translucens, Biochem. Biophys. Acta,* 126, 94, 1966.
165. **Blum, J. J.,** Phosphate uptake by phosphate starved *Euglena, J. Gen. Physiol.,* 49, 1125, 1966.
166. **Mohleji, S. C. and Verhoff, F. H.,** Sodium and potassium ion effects on phosphorus transport in algal cells, *J. Water Pollut. Control Fed.,* 52(1), 110, 1980.
167. **Niemeyer, R. and Richter, G.,** Seh Nellmarkierte phosphate und Metaphosphate bei der Blaualge Anacystis nudulans, *Arch. Microbiol.,* 69, 52, 1969.
168. **Niemeyer, R. and Richter, G.,** Rapidly labelled polyphosphates in acetabularia, in *Biology and Radiobiology of Anucleate Systems,* Vol. 2, *Plant Cells,* Bonotto, S., Ed., Academic Press, New York, 1972, 225.
169. **Healey, F. P.,** Phosphate, in *The Biology of Cyanobacteria,* Carr, N. G. and Whitton, B. A., Eds., Blackwell Scientific, Oxford, 1982.
170. **Holm-Hunsen, O.,** ATP levels in algal cells as influenced by environmental conditions, *Plant Cell. Physiol.,* 11, 689, 1970.
171. **Aitchison, P. A. and Butt, V. S.,** The relative between the synthesis of inorganic polyphosphate and phosphate uptake by chlorella vulgaris, *J. Exp. Bot.,* 24, 497, 1973.

172. **Sicko-Goard, L. and Jensen, T. E.**, Phosphate metabolism in blue-green algae. II. Changes in phosphate distribution during starvation and the "polyphosphate overplus" phenomenon in Plectonema boryanum, *Am. J. Bot.*, 63, 183, 1976.

173. **Stewart, W. D. P. and Alexander, G.**, Phosphorus availability and nitrogenase activity in aquatic blue green algae, *Freshwater Biol.*, 1, 389, 1971.

174. **Stevenson, R. J. and Stoermar, E. F.**, Luxury consumption of phosphorus by five Cladophora epipites in lake Huron, *Trans. Am. Microsc. Soc.*, 101, 151, 1982.

175. **Jensen, T. E. and Sicko, L. M.**, Phosphate metabolism in blue green algae. I. Fine structure of the "polyphosphate overplus" phenomenon in Plectonema boryanum, *Can. J. Microbiol.*, 9, 1235, 1974.

176. **Stevens, E. S., Jr. and Paone, D. A. M.**, Accumulation of cyanophycin granules as a result of phosphate limitation in Agmenellum quadruplictum, *Plant Physiol.*, 67, 716, 1981.

177. **Allen, M. M., Hutchinson, F., and Weathers, P. J.**, Cyanophycin granules polypeptide formation and degradation in the cyanobactrium Aphanocapsa 6308, *J. Bacteriol.*, 1141, 687, 1980.

178. **Stewart, W. D. P., Fitzgerald, G. P., and Burris, R. H.**, Acetylene reduction assay for determination of phosphorus availability in Wisconsin lakes, *Proc. Natl. Acad. Sci. U.S.A.*, 66, 1104, 1970.

179. **Vanderhoef, L. N., Dana, B., Emerich, D., and Burris, R. H.**, Acetylene reduction in relation to levels of phosphate and fixed nitrogen in Green Bay, *New Phytol.*, 71, 1097, 1971.

180. **Ullrich, W. R.**, Die wirkung von O_2 and CO_2 auf die ^{32}P-markierung der polyphosphate von Ankistrodesmus braunii bei der photosynthesis, *Ber. Dtsch. Bot. Res.*, 83, 435, 1970.

181. **Ulrich, W. R.**, Zur wirkung von sanertoff auf die ^{32}P-markierung von polyphosphaten und organischen phosphaten bei Ankistrodemus im. licht, *Planta*, 90, 272, 1970.

182. **Ulrich, W. R.**, Der einflub von CO_2 und pH auf die ^{32}P-markierung von polyphosphaten und organischen phosphaten bei Ankistrodesmus braunii im licht, *Planta*, 102, 37, 1972.

183. **Wedding, R. T. and Black, M. K.**, Uptake and metabolism of sulfate by *Chlorella*. I. Sulfate accumulation and active sulfate, *Plant Physiol.*, 35, 72, 1960.

184. **Sinesky, M.**, Specific deficit in the synthesis of 6-sulfoquinosyl diaglyceride in *Chlorella pyrenoidosa*, *J. Bacteriol.*, 129, 516, 1977.

185. **Bhetow, D. E.**, Growth survival and biochemical alteration of *Euglena gracilis* in medium limited in sulfur, *J. Cell. Comp. Physiol.*, 66, 235, 1965.

186. **Kylin, A.**, The effect of light, carbon dioxide, and nitrogen nutrition on the incorporation of S from external sulfate into different S-containing fractions in *Scenedesmus*, with special reference to lipid S, *Physiol. Plant.*, 19, 883, 1966.

187. **Schmidt, A. and Trebst, A.**, The mechanism of photosynthetic sulfate reduction by isolated chloroplasts, *Biochim. Biophys. Acta*, 180, 529, 1969.

188. **Cohen, Y., Paden, E., and Shilo, M.**, Facultative anoxygenic photosynthesis in the cyanobacterium *Oscillatoria limnetica*, *J. Bacteriol.*, 123, 855, 1975.

189. **Wiedeman, V. E. and Bold, H. C.**, Heterotrophic growth of selected waste-stabilization pond algae. *J. Phycol.*, 1, 66, 1965.

190. **Knobloch, K.**, Sulfide oxidation via photosynthesis in green algae, *Progress in Photosynthesis Research*, Vol. 2, Mefines, H., Eds., International Union of Biological Sciences, Tubirgen, 1969, 1032.

191. **Knobloch, K.**, Photosynthetische Sulfidoxydation gruner Pflanzen. I. Mitteilung, *Planta*, 70, 73, 1966.

192. **Knobloch, K.**, Photosynthetische Sulfidoxydation gruner Pflanzen, II. Wirkung von Stoffwechselinhibitoren, *Planta*, 70, 172, 1966.

193. **Krumbein, W. H., and Cohen, Y.**, Biogene, Klastische und evaporitische Sedimentation in einem mesothermen monomiktischen ufernahen, *Geol. Rundsch.*, 63, 1035, 1974.

194. **Gerloff, G. C. and Fishbeck, K. A.**, Quantitative cation requirements of several green and blue-green algae, *J. Phycol.*, 5, 109, 1969.

195. **Allen, M. B. and Arnon, D. I.**, Studies on nitrogen-fixing blue-green algae. I. Growth and nitrogen fixation by *Anabaena cylindrica* Lemm., *Plant Physiol.*, 30, 366, 1955.

196. **Frank, E.**, Vergleichende Untersuchungen zum Calcium-, Kalium-, und Phosphathaushalt von Grunalgen. II. Calciummangel bei *Hydrodictyon, Sphaeroplea* und *Chlorella, Flora*, 152, 157, 1962.

197. **Kylin, A. and Das, G.**, Calcium and strontium as micronutrients and morphogenic factors for *Scenedesmus, Phycologia*, 6, 201, 1967.

198. **Lewin, J. C.**, Calcification, in *Physiology and biochemistry of Algae*, Lewin, R. A., Ed., Academic Press, New York, 1962, 457.

199. **Walker, J. B.**, Inorganic micronutrient requirements of *Chlorella*. I. Requirements for calcium (or strontium), copper and molybdenum, *Arch. Biochem. Biophys.*, 46, 1, 1953.

200. **Bostwick, C. D., Brown, L. R., and Tischer, R. G.**, Some observations on the sodium and potassium interactions in the blue-green alga *Anabaena flos-aquae* A-37, *Physiol. Plant.*, 21, 466, 1968.

201. **MacRobbie, E. A. C.**, Ionic relations of *Nitella translucens*, *J. Gen. Physiol.*, 45, 861, 1962.

202. **Chang, F. C. and Kahn, J. S.**, Isolation of a possible coupling factor for photophosphorylation from chloroplasts of *Euglena gracilis*, *Arch. Biochem. Biophys.*, 117, 282, 1966.

203. **Sugiyama, T., Matsumoto, C., and Akazawa, T.,** Structure and function of chloroplast proteins. VII. Ribulose-1,5-diphosphate carboxylase of *Chlorella ellipsoidea, Arch. Biochem. Biophys.,* 129, 597, 1969.
204. **Holm-Hansen, O., Gerloff, G. C., and Skoog, F.,** Cobalt as an essential element for blue-green algae, *Physiol. Plant.,* 7, 665, 1954.
205. **Hewitt, E. J., Hukelsby, D. P., and Notton, B. A.,** Nitrogen metabolism, in *Plant Biochemistry,* Bonner, J. and Varner, J. E., Eds., Academic Press, New York, 1976, 633.
206. **Versteate, D. R. T., Storch, A., and Dunham, V. L.,** A comparison of the influence of iron on the growth and nitrate metabolism of *Anabaena* and *Scenedesmus, Physiol. Plant.,* 50, 47, 1980.
207. **Oquist, G.,** Changes in pigment composition and photosynthesis induced by iron-deficiency in the blue-green alga *Anacystic nidulans, Physiol. Plant.,* 25, 188, 1971.
208. **Lankford, C. E.,** Bacterial assimilation of iron, *CRC Crit. Rev. Microbiol.,* 2, 273, 1973.
209. **Hardie, L. P., Balkwill, D. L., and Stevens, S. E., Jr.,** Effects of iron starvation on the physiology of the cyanobacterium *Agmenellum quadruplicatum, Appl. Environ. Microbiol.,* 3, 999, 1983.
210. **Hardie, L. P., Balkwill, D. L., and Stevens, S. W., Jr.,** Effects of iron starvation on the ultrastructure of the cyanobacterium *Agmenellum quadruplicatum, Appl. Environ. Microbiol.,* 3, 1007, 1983.
211. **Meisch, H. U., Becker, L. J. M., and Schwab, D.,** Ultrastructural changes in *Chlorella fusca* during iron deficiency and vanadium treatment, *Protoplasma,* 103, 273, 1980.
212. **Lewin, J. C.,** Physiological studies of the boron requirement of the diatom Cylindrotheca fusiformis Reimann and Lewin, *J. Exp. Bot.,* 17, 473, 1966.
213. **Werner, D.,** Silicoborate als erste nicht C-haltige Wachstumsfaktoren, *Arch. Mikrobiol.,* 65, 258, 1969.
214. **Jahnke, L. S. and Soulen, T. K.,** Effects of manganese on growth and restoration of photosynthesis in manganese deficient algae, *Z. Pflanzonphysiol.,* 88, 83, 1978.
215. **Bates, S. S., Letorneau, M., Tessier, A., and Campbell, P. G. C.,** Variation in zinc adsorption and transport during growth of *Chlamydomonas variabilis* (Chlorophyceae) in batch culture with daily addition of zinc, *Can. J. Fish. Aquat. Sci.,* 40, 895, 1983.
216. **Sandmann, G. and Boger, P.,** Copper deficiency and toxicity in *Scendesmus, Z. Pflanzenphysiol.,* 98, 53, 1980.
217. **Petersen, R.,** Influence of copper and zinc on the growth of a freshwater alga, *Scenedesmus quadricauda:* the significance of chemical speciation, *Environ. Sci. Technol.,* 16(8), 443, 1982.
218. **Dalton, H. and Mortenson, L. E.,** Dinitrogen (N_2) fixation (with a biochemical emphasis), *Bacteriol. Rev.,* 36, 231, 1972.
219. **Arnon, D. I. and Wessel, G.,** Vanadium as an essential element for green plants, *Nature (London),* 172, 1039, 1953.
220. **Fries, L.,** Vanadium an essential element for some marine macroalgae, *Planta,* 154, 393, 1982.
221. **Fezy, J. S., Spencer, D. F., and Greene, R. W.,** The effect of nickel on the growth of freshwater diatom *Navicula pelliculosa, Environ. Pollut.,* 131, 9327, 1979.
222. **Van Baalen, C. and O'Donnell, R.,** Isolation of a nickel-dependent blue-green alga, *J. Gen. Microbiol.,* 105, 351, 1978.
223. **Rees, T. A. V. and Bekheet, I. A.,** The role of nickel in urea assimilation, *Planta,* 156, 385, 1982.
224. **Darley, W. M.,** Silicon and the division cycle of the diatoms *Navicula pelliculosa* and *cylindrotheca fusiformis, Proc. N. Am. Paleontol. Convention,* Part G, September, 1969, Clinical Agal Physiology and Biochemistry, 1969, 994.
225. **Lewin, J. C.,** Silicon metabolism in diatoms. I. Evidence for the role of reduced sulfur compounds in silicon utilization, *J. Gen. Physiol.,* 37, 589, 1954.
226. **Lewin, J. and Chen, C. H.,** Silicon metabolism in diatoms. VI. Silicid acid uptake by a colorless marine diatom *Nitzschia alba* Lewin and Lewin, *J. Phycol.,* 4, 161, 1968.
227. **Lewin, J. C.,** Silicon metabolism in diatoms. III. Respiration and silicon uptake in *Navicula pelliculosa, J. Gen. Physiol.,* 39, 1, 1955.
228. **Coombs, J., Halicki, P. J., Holm-Hansen, O., and Volcani, B. E.,** Studies on the biochemistry and fine structure of silica shell formation in diatoms. II. Changes in concentration of nucleotide triphosphates in silicon-starvation synchrony of *Navicula pelliculosa* (Breb.) Hilse, *Exp. Cell. Res.,* 47, 315, 1967.
229. **Coombs, J., Spanis, C., and Volcani, B. E.,** Studies on the biochemistry and fine structure of silica shell formation in diatoms. Photosynthesis and respiration in silicon-starvation synchrony of *Navicula pelliculosa, Plant Physiol.,* 42, 1607, 1967.
229a. **Coombs, J., Halicki, P. J., Holm-Hansen, O., and Volcani, B. E.,** Studies on the biochemistry and fine structure of silica shell formation in diatoms. Changes in concentration of nucleoside triphosphates during synchronized division of *Cylindrotheca fusiformis* Reimann and Lewin, *Exp. Cell Res.,* 47, 302, 1967.
230. **Moede, A., Greene, R. W., and Spencer, D. F.,** Effects of selenium on the growth and phosphorus uptake of *Scenedesmus dimorphus* and *Anabaena cylindrica, Environ. Exp. Bot.,* 20, 207, 1980.
231. **Patrick, R., Bott, P. R. and Larson, R.,** The Role of Trace Elements in Management of Nuisance Growths, U.S. Environmental Protection Agency, Corvallis, Ore., 1975.

232. **Spencer, D. F., Greene, R. W., Their, T. L., Yeung, H.-Y., Ross, Q. E., and Dodge, E. E.,** A study of the relationship between phytoplankton abundance and trace metal concentrations in eutrophic lake Charles East, using correlation techniques, *Proc. Ind. Acad. Sci.,* 87, 204, 1978.

233. **Pintner, I. J. and Provasoli, L.,** Heterotrophy in subdued light of 3 *Chrysochromulina* species, *Bull. Misaki Mar. Biol. Inst.,* 12, 25, 1968.

234. **Fries, L.,** Selenium stimulates growth of marine macroalgae in axenic culture, *J. Phycol.,* 18, 328, 1982.

235. **Allen, M. M. and Smith, A. J.,** Nitrogen chlorosis in blue-green algae, *Arch. Mikrobiol.,* 69, 114, 1969.

236. **Guerin-Dumartrait, E., Mihara, S., and Moyse, A.,** Composition de *Chlorella pyrenoidosa,* structure des cellules et de leurs lamelles chloroplastiques, en fonction de la carence en azote et de la levee de carence, *Can. J. Bot.,* 48, 1147, 1970.

237. **Thomas, W. H. and Dodson, A. N.,** On nitrogen deficiency in tropical Pacific Oceanic phytoplankton. II. Photosynthetic and cellular characteristics of a chemostat-grown diatom, *Limnol. Oceanogr.,* 17, 515, 1972.

238. **Fuhs, G. W., Demmerle, S. D., Canelli, E., and Chen, M.,** Characterization of phosphorus-limited plankton algae, in *Nutrients and Eutrophication,* Special Symposia, Vol. 1, Likens, G. E., Ed., American Society of Limnology and Oceanography, 1972, 113.

239. **Das, G.,** Growth and appearance of *Scenedesmus* as influenced by deficient inorganic nutrition, *Sven. Bot. Tidskr.,* 62, 457, 1968.

240. **Prakash, G. and Kumar, H. D.,** Studies on sulphur-selenium antagonism in blue-green algae. I. Sulphur nutrition, *Arch. Mikrobiol.,* 77, 196, 1971.

241. **Coombs, J., Darley, W. M., Holm-Hansen, O., and Volcani, B. E.,** Studies on the biochemistry and fine structure of silica shell formation in diatoms. Chemical composition of *Navicula pelliculosa* during silicon-starvation synchrony, *Plant Physiol.,* 42, 1601, 1967.

242. **Werner, D.,** Die Kieselsaure im Stoffwechsel von *Cyclotella cryptica* Reimann, Lewin, and Guillard, *Arch. Mikrobiol.,* 55, 278, 1966.

243. **Smith, F. A.,** Active phosphate uptake by *Nitella translucens, Biochim. Biophys. Acta,* 126, 94, 1966.

244. **Eley, J. H.,** Effect of carbon dioxide concentration on pigmentation in the blue-green alga *Anacystis niculans, Plant Cell Physiol.,* 12, 311, 1971.

245. **Haxo, F. and Strout, P.,** Nitrogen deficiency and coloration in red algae, *Biol. Bull.,* 99, 360, 1950.

246. **Healey, F. P., Coombs, J., and Volcani, B. E.,** Changes in pigment content of the diatom *Navicula pelliculosa* (Breg.) Hilse in silicon-starvation synchrony, *Arch. Mikrobiol.,* 59, 131, 1967.

247. **Provasoli, L. and Carlucci, A. F.,** Vitamins and growth regulators in *Algal Physiology and Biochemistry,* Stewart, W. D. P., Ed., University of California Press, Berkley, 1974, 781.

248. **Pringsheim, E. G.,** Assimilation of different organic substances by saprophytic flagellates, *Nature (London),* 136, 196, 1937.

249. **Chibnall, A. C., Williams, E. F., Latner, A. L., and Piper, S. H.,** The isolation of n-triacontanol from lucerne wax, *Biochem. J.,* 27, 1885, 1933.

250. **Ries, S. K., Wert, V., Sweeley, C. C., and Leavitt, R. A.,** Triacontanol: a new naturally occuring plant growth regulator, *Science,* 195, 1339, 1977.

251. **Haugstad, M., Ulsaker, L. K., Ruppel, A., and Nilsen, S.,** The effect of triacontanol on growth, photosynthesis and photorespiration in *Chlamydomonas reinhardtii* and *Anacystis nidulans, Physiol. Plant.,* 58, 451, 1983.

252. **Komoda, Y., Isogai, Y., and Satoh, K.,** Isolation from humus and identification of two growth promoters, adenosine and 2'-deoxyadenosine, effective in culturing the diatom *Phaeodactylum tricornutum. Chem. Pharm. Bull.,* 31(10), 3771, 1983.

MICROALGAE OF ECONOMIC POTENTIAL

A. Richmond

INTRODUCTION

Algae are photosynthetic, nonvascular plants that contain chlorophyll-a and have simple reproductive structures. Algae exhibit a remarkable diversity of form and size, and exist in nearly every environment. The brown kelps up to 70 m in length, the flagellated swimming green cells 1 μm in diameter, green scum floating in a pond, sargasso weed in the massive Atlantic gyre, and organisms responsible for coloration on mountain snow are all algae.[1] Algae, as well as other organisms, can be classified both for convenience and to show relationships. The plant kingdom, to which the algae belong, is divided into equivalent categories (divisions). Further subgroups, in descending order, are family, genus, and species.

Trainor[1] summarized the major characteristics which help distinguish the various classes of algae as follows:

1. *Pigmentation.* All of the groups contain chlorophyll and several carotenoids. The carotenoids include carotenes and xanthophylls. In one or more divisions one can find chlorophylls a and b, α- or β-carotene, and a selection from a group of more than a dozen xanthophylls. In addition to the above pigments, which are soluble in organic solvents, there are the water-soluble phycobiliproteins (sometimes called phycobilins), which are found in blue-green algae, red algae, and a small group of flagellates.[1]
2. *The stored reserve of photosynthesis.* Reserve food material is usually stored within the cell and frequently within the plastid in which photosynthesis occurred. Starch, starch-like compounds, fats, or oils are the most common forms. Apparently some organisms release a portion of their excess material and in a sense, might be using the environment as their storage area. The released material may then be brought back into the cell when needed.[1]
3. *Motility.* Some organisms are motile during much of their lives, whereas other genera lack motility, or any motile reproductive stages. Most algae do not move about by active means, but often have some reproductive stages that are motile. Some filamentous blue-green algae and certain pennate diatoms move in a steady gliding manner without flagella.[1]
4. *Wall composition.* Although most algae have a conspicuous cell wall, some genera and certain reproductive cells do not. The cell wall may be a simple outer covering around the protoplast or an elaborately ornamented structure. There are marked differences in wall composition, and at times, structure, among the various groups of algae. Among the materials isolated from algal walls are cellulose, xylans, mannans, sulfated polysaccharides, alginic acid, protein, silicon dioxide, and calcium carbonate. A typical algal wall is not constructed of one compound but instead is a matrix of one material interlaced with another or is formed by layers of different materials.[1]
5. *Gross structure and plant body types.* Groups of algae can be distinguished on the basis of morphological features. Indeed, a broad range of cell types can be readily observed. Some algae are unicells, but most algae are more complex and exist in colonies such as filaments *(Spirulina)* or round bodies *(Chlorella)* or in groups of cells such as Scenedesmus.[1]

For the more detailed classification, refined methods are employed that include cytological means by which the most accurate details of cellular structure may be elucidated. Also biochemical tools are employed to identify differences in enzymes or nucleic acids. Finally,

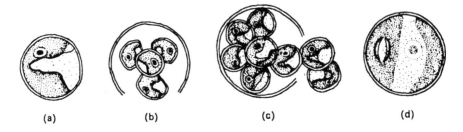

FIGURE 1. *Chlorella* sp. (a) vegetative cell (b, c) autosporogenesis, (d) *C. vulgaris Beij. var. vulgaris* Fott. × 1875. (Modified from Fott and Novakova in Bold and Wynee[3].)

physiological features which relate to the optimal temperature of the organism or pH and which test the response of the cell to specific nutrients or chemicals are all used to classify algae.

From the standpoint of production of algal biomass, algae are grouped in two major categories, macroalgae and microalgae. The former are more complex organisms, most of which are commonly known as seaweeds and are not dealt with in this book. Microalgae are unicells or colonies, with the common feature of their minute size which does not permit resolution with the naked eye and requires the aid of a microscope.

Most of the microalgal species that are commercially grown today or that seem to have an economic potential belong to two classes, Cyanophyceae and Chlorophyceae, commonly named blue-green and green algae, respectively. Blue-green algae are also termed cyanobacteria and these two names are used interchangeably in this book. Recently, there is evidence that species belonging to other classes, such as Rhodophyceae, Bacillariophyceae, and Chrysophyceae may also become commercially interesting and species belonging to these classes shall be described later.

CHLORELLA

The order Chlorellalis include unicellular and colonial green algae. Two families include species of commercial importance, the chlorellaceae which contain unicellular genera and the Scenedesmaceae with the clonial, coenobic genera.

Ecology and Reproduction

The genus *Chlorella* occupies a special position among the other genera in the order Chlorococcales. Its species possess spherical or ellipsoidal cells, exhibit a simple life cycle, and have simple nutritional requirements. In culture they grow more quickly than other microorganisms, rapidly overgrowing them. The smallest *Chlorella* species are similar to bacteria and were the first algae to be isolated like bacteria and to be grown in pure cultures.[2] In reproduction, which is exclusively asexual, each mature cell divides, producing 4, 8, or, more rarely, 16 autospores (Figure 1). These are freed by rupture or dissolution of the parental walls, the latter persisting sometimes in axenic cultures. *Chlorella* species are encountered in all water habitats and they exhibit a cosmopolitan occurrence, having been isolated from natural habitats differing widely in amounts of major nutrients, the presence or absence of growth substances and in fresh as well as marine waters. A very great breadth of nutritional requirements and habitats of *Chlorella* species is thus indicated.

Morphology and Taxonomy

Due to the little morphological differentiation of their more-or-less spherical cells, the taxonomy of the members of the section Euchlorella, which comprise the most common species *(Chlorella vulgaris Beijerinck)* has been rather confused. The criteria for their iden-

Table 1
BIOCHEMICAL AND PHYSIOLOGICAL PROPERTIES
OF 25 DIFFERENT STRAINS OF THE GENUS
CHLORELLA, SECTION *AUCHLORELLA*

Hydrogenase activity	Color of N-deficient cells	Reaction with ruthenium red	Species
Strong	Orange	None	*C. pyrenoidosa*
Weak	Pale	Weak	
None	Pale	Strong	*C. vulgaris Beijerinck*
None	Pale	None	*C. ellipsoidea*

From Kessler, E. and Soeder, C. J., *Nature (London)*, 1069, 1962. With permission.

tification based on size and shape of cells and chloroplasts, visibility of the pyrenoid, etc., are highly variable and ambiguous, and depend to a large extent on culture conditions and the developmental stage of the cells. In most cases it is therefore very difficult, if not impossible, to identify these species by morphological criteria. Kessler and Soeder[4] proposed to use mainly physiological and biochemical rather than morphological criteria for the identification of these *Chlorella* species.

They chose three biochemical properties known to be present in some, but not all, strains of chlorella. These included hydrogenase activity, formation of secondary carotenoids under conditions of nitrogen deficiency, and presence in the cells and in the cell wall of carbohydrates (probably pectin or related substances) which give a red color reaction with ruthenium red. The results are summarized in Table 1.

It is evident that the 25 strains of *Chlorella* studied fall into four well-defined groups. Kessler and Soeder[4] concluded that the taxonomy of the critical section Euchlorella of the genus *Chlorella* can be soundly based on some physiological and biochemical characteristics which can easily be determined.

In principle, growth responses could provide good taxonomical criteria if they are clearly different quantitatively. For example, a species may be distinguished on the basis that it requires thiamin and does not grow in its absence. Shihira and Krauss[5] attempted a comprehensive treatment of the genus *Chlorella* based primarily on physiological characters and distinguished species by the responses to various sugars, nitrogen sources, and other compounds. They demonstrated a great diversity of nutritional reqrirements, recognizing 20 species and 8 varieties for 41 isolates. From these 28 taxa, 75% (22) were defined by the observation of a single strain. By means of morphological differences and by using a simple idenitification key, freshwater strains in the genus *Chlorella* were grouped by Fott and Novakova[6] into nine species. The main morphological and structural characteristics they used as taxonomical criteria for species identification in *Chlorella* were as follows: form of cells, cell wall, fate of the mother-cell wall, chloroplast morphology, presence or absence of pyrenoid, number and size of autospores, mode of autospore release, and cell dimensions (Figure 2).

On the basis of their external morphology *Chlorella* species were placed in four groups: (1) spherical cells (ratio of the two axes equals 1); (2) ellipsoidal cells (ratio of the long axis to the shortest axis being 1.45 to 1.60); (3) spherical or ellipsoidal cells; (4) globular to subspherical cells.

The cell wall, which is perfectly smooth, is well defined in all *Chlorella* species, being distinctly thick in some species and very thin in others. The empty cell walls or the remains

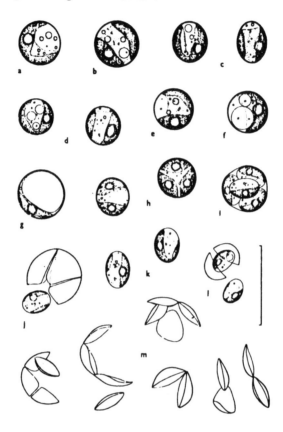

FIGURE 2. *Chlorella vulgaris Beijerinck var. vulgaris.* (a, b) Spherical cells with girdle-shaped chloroplasts; (c, d) young ellipsoidal cells, pyrenoid lateral in position, with saucer-shaped starch grains, inside the chloroplast are vacuoles and lipid droplets; (e) cell with a cup-shaped chloroplast; (f, g) older cells with chloroplasts; (i) sporangium enclosing four elliptical autospores; (j) empty cell wall with one remaining autospore, cell wall pouch-like, split into several fragments; (k) young cells; (l) release of autospores; (m) remains of burst mother cell walls — they are triangular or irregularly elongated, usually four in number. All figures drawn according to the strain Beijerinck no. 211—11b in the Cambridge Coll. 211—11b. The given abscissa equals 10 μm. (From Fott, B. and Novakova, M., A monograph of the genus *Chlorella* - The freshwater species, Acedemia, Prague, 1969. With permission.)

of them after the release of autospores exhibit various features which are peculiar to the species. In all species the chloroplast is parietal, adhering to the cell wall and covering most of the periphery. It may be saucer-shaped (mostly encountered in autospores and in young cells) or band-shaped, girdle-shaped, cup-shaped, and mantel-shaped. The pyrenoid is seen in some species.[6]

Fott and Novakova[6] suggested a key for distinguishing the species and varieties of *Chlorella*. They described in great detail the following species: *C. vulgaris* and var. autothrophica; *C. kessleri* nomen novum; *C. fusca* Shihira et Krauss 1965, including var. fusca and var. vacuolata; *C. Luteoviridis Chodat 1912; C. saccharophila* (Kruger) Migula 1907, including var. saccharophila and var. ellipsoidea; *C. zofingiensis* Donz 1934; *C. prototecoides* Kruger 1894; *C. homosphaera* Skuja 1948; and *C. minutissima* species nova.

Da Silva and Gyllenberg[7] adapted computerized numerical methods for the automated reduction of data for the taxonomic treatment of the genus *Chlorella*. Isolates, 41, were grouped into 10 clusters on the following list of characters based on several aspects of *Chlorella* physiology: effect of each of the following sugars, glucose, galactose, mannose, fructose, and sucrose on growth in light and the effect of the same sugars on growth in dark;

thiamin requirement for growth; growth only in inorganic medium; effect of acetate (0.1%) on growth in light or in dark; effect of nitrate as sole nitrogen source; effect of aminoacids as sole nitrogen source; effect of yeast extract (0.01%) on growth in light; growth on medium with nitrate, yeast extract and glucose; growth on medium with amino acids, glucose and yeast extract; occurrence of cells as spheres or ellipsoids; cup-shaped, dumbbell-shaped, mantle-shaped, or net-like chromatophore; cell color; cell diameter less than 10 μg or cell diameter greater than 10 μm.[7]

Kessler[8] considered many other physiological and biochemical characters to group 77 strains in 12 taxa (Table 2). He concluded that certain nutritional characters, i.e., utilizations of organic carbon and nitrogen compounds were highly variable and therefore unsuitable for taxonomy.

Of particular relevance to mass cultivation of algae is salt tolerance of the various *Chlorella* strains, this because a constant process of salinization of the growth medium takes place in outdoor cultivation, due to evaporative losses. Also, the relative salt tolerance of a particular strain could be used advantageously to check the growth of an unwanted contaminant. Kessler[9] reported that salt tolerance was a species-specific character in the genus *Chlorella*. The most resistant species, *C. luteoviridis,* was able to grow nutrient media containing up to 5% NaCl. The limit for *C. protothecoides* and *C. saccharophila* was at 4% NaCl, whereas *C. vulgaris var. vulgaris, C. fusca var. vacuolata,* and *C. fusca* var. rubescens grow in the presence of 3%; *C. zofingiensis* and *C. minutissima* were only capable of growing in media containing up to 1% NaCl. *C. homosphaera* did not tolerate even 1% NaCl. Those strains, which are able to synthesize secondary carotenoids, turn orange in salt concentrations close to the limit of their tolerance.

Optimizing Growth Conditions Outdoors

Chlorella had been for many years the most widely studied algae and many basic works on photosynthesis and cell metabolism have been carried through on *Chlorella*. Perhaps the most significant published works on the mass production of *Chlorella* were out on the Carnegie Institution at Cambridge, Mass., in relation to the *Chlorella* pilot plant constructed for the Institute by Arthur D. Little in 1951 (see "Outdoor Mass Cultures of Microalgae"). As documented by some of the early workers in the area of mass algaculture, French and Spoehr,[10] only scarce information was available concerning the effect of the form or design of culture chambers on yields and growth rates of *Chlorella*. Accordingly, a number of algal culture chambers of different design were operated outdoors with the minimum necessary control of growing conditions. It was intended that these experiments should provide information as to the practicality of the principles underlying the different designs, so that it could be determined which of the culture chambers could profitably be expanded in scale. The model culture chambers actually operated were (1) a 4 × 8 ft rocking tray; (2) a vertical, slightly tapered glass tube, 5 in. mean diameter; (3) a 7/16-inch plastic tube, 40 ft long, and a similar tube of glass.

Results indicated that the different culture devices all gave roughly the same daily yield of *Chlorella* per unit area of intercepted sunlight.

The low rates of photosynthesis obtained in early experiments raised the queries whether the fact that bright light is used less effectively than weak light was primarily responsible for the low growth rate. If so, can yields be increased by producing intermittent illumination for each cell by creating turbulence in the culture? Also, is the growth rate limited by the potential rate of some other limiting process when photosynthesis is proceeding rapidly? Do both large and small cells of *Chlorella* increase in weight at an equal rate per gram of cell material?

Another pressing issue arose from the work of Cook[11] who reported a maximum yield at the low density of 0.36 g dry weight of *Chlorella* per liter, with a rapid drop in yield above

Table 2
BIOCHEMICAL AND PHYSIOLOGICAL CHARACTERS OF 12 *CHLORELLA* TAXA (77 STRAINS)

	Characters										
	Hydr.	S. car.	Gel liqu.	pH	NaCl %	Therm.	Lact. ferm.	NO$_3$ red.	B$_1$ requ.	DNA % GC	Strains
C. *fusca* var. *vacuolata* Shihira et Krauss	+	+	+	3.5	3	−	+	+	−	50	10
C. *fusca* var. *fusca* Shihira et Krauss	+	+	+	4.0	2	−	+	+	−	55	1
C. *fusca* var. *rubescens* (Dangeard) Kessler et al.	+	+	−	4.0	3	−	+	+	−	55	1
C. *homosphaera* Skuja	+	+	−	6.0	0	−	−	+	−	69	1
C. *zofingiensis* Donz	−	+	−	5.0	1	−	+	+	−	49	3
C. *minutissima* Fott et Novakova	−	−	−	5.5	1	−	−	+	−	43	1
C. *saccharophila* (Kruger) Migula	−	−	−	2.0	4	−	−	+	−	50	6
C. *Luteoviridis* Chodat	−	−	−	3.0	5	−	−	+	−	44	6
C. *kessleri* Fott et Novakova	+	−	−	3.0	2	−	−	+	−	56	7
C. *sorokiniana* Shihira et Krauss (=C. *vulgaris* f. tertia Fott et Novakova)	+	−	−	4.0	2	+	+	+	−	65	10
C. *vulgaris* Beijerinck	−	−	−	4.0	3	−	+	+	−	62	17
C. *protothecoides* Kruger	−	−	−	4.0	4	−	+	−	+	62	14

[a] Abbreviations of the characters as listed in the table are hydrogenase, secondary carotenoids, liquefaction of gelatin, acid tolerance (pH limit), salt tolerance (% NaCl), thermophily, lactic acid fermentation, nitrate reduction, thiamin (B$_1$) requirement, base composition of DNA (% GC), and number of strains studied.

From Kessler, E., *Plant Syst. Evol.*, 125, 129, 1976. With permission.

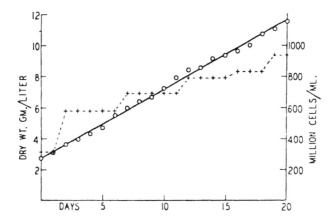

FIGURE 3.   Growth of a *Chlorella* culture in the rocking tray without harvesting. Circles and straight line show increase in cell mass; crosses and broken line show daily cell counts. The average yield for 20 days was 8.2 g/daywt/ℓ. During this period, the growth rate fell from 18%/day to 4.2%. (From Davis, E. A., et al., *Algal Culture: From Laboratory to Pilot Plant,* Burlew, J. S., Ed., The Carnegie Institution, Washington, D.C., 1953. With permission.)

and below this cell density. High cell densities, however, if they could only be maintained without significantly reducing the growth rate, were thought to greatly reduce harvesting costs. It was thus queried whether high growth rates could be maintained under high cell densities.

Valuable information concerning these questions was obtained from studies with the rocking tray. *Chlorella* cells in several cultures bleached and died soon after inoculation, this difficulty being largely due to overexposure of the cells to sunlight. Because the end of the tray had a vertical path of 4 cm on each rocking cycle, the culture drained from the high end, and the cells in the remaining thin film of culture were exposed too long to direct sunlight. Moreover, even dense cultures bleached slowly when exposed to the summer sun. Only when the rocking was adjusted to expose no part of the bottom of the tray was it possible to obtain good growth of the culture for a long time without cell bleaching.

In the early stages of growth of the culture shown in Figure 3 there was a daily increase in both cell count and cell weight. As cell density increased to about 2 g/ℓ, cell division became stepwise instead of a continuous process. There were periods of 2 to 5 days without significant change in the cell count, followed by a material increase. This stepwise increase in cell count is shown by the broken line in Figure 3.

The statistical cell size was computed by dividing the centrifugal cell volume (packed cells) by the cell count per millimeter. In young *Chlorella* cultures, where nearly all the cells are small, the statistical volume is 1.8×10^{-11} mℓ/cell. The volume of a cell was found to vary from 2.5 to 5.2×10^{-11} mℓ, with a very heterogeneous distribution of individual cell sizes. When growth ceased in the rocking tray (after 30 days) the statistical volume per cell was 6.3×10^{-11} mℓ.

The Effect of Temperature

The yield of *Chlorella* in relation to day and night temperatures was also studied by these early workers.[12] The solid line in Figure 4 shows the yields at various temperatures which were maintained continuously. The dashed line shows the yields when the daytime temperature is shown on the graph and the night temperature is 20°C in each case. The ordinate is the percentage yield for each condition, the respective control at 25°C continuous temperature being taken as 100%.

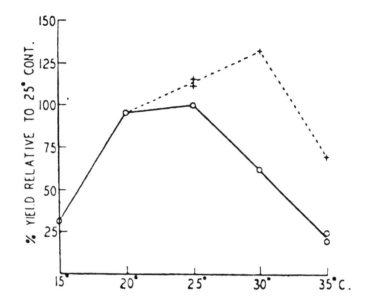

FIGURE 4. Effect of temperature on yield *Chlorella* grown outdoors in sunlight (solid line). Continuously maintained temperatures; (dashed line) indicates temperature in daytime; 20°C at night in each case. (From Davis, E. A., et al., *Algal Culture: From Laboratory to Pilot Plant*, Burlew, J. S., Ed., The Carnegie Institution, Washington, D.C., 1953. With permission.)

The effect of night temperature on the growth of *Chlorella* was investigated in detail. Fresh weight was not found to be greatly affected by night temperatures of 10, 15, 20, or 25°C. The greatest growth occurred, however, with a night temperature of 15°C, and declined with night temperatures above and below this value, night temperatures of 5, 30, and 35°C being definitely unfavorable.

On the basis of its response to temperature, strains of *Chlorella* may be regarded as "mesophilic" and "thermophilic" which vary significantly in their growth response to the temperatures prevailing outdoors throughout the year. A mesophilic strain, *C. ellipsoidea*, showed positive growth in all months throughout the year, although its growth was considerably suppressed in the hottest month, August. In contrast, the thermophilic strain grew actively in the summer months, but showed no growth in the winter months (December, January, and February). These observations carry important implications on outdoor production of algal biomass. To obtain a high yield of algae in locations where temperature varies in a wide range during the year, algal strains of different responses to temperature should be employed to secure maximal yields on a yearly basis.[13]

The works of Sorokin[14] on the isolation and characterization of thermophilic strains of algae pointed at the attractive potential of thermophilic strains for production of biomass under high temperatures. This is revealed in comparing the response to temperature of two strains of *Chlorella pyrenoidosa:* Emerson strain, representing the group of low-temperature algae, and *Chlorella* 7-11-05, representing the group of high-temperature algae (Table 3).

In outdoor cultures exposed to different degrees of illumination, i.e., sunlight, skylight, and shade or clouding, the highest yields were obtained under full sunshine when the temperature was kept at 25°C continuously. Under skylight and shade, the yields were significantly lower — 72 and 55% of the highest value for skylight and shade, respectively. The culture was clearly light-limited and its was perceived that the most promising way to increase *Chlorella* yields in sunlight was to supply the bright light to the individual cells in short flashes which would follow by dark periods of about ten times the length of the light

Table 3
TEMPERATURE CHARACTERISITCS OF TWO STRAINS OF
CHLORELLA PYRENOIDOSA

Characteristic	Unit of measurements	Strain of *emerson*	*Chlorella* 7-11-05
Temp. optimum for growth	°C	25—26	38—39
Photosynthesis		32—35	40—42
Endogenous respiration		30	40—42
Glucose respiration		30	40—42
Upper temp. limit for	°C		
Growth		29	42
Photosynthesis		39	45
Endogenous respiration		>45	>50
Glucose respiration		>45	>50
Growth rate at light saturation:			
At 25°C	No. doublings/day	3.1	3.0
At 39°C		—	9.2
Growth rate at $^1/_2$ saturating light intensity	No. doublings/day	2.4	2.3
At 25°C			
At 39°C		—	7.0
Rate of apparent photosynthesis at saturating light intensity	mm^3/O$_2$/mm^3/cells/hr	43, Rapid	47
At 25°C		decline	170
At 39°C			
Rate of apparent photosynthesis at $^1/_2$ saturating light intensity	mm^3/O$_2$/mm^3/cells/hr	32	32
At 25°C		—	115
At 39°C			
Rate of endogenous respiration			
At 25°C	mm^3/O$_2$/mm^3/cells/hr	1.8	2.0
At 39°C		1.4	5.5
Rate of glucose respiration			
At 25°C	mm^3/O$_2$/mm^3/cells/hr	4.5	8
At 39°C		1.6	18
Saturating light intensity for growth			
At 25°C	fc	500	500
At 39°C		—	1,400
Saturating light intensity for photosynthesis	fc	400	500
At 25°C			
At 39°C		—	1,600

From Sorokin, C., *Nature (London)*, 1959. With permission.

flashes. The most practical way to achieve this was judged by Davis et al.[12] to be by creating turbulence (Table 4).

The early investigators concluded that the yield of *Chlorella* can be increased at least 70% and perhaps even 300% by creating turbulence in the culture.

Cell Division and Enlargement

Davis et al.[12] observed that cell division in *Chlorella* took place during the night, the increase in cell volume occurred during the day.

Night temperature did not influence the time of cell enlargement, but had profound influence on the time of cell division. With night temperatures of 20, 25, and 30°C, cell division occurred chiefly at night, but when night temperature was 5°C, it occurred during

Table 4
THE INFLUENCE OF TURBULENCE ON
THE GROWTH OF *CHLORELLA*

| | Yield | | |
| | | | |
Rotor speed	Fresh wt., g/ℓ/12 hr	Dry wt. g/m²/12 hr	Increase (%)
0	15.1	25.2	—
16	20.5	34.1	35.3
208	25.6	42.8	69.8
475	25.9	43.2	71.4

From Davis, E. A., et al., *Algal Culture: From Laboratory to Pilot Plant*, The Carnegie Institution, Washington, D.C., 1953. With permission.

the day. With night temperatures of 10 and 15°C, cell division occurred during both day and night.

The time of cell division could be restricted to the night hours by controlling the night temperature only, so long as the cells received ample light for photosynthesis during the day. When the culture density increased to the point where all the cells in the culture were receiving too little light, cell division occurred during day and night. Under such conditions many of the cells were in effect living under night conditions during the day. By varying night temperature it was possible to obtain cell division only at night, only during the day, or during both day and night. Regardless of night temperature, cell enlargement occurred chiefly during the day.

The dependence of cell division in *Chlorella* on temperature and light intensity were studied in synchronized suspensions of *Chlorella*, with the thermophilic strain 7-11-05. The time for incipient cell division, the progress in the process after it started, and the number of cells produced are influenced by temperature and light intensity. Within limits, cell division is generally favored by the increase in temperature. The increase in light intensity first favors cell division. Then, after the optimal light intensity is attained, a further increase in light intensity inhibits cell division.[15]

"Light" and "Dark" Cells

Tamiya et al.[16] summarized their observations on the culture of *Chlorella*. Accordingly, chlorella cells assume two distinct forms in the course of their growth. One form, the "dark cells", is smaller in size, richer in chlorophyll content, and stronger in photosynthetic activity than the other, which was referred to as "light cells". When illuminated, dark cells grow, and with a substantial increase in mass, turn into light cells; the latter, when ripened, bear autospores in themselves (on the average 6 to 7 per cell) and eventually burst, setting free the autospores which then become individual dark cells. The transformation of light cells into dark cells involves no increase of cell mass and occurs only under aerobic conditions, irrespective of whether the cells are in the light or in the dark. Freshly born dark cells are somewhat smaller in size and contain less chlorophyll than the "active" dark cells, into which the former turn rapidly under the influence of light.

Tamiya et al.[16] concluded that (1) the transformation of dark cells into light cells is dependent on photosynthesis, although it is accompanied by some other metabolic processes, which are to be distinguished from the photosynthetic process in the ordinary sense; and that (2) the transformation of light cells into dark cells involves a light-independent and aerobically endergonic anabolic metabolism, although when illuminated, the transformation is also accompanied by some photosynthetic processes. Under weak light the rate of growth is exclusively determined by the photosynthetic process, whereas under strong light the light-independent metabolic processes become more significant in determining the rate of overall growth.

Based on the above, the following formulae were set forth:

$$\begin{array}{cc} \text{light} & \text{dark} \\ D \rightarrow L \;;\; L \rightarrow nD \\ k_p & k_D \end{array}$$

where D and L represent dark and light cells, respectively, k_p the rate constant of photosynthesis by which dark cells are changed into light cells, k_D the rate constant of increase in cell number in the dark process, and n the number of dark cells arising from one light cell. The overall growth rate and the relative abundance of dark and light cells in the culture are functions of light intensity, the rate of dark process, and the light-saturated and light-limited rates of photosynthesis. The dark process involved in *Chlorella* growth was revealed to have a remarkably large temperature coefficient, especially in the range of lower temperatures.[16]

Principles for Production of Autotrophic *Chlorella* Biomass

David and Milner[13] summarized the optimum conditions for growth of autotrophic *Chlorella* biomass, as it emerged from the work of the early investigators as follows.

Sunlight has at least ten times the intensity that can be utilized by *Chlorella* for maximum growth. Therefore, to obtain the highest possible yield of *Chlorella* it is important that as much as possible of the energy of sunlight be used for photosynthesis. This may be partially accomplished by means of culture turbulence, making use of the intermittent-light effect, i.e., each cell in the culture receiving the highly intensive solar irradiation in short flashes, follwed by longer periods of darkness; at this time other cells enter the intensely-illuminated photic layer.

There is an advantage in mass culture to maintain the highest cell density possible without decreasing the yield, so as to lessen the quantity of culture to be handled during harvesting. It is more economical to use a shallow culture with high cell density than a deep culture with low cell density.

Growth rates of *Chlorella* were not significantly different when the medium contained 0.56 to 4.43% by volume CO_2, a mixture of 5% CO_2 in air usually being adequate. The CO_2 supply may be cut off at night without significant effect.

The yield of *Chlorella* is influenced by both day and night temperatures. When different temperatures are maintained continuously, 25°C is most favorable for rapid growth. The yields at 25°C continuously are exceeded by using a 25°C day and 15°C night temperature, with both laboratory and outdoor cultures. An even higher yield may be obtained in outdoor cultures with 30°C day and 20°C night temperatures. The requirement for cooling overheated outdoor cultures may be lessened by using a high temperature strain of *Chlorella*.

The maintenance of an adequate supply of nitrogen in the medium is mandatory for rapid growth. Urea as a nitrogen source has been found superior in several respects to the commonly used potassium nitrate. For an equivalent nitrogen concentration urea gives greater yields. It can be used in higher initial concentration without harmful effect and it causes smaller pH fluctuations in the medium during cell growth than is the case with potassium nitrate. A favorable concentration of available iron in the medium should be maintained at all times, the chelating agent ethylenediamine tetraacetic acid (EDTA) being satisfactory.

Table 5
YIELDS OF *CHLORELLA PYRENOIDOSA* AND *C.*
ELLIPSOIDEA GROWN ON VARIOUS CARBON AND
NITROGEN SOURCES

Carbon source	Nitrogen source	Amount C source (mg/ℓ)	Yield (mg)	Organisms k^a (log10/day)	Carbon[b] assimilation (%)
Chlorella pyrenoidosa					
Glucose	Nitrate	500	180.1	0.44	43
		250	92.7	0.45	45
Galactose	Nitrate	250	77.3	0.2	37
Na acetate	Nitrate	100	9.7	0.16	26
Glucose	Ammonia	75	30.8	0.4	48
		250	106.5	0.5	51
	Urea	500	207.3	0.43	50
		250	102.9	0.47	49
C. ellipsoidea					
Glucose	Nitrate	250	109.2	0.5	53
	Urea	250	122.9	0.40	59

a Specific growth rate; varying degrees of reliability reflect differences in number of samples taken during exponential period of growth.
b Approximate: calculated for average cell product of 48% carbon.

From Samejima, H. and Meyers, J., *J. Gen. Microbiol.*, 18, 107, 1958. With permission.

Heterotrophic Growth

The heterotrophic growth of *Chlorella* has been thoroughly studied. Samejima and Myers[17] (Table 5) reported that cellular synthesis may proceed with high efficiency, but only from a very limited number of substrates (the lack of versatility in the use of organic substrates apparently results from restrictions on permeability). The yields of *Chlorella pyrenoidosa* and *C. ellipsoidea* grown on various carbon and nitrogen sources is shown below.

Acetate-limited chemostat cultures of *C. vulgaris* with acetate as a sole carbon source yielded the growth characteristics of the alga under dark-heterotrophic conditions.[18] The energic analyses gave values of true growth yields, $Y_G = 25.6$ (g cell/mol acetate, $Y_{Go} = 23.8$ (g cell/mol O_2), and $Y_{G'} = 0.12$ (g cell/kcal) and values of maintenance coefficients, $m = 0.2$ (mmol acetate/g cell/hr), $m_o = 0.4$ (mmol O_2/g cell/hr) and $m' = 0.042$ (kcal/g cell/hr), respectively.

Chemical Composition

Various environmental factors influence the chemical composition of many microalgae, *Chlorella* being among the species in which most extreme variations in composition were reported in response to several environmental and nutritional factors. Spoehr and Milner[19] who pioneered this field, described a method for determining the degree of reduction of the total organic matter of plant material from its elementary chemical composition. The degree of reduction was designated the R-value, which is proportional to the heat of combustion and is an expression of the energy content of the material. From the elementary analysis and R-value, it was possible to calculate the approximate carbohydrate, protein, and lipid contents of the algal material. Different environmental conditions affect these components in *Chlorella* very widely, e.g., the lipid content varied from 4.5 to 85.6%. The CO_2 concentration, aerobic and anaerobic atmospheres, mineral nutrients and fixed nitrogen, illumination, and temperature all effect the yield of *Chlorella* and the R-value of its composition. In general, cells having low R-value are produced when the fixed nitrogen in the medium is above 0.001 *M;* below this concentration, cells of higher R-value can be obtained. High light intensity also favors the production of cells of high R-value. In addition, Spoehr and Milner[19] found very striking changes in chlorophyll content of *Chlorella* cells with increase in their R-value, indicating that cells with very high lipid content carry on photosynthesis with a chlorophyll content 1/500 to 1/2000 that of cells of low lipid content.

Effect of Nitrogen Starvation

Many studies that concerned the metabolic response of *Chlorella* to extreme conditions were focused on identifying inherent differences between strains of *Chlorella* in the response to nitrogen starvation and high light intensity. Vladimirova et al.[20] compared *C. pyrenoidosa Pringsheim 82T* (A thermophylic strain) and *C. pyrenoidosa* Chick 82, with respect to growth rates, photosynthetic responses, and accumulation of storage material. *C. pyrenoidosa Pringsheim 82T* responsed to nitrogen starvation by accumulating carbohydrates, while *C. pyrenoidosa* Chick 82 accumulated lipids.

Under nitrogen starvation, *Chlorella* sp. K accumulated starch up to 80% of the total carbohydrates, *C. pyrenoidosa* 19 N/V more than 60%, while in *C. vulgaris* LARG-1, starch comprised only 27.6% of the carbohydrates. In strains with a low carbohydrate content, fatty acids accumulated to a greater degree. An especially large accumulation of fatty acids was observed in the strains *C. pyrenoidosa* 82 and *C. ellipsoidea* SK., fatty acids consituting 52.0 and 65.0% of the total increase in dry weight, respectively.[21]

Semenko and Rudova[22] observed that the lipid-accumulating strain of *C. pyrenoidosa* 82 lost the ability to synthesize lipids when exposed to cycloheximide (3 mg/ℓ). The total amount of protein decreased and the biomass increased only through carbohydrate accumulation. On the other hand, cycloheximide did not inhibit the biosynthesis of substances in *Chlorella* sp. K, the carbohydrate-accumulating strain. The conclusion was that accumulation of lipids in strain *C. pyrenoidosa* 82 under nitrogen starvation required preliminary *do novo* synthesis of proteins, probably reflecting the induction of adaptive enzymes.

The effect of nitrogen deprivation on mass and lipid yield of *Chlorella* was recently reported by Lien.[23] An increase in total lipid biosynthesis is not a prerequisite for storage-lipid accumulation in this organism. In fact, the rate of total lipid synthesis decreases in response to nitrogen depletion. The rate of triacylglycerol synthesis, however, more than doubles and accounts for 96% of newly synthesized total lipids starting within 12 hr after nitrogen deprivation. Consequently, a major shift in the compositions of algal lipid occurs following nitrogen depletion. In contrast to the polar lipids of nitrogen-sufficient cells, neutral lipids in the form of triacylglycerols become the predominant components of lipids from N-depleted *Chlorella* cells.

Lien[23] pointed out that the light requirement for lipid synthesis and the concomitant photodestruction of the photosynthetic capacity in nitrogen-depleted cells meant that there was a limited interval following nitrogen starvation during which this alga can synthesize storage lipid from new photosynthate. The results indicated that although nitrogen starvation greatly increases the percent of lipid in the cells, this treatment cannot increase the overall lipid yield per unit of illuminated area.

Commercial Production of *Chlorella*

The massive experimentation on the growth physiology of mass cultures of *Chlorella* which took place in the early 1950s was discontinued to a large extent in the U.S. and Europe. In Southeast Asia, however, and particularly in Japan and Taiwan, a *Chlorella* industry was developed.

The first Taiwan *Chlorella* company was established in 1964. Two years later this firm

Table 6
EFFECT OF CARBON SOURCES ON INDUSTRIAL
PRODUCTION OF *CHLORELLA*

Carbon source	Yield (g/m²/day)	Comments
C1 CO_2	10—15	
C2 CH_3COOH	20—40	Also used for pH adjustment
C6 $C_6H_{12}O_6$		
Hog manure after fermentation	15—25	Also used for N supply and pH adjustment
Amino acids	15—20	Also used for N supply and pH adjustment

From Soong, P., *Algae Biomass*, Elsevier/North Holland, Amsterdam, 1980. With permission.

started to produce *Chlorella* on a commercial scale. According to Soong,[24] there were 30 *Chlorella* factories in Taiwan as of 1977, which together have a total production capacity of 200 ton/month producing more than 1000 ton/year of *Chlorella* for use in health food products.

Some companies expanded very quickly. For instance, the China *Chlorella* Company has extended the culture area from 1700 m² to 24,000 m² and the production from 3 ton/year up to 320 ton/year from the period 1970 to 1977. Four types of culture systems are used to produce *Chlorella:* (1) open pond with agitation, (2) open pond with circulation, (3) deep culture in closed fermenter, and (4) combination of types 3 and 1.

In a precultivation step, the inoculum is produced heterotrophically in closed fermenters and after transfer to open ponds, the chlorophyll content increases significantly.[24]

The *Chlorella* industry is producing *Chlorella* mass both autotrophically and heterotrophically. Thus, in addition to CO_2 which is supplied during daytime, acetic acid or glucose are also supplied, thereby greatly increasing production.

The w/w efficiency of acetic acid and glucose utilization by *Chlorella* is approximately 30 to 50%; i.e., 2 kg of glucose or 3.5 kg of acetic acid are needed to produce 1.0 kg of *Chlorella* dry matter. The concentration of *Chlorella* in the open pond according to Soong[24] is very high: 1 to 5 g/ℓ. When glucose is used as carbon source in a closed fermenter, the concentration of *Chlorella* may be 10 times higher. Such high cell densities are not possible for autotrophic carbon nutrition and in the closed fermenter, light is obviously limiting chlorophyll formation. To increase the chlorophyll content of the cells, the *Chlorella* is removed from the fermenter by centrifugation and after being washed with water to remove the residual sugar, the *Chlorella* mass is then poured into an open pond where it is kept for a few days to increase the chlorophyll content. Instead of "greening" fermenter-grown *Chlorella* in an open pond, transparent plastic tubes may be used through which the culture is run from the fermenter and back.[24] (Table 6.)

The *Chlorella* product is distributed as a powder or as pills for the health food market. An additional product is termed "Chlorella Growth Factor" (CGF) which is known to improve the growth of lactic bacteria. A summary of the chemical properties of CGF and a flow sheet describing its production are described by Soong.[24]

SPIRULINA

Morphology and Taxonomy

Spirulina is a multicellular, filamentous cyanobacterium. Under the microscope, *Spirulina* appears as a blue-green filaments composed of cylindrical cells arranged in unbranched,

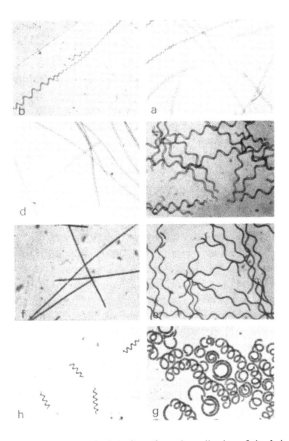

FIGURE 5. Different morphological types in *Spirulina* (from the collection of the Laboratory of Micro-algal Biotechnology at the Jacob Blaustein Institute for Desert Research at Sede Boqer, Israel). (a) Isolated from a local oxidation pond; (b) morphological similar trichome as in a, developing from *Spirulina platensis* typical trichome; (c) *Spirulina platensis*, nonvacuolated from Lake Chad; (d) straight nonvacuolated trichomes, isolated from pure culture c, from which they have been apparently transformed; (e) *Spirulina platensis*, vacuolated; (f) straight vacuolated trichomes, isolated from pure culture b, e, from which they have been apparently transformed (a reversion from d-type to c-type and from f-type to e-type was never observed; (g) *Spirulina sp.*, apparently platensis, isolated from Lake Bogoria in Kenya (courtesy of Dr. S. G. Njugana); (k) *Spirulina* (unidentified), gas vacuolated, appearing during the winter in a *Spirulina platensis* outdoor (100 m²) pond. (Magnification × 100.)

helicoidal trichomes (Figure 5). The filaments are motile, gliding along their axis. Heterocysts are absent.

The helical shape of the trichome is characteristic of the genus, but the helical parameters (i.e., pitch, length, and helix dimensions) vary with the species, and even within the same species. The helical shape is maintained only in liquid media, the filaments becoming true spirals in solid media. The transition from a helix to a flat spiral is slow, whereas the reverse occurs almost instantly.[25]

The diameter of the cells ranges from 1 to 3 μm in the smaller species and from 3 to 12 μm in the larger. According to Ciferri[25] an authentic isolate of *S. platensis* from Cahd and *S. maxima* from Mexico grown in the laboratory under identical conditions showed that *S. maxima* is characterized by a diameter of the helix of 50 to 60 μm and a pitch of 80 μm. For *S. platensis* these parameters were >35 to 50 and 60 μm, respectively. On the other hand, cell dimensions were greater in *S. platensis* than in *S. maxima* (diameter 6 to 8 μm in the former and 4 to 6 μm in the latter).[26]

The larger species of *Spirulina* such as *S. platensis* and *S. maxima* have a granular cytoplasm containing gas vacuoles and easily visible septa. Electron microscopy of *S. platensis* revealed that the cell wall is composed of possibly four layers.[27]

As summarized by Ciferri,[25] the septum appears as a thin disk, folded in part. This fold covers a portion of the septum surface and its extent seems to be related to the pitch of the trichome; the larger the pitch, the smaller the folded area and vice versa.

At low temperatures, when the demand for amino acids is limited by the reduced growth rate, cyanophycin granules are the most abundant organelles, occupying up to 18% of the cell volume. On increasing the temperature, the content in these granules progressively decreases, and at 25 to 30°C they are practically undetectable. Polyglucan granules were most prominent at low temperatures (15 to 17°C), decreasing in concentration at higher temperatures. The relative concentration of other cell organelles was not influenced significantly by light or temperature.[25]

Spirulina can grow rapidly, reaching high filament densities in warm, shallow, brackish lakes. It is one of the most commonly found and abundant algae in many alkaline saline lakes in Africa and the Americas.[28] Physiological races with higher and lower salinity optima have been described. Lewin[29] pointed out that while the genera of blue-green algae have generally been distinguished on the basis of morphological features, only a limited number of characters can be used. Most members of the Oscillatoriaceae have unbranched, cylindrical filaments of indefinite length, with no cellular differentiation except for slight variations in the shape of the end cells. Since trichome color and sheath characteristics vary considerably according to environmental conditions, a classical taxonomist of this group may be almost confined to considerations of cell size and shape; however, only the genus *Spirulina* is characterized by regular helical coiling. In fact, this is the only distinguishing feature of the genus. Lewin[29] commented that in the past, Arthrospira was distinguished from *Spirulina* by the presence of intercellular cross-walls, visible by ordinary light microscopy, but since phase contrast and electron microscopy reveal cross-walls even in small species of *Spirulina*,[30] the distinction ceased to be tenable.

Even though helicity is still being used as a taxonomic criterion at the species level, Lewin[29] brings convincing evidence that it may not be a dependable generic character. He described a pure culture initiated from a single filament of *Spirulina platensis* found in an alkaline salt-flat area of a coastal lagoon in California. The filaments were 8 to 10 μm wide, the cells about 4 to 10 μm long and the helix pitch about 110 μm long. The pitch ratio of the helices (gyre length to gyre width) was 1 to 2, the coils being generally of constant width and the cells containing abundant vacuoles. After some months, a considerable variation in helix form which had not been apparent when the alga was first isolated was observed in some subcultures. Filaments, 3 kinds, were isolated and maintained their character in subculture: the helical subclones were distinguished by helix widths of 55 and 25 μm, respectively, while the straight or slightly undulant variants showed no tendency to revert to a helical form.[29] Essentially, the same phenomenon was observed in the author's laboratory *Spirulina* culture collection (Figure 5).

Bai and Seshadri[31] recognized three distinct morphological variants in an isolate of *Spirulina fusiformis* from Madurai: S-type variant with trichomes having more or less regular and distant coils, C-type variant with trychomes having a distinct spindle shape with close coils, and H-type variant with trichomes forming a dumbbell shape with very close to tight coils. They observed that these distinct types readily transformed from one type to another and attempted to identify the environmental factors which influence this reversion. According to their summary, high light intensity and high nutrient concentration affected the transformation of the S-type to C-type, whereas high light intensity and low nutrient concentration enhanced the formation of the H-type variant from C-type. The phenomenon of these morphological transformations is recognized as polymorphism. Indeed, much confusion exists in species taxonomy of *Spirulina,* because of the great morphological variability which exists in this genus.

Fott and Karim[32] made a most detailed study of *Spirulina* species. According to them, the synonomy of the major *Spirulina* species runs as follows:

•*Spirulina platensis* (Nordstedt) Teitler 1925
Synonyms:
Spirulina Jenneri (Hass.) Kutzing var. platensis Nordstedt 1889 in Wittrock Nordstedt: Algae aquae dulcis exsiccatae, Fasc. 21, No. 679, p. 59 (basionym, type material).

Arthrosphira platensis (Nordstedt) Gomont 1982. Monogr. des Oscillariees, Ann. Sc. nat. 7, Botanique, P. 247—248, pl. VIII: 27 (iconotypus, diagnosis prima).

Description (according to diagnosis and figure of Gomont): plant mass bright green. Trichomes blue-green, slightly constricted at the cross walls, forming regular spirals. Turns of spirals 26—36 μm broad, 43—47 μm distant. Ends of trichromes not or very slightly attenuated, terminal cell broadly rounded. Cells of trichomes 6—8 μm in breadth, 2—6 μm long. Locus classicus: Montevideo, Uruguay, America merid., 1889.

•*Spirulina geittleri* De Toni 1939
Synonyms:
Spirulina maxima (Setchel et Gardner) Geitler 1932, Rabenhorsts Kryptogamen-Flora 14, p. 923, Figure 591b.

Spirulina platensis auct. in Rich,[28] Leonard and Compere,[33] Thomasson,[34] Marty and Busson,[26] Baxter and Wood.[35]

Arthrospira maxima Setchel et Gardner in Gardner 1917, New Pacific Coast marine Algae Vol. 6/14:377—379, pl. 22:3 (basionym, diagnosis latina, iconotypus).

Oscillatoria platensis (Nordst.) Bourelly 1970, Algues d'eau douce III, P. 434, pl. 128:5.

The description of *Spirulina geitleri* De Toni based on Fott and Karim[32] examinations of specimens from Jebel Marra, Republic of Tchad and from Ethiopia runs as follows: "trichomes forming regular spirals that diminish gradually towards both ends. Trichomes usually not constricted at the cross walls, but in some forms slight constriction at the cross walls may occur. Ends of trichomes slightly tapering (about 1/10—1/5 of the width of trichome measured in the middle of the spiral). Terminal cell broadly rounded and some forms with moderately thickened cell wall. Dimensions: cells of trichomes 3.4—15 μm breadth, their length 1/2—1/3 of the breadth. Turns of spirals: their breadth (20)-67-(70) μm, their distance (4)-33-76-(80) μm. Occurrence: In salt water ponds in Oakland (California, locus classicus) and in plankton of inland lakes with saline and alkaline water in Africa. Probably in tropical Asia."

Fott and Karim[32] distinguish two forms of *Spirulina geitleri* De Toni, as follows:

•*Forma geitleri*
Synonym:
Spirulina platensis auct. from Africa
Terminal cell broadly rounded, sometimes with cup-like thickened outer wall. Trichomes 6—13.5 μm broad, turns of spirals 42-67-(70) μm broad, their distance 33-76-(80) μm.

Occurrence: Saline pond in Oakland, U.S.A.[36] (holotypus): lakes in Africa: Republic of Tchad,[33—37] (Lake Chilwa in Malawi), Bishoftu Crater Lake Addis Abeba in Ethiopia; Green Lake in Ethiopia,[34] the large Dariba Lake, Jebel Marra, Sudan.[32]

•*Forma minor (Rich) Fott & Karim comb. nova*[32]
Synonyms:
Arthrospira platensis (Nordst.) Gomont f. minor Rich, Revue Algologique 6:77, 1931, (basionym).

Oscilatoria platensis (Nordst.) Bourrelly var. minor Rich in Iltis 1970, O.R.S.T.O.M., ser. Hydrobiol. 4, 3/4: 133, Figure 2:7. Erroneously designated by Iltis as variety.

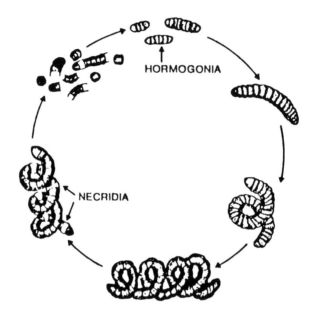

FIGURE 6. Life cycle of *Spirulina*. (From Materassi, R., *Prospective de lella Coltura di Spirulina in Italy*, Consiglio Nazionale delle Riche, Rome, 1980. With permission.)

Terminal cell of trichome capitate, with thickened cell wall, cells of trichomes 3.4—6 μm broad, turns of spirals 37—67 μm broad, their distance 38—76 μm.

Occurrence: Rift Valley in Kenya[28] (locus classicus), natron Kanen Lakes in Tchad,[37] the large Dariba Lake in Jebel Marra, Sudan (our observation), and probably elsewhere in Africa.

Fott and Karim[32] believe that the independence of both forms, *F. geitleri* and *F. minor* (Rich) Fott et Karim comb. nova is substantiated by their common occurrence in the same plankton community of the large Dariba Lake in Jebel Marra and without any transitional forms between them. Moreover, they claim that both taxa are sharply distinguished by the different breath of their trichomes and by the terminal cell, being capitate in *F. minor* and broadly rounded in *F. geitleri*.

Yanagimoto and Saitoh[38] reported that two common stock cultures of *Spirulina*, "Chad" and "Mexico" were wrongly classified. "Chad A" was considered by them to be *Oscillatoria mougeotii*, whereas "Mexico C" was thought to be *spirulina masartii*. The authors observed that the diameter and the spiral pitch of "Chad A" were much larger than those of *Spirulina platensis* and suggested it might be a more suitable strain for mass culture.

Summing up, it is clear that the taxonomic classification of *Spirulina* sp. is complicated and controversial. Clearly, many forms of *Spirulina* and *Oscillatoria* exist that should be tested for their suitability for mass culture. It seems reasonable to believe that in different locations, different forms may be found experimentally to be the most suitable.

Life Cycle

The life cycle of *Spirulina* in laboratory culture is rather simple[39] (Figure 6). A mature trichome is broken in several pieces through the formation of specialized cells, necridia, that undergo lysis, giving rise to biconcave separation disks. The fragmentation of the trichome at the necridia produces gliding, short chains (two to four cells), the hormogonia, which give rise to a new trichome. The cells in the hormogonium lose the attached portions of the necridial cells, becoming rounded at the distal ends with little or no thickening of the walls. During this process, the cytoplasm appears less granulated and the cells assume a pale blue-green color. The number of cells in hormogonia increases by cell fission while the cytoplasm becomes granulated and the cells assume a brilliant blue-green color. By this process trichomes increase in length and assume the typical helicoidal shape. Random but spontaneous breakage of trichomes together with the formulation of necridia assure growth and dispersal of the organism.[25]

Ecology and Habitat

Spirulina is a ubiquitous organism and species of *Spirulina* are found in greatly differing environments, soil, marshes, brackish water, seawater, thermal springs, and fresh water. Clearly, the organism may colonize extreme environments in which life is very difficult for other organisms. Typical are *Spirulina platensis* and *S. maxima,* which profusely populate certain alkaline lakes in Africa and Mexico.[25] As summarized by Ciferri,[25] an extensive study of the phytoplankton of the alkaline lakes was conducted by Iltis.[40] These bodies of water were classified into three categories according to the salt content which was mostly carbonates and bicarbonates. Lakes with a salt concentration below 2.5 g/ℓ support a varied population of microorganisms belonging to chlorophyceae, cyanobacteria, and diatoms. In the mesohaline lakes with a salt concentration ranging from 2.5 to 30.0 g/ℓ, a cyanobacterial population comprising many species *(Synechocystis, oscillatoria, Spirulina, Anabaenopsis)* predominated. In alkaline lakes containing salt concentrations over 30 g/ℓ, the cyanobacterial population was practically monospecific, *Spirulina* being the only organism present in these lakes. According to Ciferri,[25] *Spirulina platensis* was found in waters containing from 85 to 270 g of salt per liter, growth being optimal (at salt range of 20 to 70 g/ℓ).

In Table 7, measurements of pH salinity and alkalinity in several alkaline lakes in Central Africa are recorded in relation to the species specificity. Clearly, the higher the pH and the conductivity, the more marked, in general, the predominance of *Spirulina.*

Ciferri[25] summarized a detailed investigation of two lakes, Rombou and Bodou, both characterized by a very alkaline pH (10 to 10.3 for Rombou and 10.2 to 10.4 for Bodou) but different salt concentrations (13 to 26 g/ℓ for the former and 32 to 55 g/ℓ for the latter). These investigations seemed to confirm that salt concentration plays a direct role in establishing predominance of *S. platensis.* In Lake Rombou, cyanobacteria represent <50% of the phytoplankton population, whereas in Lake Bodou they account for at least 80% of the total population. In addition, in the former, *S. platensis* represented major, but not the only, phytoplankton component with extensive quantitative seasonal variations, whereas in the latter *S. platensis* was practically the only cyanobacterium present. Indeed, with the exception of the months of November and December, when *S. platensis* accounted for 80% of the plankton, in all other months this species represented the totality of the biomass in Lake Bodou.

An analogous situation appears to exist in the alkaline lakes of the Rift Valley in East Africa, lakes characterized by very high pH, reaching, in certain cases, values close to pH 11, and very high salt concentrations, particularly sodium carbonate.[25] Thus, in two crater lakes in Ethiopia, Lakes Kilotes and Aranguadi, both characterized by a high salt content and a highly alkaline pH, support a dense population of *Spirulina* (150). In Lake Kilotes (pH 9.6), *S. platensis* is the predominant organism, although it is accompanied by an unidentified species of Chroococcus. In Lake Aranguadi, characterized by a more alkaline pH (10.3), *S. platensis* is the only microorganism present and its abundance is such that waters appear deep green (in Abyssinian, aranguadi means green). The high concentration of *S. platensis* was responsible for the extremely high photosynthetic rates (1.2 to 2.4 g of O_2 produced/m²/hr).[25]

Melack[41] measured the photosynthetic activity and growth of *Spirulina platensis* in an equatorial soda lake in Kenya, in which a unialgal bloom of *Spirulina platensis* (Nordst.)

Table 7
SPIRULINA IN LAKES OF AFRICA

Country	Year of observation	Area (km²)	Maximum depth (m)	pH	Conductivity (μmhos/cm) at 20°C	Alkalinity HCO₃⁻ + CO₂ (meq/ℓ)	Species *Spirulina* present	Relative abundance	Other phytoplankton components (in decreasing order of abundance)
Ethiopia	1964—1966	0.8	6.4	9.6		51—67	*Spirulina* sp.	Main component, but showing quantitative seasonal variations	*Chroococcus* sp.
Chad	1967—1968		2.5	9.7—10.2	13.000—16.000	146—217	*S. platensis var. minor S. laxissima*	Predominant species (up to 93% of the biomass) all year	*Anabaenopsis arnoldii*
Chad	1967—1968		2	9.7—10.2	4.100—7.200	45—67	*S. platensis var. minor*	Predominant species only in certain months (2 to 70% of the biomass)	*Anabaenopsis arnoldii* *Synecocystis* spp. *Chroococcidiopsis* sp.
Kenya	1974—1976	33	8.5	9.8—10.3		480—800	*S. platensis S. laxissima*	Extensive seasonal or yearly variations, but always the predominant species	Diatoms
Chad	1967—1968	0.75	1.5	10.2—10.4	34.400—47.800		*S. platensis*	From a minimum of 82 up to 100% of the biomass	
Chad	1967—1968			10—10.8	4.260—88.250		*S. platensis*	Predominant species (up to 97% of the biomass) only in certain months	*Cryptomonas sp.*, diatoms, volvocales
Ethiopia	1964—1966	0.8	25	10.3		51—67	*S. platensis*	Only species present	
	1979	49	4.5	10.5—11	14.000—26.000	5.000—90.000 mg/ℓ	*S. platensis var. minor Spirulina sp.*	Predominant species (up to 98% of the biomass) in certain years; When *Spirulina* spp. population decreases, increase in unicellular cyanobacteria	*Synecocystis sp. Chroococcus minutus, Monoraphidium minutum,* diatoms

From Ciferri, O., *Microbiol. Rev.*, 47, 551, 1983. With permission.

Table 8
EFFECT OF MODIFYING WINTER TEMPERATURE ON THE GROWTH OF *SPIRULINA PLATENSIS*

Experiment no.	Conditions	Specific growth rate/day
1	Temp. in the pond not modified	0.3
2	Day temp. raised to 25°C, night temp. not modified	0.16
3	Night temp. prevented from falling below 10°C, day temp. not modified	0.3
4	Day and night temp. raised to 25°C	0.11

From Richmond, A. et al., *Algae Biomass*, Elsevier/North-Holland, Amsterdam, 1980. With permission.

Geitl. was very abundant in August 1973. Their measurements revealed particularly high activities of a phytoplankton in natural environments: photosynthetic rates of bottled phytoplankton reached 12,900 mg O_2 m^{-3}/hr and areal rates ranged from 620 to 5220 mg O_2 m^{-2}/hr. Free water changes in dissolved oxygen varied from 2800 to 12,000 mg O_2 m^{-2}/hr. A principal reason for these exceptionally high rates of photosynthesis was the extraordinary chlorophyll-a content of the euphotic zone: 200 to 650 mg chl-a m^{-2}. Growth rate measurements based on actual weight changes yielded turnover times of 8.9 to 18.9 hr.

The basic biological parameters that concern the growth and biomass production of *Spirulina* sp. were investigated by Ogawa and Aiba,[42] who measured the quantum requirements for CO_2 assimilation, conversion efficiency of energy to biomass, and photorespiration in *Spirulina* sp. They estimated the quantum requirements (defined by Q/QCO$_2$ Einstein/mol CO_2) to be around 20, as compared with 11 to 16, the quantum requirements of *Chlorella* cells exposed to visible light.[43] The conversion efficiency, Y, (defined by energy retained as cell material in kcal per light energy absorbed) was in the order of 10%. By impregnating cells with DCMU (3-(3,4-dichlorophenyl)-1,1-dimethyl urea), which was shown to completely inhibit oxygen evolution through photosynthesis without any ill effects on respiration of *S. platensis*, Ogawa and Aiba[42] concluded that photorespiration in this algae was negligible.

Response to Temperature

Spirulina is a thermophilic alga, i.e., the optimal temperature for its growth being relatively high. According to Zarrouk,[44] the optimal temperature range is between 35 to 40°C. In the author's laboratory, numerous observations with different strains *Spirulina* indicated the optimal temperature for growth was between 35 to 37°C, 40°C being definitely injurious. In outdoor cultures, however, a rise in temperature to 39°C for a few hours did not cause any detectable harmful effects.

The minimal temperature that still permits some growth in *Spirulina* sp. is approximately 18°C, i.e., some 20°C below the optimal. Outdoors, when maximal day temperature declines below 12°C, the culture deteriorates. In contrast to day temperature, *Spirulina* can tolerate relatively low night temperatures. Richmond et al.[45] described an experiment which is summarized in Table 8.

Average maximum day temperatures and minimum night temperatures during the period of experimentation were 18 and 5°C, respectively. Clearly, *Spirulina* could tolerate relatively low night temperatures, the limitation of temperature on the growth of the culture being apparent during the day only. Low night temperature, however, exerts a potentially dangerous indirect effect in that following a cold night, water temperature in the morning is still low (e.g., 4 to 6°C), while the extent of irradiance may be relatively high, forming conditions which promote photooxidation.

Table 9
THE EFFECT OF THE CONCENTRATION OF
UREA ON THE GROWTH (0.D.560) OF
***SPIRULINA*[24]**

	Urea (mol/ℓ) control[a]				
Days	**0**	**1 × 10⁻³**	**5 × 10⁻³**	**2 × 10⁻²**	**Control**
0	0.204	0.204	0.204	0.204	0.204
3	0.229	0.482	0.523	0.523	0.432
5	0.119	0.357	0.783	0.721	0.620
7	0.092	0.319	1.155	0.509	0.854

[a] 1.5×10^{-2} M NaNO$_3$ in medium.

From Shelef, G. and Soeder, G. J., *Algae Biomass — Production and Use*, Elsevierl North Holland, Amsterdam, 1980. With permission.

Nutrition of *Spirulina*

Zarouk[44] studied the effect of salinity on culture grown under laboratory conditions. He reported that *Spirulina* can tolerate approximately 7 g of NaCl and 50 g NaHCO$_3$/ℓ without measurable ill effects, this tolerance reflecting the conditions in the natural habitats of this alga. Alkalinity is mandatory for the growth of *Spirulina* as reflected in the pH optimum for its growth, which according to Zarouk,[44] ranges from about 8.3 to 11.0. The experience in the author's laboratory indicates that outdoors, a pH of 10.4 is yet not limiting to growth, but pH 10.8 is limiting. *Spirulina* can readily tolerate progressive changes in the pH. The culture, however, could quickly deteriorate when the pH is changed abruptly, as may happen in a growth medium which is not well buffered. The O.2 M in NaHCO$_3$, which is the major salt component in the growth medium for *Spirulina*,[44] provides a good buffering capacity for the growth medium. No limitation to growth would take place even when this concentration is radically reduced, e.g., to 0.05 M NaHCO$_3$. In such low alkalinity, however, the culture may become readily contaminated by other algae.

According to Zarouk,[44] mainly nitrates are assimilated by *Spirulina*, but ammonium salts may be used as long as the NH$_4^{2+}$ is relatively low, i.e., about 100 mg N/ℓ. Urea could be used with no ill effects at pH 8.4 as long as its concentration is kept below approximately 1.5 g/ℓ. This is born out in Snoog's report,[24] summarized in Table 9.

Zarouk[44] provides good evidence that both Na and K are indispensable in the *Spirulina* growth medium. Inhibition of growth takes place when the K to Na ratio is >5. As long as this ratio is below 5, growth is uninhibited even at very high concentrations of Na^{+2}, i.e., 18 g/ℓ.

Like most cyanobacteria, *Spirulina* is an obligate photoautotroph, i.e., cannot grow in the dark in media containing organic sources of carbon. Nevertheless, in the light it may utilize carbohydrates since, for instance, the addition of 0.1% glucose to the growth medium enhances growth rate and cell yield.[46,47]

Ogawa and Terui[47] demonstrated that glucose could be used by *S. platensis*. ¹⁴C-labeled glucose was supplied to *S. platensis* and within 80 hr, no glucose was left in the medium. Fifty percent of the label was recovered with the cells, the rest being released either as CO$_2$ (34%) or as organic by-products excreted into the medium (19%).[25]

Phycocyanins, biliproteins involved in the light harvesting reactions, may also serve as a nitrogenous storage material since the phycocyanin concentration was highest when *S. platensis* was cultivated under favorable nitrogen concentrations. When the cultures were completely deprived of nitrogen, a corresponding specific decrease in the cell content of phycocyanin was observed.[48] No other nitrogen-containing compounds decreased under these conditions and the decrease of phycocyanin concentration was associated with an increase in the activity of a protease acting on purified c-phycocyanin. If, under these conditions as well as after inhibition of protein synthesis, the cellular concentration of phycocyanins decreased, severe inhibition of photosynthesis and growth was observed.[25]

Spirulina as a Food Supplement for Humans

The history of *Spirulina* as a staple for the human diet is fascinating. As recounted by Furst,[49] Fray Toribio de Benavente reached the Valley of Mexico in 1524, 3 years after the fall of the Aztec. He described a harvest of tecuitlatl: "There breeds upon the water of the lake of Mexico a kind of very fine mud and at a certain time of year when it is thickest the Indians collect it with a very fine-meshed net until their acales or canoes are filled with it; on shore they make on the earth or the sand some very smooth beds, two or three brazas [3.4—5.1 m] wide and a little less in length, and they cast it down to dry, sufficient to make a case two dedos [3.6 cm] thick. In a few days it dries to the thickness of a worn ducat and they slice this cake like wide bricks; the Indians eat much of it and enjoy it well, this product is traded by all the merchants of the land, as cheese is among us; those who share the Indians' condiments find it very savory, having a slightly salty flavor."[49]

Identification of tecuitlatl, however, came not from Mexico but from Africa. During the 1950s a world-wide interest in novel sources of protein to feed the growing human population led researchers to investigate the possibilities of large-scale algaculture. In 1963, the French Petroleum Institute became interested in reports of a dried alga cake called dihe that was eaten by people along the shores of Lake Chad in central West Africa.

While the French were experimenting with *Spirulina* cultures under laboratory conditions, a Belgian independently discovered dihe and reported on the indigenous techniques of its production. Jean Leonard, a civilian botanist who at the time was on a Belgian military expedition crossing the Sahara from the Atlantic to the Red Sea, became interested in the dihe cakes eaten by the Kanembu of Lake Chad and the blue-green alga from which the cakes were made.[49]

According to Furst[49] the Kanembu, who at times get much of their protein from *Spirulina*, wait for winds to push the alga toward the shore, where it collects and becomes concentrated into a thick mash. Women with calabashes ladle the algal mass into circular depressions in the sand where it is dried by the hot sun. As the blue-green sheet gels, the glossy surface is smoothed by hand and marked off into squares. When most of the water has evaporated or seeped into the sand, the squares are pulled up, dried further on mats, and cut into small, flat, brittle cakes. The Kanembu eat dihe in a thick, pungent sauce made of tomatoes, chili peppers, and various spices poured over millet, the staple of the region.

As summarized by Ciferri,[25] a careful evaluation of the consumption of *S. platensis* in the Chad area was reported in 1976 by Delpeuch et al.[50] It appears that the area in which dihe is eaten regularly is restricted to a fairly limited region, east and northeast of Lake Chad, with a total population of approximately 300,000. Of this population only the quantitatively most important ethnic group, the Kanenbou, consumes dihe regularly, whereas its consumption is nil among the fishermen living around the lake and the nomads north of it. Depending on the season, dihe is present in seven of ten meals. Direct consumption of the dihe biscuits takes place only for superstitious reasons among pregnant women. The reason is that its dark color will supposedly screen the unborn baby from the eyes of sorcerers. In general, dihe is eaten as a constituent of a number of sauces that always accompany the standard millet meal. The dry dihe is pounded in a mortar and the powder is suspended in water. Salt, pimento, tomatoes, and, if available, beans, meat, or fish are added to complete the sauce. In a meal, a person eats approximately 10 to 12 g of dihe which satisfies at most 8% of the caloric need and little more than 10% of the protein requirement. That dihe may represent an "emergency" sort of food may be inferred from the finding that its consumption decreases when the economic conditions, or the local availability, allow consumption of meat or fish. However, during periods of severe famine, dihe is still consumed extensively although one expects that these periods may often be the result of severe droughts that may cause drying of many temporary lakes, reducing the supply of *Spirulina*.[25]

When the Spanish first came to Mexico, *Spirulina* was harvested regularly from Lake Texcoco. The trade in the tecuitlatl died out in time, but *Spirulina* continued to survive in the shrunken remains of Lake Texcoco. Sosa Texcoco, the company that holds the concession for the production of caustic soda and soda ash from the remnants of Lake Texcoco, was aware of the existence of *Spirulina* during the 1960s. They also knew that as recently as a generation ago some Mexican Indians who lived along the shores of the lake were still harvesting tecuitlatl.

When tests conducted by the Mexicans confirmed the French report that *Spirulina* has nearly the high quality of whole egg protein and is rich in vitamins and minerals, the Mexicans decided to produce tecuitlatl for commercial sale.[25]

Cultivation of Mass Cultures

The basis for the growth medium *Spirulina* is the formula proposed by Zarouk,[44] with modifications in the relative proportions of carbonate and bicarbonate. The higher the pH that is to be maintained in the pond, the greater the proportion of the former. Also, ammonia or urea, which are less expensive than nitrate, are used in large-scale culture and phosphorous could be well supplied as phosphoric acid. If the water contained relatively high Ca^{++} and Mg^{++}, it would be advantageous to decrease their concentration in the water to be used for the growth medium either by passage through a proper ion exchanger or by precipitation as carbonates. The outdoor cultivation of *Spirulina* today is carried out in raceways, mixing being provided by a paddlewheel (see "Elements of Pond Design and Construction"). Ciferri[25] describes an interesting alternative — growing *S. platensis* and *S. maxima* in polyethylene tubes. The tubes obviously function as solar collectors which extend the growing season in Italy well into the winter, but causes overheating in summer. Daily productivity in tubes has reached 15 g (dry weight)/m^2 in summer and 10 g (dry weight)/m^2 in winter to give a yearly production estimated, perhaps a bit optimistically, to be about 40 to 50 tons/ha. Other advantages of the tubular system are the considerable reduction of water loss by evaporation, the possibility of utilizing sloping (up to 10%) terrains, and the screening from external contaminants (biological or otherwise). The quality of the biomass produced in tubes, in terms of protein content, seems to be lower than that of the biomass produced in open ponds.[25]

A basic issue in the production of photoautotrophic organisms in general and *Spirulina* in particular is to maintain a continuous culture with an optimal population density. In all seasons, maintaining the culture at a concentration of approximately 300 to 500 mg of dry weight per liter at depth of 12 to 15 cm resulted in a very substantial increase in the output rate as compared to lower or higher densities. In winter, the production is greatly reduced and the dependence of the output on the population density decreases. This is because temperature rather than light becomes growth limiting. In summer, when temperature limitation may exist only in early morning, the dependence of the output on the population density is most acute. Significantly, peak outpu in summer is achieved at a lower population density than in winter, as expected in a system in which light availability for the cells is the sole limitation to growth. At optimal density, defined as that density which results in the highest output rate per unit area, light limitation is extremely keen. At this density, solar irradiance penetrates to only a fraction of the pond's profile, leaving most of the cells in the culture in complete darkness at any given instant. This is why a turbulent flow in the pond is such an important issue; indeed, this is the key factor for obtaining high output through a better utilization of the irradiant flux which reaches the pond surface. Slow, essentially laminar flow in the pond results in low efficiency of solar energy utilization because at a given time interval, most of the culture is exposed to insufficient irradiance. The rest of the culture is exposed to a radiation intensity which at best cannot be efficiently utilized, and at worst, as in the case of *Spirulina* strains sensitive to high radiation intensity, causes damages through photoinhibition and photooxidation. The thesis that turbulence is mainly necessary to eliminate the boundary layers interfering with nutrient flux and gaseous exchange is erroneous and leads to conceptual fallacies in designing tactics for the production of mass cultures. How to maintain optimal turbulence in the pond is a basic technological and economical problem which is yet far from having a satisfactory solution (see "Technological Aspects of Mass Cultivation — A General Outline" and "Outdoor Mass Cultures of Microalgae").

Successful maintenance of *Spirulina*, as well as other microalgae in an outdoor mass culture requires a constant vigil, i.e., up-to-the-minute information from which to assess the state of the culture. The concentration of dissolved oxygen in the pond was repeatedly found in the author's laboratory, as well as by other workers, to be a most sensitive and reliable parameter from which to assess the relative well-being of the culture. The O_2 concentration which is below saturation at night climbs quickly at sunrise, reaching on a hot, sunny, early afternoon a value of over 400% saturation. The lower the turbulence in the pond, the higher concentration of dissolved O_2. The concentration of O_2 in the pond relates to two contrasting aspects: when it is increasing rapidly during the day, the culture is judged to be active and well, yet the rise in O_2 concentrations to extreme values is potentially very dangerous. It inhibits the process of photosynthesis and forms prerequisite for the development of photooxidation. Photooxidation (see "Outdoor Mass Cultures of Microalgae") causes deterioration and could lead to the complete loss of the culture. In contrast, a relatively slow rise in dissolved O_2 which is accompanied with a decline in the maximal concentration of pond O_2 is in itself harmless, but should cause alarm because it may reflect the onset of deterioration in the culture. Indeed, any unexplainable decline in O_2 concentration indicates some weakness in the culture. In the author's laboratory, *Spirulina* cultures were saved in time from complete deterioration due to the early warning provided by a definite, albeit initially small, decline in dissolved oxygen.

The optimal pH for maintenance of a *Spirulina* culture is not yet clear. It seems obvious that the pH should be maintained as high as possible to creat exclusive surroundings for *Spirulina* without limiting growth. At both extremes of the growing season, i.e., in early spring and autumn, an output of up to 10 g dry weight/m²/day may take place without any change in pH. This indicated that some 5 g of carbon/m²/day must have been taken up by the system from the air. Up to pH 9.6, a loss of CO_2 from the culture takes place with the rise in the concentration of dissolved O_2. This phenomenon alone bestows an advantage on maintaining the cultures on as high a pH and on as low total carbon concentration as will not limit growth and will not endanger steady state in the culture.

The most crucial challenge in the commercial production of *Spirulina* seems to maintain a monoalgal culture throughout the year. The basic demand in this respect is to provide growth conditions which will not sway too much from the optimal for *Spirulina*. Since nutritional deficiencies can be easily controlled in mass cultures, the most important factor to regulate is the temperature. The farther away the temperature declines from the 35 to 37°C optimal for *Spirulina*, the slower the growth rate and the easier for contaminants to take over the culture. *Chlorella* sp. seems to be the most formidable organism that contaminates *Spirulina* cultures. However, when the culture is harvested by filtration, any organisms small enough to pass through the filter, such as a short *Spirulina* strain that has arisen in the ponds operated by the author's laboratory, may develop into a serious contaminant. Thus, the best mode of harvesting the algal biomass should answer two criteria: it should preferably completely remove all the biomass during harvesting and it should be such that no cell breakage takes place. Such a harvester is not yet available for the *Spirulina* industry, a factor which curtails commercial production. A point to stress is that when the harvesting process is accompanied by even a relatively small amount of cell breakage, the returning flow is sufficiently enriched in inorganic matter to provide an advantage to mixtrophic or heterotrophic competitors.

From an economic standpoint, the major handicap which today impedes large-scale mass production of *Spirulina* is the high cost of production which results predominantly from the low yields per unit area obtained at the present. Close to 40% of the cost of production stems from capital costs, in which the cost of pond construction is prominent; hence, the emphasis on the areal yield. Theoretically, in well-mixed ponds grown in optimal temperature and maintained correctly, an output of some 40 g m²/day or more is no doubt possible. One half this output on an annual basis would reduce the cost of production significantly, permitting expansion of production, a major prerequisite for establishing a stable consumption of *Spirulina* in various forms and for several purposes.

Commercial Production of *Spirulina*

Commercial production of *Spirulina* today takes place in Mexico, Taiwan, Thailand, California, Japan, and Israel. In each country, local conditions dictate the choice of materials for pond construction and alga nutrition. As $NaHCO_3$ is rather expensive in Taiwan,[24] Taiwanese use NaOH as a substitute, adjusting the pH with CO_2. Nitrate, which is very stable at high pH, is also costly in Taiwan. Thus, they use urea, ammonium, and ammonium sulfate, taking care to proceed with the feeding in a continuous manner maintaining the nutrients at constant but low concentrations. Soong[24] commented that under these circumstances, the yield of *Spirulina* was even greater than the yield obtained with nitrate as N source.

In Mexico, the largest single plant for the production of *Spirulina* had been in operation in Lake Texcoco, in the Valley of Mexico, located 2200 m above sea level but in a semitropical climate (average yearly temperature, 18°C). *S. maxima* grows naturally in the lake and the firm that operates a plant to extract soda from the lake is now recovering and commercializing the cyanobacterial biomass. *S. maxima* is harvested from the most external portion of a giant solar evaporator of spiral shape with a diameter of 3 km and a surface area of 900 ha.[51] The biomass is recovered by filtration and, after homogenization and pasteurization, spray dried. Daily production of the plant has been reported to approach 2 tons (dry weight), with a yield of 28 tons of protein per hectare per year.[52] Although the long-term goal is that of a protein source for general human consumption, so far it appears that *S. maxima* biomass is commercialized mostly as a health food or as a food for special purposes (including some fancy uses such as to enhance the color of certain Japanese ornamental fish).[25]

In Thailand, the Siam Algae Co. Ltd., a subsidiary of Greater Japan Ink and Chemicals, Inc. (DIC) in Japan, commenced operation in October 1978. Thailand was chosen after a long search for the ideal climate. DIC manufactures a health food, a food coloring, and fish feeds for fancy koi (carps). The system of ponds is different from the spiral stream system. DIC produced about 100 tons/year in 1980 by growing an Ethiopian strain of *Spirulina*.[53] by Nakamura.[53]

In the U.S., the major usage of *Spirulina* is as a health food, taking root mainly through the efforts of Switzer and Hills. As defined in a document issued by the Federal Drug Administration (FDA) "Spirulina is a source of protein and contains various vitamins and minerals. It may be legally marketed as a food or food supplement so long as it is labeled accurately and contains no contaminated or adulterated substances."[54]

In the U.S., the pioneer in the production of *Spirulina* is Earthrise Farms, established in

FIGURE 7. The Earthrise plant in Imperial Valley, California. The ponds are 5000 m² in area, lined with food-grade PCV and stirred by a paddlewheel approximately 2 m in diameter, providing a flow of about 20 cm/sec. (Courtesy Dr. A. Jassby, Earthrise.)

the Imperial Valley of California (Figure 7). The company operates 10 to 5000 m² and a whole array of smaller ponds (e.g., two of 1000 m², two of 200 m², two of 40 m², and a dozen smaller ponds for experimentation). Plans are to reach a net yield of approximately 100 tons with the present installation.

In Israel, two commercial plants have recently begun production. The Koor-Hills Co. established a plant near the city of Eilat, on the gulf of Aqaba. At the present, their annual production capacity is between 10 to 20 tons. The other plant, Ein Yahav Algae, is being built at the present at Ein Yahav, in the Arava Valley which connects the Dead Sea and the Red Sea, and will have a similar capacity at the initial phase of operation (Figure 8). Both production plants have plans for expansion.

The world production of *Spirulina* is summarized in Table 10. Even a small quantity of the unprocessed dried algae contains significant nutritional elements aside from the high protein content (Table 11). *Spirulina* appears to have the highest vitamin B_{12} content of any unprocessed plant or animal food. It was assumed that B_{12} occurred in substantial amounts only in animal sources: the availability of a natural plant source of B_{12} is a boon for vegetarian diets. According to Switzer[54] the *Spirulina* powder from Mexico contains about 30 μg of B_{12} in 2 heaping tablespoons (20 g), compared to 16 μg in an equal amount of liver, which usually is touted as the best source. The U.S. Recommended Daily Allowance (RDA) is 6 μg. In addition, two heaping tablespoons of *Spirulina* provide significant quantities of other B-complex vitamins, including 70% of the RDA for B_1 (thiamine), 50% for B_2 (riboflavin), and 12% for B_3 (niacin).

The blue-green color of *Spirulina* (and of other blue-green algae) is due to two pigments: phycocyanin, the blue color, and chlorophyll, the green. These two pigments mask the presence of yet another group of pigments, the carotenoids, which typically are red, orange, and yellow in color. Of interest to humans is the main carotenoid pigment, β-carotene, which is absorbed by the body and converted to vitamin A. Researchers for the World Health Organization (WHO) estimate that approximately 16% of the β-carotene ingested is converted to usable vitamin A. On this basis 1 tablespoon (10 g) of *Spirulina* contain about 25,000

FIGURE 8.   At the commercial plant of Ein-Yahav Algae, Israel. A 1000 m², PVC-lined *Spirulina* pond covered with polyethylene sheet for solar heating in winter. The paddlewheel may be seen in the far right. (Photo by Dr. A. Vonshak.)

Table 10
COMMERCIAL PRODUCTION OF *SPIRULINA* FOR FOOD (1984)[55]

Country	No. of production plants	Area (ha)	Annual production	Comments
Mexico	1	10	300	Yield based on published reports of 10 g/m²/day for 11 months, and a total of 3000 tons for 9 years (1974—1982); production began in 1974
Taiwan	4	16	300	Annual product of 300 was assumed; Taiwan farms are concrete, owned by blue continent *Chlorella* (3 ha in north) = tung HAI *Chlorella;* (3 ha in central) = nan pao chemicals; (6.6 ha in south) = Far East microalgae
U.S.	1	5	90	Production began in 1983; ponds are lined with food grade PVC; owned by Proteus Corporation and Dainippon Ink G Chemicals
Thailand	1	1.8	60	Production began in 1978; ponds are concrete; Siam Algae Co. is owned by Dainippon Ink G Chemicals
Japan	1	1.3	40	Japan farm on Miyako Island consists of concrete ponds under greenhouse; Japan Spirulina Co. is owned by Nichimen
Israel	2	1.5	30	Production in one plant began in 1983, in the other in 1984; plants owned by Koor-Hills and Ein-Yahav algae; ponds lined with food-grade PVC and the production figures are estimations
World	10	35.6	850	

IU or over 500% of the RDA for preformed or true vitamin A. β-Carotene is converted to vitamin A only as needed in the body.[54]

Many other attributes of *Spirulina* are also of nutritional significance, e.g., iron and essential unsaturated fatty acids, the most important of which is γ-linolenic acid. *S. platensis* is unique among photoautotrophic organisms so far studied, containing major quantities of

Table 11
CHEMICAL ANALYSIS

Chemical composition		Sodium (Na)	412 mg/kg
Moisture	7.0%	Chloride (Cl)	4,400 mg/kg
Ash	9.0%	Magnesium (Mg)	1,915 mg/kg
Proteins	71.0%	Manganese (Mn)	25 mg/kg
Crude fiber	0.9%	Zinc (Zn)	39 mg/kg
Xanthophylls	1.80 g/kg	Potassium (K)	15,400 mg/kg
	of product	Others	57,000 mg/kg
Carotene	1.90 g/kg		
	of product	Sterols	
Chlorophyll-a	7.60 g/kg	Cholesterol	325 mg/kg
	of product	Sitosterol	196 mg/kg
		Dihidro-7-cholesterol	
Total organic nitrogen	13.35%	Cholesten-7-ol-3	32 mg/kg
Nitrogen from proteins	11.36%	Stigmasterol	
Crude protein (% Nx6.25)	71.0%	Others	
		Nutritional value	
Essential amino acids		Protein Efficiency Ratio (PER) of 2.2 to 2.6	
		(74—87% that of casein)	
Isoleucine	4.13%	Net Protein Utilization (NPU) of 53 to 61%	
Leucine	5.80%	(85—92% that of casein)	
Lysine	4.00%	Digestibility of 83 to 84%	
Methionine	2.17%	Available lysine	(average 85%)
Phenylalanine	3.95%		
Threonine	4.17%	Nitrogen from	1.99%
Tryptothan	1.13%		
Valine	6.00%	Nucleic acids	
		Ribonucleic acid (RNA)	3.50%
Nonessential amino acids		RNA = N x 2.18	
Alanine	5.82%	Deoxyribonucleic acid (DNA)	1.00%
Arginine	5.98%	DNA = N x 2.63	
Aspartic acid	6.43%		
Cystine	0.67%	Carotenoids	4,000 mg/kg
Glutamic acid	8.94%	α-Carotene	Traces
Glycine	3.46%	β-carotene	Av. 1,700 mg/kg
Histidine	1.08%	Xantophylis	Av. 1,000 mg/kg
Proline	2.97%	Cryptoxanthin	Av. 556 mg/kg
Serine	4.00%	Echinenone	Av. 439 mg/kg
Tyrosine	4.60%	Zeaxanthin	Av. 316 mg/kg
		Luthein and Euglenanone	Av. 289 mg/kg
Vitamins			
Biotin (H)	Av. 0.4 mg/kg	Total lipids	7.0%
Cyanocobalamin (B$_{12}$)	Av. 2 mg/kg	Fatty acids	5.7%
d-Ca-Pantothenate	Av. 11 mg/kg	Lauric (C-12)	229 mg/kg
Folic acid	Av. 0.5 mg/kg	Myristic (C-14)	644 mg/kg
Inositol	Av. 350 mg/kg	Palmitic (C-16)	21,141 mg/kg
Nicotinic acid (PP)	Av. 118 mg/kg	Palmitoleic (C-16)	2,035 mg/kg
Pyridoxine (B$_6$)	Av. 3 mg/kg	Palmitolinoleic (C-16)	2,565 mg/kg
Riboflavin (B$_2$)	Av. 40 mg/kg	Heptadecanoic (C-17)	142 mg/kg
Thiamine (B$_1$)	Av. 55 mg/kg	Stearic (C-18)	353 mg/kg
Tocopherol (E)	Av. 190 mg/kg	Oleic (C-18)	3,009 mg/kg
		Linoleic (C-18)	13,784 mg/kg
Moisture	7.0%	γ-linolenic (C-18)	11,970 mg/kg
		α-Linolenic (C-18)	427 mg/kg
Ash	9.0%	Others	699 mg/kg
		Insaponifiable	1.3%
Calcium (Ca)	1,315 mg/kg	Sterols	325 mg/ᵏ
Phosphorous (P)	8,942 mg/kg	Titerpen alcohols	800 mg/kg
Iron (Fe)	580 mg/kg	Carotenoids	4,000 mg/kg

Table 11 (continued)
CHEMICAL ANALYSIS

Total lipids (continued)		*Coliforms*	20/g
Chlorophyll-a	7,600 mg/kg	*Salmonella*	None
Others	150 ≫/kg	*Shigella*	None
3-4,Benzypyrene	3.6 mg/kg	*E. coli* enteropathogene	None
Toxicology	Nontoxic	Toxics and pesticides (all negative)	
Heavy metals	Typical	α-BCH	
Arsenic (as As203)	1.10 ppm	1,2,3,4,5,6 Hexachlorocyclohexane (neg.)	
Cadmium (Cd)	<0.10 ppm	β-BHC	
Lead (Pb)	0.40 ppm	1,2,3,4,5,6 Hexachlorocyclohexane (neg.)	
Mercury (Hg)	0.24 ppm	γ-BHC	
Selenium (Se)	0.40 ppm	1,2,3,4,5,6 Hexachlorocyclohexane (neg.)	
		δ-BHC	
		1,2,3,4,5,6 Hexachlorocyclohexane (neg.)	
Total carbohydrates	16.5%	DDT	
Ramnose Av.	9.0%	1,1,1 Trichloro-2,2 bis 9p chlorophenyl)ethane (neg.)	
Glucane Av.	1.5%	op'DDD	
Phosphoryled Av.	2.5%	1,1 Dichloro 2-(O-Chlorophenyl)ethane (neg.)	
cyclitols		1,1 dichloro 2-(p-chlorophenyl) 2-(p-chlorophenyl)	
Glucosamine and Av.	2.0%	ethane (neg.)	
muramic acid		op'DDE	
Glycogen Av.	0.5%	2,2 bis (p-chlorophenyl) 2-2 p Dichloroethylene (neg.)	
Sialic acid and Av.	0.5%		
others		Physical properties	
		Appearance	Fine powder
Microbiological analysis of	Max. value	Color	Dark green
Spirulina powder		Odor and taste	Mild, resembling
Standard plate count	20,000/g		sea vegetables
Fungi	10/g	Bulk density	0.5 g/ℓ
Yeasts	10/g	Particle size	9—25 μm

From Hills, C., Ed., *The Secrets of Spirulina, Medical Discoveries of Japanese Doctors*, University of the Tree Press, Boulder, Colorado, 1980. With permission.

γ-linolenic acid (6,9,12-octadecatrienoic acid). This acid is synthesized by the alga by direct desaturation of linoleic acid and is primarily located in the mono- and digalactosyl diglyceride fractions.[56]

Spirulina "Health Food"

Today, the natural food consumer is being offered an ever-expanding variety of new *Spirulina* recipes and products, using the raw, unprocessed blue-green *Spirulina* powder as the unique ingredient. Some other innovative *Spirulina* products are an orange-flavored chewable wafer; protein powders with 10% *Spirulina*, one soy-based, the other milk-egg based; and *Pastalina*, a soy-whole wheat noodle green with *Spirulina*. Switzer[54] describes many recipes for food based on *Spirulina* powder, the following being one example:

Dihe (based on the original Lake Chad recipe)

1 T. spirulina	¹/₄ c. onion, chopped
1 clove garlic, finely chopped	¹/₂ c. vegetable stock or water
¹/₄ c. pimento, chopped	1 c. millet, raw
¹/₂ c. red and green bell peppers, chopped	

Gently boil the millet in a covered pan with 3 c. water for 30 min, or until all liquid is dissolved. Remove and drain if necessary. Premix vegetable stock or water with *Spirulina* powder in a blender separately. In a separate saucepan, saute the onions and garlic in oil. Add pimentos and red and green bell peppers to the saute, and after a few moments pour in the *Spirulina* mix and stir until it has a uniform sauce consistency. Ladle sauce over individual servings of steaming millet. Salt and pepper to taste. Serves 4.

According to Switzer,[54] the possibility of preparing fermented foods similar to cheese, yogurt, and tofu also offers many exciting new possibilities for *Spirulina*. In the future, extraction methods will provide a decolored *Spirulina*, a yellow-white, high-protein powder that is virtually odorless and tasteless, suitable for widespread use.

Therapeutic Properties of *Spirulina*

Certain features of *Spirulina* suggest clinical applications in addition to its unique food value, but there are very few clinical studies as yet. According to Hills,[53] numerous tests that have been conducted in Japan revealed therapeutic effects of *Spirulina* on patients suffering from a long list of diseases. It should be stressed, however, that the number of cases cited by Hills for each effect was very small and rigorous scientific controls were seemingly not applied. Nevertheless, some reliable scientific reports are available and were compiled by Jassby[57] as follows.

Therapeutic Feeding

Whole *Spirulina* has been fed to undernourished children[58] and adults[59] with satisfactory results (see also ''Nutritional Properties of Microalgae — Potentials and Constraints'').

Wound Treatment

Pharmaceutical compounds containing *Spirulina* as the active ingredient produced accelerated cicatrization of wounds.[60] The effect was observed with whole *Spirulina*, culture juice, and extracts. Treatment was effected with creams, ointments, solutions, and suspensions. A separate study showed *Spirulina* and its enzymatic hydrolyzates promote skin metabolism and prevent keratinization.[61]

Thyroid Stimulation

Babaev et al.[62] showed that the iodine in *Spirulina* occurs mostly in thyroid hormones, about 52% as T4 and 18% as T3 in their study. Their results may explain growth stimulation observed in animal feeding tests with live suspensions of *Spirulina*.[63,64]

Cancer Treatment with Phycocyanin

The blue pigment phycocyanin, which constitutes sometimes more than 20% of *Spirulina* dry weight, was extracted and given orally to laboratory mice that had been injected with liver tumor cells.[65] The survival rate of the treatment group was significantly higher than that of the controls. After 8 weeks, none of the control group of 20 mice had survived, while 25% of the treatment group was alive. After 5 weeks, 25% of the control group was alive compared to 90% of the treatment group. In further studies, lymphocyte activity of the treatment group was higher than that of the control group and of a normal group of mice. According to Iijima et al.,[65] phycocyanin may generally stimulate the immune system, providing protection from a variety of diseases.

Cancer Protection from β-Carotene Content

Epidemiological studies suggest that high dietary vitamin A intake decreases cancer risks.[66] A recent study showed that provitamin A β-carotene intake, and not the preformed vitamin A intake from animal sources, correlates with lower cancer rates.[67] Although no specific studies have been done with *Spirulina,* its high β-carotene content (usually greater than 2000 IU/g) suggests that it too may decrease certain cancer risks when ingested in suitable amounts.

γ-Linolenic Acid (GLA) and Prostaglandin Stimulation

Spirulina is a concentrated source of GLA, which constitute approximately 1% of its dry weight. Prostaglandin PGE_1 is involved in many essential tasks in the body, including regulation of blood pressure, cholesterol synthesis, inflammation, and cell proliferation. PGE_1 is formed usually from dietary linoleic acid. The linoleic converts first to GLA with the enzyme δ-6-desaturase and the GLA in turn is converted to dihomo-GLA (DGLA). DGLA gives rise to PGE_1. Dietary saturated fats and alcohol as well as other factors easily inhibit δ-6-desaturase, resulting in GLA deficiency and suppressed PGE_1 formation.[68]

GLA (and subsequent PGE_1 deficiency) may figure in many degenerative diseases. Clinical studies demonstrate that dietary GLA can help arthritis,[69] heart disease,[70] obesity,[71] and zinc deficiency.[72] Alcoholism, manic-depression, certain aging symptoms and schizophrenia also have been ascribed partially to GLA deficiency.[73-75]

Human Enzyme Reactivation

Human erythrocyte-cholinesterase was extracted and inhibited in vitro by organophosphoric pesticides (methyl parathion). Both *Spirulina* and *Chlorella* extracts restored a significant amount of the cholinesterase activity.[76]

Jassby[57] concluded his list of scientific evidence for the existence of therapeutic properties in *Spirulina* by supplying anecodtal evidence which may yet prove to be true. Accordingly, consumers of *Spirulina* reported several consistent effects. These include the ability to postpone or skip a meal with quantities as little as 3 g, reduction of premenstrual stress symptoms and freedom from "hangovers" when 3 g of *Spirulina* are ingested the previous evening.[57]

DUNALIELLA

Taxonomy and Habitat

The genus *Dunaliella* of the order *Volvocales* includes a variety of ill-defined species of unicellular green algae. Members of the genus *Dunaliella* are generally ovoid in shape, 4 to 10 μm wide and 6 to 15 μm long. The cells are motile due to the presence of two equal, long flagella in each cell, and contain one large, cup-shaped chloroplast which occupies about one half the cell volume. The chloroplast contains a large pyrenoid surrounded by polysaccharide granules, the storage product. The chief morphological characteristic of *Dunaliella,* in contrast to other members of the chlorophyta, is the lack of a rigid polysaccharide wall. Instead, the cell is a natural protoplast, enclosed by only a thin elastic membrane. This permits rapid changes in cell shape and volume in response to osmotic changes, and causes it to burst if subjected to extreme hypotonic stress.

Asexual reproduction which is isogamous involves longitudinal division of the motile cells. Meiosis occurs at zygote germination resulting in the formation of motile individuals (Figure 9).

Dunaliella demonstrates a remarkable degree of environmental adaptation to salt and is widely distributed in natural habitats. It has been found in oceans, brine lakes, salt marshes, and salt water ditches near the sea and is the only algal genus to be detected in significant numbers in the Dead Sea or the salt-saturated part of Utah's Great Salt Lake. The algae are

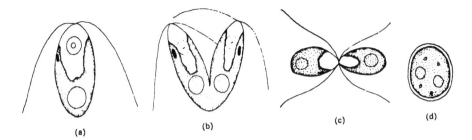

FIGURE 9. *Dunaliella salina.* (a) Vegetative cell ($\times$ 1331); (b) cell division ($\times$ 1331); (c) union of isogametes ($\times$ 675); (d) zygote ($\times$ 675). (From Bold, H. C. and Wynne, M. J., *Introduction to the Algae: Structure and Reproduction,* 1978, p. 73. With permission of Prentice-Hall, Englewood Cliffs, N.J.)

very well adapted to propagate in media ranging from less than seawater concentration (0.1 M NaCl) to saturated salt solutions. (>5 M NaCl).[77]

Halotolerant and Halophilic Types

Ginzburg and Ginzburg[78] investigated the interrelationships of light, temperature, sodium chloride, and carbon source on the growth of *Dunaliella* sp.

No one optimum set of conditions of growth for *Dunaliella* could be defined, since the effects of the major factors affecting cell growth, i.e., temperature, light intensity, carbon dioxide, and NaCl concentrations, all depend on each other. In general, the genus can be divided into two subgroups, one halotolerant and the other halphilic, the two resembling each other in many respects.

A difference was observed between the two subgroups in respect to carbon nutrition. When carbon was supplied in the form of 0.1% sodium bicarbonate, the halotolerant varieties grew almost as well as with 2% CO_2 while the halophilic ones grew more slowly. The halophilic species seem to require a higher concentration of carbon for maximum growth than the halotolerant species.

Halotolerant varieties growing on atmospheric CO_2 or on 0.1% $NaHCO_3$ were very sensitive to light intensity. This factor determined the NaCl concentration and the temperature which they could tolerate. At low intensity (4000 lux) and at 0.5 M NaCl the maximum growth rate was found at 26°C; the growth rate fell off fast as the temperature either rose or fell. As light intensity was increased, the algae became less sensitive both to NaCl and to temperature. Halotolerant varieties remained green no matter what the light intensity to which they were subjected, this inability to synthesize carotenes to any large extent accounting perhaps for their failure to grow satisfactorily outdoors where light intensity reached over 100 klux for several hours each day.

Halophilic varieties were equally sensitive to light intensity though in different ways. They grew little or not at all at low light intensity unless supplied with high concentrations of CO_2. They responded to high light intensities and high temperatures by rapid growth and by the production of carotenes; thus, variety D13 remained green at 400 lux, turned bright yellow at 16,000 lux, and dark brown outdoors.[81]

Osmoregulation

Ben Amotz and Avron,[77,79,80] who have made significant contributions to the understanding of the osmoregularity mechanism in *Dunaliella* provided several lines of evidence regarding the underlying mechanism which enables *Dunaliella* to display its unique halotolerance. Microscopic observations show that *Dunaliella* cells behave like perfect osmometers rapidly shrinking or swelling under hypertonic or hypotonic conditions, respectively (Figure 10).

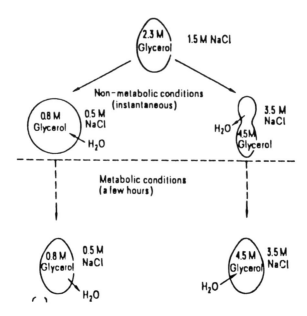

FIGURE 10. Osmoregulation in halotolerant wall-less algae. Schematic representation of the adjustment of *Dunaliella* to hypertonic and hypotonic conditions. (From Ben-Amotz, A. and Avron, M., *Plant Physiol.*, 72, 593, 1975. With permission.)

Clearly, the absence of a rigid polysaccharide cell wall in *Dunaliella* permits a rapid adjustment of the intracellular osmotic pressure by fluxes of water through the cytoplasmic membrane. Thereafter the cells slowly return to their original ellipsoid-like shape through a phase of metabolic adjustment. During this metabolic adjustment period, the algae produce and accumulate glycerol above the original level in hypertonic conditions, whereas upon transfer to hypotonic conditions the algae reduce the glycerol content below the original level. In either case, water flows through the cytoplasmic membrane in response to the new level of intracellular glycerol so that at the steady state the original cell volume is regained.[79.81] Cellular osmoregulation in *Dunaliella* can be defined, therefore, as the ability of the cell to maintain approximately constant volume in the face of changing water potential.

The kinetics of synthesis and elimination of glycerol upon transition from low to high salt concentration or vice versa have been studied in detail,[79.82] and indicate that (1) the process is very rapid (glycerol synthesis or elimination can be detected within minutes after the transition), and (2) such synthesis is independent of protein synthesis or of illumination. Thus, the mechanism of response is ever present and rapid in responding.

A basic issue of osmoregularity in *Dunaliella* concerns the fate of the glycerol. Does it stay within the cell or does it leak out through the plasmalemma, necessitating continuous synthesis to maintain the osmotic potential in the cell? On the one hand, there is evidence that *Dunaliella* retains glycerol very well especially at high salinities,[79.83] but on the other hand there is evidence for a distinct permeability of the plasmalemma of *Dunaliella* for glycerol.[84] Enhuber and Gimmler[85] examined the glycerol permeability of the plasmalemma of *D. parva* in an attempt to learn whether *Dunaliella* minimizes glycerol efflux, following the strategy of avoidance or replacing glycerol leakage from the cells into the medium by a continuous synthesis of new glycerol, and paying the energetic costs for this in a strategy of tolerance.[86] Their conclusion was that the plasmalemma of *D. parva* does not exhibit a particularly low permeability toward glycerol as would be expected from a glycerol-accumulating alga. Rather, significant amounts of glycerol diffuse continuously into the medium

Table 12
THE EFFECT OF SALT CONCENTRATION ON THE PRODUCTIVITY OF GLYCEROL IN OUTDOOR CULTURES OF DUNALIELLA BARDAWIL[87]

NaCl conc (M)	Av. level of glycerol at harvest (g/ℓ)	Productivity of glycerol (g/m²/day)
3.5	0.18	4.4
4.0	0.14	3.4
4.5	0.16	3.0
5.0	0.16	2.4

From Ben-Amotz, A. et al., *Experientia*, 38, 1982. With permission.

following the glycerol concentration gradient between the cells and the medium. Efflux rates vary between 0.1 and 2 mol glycerol per milligram of chlorophyll per hour depending on the external NaCl concentration. After 1 day up to 25% of the total glycerol of the algal suspension was found in the medium. Within 10 days this value can increase to 60%, depending on the growth constant of the culture.

Production of Glycerol

The potential production of glycerol by *Dunaliella* is of obvious economic interest and was analyzed by Ben Amotz et al.[87] Assuming a solar conversion efficiency of 8% for calculating the potential for production of glycerol by *Dunaliella*, algae containing 40% glycerol on a dry weight basis can yield 16 g glycerol/m²/day. The actual production of glycerol in outdoor culture of *Dunaliella* was tested in 10-cm depth miniponds for about 60 days between the months of May and June. When the algae content of the pond reached the desired level of glycerol, one half of the culture volume was harvested by centrifugation and the remaining algae diluted with fresh medium to the original volume.

Table 12 shows that maximal long-term productivity of about 4.5 g glycerol m^{-2}/day has been obtained at a salt concentration of 3.5 M. Since the conditions for optimal growth are not necessarily those which maximize glycerol production, laboratory experiments have been undertaken to check the effect of salt on the production of both glycerol and algal biomass. The highest yield of glycerol was obtained in about 2 M NaCl while the conditions favoring maximal biomass production were in the low range of salt concentration (Figure 11).

Another factor was the contamination by predators and other organisms, which was inversely proportional to the salt concentration. This clearly illustrates a cardinal issue in the biotechnology of mass algaculture, i.e., evaluation of the optimal production capability of a product requires that a whole array of factors be considered.

β-Carotene

Another product of economic importance from *Dunaliella* is β-carotene. It is a natural product of high value as provitamin A and as a natural pigment for the food industry. In the near future for mass production of *Dunaliella* for extraction of β-carotene will be realized Ben-Amotz and Avron[80] studied the factors which determine massive β-carotene accumulation in *Dunaliella bardawil*. Increasing light intensity and duration as well as inhibition of growth by various stress conditions such as nutrient deficiency or high salt concentration

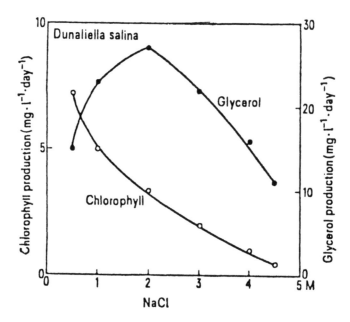

FIGURE 11.   Effect of extracellular salt concentrations on the chlorophyll and glycerol production by *Dunaliella*. Algae were grown in a constant temperature growth room as described previously. (From Ben-Amotz, A. et al., *Experientia*, 38, 1982. With permission.)

caused a decrease in chlorophyll and an increase in β-carotene. As a result, the β-carotene to chlorophyll ratio increased from about 0.4 to 13 (g/g) and the alga changed its visual appearance from green to deep orange. In contrast, *D. salina* grown under similar conditions decreased in cell content of both chlorophyll and β-carotene, the culture turning from green to yellow. Low chlorophyll-containing cells of *D. bardawil* or *D. salina* exhibited very high photosynthetic rates when expressed on a chlorophyll basis (600 μmol O_2 evolved per milligram chlorophyll per hour).

Variation of pigment content in *D. bardawil* by a large variety of environmental agents has been correlated with the integral irradiance received by the algal culture during a division cycle.[80] The higher the integral irradiance per division cycle, the lower the cell chlorophyll and the higher the β-carotene content per cell, and therefore the higher the β-carotene to chlorophyll ratio. The results are interpreted as indicating a protecting effect of β-carotene against injury by high irradiance under conditions that impair the content of cell chlorophyll.

SCENEDESMUS

Taxonomy and Reproduction

Scenedesmus belongs to the family Scenedesmaceae, division Chlorophycophyta. As described by Bold and Wynne,[3] species of *Scenedesmus* are widely distributed in freshwater and soil. The cylindrical cells, with rounded or pointed ends, are laterally joined in groups of 4 or 8 or more rarely, 16. The terminal cells and some of the others in some species (e.g., *S. quadricauda*), have spines. Some species also have tufts of fine bristles[88] which have buoyancy. The cells are uninucleate and have a laminate chloroplast that contains a pyrenoid.

Trainor et al.[89] found that the composition of the medium may have great influence on the form of *Scenedesmus* cultures. When an axenic strain of (probably) *S. dimorphus* (Turp.) Kutz. was grown in the presence of yeast extract, the cells of the coenobia were joined only

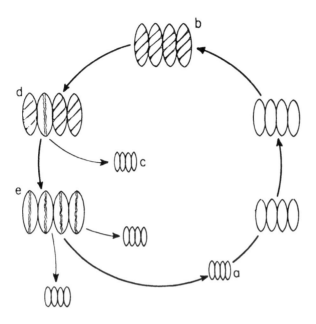

FIGURE 12.   Asexual reproduction. A young colony (a) grows during the light phase. When mature (b) each protoplast divides twice. A new colony (c) is released from within the parent cell (d). Soon all parent cells have released four-celled colonies. The wall of the parent colony (e) remains. (From Trainor, F. R. et al., *Bot. Rev.*, 42(1), 5, 1976. With permission.)

at their apices as in the genus *Dactylococcus nageli;* while in a culture medium containing 0.1% glucose, the cells were coherent laterally to form typical coenobia. Another example for polymorphism in *Scenedesmus* was revealed in *S. longus* Meyen (which resembled the unicellular *Chodatella subsalsa* Lemm. in liquid medium), whereas typical coenobia developed in the same medium solidified with agar. Likewise, *S. dimorphus*, which produced typical coenobia in axenic culture, was entirely unicellular in the presence of a soil-inhabiting bacterium.[90]

Reproduction of *Scenedesmus* is by autocolony formation in which each parental cell forms a miniature colony that is liberated through a tear in the parental wall (Figure 12). Sexual reproduction has been described only for *S. obliquus*.[3]

Mass Culture of *Scenedesmus*

The most successful operation in the outdoor culture of *Scenedesmus* took place in Peru. A Peruvian-German[91] project has been in operation since 1973 culturing *S. acutus* var. alternans Nr. 276-3a at the Pilot Plant in Casa Grande, situated near Trujillo in northern Peru.

The pilot plant is equipped with several small ponds totaling a net cultivation area of 100 m² agitated by paddlewheels. Also, a pond of 100 m² was built in which the algal suspension ran by gravity and was recirculated by a centrifugal pump. The algae are harvested from the suspension by a continuous centrifuge and dried on a drum drier.

The influence of temperature and of solar irradiance on the output rate of *Scenedesmus* are studied in detail. The marked decline in output during the winter (June through September) seems to be primarily associated with the decline in temperature, although the decrease in the number of hours of sunshine must have also exerted a negative effect on the output (see "Elements of Pond Design and Construction"). The yields obtained in the summer months indicated that under appropriate conditions, yields of 25 g/m²/day may be expected on an annual basis, equivalent to approximately 90 tons of dry matter containing 50% protein per hectare per year.

Van Vuren and Grobbelaar[92] also described the close relationship between temperature and light intensity in their effect on the growth rate of *Scenedesmus*. *S. bijugates* was found to have a higher optimal temperature (37°C) at high light intensities (100 to 150 $\mu E/m^2/sec$) and a lower optimal temperature (32°C) at a lower light intensity of 50 $\mu E/m^2/sec$.

Castillo et al.[91] concluded that the optimal temperature range for *Scenedesmus* was between 30 and 32°C. Nevertheless, even though yields decreased at higher temperatures, *Scenedesmus* can grow in temperatures at least as high as 38°C. Similar findings were reported by Van Vuren and Grobbelaar,[92] who projected an Arrhenius plot for the influence of temperature on the growth rate of *S. bijugates*. Their conclusion was that *S. bijugates* has an optimum temperature at 33.5°C and a maximum of about 43°C at light intensities of 150 $\mu E/m^2/sec$. Castillo et al.[91] stated that no saturation effect could be identified for solar irradiances since the output seemed to have increased linearly even at the highest irradiation values obtained, 800 cal/cm^2.

The population density had a marked effect on the yield of *Scenedesmus*, the maximal yield of approximately 25 g/dry wt/m²/day being obtained at a density of approximately 35 g dry wt/m². When the population density was maintained at 10 g/m² or at 80 g/m², the daily yield dropped to approximately 10 g/dry wt/m²/day.[92]

Nutritional Requirements

Van Vuren and Grobbelaar[92] found that in addition to the well - established effect of the N-content in the medium on the percent of crude protein, the protein content could be manipulated from 30 to 47% over a range of potassium concentration of 22 to 254 mg/ℓ. Venkataraman et al.[93] studied the effect of different carbon sources on the growth of *Scenedesmus* (Figure 13). A combination of CO_2 and molasses yielded the best results and may represent a valuable formula for industrial production of *Scenedesmus*.

Scendesmus may be grown heterotrophically, in darkness, particularly on glucose, as shown in Table 13.

In addition to glucose and galactose, slow but continued growth was also supported by mannose, fructose, maltose, and sucrose.

PHAEODACTYLUM

Taxonomy and Habitat

Phaeodactylum belongs to division Chrysophycophyta. According to Lewin et al.,[94] cells of *Phaeodactylum tricornutum* are of two characteristic types, oval and fusiform, each of which remains constant for many cell divisions in clonal cultures. Triradiate cells arose rarely as atypical forms of the fusiform variety; also, oval cells could arise as endospores within a fusiform cell. Oval and fusiform cells both contained approximately the same amount of silica (0.4 to 0.5% dry weight). In each case, most of this silica could be recovered as a particulate fraction resistant to digestion in hot nitric acid. The silica obtained from oval cells was in the form of diatom valves, whereas that from fusiform cells consisted of irregular particles. In comparison with other diatoms, *Phaeodactylum* is very weakly silicified and enough silicon dissolves from Pyrex® glass vessels in alkaline culture media to fulfill its meager requirements.[94]

Lipid Productivity

P. tricornutum is capable of producing and accumulating high levels of lipids, up to 34% of its dry weight when nitrogen is limited.[94,95] The physiology of *P. tricornutum* with respect to lipids differs considerably from that of other phytoplankton which have been investigated.

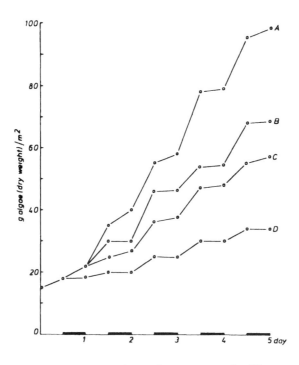

FIGURE 13. Growth of *Scenedesmus acutus* under different conditions, outdoors. (A) Air — 100 ℓ/min with 30 ℓ CO_2/hr and 140 mg/ℓ molasses; (B) air — 100 ℓ/min with 30 ℓ CO_2/hr; (C) air — 100 ℓ/min and 140 mg/ℓ molasses; (D) air — 100 ℓ/min; day; night. (From Venkataraman, V. L. et al., *Life Sci.*, 20, 223, 1977. With permission.)

Table 13
GROWTH RATE OF
SCENEDESMUS SP. D3 ON
VARIOUS CARBON SOURCES[17]

	Specific growth rate/day
In darkness	
Substrate (% w/v)	
Glucose 1	0.26
Galactose 1	0.04
Basal medium alone	0.00
In light	
Basal medium alone	0.65

From Samejima, H. and Myers, J., *J. Gen. Microbiol.*, 18, 107, 1958. With permission.

The percentage of carbon found in the lipid fraction is substantially higher for *P. tricornutum* than has been found in incorporation studies with other phytoplankters.

Like most diatoms, *Phaeodactylum* accumulated lipids that contain mainly palmitic (16:6), hexadecanoic (16:1), and C_2O polyenoic acids, linolenic acid being a minor constituent. The lipid functions as a constitutive component and also serves as a storage product. However, it is not a storage product which is utilized during short-term light deprivation, since it is not consumed during the night.

Phaeodactylum may develop very dense cultures. According to one report, *Phaeodactylum* produced 3.37 to 3.76 mg chlorophyll per liter per day, with 0.056 cal/cm²/min at 12°C reaching a maximum of 80 mg chlorophyll per liter and a cell density of 1.45×10^8 cells/mℓ. Studies of the mass culture of *Phaeodactylum* in outdoor tanks of 1000 ℓ in fertilized seawater indicated that the growth of the algae followed a constant pattern. An active growth period was followed by a period of greatly reduced growth rate (associated with the accelerated growth of a heterotroph — *Monas* sp.). The mean value of solar energy utilization for the *Phaeodactylum* culture in these experiments was 3.7%.[96] Under laboratory conditions the percentage of carbon in the lipid fraction increased linearly with increasing irradiance up to 600 μE/m²/sec, reaching a maximum of about 35%.

The rate of production of *Phaeodactylum* biomass could reach very high levels. According to Thomas et al.,[97] outdoor cultures of *Phaeodactylum* in shallow raceways in Hawaii yielded as much as 41 g/m²/day.

PORPHYRIDIUM

Taxonomy and Habitat

Porphyridium belongs to the order *Porphyridiales*, division Rhodophycophyta. The taxonomy and nomenclature of *Porphyridium* have been investigated by many workers. According to Ott,[98] color is the only valid criterion for specific separation in this genus, environmental changes causing only slight color variations. The correct name of the type species is *P. purpureum* (Bory) Drew et Ross, *P. cruentum* (S.F. Gray) Nag. being a synonym.[3] As described by Bold and Wynne,[3] a diverse spectrum of habitats is occupied by the genus *Porphyridium*, freshwater, brackish, and marine water, as well as the surface of moist soils or pots in greenhouses upon which growths may form reddish coatings. A single prominent stellate chloroplast is present in each cell, with a central pyrenoid region within the chloroplast. The ultrastructure of *Porphyridium* has been reported in great detail (e.g., Chapman and Lang[99]) and it has served as a useful tool in investigations of phycobilisomes[100] and of polysaccharide production.[101]

Polysaccharide Formation

Cells of *Porphyridium* are essentially without walls, in that a skeletal or microfibrillar component is lacking. It produces profusely an amorphous mucilaginous polysaccharide which is an acidic heteropolymer composed of sulfated sugars. The polysaccharide which forms ionic bridges through divalent cations and reaches a very high molecular weight, is constantly being excreted from the cell, forming an encapsulating halo around the cell. It is produced in particularly copious amounts during the stationary phase of growth. The polysaccharide is solubilized as it moves away from the cell into the surrounding medium and is constantly being replenished at the cell surface by dictyosome-mediated processes.[101]

Growth Physiology

P. cruentum has been successfully grown in axenic cultures in a completely artificial seawater medium. It appears to be a complete photoautotroph and the relative growth constant was found to vary between 0.56 and 1.17 log units per day depending upon culture conditions.[102]

Golueke and Oswald[103] investigated the effect of temperature, detention period, light intensity, and salinity on the growth rate and overall light energy conversion efficiency of *P. cruentum* cultured on a medium consisting of concentrated seawater and sewage enriched with urea, chelated iron, and other additives. The found that the optimal temperature was within the range of 21 to 26°C, growth being retarded at temperatures less than 13°C, and completely inhibited above 31°C. Overall light energy conversion efficiency increased from 2.24% at the 4-day detention period to 2.76% at the 10-day period. Conversion efficiency ranged from 5.8% at a light energy absorption rate of 8.2 cal:liter:min to 2.3% at 35 to 39 cal/ℓ/min.

At salt concentrations less than 3.5%, *Porphyridium* in open cultures could not successfully compete with other algae. Salt concentrations as high as 4.6% had no inhibitory effect on its growth. Similarly, Jones et al.[102] reported that waters with salinities from 4.5 to 35.0% S (salinity) supported optimal growth. They also found that the algae grew well in a range of pH from 5.2 to 8.3.

Arachidonic Acid

Recently, *P. cruentum* was assessed by Ahern et al.[104] to be a potentially competitive source of arachidonic (5,8,11,14-eicosatetraenoic) acid. Measurement of the production of this prostaglandin precursor at various temperatures and light intensities revealed that increasing the light intensity within the range of 1700 to 8000 lux promoted the growth rate without affecting the arachidonic acid yield per cell. Also, whereas raising the temperature from 18 to approximately 32°C lowered the yield of arachidonic acid per cell, the rate of its production per unit volume and time was increased. The rates of production of *P. cruentum* cells and of arachidonic acid grown in large-scale batch cultures of 8.5 ℓ at 25 to 26°C were approximately the same as obtained in shaking batch cultures. Continuous cultivation at 25 to 26°C for 10 days and 31°C for 3 days yielded productivity rates as high as 0.4 and 0.88 mg arachidonic acid per liter per hour, respectively.

BOTYROCOCCUS BRAUNII

This is a green, colonial algae which is unique among the algae in its ability to accumulate substantial quantities of hydrocarbons. According to a recent review by Wolf,[105] record hydrocarbon contents of over 80% of the cell dry weight were reported, the typical hydrocarbon content ranging between 25 to 40%. *Botyrococcus* is widely distributed in temperate and tropical regions where it occurs primarily in fresh, though occasionally brackish, and salt waters. Colonies appear in different colors, green, orange, red, or brown, depending on their physiological state and environmental conditions.

The colonies of *Botyroccus* often form floating scum on the surface of undisturbed waters, the buoyancy resulting from the reduction of specific gravity by accumulated hydrocarbons. The type of hydrocarbon produced is apparently related to physiological status. Active state colonies produce unbranched olefins (large $C_{27:2}$, $C_{29:2}$, $C_{29:3}$, and $C_{31:2}$) that have been reported to comprise up to 32% of the dry weight. In contrast, resting state colonies produce unusual branched olefins (general formula C_nH_{2n-10}; n = 30 to 37) that appear to be of terpenoid origin. This "botyrococcene fraction" has been shown to comprise from 27 to 86% of the dry weight in natural collections.[105]

Because of its ability to accumulate large quantities of hydrocarbon and form prodigious natural blooms, *B. braunii* has been proposed as a source of renewable liquid fuel. Just how much oil can be generated per unit pond area in mass culture is definitely uncertain because of the currently observed very slow growth rates *Botyrococcus*. Under controlled laboratory conditions, the doubling time at 26°C with 1% CO_2 was reported to be over 2 days, compared with a few hours that represent the generation time of most algal species maintained under optimal conditions. Thus, a basic physiological issue with regards to the possibility for mass culturing *Botyrococcus* related to the extent to which the observed growth rate could be increased. Whether this could be achieved by fully identifying the limiting growth conditions or by the proper selection, breeding, and perhaps genetic modifications, a faster growing strain of *Botyrococcus* that will yet retain its unusual capacity to produce and store olefins which comprise a substantial amount of the cell dry weight must be singled out before mass cultivation could be considered.

CHLAMIDOMONAS

The unicellular green alga *Chlamidomonas* belongs to the family Chlamydomonadaceae. This ubiquitous, biflagellate alga, with several hundred species, is one of the largest algal genera, occurring in soils and aquatic habitats including brackish and marine.[3]

Chlamidomonas is at present one of the few algae which are grown commercially, albeit at yet a very small scale, being used as a soil conditioner in loamy soils.[106] Metting and Rayburn[106] reported that carbohydrate concentrations were consistently greater in both the surface 10 mm and the upper 300 mm of treated portions of loamy sand and silt loam, compared to untreated portions from the same soil. Increases in water retention from treatments were significant (at the 0.05% level), ranging from 2 to nearly 5% above untreated soils, the innoculation rate being 3 to 5 $\times$ 10[11] cells per hectare.

Differences in chemical and physical properties could not be measured between treated and untreated samples from a dryland silt loam. Moisture was therefore implicated as the factor most likely to control growth and polysaccharide production by the microalgal inoculum.

REFERENCES

1. **Trainor, F. R.,** *Introductory Phycology,* John Wiley & Sons, New York, 1978.
2. **Beijerinck, M. W.,** Kulturversuche mit Zoochlorellen, Lichenogonidien und anderen niederen. *Alg. Bot. Z.,* 48, 725. 741, 757, 781, 1890.
3. **Bold, H. C. and Wynne, M. J.,** *Introduction to the Algae, Structure and Reproduction,* Prentice-Hall, Englewood Cliffs, N.J., 1978.
4. **Kessler, E. and Soeder, C. J.,** Biochemical contributions to the taxonomy of the genus *Chlorella, Nature (London),* 4833, 1069, 1962.
5. **Shihira, I. and Krauss, R. W.,** *Chlorella: the Physiology and Taxonomy of 41 Isolates,* Port City Press, Baltimore, Md., 1965.
6. **Fott, B. and Novakova, M.,** Studies in phycology — A monograph of the genus *Chlorella,* in *The Freshwater Species,* Fott, B., Ed., Academia, Prague, 1969.
7. **DaSilva, E. J. and Gyllenberg, H. G.,** A taxonomic treatment of the genus *Chlorella* by the techniques of continuous classifications, *Arch. Mikrobiol.,* 87, 99, 1972.
8. **Kessler, E.,** Comparative physiology, biochemistry and the taxonomy of *Chlorella* (Chlorophyceae). *Plant Systems Evol.,* 125, 129, 1976.
9. **Kessler, E.,** Physiologische und biochemische Beitrage zur Taxonomie der Gattung *Chlorella.* IX. Salzresistenz als taxonomisches Merkmal., *Arch. Mikrobiol.,* 100, 51, 1974.
10. **French, C. S. and Spoehr, H. A.,** Introduction to Davis, E. A., Dedrick, J., French, C. S., Milner, H. W., Myers, J., Smith, J. H. C., and Spoehr, H. A., Laboratory experiments on *Chlorella* culture at the Carnegie Institution of Washington, Department of Plant Biology, in *Algal Culture: From Laboratory to Pilot Plant,* Publ. no. 600, Burlew, J. S., Ed., The Carnegie Institution, Washington, D.C., 1953.
11. **Cook, P. M.,** Large-scale culture of *Chlorella,* in *The Culturing of Algae: A Symposium,* The Charles F. Kettering Foundation, Yellow Springs, Oh., 1950, 53.
12. **Davis, E. A., Dedrick, J., French, C. S., Milner, H. W., Myers, J., Smith, J. H. C., and Spoehr, H. A.,** Laboratory experiments on Chlorella culture at the Carnegie Institution of Washington Department of Plant Biology, in *Algal Culture: From Laboratory to Pilot Plant,* Publ. no. 600, Burlew, J. S., Ed., The Carnegie Institution, Washington, D.C., 1953.
13. **Davis, E. A. and Milner, H. W.,** Conclusions: Optimum Conditions for Growth of Chlorella, in Davis, E. A., Dedrick, J., French, C. S., Milner, H. W., Myers, J., Smith, J. H. C., and Spoehr, H. A., Laboratory experiments on Chlorella culture at the Carnegie Institution of Washington, Department of Plant Biology, in *Algal Culture: From Laboratory to Pilot Plant,* Publ. No. 600, Burlew, J. S., Ed., The Carnegie Institution, Washington, D.C., 1953.
14. **Sorokin, C.,** Tabular comparative data for the low- and high-temperature strains of *Chlorella. Nature (London),* 1959.
15. **Sorokin, C. and Krauss, R. W.,** The dependence of cell division in Chlorella on temperature and light intensity, *Am. J. Bot.,* 52(4), 331, 1965.

16. **Tamiya, H., Iwamura, T., Shibata, K., Hase, E., and Nihei, T.,** Correlation between photosynthesis and light-independent metabolism in the growth of *Chlorella, Biochim. Biophys. Acta,* 12, 23, 1953.

17. **Samejima, H. and Myers, J.,** On the heterotrophic growth of *Chlorella pyrenoidosa, J. Gen. Microbiol.,* 18, 107, 1958.

18. **Endo, H., Hosoya, H., and Koibuchi, T.,** Growth yields of Chlorella regularis in dark-heterotrophic continuous cultures using acetate, *J. Ferment. Technol.,* 55, 369, 1977.

19. **Spoehr, H. A. and Milner, H. W.,** The chemical composition of *Chlorella;* effect of environmental conditions, *Plant Physiol.,* 24, 120, 1949.

20. **Vladimirova, M. G., Rudova, T. S., Shatilov, V. R., Salamatova, L. V., and Nazarova, G. D.,** Comparative characteristics of *Chlorella pyrenoidosa pringsheim 82T* and *Chlorella pyrenoidosa Chick 82* grown under conditions of intense culture, *Fiziol. Rast.,* 26(6), 1125, 1969.

21. **Klyachko-Gurvich, G. L., Zhukova, T. S., Vladimirova, M. G., and Kurnosova, T. A.,** Comparative characterization of the growth and direction of biosynthesis of various strains of *Chlorella* under nitrogen deficiency conditions. III. Synthesis of fatty acids, *Fiziol. Rast,* 16(2), 205, 1969 (translation).

22. **Semenko, V. E. and Rudova, T. S.,** Effect of cycloheximide on changes of the biosynthesis in *Chlorella pyrenoidosa* cells caused by nitrogen starvation, *Fiziol. Rast,* 22(5), 958, 1975.

23. **Lien, S.,** Effect of nitrogen deprivation on mass and lipid yield of *Chlorella,* in Algal Biomass Workshop, University of Colorado, Boulder, Solar Energy Research Institute, 1984.

24. **Soong, P.,** Production and Development of *Chlorella* and *Spirulina* in Taiwan, in *Algae Biomass,* Shelef, G. and Soeder, C. J., Eds., Elsevier/North Holland, Amsterdam, 1980.

25. **Ciferri, O.,** *Spirulina,* the edible microorganism, *Microbiol. Rev.,* 47(4), 551, 1983.

26. **Marty, F. and Busson, F.,** Donnees cytologique, et systemtiques sur *Spirulina platensis* (Gom.) Geitler et Spirulian Geitleri J. de Toni (Cyanophyceae-Oscillatoriaceae), *C. R. Acad. Sci. Ser. D:,* 270, 786, 1970.

27. **Van Eykelenburg, C.,** On the morphology and ultrastructure of the cell wall of *Spirulina platensis, Antonie van Leeuwenhoek J. Microbiol. Serol.,* 43, 89, 1977.

28. **Rich, F.,** Notes on Arthrospira platensis, *Rev. Algal.,* 6, 75, 1931.

29. **Lewin, R. A.,** Tasks for the microbial geneticist in phycotechnology, private communication, 1980.

30. **Holmgren, P. R., Hostecter, H. P., and Scholes, V. E.,** Ultrastructural observation of cross-walls in the blue-green alga *Spirulina major, J. Phycol.,* 7, 309, 1971.

31. **Bai, N. J. and Seshadri, C. V.,** On coiling and uncoiling of trichomes in the genus *Spirulina, Arch. Hydrobiol. Suppl.,* 60, 32, 1980.

32. **Fott, B. and Karim, A. G. A.,** Spirulina plankton community in a lake in Jebel Marra, Sudan, *Arch. Protestenkd.,* 115, 408, 1973.

33. **Leonard, J. and Compere, P.,** *Spirulina platensis* (Gom.) Geitl., algue bleue de grande valeur alimentaire par sa richesse en proteines, *Bull. Jard. Bot. Natl. Belg.,* 37, 1, 1967.

34. **Thomasson, K.,** Ett fall av tropisk vattenblomning, *Bot. Not.,* 113, 214, 1960.

35. **Baxter, R. M. and Wood, R. B.,** Studies on stratification in the Bishoftu Crater Lakes, *J. Appl. Ecol.,* 2, 416, 1965.

36. **Gardner, N. L.,** New Pacific Coast marine algae I, *Univ. Calif. Berkeley Publ. Bot.,* 6, 337, 1917.

37. **Iltis, A.,** Phytoplankton des eaux natronees du Kanem (Tchad). IV. Note sure les especes du genre *Oscillatoria,* sous-genre *Spirulina* (Cyanphyta), *Cahiers O.R.S.T.O.M. Ser Hydrobiol.,* 4, 129, 1970.

38. **Yanagimoto, M. and Saitoh, H.,** Blue green algae coexistent in stock culture of *Spirulina, Rep. Natl. Food Res. Inst.,* 38, 96, 1981.

39. **Balloni, W., Tomaselli, L., Giovannetti, L., and Margheri, M. C.,** Biologia fondamentale del genere *Spirulina,* in *Prospettive della Coltura di Spirulina in Italia, Materassi, R.,* Ed., Consiglio Nazionale delle Ricerche, Rome, 1980.

40. **Iltis, A.,** Phytoplankton des eaux natronees du Kanem (Tchad). X. Conclusions, *Cahiers O.R.S.T.O.M. Ser. Hydrobiol.,* 9, 13, 1975.

41. **Melack, J. M.,** Photosynthesis and growth of *Spirulina platensis* (Cyanophyta) in an equatorial lake (Lake Simbi, Kenya), *Limnol. Oceanogr.,* 24, 753, 1979.

42. **Ogawa, T. and Aiba, S.,** CO_2 assimilation and growth of a blue-green alga, *Spirulina platensis* in continuous culture, *J. Appl. Chem. Biotechnol.,* 28, 515, 1978.

43. **Emerson, R. and Lewis, C. M.,** The dependence of the quantum yield of *Chlorella* photosynthesis on wave length of light, *Am. J. Bot.,* 30, 165, 1943.

44. **Zarrouk, C.,** Contribution a l'Etude d'une Cyanophycee. Influence de Divers Facteurs Physiques et Chimiques sur las Croissance et la Photosynthese de *Spirulina maxima,* Thesis, University of Paris, France, 1966.

45. **Richmond, A., Vonshak, A., and Arad, S.,** Environmental limitations in outdoor production of algal biomass, in *Algae Biomass,* Shelef, G. and Soeder, C. J., Eds., Elsevier/North-Holland, Amsterdam, 1980.

46. **Kenyon, C. N., Rippka, R., and Stanier, R. Y.,** Fatty acid composition and physiological properties of some filamentous blue-green algae, *Arch. Mikrobiol.,* 83, 216, 1972.

47. **Ogawa, T. and Terui, G.,** Growth kinetics of *Spirulina platensis* in qutotrophic and mixotrophic cultures, in *Fermentation Technology Today,* Terui, G., Ed., Society of Fermentation Technology, Tokyo, 1972.

48. **Boussiba, S. and Richmond, A. E.,** C-phycocyanin as a storage protein in the blue-green alga *Spirulina platensis, Arch. Microbiol.,* 125, 143, 1980.

49. **Furst, P. T.,** *Hum. Nature,* 60, March 1978.

50. **Delpeuch, F., Joseph, A., and Cavelier, C.,** Consumation alimentaire et apport nutritionnel des algues bleues *(Oscillatoria platensis)* chez quelques population du Kanem (Tchad), *Ann. Nutr. Aliment.,* 29, 497, 1976.

51. **Durand-Chastel, H.,** Production and use of *Spirulina* in Mexico, in *Algae Biomass,* Shelef, G. and Soeder, C. J., Eds., Elsevier/North-Holland, Amsterdam, 1980, 51.

52. **Santillan, C.,** Mass production of *Spirulina, Experientia,* 38, 40, 1982.

53. **Hills, C., Ed.,** *The Secrets of Spirulina, Medical Discoveries of Japanese Doctors,* University of the Tree Press, Boulder Creek, Calif., 1980.

54. **Switzer, L.,** *Spirulina, the Whole Food Revolution,* Bantam Books, Toronto, 1982.

55. **Jassby, A.,** Proteus Corporation, San Rafael, Calif., personal communication, 1984.

56. **Nichols, B. W. and Wood, B. J. B.,** The occurrence and biosynthesis of gamma-linolenic acid in a blue-green alga, *Spirulina platensis, Lipids,* 3(1), 46, 1967.

57. **Jassby, A.,** Proteus Corporation, P.O. Box 1196, San Rafael, Calif., 94915, Personal communication, 1984.

58. **Ramos Galvan, R.,** Clinical experimentation with *Spirulina,* paper presented at Colloque sur la valeur nutritionelle des algues *Spirulina,* Rueil, May 1973.

59. **Sautier, C. and Tremolieres, J.,** Food value in *Spirulina* algae in humans, *J. Ann. Nutr. Aliment.,* 30, 517, 1975.

60. **Clement, G., Rebeller, M., and Zarrouk, C.,** Wound treating medicaments containing algae, *Fr. Med.,* 5279, 1967.

61. **Yoshida, R.,** *Spirulina* hydrolyzates for cosmetic packs, *Jpn. Kokai,* 77, 31, 1977.

62. **Babaev, T. A., Barashkina, Y. I., Kuchkarova, M. A., Tulyaganov, A.,** Khodzhiakhmedov, G., and Turakulov, Y. K., Iodine-containing compounds of *Spirulina platensis,* (Gom.) Geitl., *Uzb. Biol. Zh.,* 5, 6, 1979 (in Russian).

63. **Selyametov, R. A., Kuchkarova, M. A., and Tulaganov, A. T.,** *Spirulina platensis* (Gom.) Geitler in the rations of weaning piglets, *Uzv. Biol. Zh.,* 2, 73, 1977.

64. **Babaev, T. A., Karimova, K. M., Sattyev, R., Kuchkarova, M. A., Zaripov, E., and Turakulov, Y. K.,** Thyroxine-active principle of *Spirulina, Uzb. Biol. Zh.,* 7, 3, 1980 (in Russian).

65. **Iijima, N., Fugii, Shimamatsu, H., Katoh, S.,** Anti-tumor agent and method of treatment therewith, U.S. Patent pending, ref. no. P1150-726-A82679, 1982.

66. **Peto, R., Doll, R., Buckley, J. D., and Sporn, M. B.,** Can dietary beta-carotene materially reduce human cancer rates? *Nature (London),* 290, 201, 1981.

67. **Shekelle, R. B., Liu, S., Raynor, W. J., Jr., Lopper, M., Maliza, C., and Rossof, A. H.,** Dietary vitamin A and risk of cancer in the Western Electric study, *Lancet,* 8257, 1185, 1981.

68. **Tudge, C.,** Why we could all need the evening primrose, *New Sci.,* 19, 506, 1981.

69. **Kunkel, S. L., Ogawa, H., Ward, P. A., and Zurier, R. B.,** Suppression of chronic inflammation by evening primrose oil, in *Progress in Lipid Research,* Vol. 20, Holman, R. T., Ed., Pergamon Press, New York, 1982, 885.

70. **Kernoff, P. B. A., Willis, A. L., Stone, K. J., and McNicol, G. P.,** Antithrombotic potential of dihomo-gamma-linolenic acid in man, *Br. Med. J.,* 2, 1441, 1977.

71. **Vaddadi, K. S. and Horrobin, D. F.,** Weight loss produced by evening primrose oil administration, *IRCS Med. Sci.,* 7, 52, 1979.

72. **Huang, Y. S., Cunnane, S. C., Horrobin, D. F., and Davignon, J.,** Most biological effects of zinc deficiency corrected by gamma linoleic acid 18 3-omega-6 but not by linoleic acid 18 2-omega-6, *Atherosclerosis,* 41, 193, 1982.

73. **Horrobin, D. F.,** Loss of delta-6-desaturase activity as a key factor in aging, *Med. Hypotheses,* 7, 1211, 1981.

74. **Horrobin, D. F.,** The possible roles of prostaglandin E1 and of essential fatty acids in mania, depression and alcoholism, in *Progress in Lipid Research,* Vol. 20, Holman, R. T., Ed., Pergamon Press, New York, 1981.

75. **Horrobin, D. F. and Huang, Y. S.,** Schizophrenia: the role of abnormal essential fatty acid and prostaglandin metabolism, *Med. Hypotheses,* 10, 329, 1983.

76. **Matsueda, S., Nagaki, M., and Shimpo, K.,** Isolation and purification of the human erythrocyte cholinesterase and the enzymoeffect of *Chlorella* components to the activity, *Sci. Rep. Hiroski Univ.,* 23, 17, 1976.

77. **Ben-Amotz, A. and Avron, M.,** Glycerol and β-carotene metabolism in the halotolerant alga *Dunaliella:* a model system or biosolar energy conversion, *Trends Biochem. Sci.,* 6(11), 297, 1981.

78. **Ginzburg, M. and Ginzburg, B. Z.,** Interrelationships of light, temperature, sodium chloride and carbon source in growth of halotolerant and halophilic strains of *Dunaliella, Br. Phycol. J.,* 16, 313, 1981.

79. **Ben-Amotz, A. and Avron, M.,** The role of glycerol in osmotic regulation of the halophilic alga *Dunaliella parva, Plant Physiol.,* 51, 875, 1983.

80. **Ben-Amotz, A. and Avron, M.,** On the factors which determine massive β-carotene accumulation in the halotolerant alga *Dunaliella bardawil, Plant Physiol.,* 72, 593, 1983.

81. **Ben-Amotz, A.,** Adaptation of the unicellular alga *Dunaliella parva* to saline environment, *J. Phycol.,* 11, 44, 1975.

82. **Borowitzka, L. and Brown, A. D.,** The salt relations of marine and halophilic species of the unicellular green alga *Dunaliella, Arch. Microbiol.,* 96, 37, 1974.

83. **Sussman, I.,** The Metabolic Pathway of Osmoregulation in *Dunaliella,* Ph.D. thesis, Weizmann Institute of Science, Rehovot, Israel, 1982.

84. **Jones, T. W. and Galloway, R.,** Effects of light quality and intensity of glycerol content in *Dunaliella tertiolecta* (Chlorophyceae) and the relationship to cell growth/osmoregulation, *J. Phycol.,* 15, 101, 1979.

85. **Enhuber, G. and Gimmler, H.,** The glycerol permeability of the plasmalemma of the halotolerant green alga, *Dunaliella parva* (Volvocales), *J. Phycol.,* 16, 524, 1980.

86. **Levitt, J.,** in *Responses of Plants to Environmental Stresses,* Academic Press, New York, 1972, 17.

87. **Ben-Amotz, A., Sussman, I., and Avron, M.,** Glycerol production by *Dunaliella, Experientia,* 38, 49, 1982.

88. **Trainor, F. R. and Massalski, A.,** Ultrastructure of *Scendesmus* strain 614 bristles, *Can. J. Bot.,* 49, 1273, 1971.

89. **Trainor, F. R., Cain, J. R., and Shubert, L. E.,** Morphology and Nutrition of the colonial green alga *Scenedesmus:* 80 years later, *Bot. Rev.,* 42(1), 5, 1976.

90. **Trainor, F.,** A study of unialgal cultures of *Scenedesmus* indubated in nature and in the laboratory, *Can. J. Bot.,* 43, 701, 1965.

91. **Castillo, S. J., Merino, M. F., and Heussler, P.,** Production and ecological implications of algae mass culture under peruvian conditions, in *Algae Biomass,* Shelef, G. and Soeder, C. J., Eds., Elsevier/North-Holland, Amsterdam, 1980.

92. **Van Vuren, M. M. J. and Grobbelaar, J. U.,** Selection of algal species for use in open outdoor mass cultures, in Symp. Aquaculture in wastewater—S236., CSIR, Pretoria, South Africa, 1980.

93. **Venkataraman, L. V., Becker, W. W., and Tumkur, R. S.,** Studies on the cultivation and utilization of the alga *Scenedesmus acutus* as a single cell protein, in *Life Sciences,* Vol. 20, Pergamon Press, New York, 1977, 223.

94. **Lewin, J. C., Lewin, R. A., and Philpott, D. E.,** Observations on *Phaeodactylum tricornutum, J. Gen. Microbiol.,* 18, 418, 1958.

95. **Holdsworth, E. S. and Colbech, J.,** The pattern of carbon fixation in the marine unicellular alga *Phaeodactylum tricornutmum, Mar. Biol.,* 38, 189, 1976.

96. **Ansell, A. D., Raymont, J. E. G., Lander, R. F., Growley, E., and Shackley, P.,** Studies of the mass culture of Phaeodactylum. II. The growth of Phaeodactylum and other species in outdoor tanks, *Limnol. Oceanogr.,* 8, 184, 1963.

97. **Thomas, W. H., Seibert, D. L. R., Alden, M., and Edlridge, P.,** Cultural requirements, yields and light utilization efficiencies of some present saline algae, in Algal Biomass Workshop, University of Colorado, Boulder, Solar Energy Research Institute, April, 1984.

98. **Ott, F. D.,** A review of the synonyms and the taxonomic positions of the red algal genus *Porphyridim norgeli* 1849, *Nova Hedus,* 23, 237, 289, 1972.

99. **Chapman, R. L. and Lang, N. J.,** Virus-like particles and nuclear inclusions in the red alga *Porphyridium purpureum* (Bory) Drew et Ross, *J. Phycol.,* 9, 117, 1973.

100. **Gantt, E.,** Phycobilisomes: light harvesting pigment complexes, *BioScience,* 25, 781, 1975.

101. **Ramus, J. and Groves, S. T.,** Incorporation of sulfat into the capsular polysaccharide of the red Porphyridium, *J. Cell Biol.,* 54, 399, 1972.

102. **Jones, R. F., Speer, H. L., and Kury, W.,** Studies on the growth of the red alga *Porphyridium creuntum, Physiol. Plant.,* 16, 636, 1963.

103. **Golueke, C. G., and Oswald, W. J.,** *The Mass Culture of Porphyridium cruentum,* University of California Press, Berkeley, 1961.

104. **Ahern, T. J., Katoh, S., and Sada, E.,** Arachidonic acid production by the red alga *Porphyridium cruentum,* in *Biotechnology and Bioengineering,* Vol. 25, John Wiley & Sons, New York, 1983, 1057.

105. **Wolf, F. R.,** *Botryococcus braunii* an unusual hydrocarbon-producing alga, *Appl. Biochem. Biotechnol.,* 8, 249, 1983.

106. **Metting, B. and Rayburn, W. R.,** The influence of a microalgal conditioner on selected Washington soils: an empirical study, *Soil Sci. Soc. Am. J.,* 47(4), 1983.

TECHNOLOGICAL ASPECTS OF MASS CULTIVATION — A GENERAL OUTLINE

A. Richmond and E. W. Becker

Three major technical aspects are involved in developing commercial systems for mass cultivation of algae. The first relates to the construction of the reactor system — its shape and depth, the method of mixing the algal-laden water, and most important, the lining. The second concerns separating the algal mass from the medium, and the third relates to the dehydration of the algal mass to facilitate product distribution and storage. A satisfactory solution to these three technological issues is essential before any large-scale cultivation can be economically pursued.[1]

TYPES OF REACTORS FOR MASS CULTIVATION OF AUTOTROPHIC ALGA

Mass production of algae requires an inexpensive, yet reliable enclosure for growing the culture. In practice, the design of these enclosures represents a compromise between the cost of the investment in relation to the expected returns and the desire to establish conditions for the highest possible output rate, i.e., high flow velocity and turbulence, smooth long-lasting lining, and some means of elevating the temperature in the winter. A variety of designs have been constructed and tested, both on the experimental and commercial scale.[2] Among these, two basic systems can be distinguished: horizontal turbulent flow created by various types of devices and inclined baffled surfaces (cascades) where the algal suspension is constantly pumped back from the lower to the upper part of the unit.

Horizontal Ponds

Three basic designs have been put into operation: (1) oblong forms ("raceways") constructed either as one unit or as several units joined together ("meander"), with agitation by paddlewheel(s), a propeller, air injection, or an air-lift; (2) circular ponds with agitation by a rotating arm; (3) closed circulation systems in which the pond is covered by a transparent film or the culture is circulated in plastic tubes or in shallow trenches roofed with plastic.

Raceways

Oblong ponds are by far the most common style of construction used today in commercial plants which produce autotrophic algae. J. Dodd's chapter gives a detailed technical description of this system.

Circular Ponds

During the initial phase of commercial production of microalgae *Chlorella* by companies in Japan and Taiwan established between 1960 and 1970, circular cultivation ponds up to 45 m in diameter were used, some of them covered with glass domes.[3-5] Circular ponds, however, have many disadvantages, since they require expensive structures of heavy reinforced concrete and high energy consumption for continuous stirring. In addition it is difficult to obtain turbulence in the center and land use is inefficient.[6]

Closed Structures

Algal cultures in closed systems are not exposed to the atmosphere but are covered with a transparent material or contained within transparent tubing. Closed systems have the distinct advantage of preventing evaporation and elevating the temperature. In areas where water loss by evaporation far exceeds the gain from rainfall, as is typical of the warm sunny

FIGURE 1.   A covered 100 m² *Spirulina* pond operated by the Laboratory for Applied Hydrobiology at the Ben-Gurion University, Sede Boqer, Israel. The lining is 0.8 mm PVC and the cover is 0.02 mm-thick polyethylene used for greenhouses. The polyethylene stretched over the pond is supported by 20 mm filter glass rods. The paddle creates a flow of approximately 50 cm/sec.

regions most suitable for algal production, evaporative losses may be as high as 2500 mm/ year, the equivalent of some 20 pond volumes. Clearly, such high evaporation increases the costs of water and pumping, but more importantly, it increases the salinity of the medium, necessitating its replacement when algal growth becomes retarded.

In addition to decreasing evaporative losses, covering a pond will prevent radiative cooling of the system, particularly at night. The cover also acts as a greenhouse by trapping longwave radiation during the day. Covering ponds for the production of warm-water algae such as *Spirulina platensis* is mandatory in the winter in many subtropical and moderate regions, when the temperature in an open pond may decline to only a few degrees above freezing.[7] At the author's laboratory at Sede Boqer (Israel), covering a 100 m² pond in the type of polyethylene film used for greenhouses raised the daytime temperature of the medium by 6 to 8°C (Figure 1). Another advantage to closed systems is a reduction in the amount of dirt and insects contaminating the algal product.

Closed systems also have some disadvantages. Light penetration is markedly reduced because most of the commercially available materials used for tubing or for covering ponds are not completely transparent. In addition, dust accumulates on the outer surface of the covers and water condensing on its inner surface contributes to reducing the radiation reaching the culture by some 40 to 50%. Another obvious disadvantage is that a closed system entails a significant increase in the capital outlay per unit of pond area (some 15%).

The current high prices obtained for algae, such as *Spirulina* produced for the health food market, justify the cost of covering the ponds during the cold months of the year. Whether closed systems could be used for mass production of animal feeds from algae is not yet clear. Perhaps thermophilic algae with a high light saturation constant grown in reactors made of materials that reject dust and condensation could show high enough productivity to justify an expensive covered reactor.

The first pilot plant with a closed cultivation system for the production of *Chlorella* is believed to be one constructed in 1965 in Tylitsch (Poland) in a region with abundant

FIGURE 2. General view of the algal plant at Trebon, Czechoslovakia. Inclined surface (slope 3%) made from glass plates, two units of 50 m² each, one unit of 900 m².

springwater naturally rich in CO_2.[8] The unit, based on a design described by Little,[9] consisted of thin-walled polyethylene tubes placed on terraces on a hillside. The individual terraces were sloped alternatively to the left and the right. The culture flowed through the tubes from the higher to the lower end of each terrace, where it was collected in small cement tanks. CO_2 was added to the medium before it was delivered to the tubes of the next lower terrace. The total length of the meandering system was 4200 m². At the lowest point the culture was collected in a sedimentation tank where part of it was harvested by centrifugation and the remainder pumped back after enrichment with fresh medium. The average daily yield was reported to be 1 g/ℓ.

A similar system was described for the cultivation of *Spirulina*.[10] Here, polyethylene tubes (wall thickness 0.3 cm, 14 cm in diameter) were arranged as a "raceway" and the medium was circulated through the tubes by a pump. The daily production reached 15 g/m² in summer and 10 g/m² in winter. Another pilot plant based on this system was designed in Italy for the production of *Spirulina*. The total area covered was 2000 m² for the inoculum circuit and 15,000 m² for the production circuit. No reports are available on productivity in this plant, which may not be in operation.

Sloped Cultivation Units

This type of system is designed so as to create a turbulent flow while the algal culture flows through a sloping enclosure. The first set of sloping culture units was designed and operated by Setlik et al.[11] in Czechoslovakia. Their cascade system (Figure 2) consists of a sloping plane in which high turbulence is created by special small baffles. The suspension is collected at the lower end of the slope, flows to a sink, and is pumped from there back to the origin. The average culture depth is only 3 to 5 cm on the slope, which allows the density of the algal suspension to be relatively high (up to 3 g/ℓ). The original design by Setlik and associates was quite sophisticated and hence costly. It was later modified to establish a steady, uniform flow of a relatively thin layer of suspension. The suspension flowed on a slightly inclined plane (3% slope) made of glass sheets, fitted with transverse baffles. A 900-m² unit of this type was constructed (Figure 3). A specific feature of this design was the different modes of operation in the day and night. The suspension was kept

FIGURE 3. Part of a sloped meandering algal pond (Peruvian-German Algal Project, Trujillo, Peru).

circulating on the cultivation surface only during the day when there was enough light. At other times the suspension was kept in a tank where it was aerated and stirred. In this way heat loss from the medium during the night was minimized by the reduced surface/volume ratio. In addition, the culture could be drained to the tank during periods of heavy rainfall to prevent its dilution. The yields varied between 14 g/m²/day (average) to 21 g/m²/day (maximum) in the 50-m² units and 10 g/m²/day (average) to 19 g/m²/day in the 900-m² unit, indicating that the specific yield tends to decrease as the size of the culture units is increased.

Vendlova[12] described a less expensive pilot scale experimental sloping culture system constructed at Rupite in Southwest Bulgaria. *Scenedesmus* and *Chlorella* were grown in a 300 m² algal plant in concrete ponds with a 3% slope. The medium is recycled from the lowest point, where it is collected in a trough, to the highest point by pumps; turbulence of the suspension is caused by iron baffles. The average daily yields for the period April to October were reported to be 17 to 20 g/m². A modified version of the principle of sloped algal ponds was designed and constructed by the Peruvian-German algal project at Sausal, Peru.[13] The Peruvian ponds, made from reinforced concrete, consist of a meandering type of raceway system on a slight slope perpendicular to the length of the channels (Figure 3). The channels are built of self-supporting segments 5.5-m long and 1-m wide, connected by joints sealed with plastic strips. The thickness of the bottom and sides is 0.1 m, and the walls of two adjacent channels form a small wall 0.2 m wide, enough for a footpath for maintenance. The suspension flows from the lowermost channel into a sump pit where it is pumped back to the uppermost channel. The average culture depth is about 20 cm. With his system Heussler et al.[13] obtained the highest yield recorded up to now for the mass culture of green autotrophic microalgae. The effluent from harvesting is returned to the inlet; thus, the flow rate in the pond is not altered during harvesting. Although the channels were plastered as smoothly as possible, thick crusts formed on their surfaces, composed mainly of lime precipitated from the hard water, diatoms, and filamentous cyanobacteria. Infected *Scenedesmus* cells became preferentially attached to these crusts, making control of parasitic infections very difficult. This problem was solved by coating all surfaces in contact with the algal suspension with an epoxy paint.

The optimum operation of such sloped ponds depends on adjustment of the pump capacity to the flow rate of medium in the channel. If the flow rate in the channel increases and

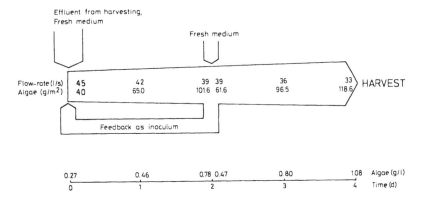

FIGURE 4. Flow diagram and growth parameters of a proposed continuously operated sloped meandering algal pond. Size: 10.4 ha, channel width: 1 m, initial suspension level: 0.15 m, flow speed: 0.3 m/sec, assumed daily production: 30 g/m². [14]

exceeds the pump capacity, the level of liquid in the lower part of the pond will rise. To avoid loss of suspension due to such imbalances or to pump failure, the sides of the pond at its lower end will rise enough to contain the whole volume of the pond. Yields of 40 g/m²/day could be obtained, and the average daily growth rates about 22 g/m² during semi-continuous operation of the plant. For one period of operation, Heussler and co-workers reported a yield of 54 g/m²/day. Based on this pilot plant larger units of up to 10 ha have been contemplated. [14] One plan is to construct a 5 ha unit with a 12.5-km long, 4-m wide channel, in which the suspension would need 8.7 hr for one circulation. By reducing the channel width to 1 m the total length would be about 50 km, and at a flow rate of 0.4 m/sec, the suspension would require about $1^1/_2$ days to complete one circulation. According to preliminary tests it seems possible to reduce the flow to 0.3 m/sec without settling. Hence, in a pond of 10.4 ha having a channel width of 1 m, the algae would take 4 days to flow from the highest to the lowest point. The yields obtained in Peru have shown that this is sufficient for the culture to reach a density suitable for harvesting. Figure 4 shows a diagram for the operation of such a large unit, based on the average yields of *Scenedesmus*. The algal suspension used as the inoculum at the upper part of the pond has an initial concentration of 40 g/m². The pond is equipped with mobile partition walls between some channels that allow a gradual increase in the cultivation area at the beginning of cultivation. After a 2-day run down the channel the algae density will be high enough to feed a part of the suspension back to the top of the channel as an additional inoculum. After 2 more days, the remaining suspension will reach the lower end of the pond with a biomass content of 120 g/m², representing a productivity of about 30 g/m²/day. It is envisaged that such a system would have several advantages: algal growth would occur in the range of its optimum density, fertilizer would be added in steps according to the requirements, growth should not be limited by parasitic infection, and the harvesting equipment is fed with a continuous flow of culture having only a few relatively small pumps for recycling the inoculum and the effluent from the harvesting process and for the supply of fresh water. It is not at all clear whether such a system has any advantage over the common horizontal ponds stirred by a paddlewheel. According to some data, substantially higher yields may be obtained in sloping raceways, however, this has not been confirmed. In one experiment specifically designed to compare the performance of two strains of *Scenedesmus* in baffled slopes and in level raceways, no final conclusion could be drawn from the results. It was evident that the different thermal regime in the two systems may have been the only factor responsible for the differences in yield. [15]

POND LINING

The lining represents the most expensive item in pond construction. Hence, a number of suggestions and trials have been made in an attempt to avoid the necessity for a watertight lining.[16] The stability of unlined ponds, however, and their ability to sustain algal cultures have yet to be demonstrated. At present, commercial algal ponds are lined with either concrete or plastic. The latter is considered best for pond lining provided it is UV-resistant. Soeder[6] reports from experience in his laboratory in Dortmund, Germany, and Becker[17] describes for tropical India that UV-resistant polyvinylchloride (PVC) is the most suitable lining material. This is also the conclusion from 5 years experience with this material in the author's laboratory at Sede Boqer. If linings of PVC are stable for say, 20 or 30 years, the construction costs for open ponds would be greatly reduced. As yet, the long-term effect of solar radiation on PVC linings has not been sufficiently documented. A wise precaution would be to use "food-grade" lining as recommended by the manufacturer.

Continued research may show that less expensive linings, such as reinforced UV-resistant polyethylene, which is very stable chemically and biologically inert, would also be suitable. Concrete linings are used in several experimental and commercial plants. They require a much larger initial investment and may not be sufficiently stable. The are not recommended for large-scale operations.

MIXING AND TURBULENCE

When the nutritional requirements are satisfied and the environmental conditions are not growth-limiting, mixing designed to create a turbulent flow constitutes the most important requisite for consistently high yields of algal mass.

The high population density, i.e., 500 to 1000 mg dry wt/ℓ or 3 to 15 mg chlorophyll per liter is required to obtain high output per unit area. Maintenance of high cell concentration in the culture, however, creates several pressures on the individual cells in the culture which are caused by vigorous mixing.

One obvious reason for mixing is to prevent the microalgal cells from sinking to the bottom of the pond. Sinking occurs when the flow is too slow and will be particularly severe in the pond areas where turbulence is smallest (See Dodd's Chapter). Clearly, accumulation of organic matter in "dead" areas of the pond will affect cell deterioration and anaerobic decomposition, which on the one hand decrease the output, and on the other hand adversely affect the quality of the product. In extreme cases, toxic materials would form with ramifications far greater than the mere effect on decreased productivity.

One additional reason to maintain high turbulence relates to the nutritional and gaseous gradients which are formed around the algal cells in the course of their metabolic activity. Such gradients impose restrictions on the growth rate and are alleviated with high turbulence. The high density of actively photosynthesizing cells also creates extremely high concentrations of dissolved O_2 which may reach over 400% saturation at midday. Vigorous mixing decreases the O_2 tension in the culture, particularly when mixing is effected by a properly designed device.

The major objective for creating a turbulent flow in the culture relates to the phenomenon of mutal shading (see Outdoor Mass Cultures of Microalgae), which in a highly productive pond is so intense as to permit only a very shallow photic zone in which the cells receive sufficient illumination. In a *Spirulina* pond maintained at the optimal population density, for example, it is estimated that approximately 85% of the cells do not receive, at any given instant, sufficient radiation to facilitate a photosynthesis rate over the compensation point.

A turbulent flow causes a continuous shift in the relative position of the cells with respect to the photic zone. Thus, turbulence, in effect, causes the solar radiation impinging on the

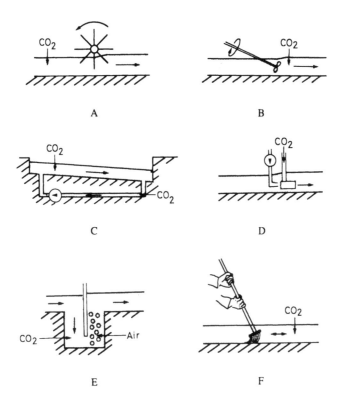

FIGURE 5. Different methods of agitation in algal cultures (A) Paddlewheel, (B) free propeller, (C) pump and gravity flow, (D) injector, (E) airlift, (F) manual stirring.

surface of the pond to be distributed more evenly to all the algae. This causes far-reaching effects on the efficiency of photosynthesis, for it is well documented that if a radiant flux of high intensity (such as solar irradiance, e.g., 2500 $\mu E/m^2/sec$) strikes the photosynthetic apparatus and is interspaced with short intervals of darkness, the efficiency of conversion of heat radiant energy may equal or exceed the maximum efficiency in continuous light. Correct modulation or flashing of the incident light on the optimal time scale is necessary in order to exploit this flashing light effect. A certain perplexity exists at present in this respect, stemming from a discrepancy between the early reports, which state the flash periods had to be extremely short (as little as 10 msec), whereas more recent reports say it is becoming clear that light and dark periods in the range of seconds may still significantly increase the overall photosynthetic efficiency.[9] Much more experimentation is required to formulate the parameters involved in inducing optimal mixing in mass algal cultures.

A turbulent flow in the pond may be created by several techniques (Figure 5), as follows

1. A paddlewheel, which is used in most experimental and commercial algal projects, is a relatively expensive device, with a power demand of about 600 W for a pond of 100 m^2. The technical design details are described in Dodd's chapter.
2. A free propeller, which has been tested in experimental ponds only, has a power demand similar to that of the paddlewheel. It may not be suitable for filamentous species.
3. A combination of pumps and gravity flow, which has a power demand between 100 and 200 W for a pond of 100 m. This principle has been adopted in a few large-scale plants.

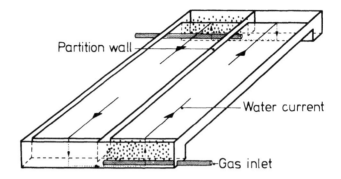

FIGURE 6.   Design of algal pond with airlift agitation system.[19]

4.   An injector. The algal suspension is injected through a nozzle and CO_2 is added so as to obtain high turbulence and high CO_2 transfer rates. High energy demands of 1000 to 2000 W for a 100 m² pond are required which prohibits the application of the system. This system cannot be used for filamentous algae.

5.   An airlift. This is very simple installation with a low power demand. Assuming a compressor efficiency of 70% and an air demand of 120 ℓ/sec, a power demand of 195 W has been calculated for a 85 m² pond. The system ensures high CO_2 utilization efficiency and reduces oversaturation with O_2. Clement and Van Landeghem[19] introduced airlift pumps to algal biotechnology for growing *Spirulina* (Figure 6). There seems to be technical difficulties for applying airlift pumps for large-scale cultivation.

6.   Low-technology devices. Various attempts have been made to decrease algal production costs, particularly in developing countries, by harnessing natural energy sources such as wind, sun, and even animals or humans (Figure 7).

Most means currently used to induce flow are inadequate to ensure attainment of the maximum photosynthetic potential of the cell, because the turbulence induced is random. Thus, all the cells in the suspension are not subjected to the same, uniform time pattern of movement into and out of the irradiated zone which would match the desirable intermittency pattern. A suitable mean periodicity of the motion is achieved but the variance from the mean value is so large that the optimal light-dark cycles are experienced only by a small percentage of the algae present.[11] What is needed is a nonrandom movement of the individual cells between the light and dark regions of the pond. This has recently been achieved by Laws et al.[18] who described a simple algal production system designed to utilize the flashing light effect. Arrays of foils similar in design to airplane wings are placed in the algal culture to create systematic mixing. This results from vortices which are produced in the culture due to the pressure differential created as water flows over and under the foils. In a flume having a flow rate of 30 cm/sec, the foil arrays produced vortices with rotation rates of approximately 0.5 to 1.0 Hz. This rotation rate was judged by the authors to be satisfactory for taking advantage of the flashing light effect if the culture is sufficiently dense. Solar energy conversion efficiencies in an experimental culture of *Phaeodactylum tricornutum* increased 2.2- to 2.4-fold with the foil arrays in place vs. controls with no foil arrays. The mechanism devised to produce systematic vertical mixing in the flume is illustrated in Figures 8 and 9.

At the tips of the foil, water flows from the high pressure region below the foil to the low pressure region above the foil, creating a vortex off each tip of the foil. If the foils are properly spaced along a suitable supporting structure (Figure 8), the vortices on adjacent foils rotate in opposite directions and thus reinforce each other. The width of each foil and the gap between foils should be equal to the depth of the culture, so that circular vortices created by the foils effectively mix the culture from top to bottom (Figure 9). The system

FIGURE 7. Stirring of algal suspension in experimental ponds by means of savonius rotos. CEXAS, Medipillo, Chile.[36]

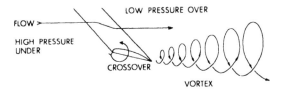

FIGURE 8. Design of a single foil indication mechanism of vortex production. (From Laws, E. A. et al., *Biomass*, 4, 1, 1982. With permission.)

of vortices with rotational axes parallel to the direction of flow is claimed to produce the sort of systematic mixing necessary to produce a flashing light effect.[18]

The use of foil arrays or similar devices seems to be an efficient, yet inexpensive way to induce nonrandom mixing, irrespective of the method used to induce flow in the culture. Such devices should increase photosynthetic efficiencies in mass cultures very significantly.

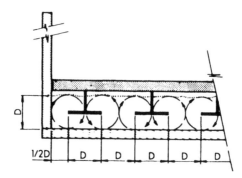

FIGURE 9. Positioning of individual foils in foil array. Lower figure indicates positioning of array in the flume. D is the depth of the water. Arrows indicate rotational direction of vortices. (From Laws, E. A. et al., *Biomass*, 4, 1, 1982. With permission.)

SUPPLY OF CO_2

Techniques for the supply of CO_2 represent an important element in the large-scale cultivation of algae, particularly for species grown near neutral pH. Several systems have been developed aimed at supplying CO_2 efficiently to shallow suspension. In most cases the gas is supplied in the form of fine bubbles. Due to the shallowness of the suspension the residence time of the bubbles is not sufficient to allow all the CO_2 to be dissolved, so that gas losses to the atmosphere are difficult to control.

A novel method for supplying CO_2 to algal cultures has been reported by Heussler et al.[13] which is based on the principle of maximizing the contact time of CO_2 with the suspension to minimize loss of the gas (Figure 10). The gas exchanger consists of a plastic frame which is covered with transparent sheeting and immersed in the suspension. CO_2 is fed under the sheeting via a float valve, so that the exchanger floats up to the surface of the suspension on the developing gas cushion, the dissolved gas flowing through beneath the frame. The float valve is adjusted so that all sides of the frame are immersed in the suspension to prevent CO_2 escape and to ensure that the exchanger floats in a stable position. The valve lets in only as much CO_2 as is simultaneously dissolving in the algae suspension.

It has to be considered that the CO_2 in the compartment is continuously diluted with other gases, especially O_2 derived from photosynthesis and N_2 from the atmosphere. This can be overcome either by increasing the volume of the gas exchanger in order to maintain a sufficient CO_2 supply over longer periods of time, which is not very economical, or by achieving a continuous de-aeration through a small hole. Depending on the rate of this leakage an equilibrium will be reached after a certain period of time. The loss of CO_2 by this alternative is relatively small (about 4% of the total CO_2 supply).

The required number of gas exchangers may be installed anywhere in a pond and thereby

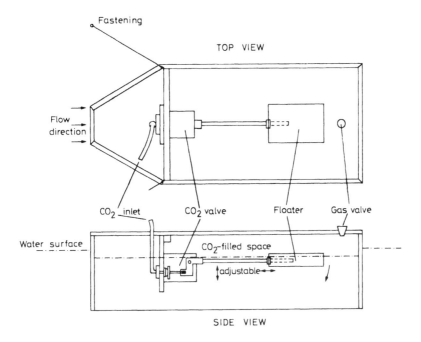

FIGURE 10.   Schematic diagram of floating gas exchanger.

facilitate even CO_2 distribution over the whole surface of the pond. A transfer area of 1.2 to 2 m² is sufficient for supplying CO_2 to 100 m² of pond surface. Depending on the temperature of the nutrient solution and the intensity of the growth of the algae, the degree of utilization of the CO_2 ranges from 25 to 65% compared with 13 to 20% if it is supplied in bubble form.[13]

Another system for the distribution of CO_2 is to let it diffuse through a porous metal or plastic pipe so as to form the smallest bubbles possible (these will not be seen on the surface). In general, the higher the pH of the culture medium, the higher the efficiency of CO_2 application. This gives a definite advantage to culturing algal species, such as *Spirulina*, which thrive at highly alkaline pH.

HARVESTING THE ALGAE

A major challenge in commercializing microalgal production relates to an efficient harvesting of the biomass, i.e., the removal of these microscopic organisms from the growth medium in which they grow as a dilute (200 to 500 mg dry wt/ℓ) suspension, and concentrating them by approximately two orders of magnitude. Many methods and machines are available for the harvesting of microalgae. These include centrifugation, electroflotation, and chemical flocculation, followed by sedimentation or air flotation, continuous belt filtration, vibrating and stationary screens, sand bed filtration, and autoflocculation. Only a few of these systems, however, appear to have potential as efficient, low-cost harvesting methods. Some are described in this section.

Centrifugation

The most direct method for thorough removal of algal cells from the growth medium is centrifugation. The self-cleaning plate separator and the nozzle and screw centrifuges effectively concentrate all type of microalgae.[20] Centrifugal fields of 500 to 1000 × for 0.25 to 1.0 min usually suffice for a high degree of removal. Centrifugation is equally applicable to filamentous and nonfilamentous microalgae. Some algae, such as *Spirulina*, which are rich in air vesicles may tend to rise in a centrifuge rather than settle and can be removed as a floating cream. The major advantage of centrifugation is its simplicity and the fact that the product contains no added chemicals.[21] It is generally accepted, however, that the high investment cost and energy demand (approximately 1 kWh/m³) make centrifuges impractical for the mass production of inexpensive algal biomass. Nevertheless, Dodd and Goh[22] recently reported advances in the design of solid bowl decanter centrifuges which made them suitable for thickening and dewatering materials such as microalgae with little or no added polymeric flocculants. A Humbolt-Wedag model S2-decanter centrifuge was tested in Singapore on an algal slurry of 1 to 2% solids harvested by a belt filter. Cakes consisting of 15 to 17% solids were achieved without adding polymers. Improvements in low differential speed scroll and backdrive design as well as concurrent flow and greater pool depth have allowed longer residence times with less floc disruption. Centrifuges previously used for algae thickening were typically of the disk type, which are more costly and difficult to maintain than the decanter type.[23]

Sedimentation with Natural Gravity

Various systems were tested by Mohn,[24] who found that all those based on the principle of sedimentation were low in efficiency. Thus, even though the energy consumed per cubic meter was approximately 1/10 that required by the plate separator centrifuge, the relative cost of algal harvesting by sedimentation was one order of magnitude higher.

Filtration

Filtration of microalgae from the pond water followed by backwashing to remove them from the filter may be useful for the harvesting of many species. Filter media such as fine sand, diatomateous earth, and cellulose fibers have been used for years in water supply and swimming pool systems. Because algae tend to clog filters quickly, backwashing must be frequent, and because so much backwash water is required, concentration factors exceeding a factor of ten are rarely obtained.[25] Filter fouling with microbial growth is also a problem. This have been overcome to some extent by Dodd and Anderson[26] who developed a continuously moving cellulose filter which becomes impregnated with the algae as it moves along. The fine-weave belt filter is part of a continuous filtration device consisting of a large filtration drum, a much smaller separation drum for suction recovery of solids, a polyester fabric belt, a tank, and a frame for supporting belt rollers and washing spray bars (Figure 11). In one operating machine described by Dodd and Goh,[22] the belt was 1-m wide, the filtration drum 2 m in diameter, and the separation drum 0.3 m in diameter.

Pressure filtration can also be used for algal harvesting. Of the many systems tested by Mohn,[20] the chamber filter press, which is used both in the brewing industry and for dewatering sludge, seems superior in its high reliability and low operating cost.

Screening and Straining

Many filamentous or colony-forming microalgae are separable by screens or sieves. Examples are the cyanobacteria *Spirulina* and *Oscillatoria*. *Oscillatoria* filaments are 30 to 150 μm in length and 3 to 5 μm in diameter. All but the smallest filaments of these algae are removable with a 375 mesh screen. Screens are available with openings as small as 1 to 2 m, but the rates at which they pass water are too slow. Screen types that are applicable to harvesting algae include vibrating screens, rotating screens, microstrainers, and cascade screens. The advantage of screening is that a pure product is obtained, while the disadvantage is that many microalgae cannot be removed by screening and fine screen filters are prone to clog. Corrosion has been largely overcome by using plastic materials such as nylon or Teflon®. Clogging by slime, however, remains a problem.[21] Vibrating screens operating with 300 to 500 mesh screens are being used today for screening *Spirulina*. If the filaments

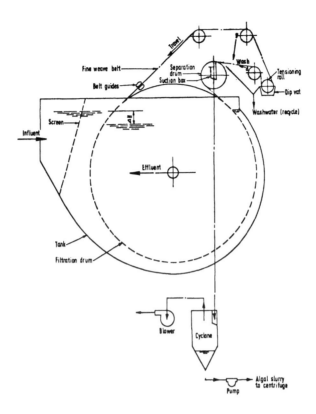

FIGURE 11. Fine weave belt filter. Schematic diagram.

are long enough, i.e., over 200 or 300 μm, efficiency of harvesting may be as high as 95%. One difficulty with vibrating screens is the rubbing effect on the delicate filaments, causing cell disruption and a harmful increase in the organic load of the pond.

Another difficulty associated with screens is that they preferentially remove the larger algal species cultivated in the pond, enriching, in effect, the culture with smaller contaminating species. Thus, in a culture of *Spirulina platensis* which becomes infected with *Chlorella* sp. repeated harvesting with a vibrating screen so greatly elevated the population of *Chlorella* that the entire culture could be discarded. This difficulty, however, may be overcome if the new type of microstrainer were used.

Until recently, microstrainers were generally unsatisfactory for the removal of suspended solids from effluents, except in special cases where large colonial or filamentous algae could be retained on the 20 to 24 μm fabric used previously. The polyester fabrics now available, with nominal ratings down to 1 μm, allow capture of small unicellular algae such as *Chlorella*, albeit at a slow filtration rate. Throughput is increased by higher pressure heads and may be facilitated by design improvements. However, the thinness of the cake, water carry-over at high drum speeds, and the need to use wash-water to dislodge the cake into the backwash-trough cause severe dilution of the removed solids. The concentration of the product is therefore only about five to ten times that in the pond effluent, requiring additional means for removal of water from the slurry.

Flocculation

Separation of the algal mass from the growth medium is greatly simplified when the single microscopic cells adhere together to form visible flocs that sediment or float. Flocculation may be spontaneous, i.e., due to environmental factors such as a sharp change in the pH or to cations associated with water hardness. Flocculation may also be induced by adding chemicals to the culture to induce the cells to aggregate into easily harvestable flocs. The costs of production of microalgae should be significantly reduced by the development of a dependable and efficient flocculation procedure.

Autoflocculation

The apparently spontaneous formation and settling of algal flocs has long been recognized and studied. Conditions that impede growth of the cultured organisms often promote flocculation. This autoflocculation can be associated with elevated pH or can be due to coprecipitation of the cells by magnesium, calcium, phosphate, or carbonate salts. In addition, the formation of algal aggregates may result from interactions between algae and bacteria or between algae and excreted organic polymers. Autoflocculation may thus be imitated in culture by manipulations related to pond management, e.g., interrupting the supply of CO_2 to the culture, which results in an immediate elevation of the pH. Species such as *Scenedesmus*, which grow optimally near neutral pH, will almost completely flocculate when the pH is raised to over 8.5.

An interaction between potassium ions and other cations in causing flocculation of *Scenedesmus* was reported from Peru by Heussler et al.[13] With the hard water available there for algaculture, addition of ammonia in the amounts required to replenish the nitrogen consumed by the harvested algae effected nearly complete flocculation of the suspended algae within a few minutes. The extent of flocculation increased when the phosphate required for fertilizing the culture was added as phosphoric acid immediately before the ammonia. Flocculation was induced by the shift from a nearly neutral to an alkaline pH; this occurred after addition of the phosphoric acid and ammonia. The effective cation was not ammonium, but the calcium present in the hard water at a concentration between 100 and 160 mg/ℓ. The coprecipitated calcium could be washed out, so that the ash content of the dry algal product was not increased.

Chemical Flocculation

Many compounds have been tested to identify the most efficient flocculant for microalgae:[27] aluminum sulfate, lime, ferro and ferric sulfate, ferric chloride, and scores of commercially available polyelectrolytes tested under a variety of conditions (e.g., changes in coagulant dose and time of application and the pH and mixing conditions). It was concluded that aluminum sulfate (alum) was the most suitable chemical agent, a dose of 150 mg/ℓ at pH 6.5 giving satisfactory results, a conclusion that had been arrived at earlier by Golueke et al.[28]

A major disadvantage of chemical separation of algae, in addition to the cost of the chemicals, is their presence in the separated biomass, as well as their contamination of the culture medium if the harvested effluent is returned to the pond. Polymeric flocculants thus seem more suitable for flocculating algal cells intended for food or feeds, because they are used at much lower concentrations than are other chemical flocculants (Figure 8).

Chemically induced algal flocculation probably occurs[29] through crosslinking of the algal cells by extended polymer chains, which forms a three-dimensional matrix that sediments under quiescent conditions. The degree of flocculation is a direct function of the extent of polymer coverage of the algal surface. To induce flocculation by this method requires that the algal surface charge be reduced to a level at which the extended polymers can bridge the distance of separation caused by electrostatic repulsion.

The requisite polymer dosage in any flocculation process is influenced by many variables, the most prominent of which are the pH, the concentration of algae, and the algal growth phase. Flocculation is most effective at low pH because of reduced electrostatic repulsion between the cells and also greater extension of the cationic polymer chains (Figure 12). A

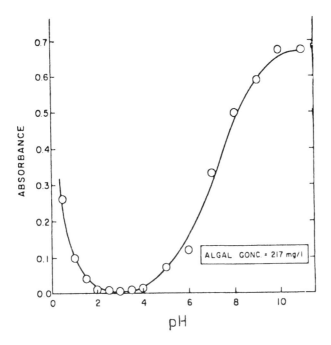

FIGURE 12.   Effect of pH on flocculation with constant dosage of polyelectrolyte (10 mg/ℓ, Dow-31). (From Tenny, M. W. et al., *Appl. Microbiol.*, 18, 965, 1969. With permission.)

stoichiometric relationship was found between the algal cell concentration and the required polymer dosage (Figure 13). Flocculation of algae with cationic polymers is most effective in the late log and early declining growth phases.

To avoid residual effects of the flocculant on the food quality of algal biomass, they should be edible and nontoxic. Both "rubifloc", a trade name for a material obtained from the sugar refining industry, and "chitosan" (deacetylated polymer of -N-acetyl-D-glucosamine), a natural carbohydrate obtained by acetylation of the chitinous exoskeleton of marine arthropods, are useful.

Chitosan has been tested in India as a cationic flocculant to concentrate algal suspension.[30] Within 10 min of adding 0.05 g/ℓ of chitosan (ph 8.4), 96.5% precipitation was obtained. Chitosan was not effective in concentrating *Spirulina*, which is irrelevant, since this alga can be harvested by filtration without the use of flocculants. The utilization of chitosan offers some advantages in comparison to that of other conventional flocculants. There has been no evidence of toxic side effects, it can be used in very low concentrations, and its production is fairly simple, as follows: chitinous shellfish wastes are separated from protein residues by treatment with 2% NaOH at 65°C, the calcium components are removed by extraction with HCl, and the remaining chitin is converted to chitosan by deacetylation with 50% NaOH at 130 to 150°C.

Electroflocculation

An electrical flotator for facilitating easier harvest of the flocculated biomass has been tested. By electrolysis, it produces fine bubbles of hydrogen and oxygen, which adhere to the algal flocs. These form a froth that can be readily skimmed from the culture surface.

Finally, direct induction of flocculation with an electric current has been reported. Passage of an electric current through the algal suspension causes the formation of buoyant algal flocs, the degree of flocculation being a function of the duration of the electric treatment. Both the rate and degree of flocculation depend on the pH of the medium, the cell density,

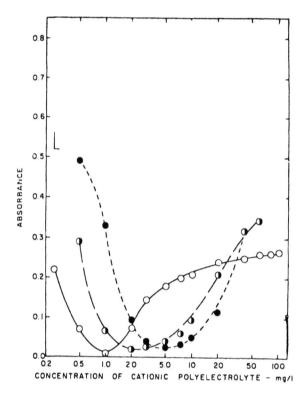

FIGURE 13. Effect of algal concentration on requisite polymer concentration at pH 0.3. The algal concentration was (O——O) 100 mg/ℓ, (◑——◑) 200 mg/ℓ, (●—●) 350 mg/ℓ. (From Tenny, M. W. et al., *Appl. Microbiol.*, 18, 965, 1969. With permission.)

and the type of counter ions added. At pH 7, up to 90% of the algal cells could be separated within 30 min by application of a 3 V/cm electric field. Both *Chlorella* and *Scenedesmus* showed a similar extent of flocculation with this method.

DEHYDRATION OF THE ALGAL MASS

Artificial Heat

Dehydration is of major economic importance, since this process can account for up to 30% of the production costs. Various systems for dehydration differ in the extents of both capital investment and energy requirements, and have a marked effect on the food value and taste of the product, especially in the case of green algae with rigid cell walls. Several methods are available to dry algae for preservation, the most prominent of which are spray drying, surface drying and sun drying.

Spray drying of an algal slurry containing 10% solids gives a uniform powder but is expensive (between 0.5 and $1.00/per kg). Pigments and vitamins may be preserved by the addition of antioxidants to the slurry prior to drying. Digestibility of spray dried green algae is only about 50% compared with 80% achieved by drum drying.[23] Most of the algal material intended for health food is processed today by spray drying, which yields powder of different characteristics, depending on the selected drying pattern.

Surface drying on drums or plates is less costly than spray drying, estimated at 0.4 to $0.6/kg. Additional grinding after drying may, however, be required to produce a usable product. Drum drying also causes more protein denaturation than spray drying.[33] Pabst[32] has shown that dehydrating *Scenedesmus* with a thin-layer drum dryer yields an excellent product. An algal suspension of 6 to 8% dry matter obtained by centrifugation is ejected onto a rotation steam-heated drum. The cells are heated for a few seconds to approximately 120°C and the ensuing dehydration causes the hard cell wall of the green algae to open, thereby greatly increasing the digestibility and biological value of the powdered product. According to Soeder and Mohn,[34] 15.7 kcal are needed for the evaporation of 18.2 kg of water in order to obtain 1 kg of dry powder with a water content of 4%. This energy input is in addition to the 1.4 kWh needed to run the machine itself. Some disadvantages of drum drying are that it requires a steam source, has relatively low thermal efficiency, and is troublesome because the algae tend to stick to the drum walls. Highly polished drums and frequently sharpened knives are necessary for satisfactory results.

A recently developed sonic dehydrator using a pulse jet suitable for direct combustion of digestor gas (65% methane) is now in commercial use for fish meal production and has been tested for drying *Spirulina*. According to Dodd and Goh[22] this type of dryer used the sonic energy of the pulse jet (about 250 cy/sec) to finely disperse the slurry, which is injected into the exhaust near the outlet, allowing rapid drying and high thermal efficiency. Further experimentation is needed to determine whether the cell wall of green algae is broken, but by adjustment of the point of injection exposure time and rate of feed, cell wall disruption comparable to drum drying may be achieved. An important advantage of the pulse jet sonic dryer is that it can be in full operation from a cold start within 15 min, with no refractory lining or significant penalties for frequent startup and shutdown as with the boiler drum dryer. High combustion temperatures exceeding 1250°C and low exhaust temperatures of about 93°C, together with direct heat exchange with the finely dispersed slurry, give optimum efficiency of fuel conversion to usable heat, which is reported to be above 99%. Heat recovered from the exhaust gases can be set to heat the slurry and the anaerobic digesters. Energy consumption per kilogram of water removed is reported to be 720 kJ/kg (1500 Btu/lb). This represents about a 50% savings over the boiler drum dryer.[22]

Sun Drying

The least expensive procedure for processing algae after concentration and dewatering is sun drying, which is the method used by the local population along some African lakes to prepare brittle, dry cakes of *Spirulina*. Since the water content of the algal concentrate is not drainable, flat plates are probably most appropriate for drying by evaporation. Systematic studies of this procedure demonstrated that if the algal layer is thicker than 0.75 cm it requires more than 1 day to dry.[35] If the initial depth is greater than 1.3 cm purification occurs during days 2 and 3. Since the preservation of protein, vitamins, etc. thus necessitates single-day drying, the depth for flat plat sun drying should be 0.5 cm. With this depth, solar irradiation of 2 KJ/cm²/day and 6% initial solids in the algal material and loading rates of about 130 g/m²/day of dry algal solids are possible, and a product with less than 10% moisture is obtained. Except for *Spirulina*, however, this simple method is not applicable for the algae used for mass cultivation. Nutritional studies have demonstrated that sun drying leaves the algae (especially *Chlorophyceae*) indigestible by nonruminants, since the rigid cellulose-like algal cell walls are not ruptured.[32]

Other Methods

Other possible methods to preserve algae as a stable concentrate involve freezing or canning. Freeze-drying has been used successfully, but is more expensive than spray drying. Freeze-dried algae are an attractive, pale green powder or a light granular material. Freezing of an algal paste allows it to be stored indefinitely in a freezer. Freezing prior to hot spray drying results in a product that has more denatured proteins and less pigmentation than the same product dried without prior freezing. Quick-frozen and freeze-dried fresh algae, on the other hand, maintain virtually all their original protein and pigmentation. Canning of dewatered algal pastes may be a reasonable storage procedure for food grade algae, but no studies on this method have been found in the literature.

REFERENCES

1. **Richmond, A. and Preiss, K.,** The biotechnology of algaculture, *Interdisciplinary Sci. Rev.,* 5, 1, 1980.
2. **Becker, E. W.,** The production of microalgae as a source of biomass, *Biomass Util.,* 67, 205, 1983.
3. **Stengel, E.,** Anlagentypen und Verfahren der technischen Massenproduktion, *Ber. Dtsch. Bot. Ges.,* 83, 589, 1970.
4. **Tamiya, H.,** Mass culture of algae, *Ann. Rev. Plant Physiol.,* 8, 309, 1957.
5. **Kanazawa, Z., Fujita, C., Yuhara, T., and Sasa, T.,** Mass culture of unicellular algae using the "open pond circulation method", *J. Gen. Appl. Microbiol.,* 4, 135, 1958.
6. **Soeder, C. J.,** Productivity of microalgal systems, *University of the Orange Free State, Publ. Ser. C.,* 3, 9, 1981.
7. **Walmsley, R. D. and Shillinglaw, S. N.,** Mass algal culture in outdoor plastic covered minipond systems, *Ann. Appl. Biol.,* 104, 185, 1984.
8. **Müntz, K.,** Die Massenkultur von Mikroalgen, bisherige Ergebnisse und Probleme, *Kulturpflanze,* 15, 311, 1967.
9. **Little, A. D.,** Pilot plant studies in the production of Chlorella, in *Algal Culture: From Laboratory to Pilot Plant,* Publ. no. 600, Burlew, J. S., Ed., The Carnegie Institution, Washington, D.C., 1964, 235.
10. **Torzillo, G.,** Sperimentazione sulla colture massiva di *Spirulina maxima* in sistema tubolare nel bienno 1979—1980, in *Prospetive della Coltura di Spirulina in Italia,* Materassi, R., Ed., Accademia dei Georgofili, Firenze, 1980, 329.
11. **Setlik, I., Veladimir, S., and Malek, I.,** Dual purpose open circulation units for large scale culture of algae in temperate zones. I. Basic design considerations and scheme of pilot plant, *Algol. Stud. (Trebon),* 1, 11, 1970.
12. **Vendlova, J.,** Outdoor cultivation in Bulgaria, *Ann. Rep. Algol. Lab. (Trebon),* 143, 1969.
13. **Heussler, P., Castillo, J., Merino, S., and Vasquez, V.,** Improvements in pond construction and CO_2 supply for the mass production of microalgae, *Arch. Hydrobiol.,* 11, 254, 1978.
14. **Heussler, P.,** Aspects of sloped pond engineering, *Arch. Hydrobiol. Beih. Ergebm. Limmol.,* 20, 71, 1985.
15. **Balloni, W., Materassi, R., Pelosi, E., Pushparaj, B., Florenzano G., Stengel, E., and Soeder, C., J.,** Comparison of two different culture devices for mass production of microalgae at Firenze (Italy) and Dortmund (Germany). I. Yields of *Scenedesmus obliquus* and *Coelastrum* at Firenze, *Algol. Stud.,* 28, 324, 1981.
16. **Berend, J., Simovitch, E., and Ollian, A.,** Economic aspects of animal food production, in *Algae Biomass,* Shelef, G. and Soeder, C. J., Eds., Elsevier/North-Holland, Amsterdam, 1980, 799.
17. **Becker, E. W.,** Biotechnology and exploitation of the green alga *Scenedesmus obliquus* in India, *Biomass,* 4, 1, 1982.
18. **Laws, E. A., Terry, K. L., Wickman, J., and Chalup, M. S.,** A simple algal production system designed to utilize the flashing light effect, *Biotechnol. Bioeng.,* 25, 2319, 1983.
19. **Clement, G. and Van Landeghem, H.,** *Spirulina:* ein günstiges Objekt für die Massenkultur von Mikroalgen, *Ber. Dtsch. Bot. Ges.,* 83, 559, 1970.
20. **Mohn, H.,** Experiences and strategies in the recovery of biomass from mass cultures of microalgae, in *Algae Biomass,* Shelef, G. and Soeder, C. J., Eds., Elsevier/North-Holland, Amsterdam, 1980, 547.
21. **Oswald, W. J.,** Pilot plant high rate pond for study of waste treatment and algae production, Final Report World Health Organization, Office for the Western Pacific, Malina, 1977.
22. **Dodd, J. C. and Goh, A.,** Algae harvesting with a fine-weave belt filter, paper presented at 6th Mid-America Conf. Environ. Eng. Design, Kansas City, Mo., June 14th to 15th, 1982.
23. **Moll, R. T. and Letki, A. G.,** New centrifuge developments that reduce operating costs for thickening municipal sludge, paper presented at N.Y. Water Pollut. Contr. Assoc. Annu. Meet., New York, January 20, 1981.
24. **Mohn, H.,** Experiences and strategies in the recovery of biomass from mass cultures of microalgae, in *Algae Biomass,* Shelef, G. and Soeder, C. J., Eds., Elsevier/North-Holland, Amsterdam, 1980, 547.
25. **Borchard, J. A. and Omelia, C. R.,** Sand filtration of algal suspension, *J. Am. Water Works Assoc.,* 53, 1493, 1961.
26. **Dodd, J. C. and Anderson, P. A.,** Integrated high-rate pond-algal harvesting system, Paper no. 55, Sydney Conference, Int. Assoc. Water Poll. Res. 1976.

27. **McGarry, M. G.,** Algal flocculation with aluminum sulphate and polyelectrolytes, *J. Water Pollut. Contr. Fed.,* 42, 191, 1970.
28. **Golueke, C. G., Oswald, W. J., and Gee, H. K.,** Harvesting and processing sewage-grown planktonic algae, *J. Water Pollut. Contr. Fed.,* 37, 4, 1965.
29. **Tenny, M. W., Echelberge, W. F., Ronald, J. R., Schuessler, R. G., and Pavoni, J. L.,** Algal flocculation with synthetic organic polyelectrolytes, *Appl. Microbiol.,* 18, 965, 1969.
30. **Nigam, B. P. and Venkataraman, L. V.,** Application of chitosan as a flocculant for the alga *Scenedesmus acutus, Arch. Hydrobiol.,* 88, 378, 1980.
31. **Kumar, H. D., Yadava, P. K., and Gaur, J. P.,** Electrical flocculation of the unicellular green alga *Chlorella vulgaris* Beijerninck, *Aquat. Bot.,* 11, 187, 1981.
32. **Pabst, W.,** Die Massenkultur von Mikroalgen, *Kraftfutter,* 58, 2, 1975.
33. **Becker, E. W., Venkataraman, L. V., and Khanum, M. P.,** Digestibility coefficient and biological value of the proteins of the alga *Scenedesmus acutus* processed by different methods, *Nutr. Rep. Int.,* 14, 457, 1976.
34. **Soeder, C. J. and Mohn, H.,** Technologische Aspekte der Mikroalgenkultur. I., *Symp. Mikrobielle Proteingewinnung, Branuschweig-Stöckheim 1975,* Verlag Chemie, Weinheim, 1975, 93.
35. **McGarry, M. G. and Tongkasame, C.,** Water reclamation and algae harvesting, *J. Water Pollut. Contr. Fed.,* 43, 824, 1971.
36. **Ayala, F., Liaz, M. E., and Bravo, R.,** Microalgae culture in salt-water media, *Arch. Hydrobid. Beih. Engebm. Limmol.,* 20, 53, 1985.

ELEMENTS OF POND DESIGN AND CONSTRUCTION

Joseph C. Dodd

INTRODUCTION

Pond systems for the production of microalgae biomass have evolved over the past 30 years from the early meandering (folded) channel designs with propeller pump mixing introduced by Oswald.[1] Ponds now in use incorporate a wide range of configurations employing various methods for mixing, feeding, control of algae populations and contaminants, enhancement of productivity, etc. Rather than try to cover all of the various options in a general way, much of which is discussed elsewhere, this chapter will concentrate on the single loop paddlewheel mixed design which has proven successful for a number of applications at relatively large scale. Elements of planning and design will be discussed beginning with considerations of site selection and layout of facilities, pond configuration, hydraulics, and size, followed by details of paddlewheel design, settleable solids control, pond bottom lining and wall construction, and utility services. Construction factors and their influence on design will also be discussed.

SITE SELECTION AND LAYOUT OF FACILITIES

The success of large-scale cultivation of microalgae biomass is strongly influenced by site-specific factors such as climate, water quality and quantity, topography and geology, availability and cost of land, and production inputs such as CO_2, nutrients, energy, and labor. These factors differ in their importance and the degree to which they can be controlled depending on the objective and scale of production. High value microalgae products such as health food can be economically viable at relatively small scale using intensive cultivation methods in competition with conventional agriculture. Climate control features such as greenhouses may be appropriate in certain cases. Where the objective is wastewater treatment with by-product recovery for animal feed, or conversion of algae biomass to fuel or chemical products, extensive cultivation using land and water resources of low value is more appropriate. Site selection factors and their influence on layout of facilities and design are discussed below.

Climate

Climate is probably the most important factor in site selection because of the predominant influence of solar radiation and temperature on productivity. Rainfall and high velocity wind patterns also influence the layout and design of facilities, as well as the scheduling and cost of construction. Tropical climate is often thought of as ideal for algae production due to the uniform temperature, but the rainfall and wind associated with tropical storm events and monsoon cloud cover can disrupt pond operations and reduce the long-term stability and productivity of cultures. On the other hand, the dryness and clear skies of desert climate are attractive, but excessive daytime temperatures in the summer and low temperatures at night in the cooler months, particularly at higher elevations, can reduce productivity or may require shutdown for a period of months. Although the use of covered ponds or geothermal heating may be economical for high value products or where special circumstances exist, the modification of climate by artificial means is unlikely to be economically feasible for extensive cultivation.

Climate affects pond layout and design in many ways. Orientation and location of the ponds can influence loss of productivity due to shading effects, as well as the susceptibility to liner damage and wind-induced wave action resulting from high velocity winds. The provision of freeboard with use of vertical above-grade walls is advantageous, particularly when the site is susceptible to storm events with risk of wind and flooding damage. In dry areas, the lack of rain combined with high evaporation rates will influence water supply and on-site water storage requirements. Climate also influences air quality, which has caused problems at a number of projects aimed at the production of food, due to atmospheric fallout of environmental pollutants such as heavy metals which found their way into the product.[2] Though not directly related to climate, another aspect of air quality that can be of concern in agricultural areas is the contamination from crop spraying which can drift into the ponds and water supplies.

Topography and Geology

The large-scale production of microalgae biomass in shallow ponds requires the precise grading of land to very flat slopes over large areas, preferably by laser-controlled leveling equipment as used for irrigated agriculture. The feasibility of such leveling is sensitive to both topography and soil characteristics. Although ponds can be terraced to conform to slightly sloping terrain, additional costs and technical constraints generally weigh against the use of land having more than about 1% prevailing slope. The very shallow depth and flat channel slopes of microalgae production ponds, typically 10 to 20 cm and 0.01 to 0.05% respectively, make them very sensitive to loss of tolerance on pond grade. Laser controlled equipment now in use can achieve construction tolerances compatible with the above values, but this is insufficient if the tolerance is subsequently lost due to relative surface movement. Relative surface movement is a problem particularly where cutting and filling has been necessary to level the pond, due to consolidation of underlying soil when surcharged in fill areas, insufficient compaction of fill, and shrinkage and swelling due to changes in moisture content, especially with expansive clays. Ground movements due to subsidence or earthquake activity can also be a problem in some areas. Risk of relative surface movement from any of the above causes can limit the maximum practical size of a single pond. This should be considered along with hydraulic limitations as further discussed in the section on pond size.

Soil characteristics influence the feasibility and cost of pond construction, and the design of pond lining and other features. Laser leveling is best adapted to relatively homogeneous soils which can be easily cut with a scraper. Soil containing rocks or other hard material, or clays which are too wet to support equipment or too dry and hard for effective cutting, can significantly increase the cost of earthwork and lining. Membrane liners generally require a smooth supporting surface free from sharp or projecting material, voids, and other defects, which can be difficult to obtain with heterogeneous or hard soils. Lining is generally required for contamination control reasons in ponds aimed at food production and has usually been of the sheet membrane type for reasons of cost. In the case of extensive ponds for wastewater treatment or energy production, membrane lining is costly and it is generally necessary to choose a site where the soil is sufficiently impervious to be self-sealing without a liner. A thin surface course of graded crushed rock, rolled to a smooth compact surface, is recommended for clay soils to control erosion and turbidity, and to facilitate pond maintenance. Expansive clay soils can be troublesome due to problems in controlling grade, weakness when wet, and induced pressures and possible movement of structures such as walls and footings which can cause leakage. These problems can be minimized with careful attention to design details such as joints, scheduling of construction, control of moisture and compaction, etc.

Water Quality and Quantity

Even when the effluent from the harvesting process is recycled to the pond, large quantities of water of suitable quality are required to make up for evaporation losses, which often amount to 1 cm/day in summer, and for necessary blowdown, wash water, and other uses. Where available at reasonable quality and cost, well water is less subject to some of the problems which can occur with surface water supplies, such as biological contamination and hazard from spills of toxic chemicals used in agricultural areas. On-site storage of water is usually a requirement to provide for interruptions in supply. This also provides storage time to allow for bioassays and other tests to verify water quality prior to use in the ponds.

In addition to the clean water storage for makeup water, it may also be necessary to provide storage for blowdown water and other waste streams which may be unsuitable for discharge. This is particularly true of ponds using high pH or high salinity, outside the limits allowed by regulatory agencies for discharge to surface waters. In some cases it may be necessary to provide evaporation area for such streams to meet a zero discharge requirement.

Production Inputs

Production inputs such as CO_2, nutrients, energy, labor, and specialized technical services can have a significant bearing on site selection and cost of production. CO_2 is the major chemical cost for many commerical plants growing microalgae, and where low-cost CO_2 is locally available as an industrial by-product or from natural CO_2 wells, major economies can be realized. Similarly, geothermal or waste heat sources can be used in special circumstances to enhance productivity during cold periods and for drying processes. Such possibilities would need to be considered at the planning and preliminary design stages to evaluate technical and economic feasibility for incorporation into the detailed design. A dependable source of electrical power is an essential requirement. Where outages exceeding 8 to 12 hr during the peak growing season can be experienced, standby power for essential services (including pond mixing) should be considered.

Layout of Facilities

Each microalgae production plant has its own special requirements depending on the type of product, site conditions, local preferences, etc., so that only generalized statements on site and facilities layout will be considered. An integrated plant capable of taking a flask culture up through production scale will be assumed. A progression of culture facilities through small, intermediate, and large ponds is therefore required. A number of ponds of each size is recommended, not less than two and preferably four, to provide sufficient flexibility and inoculation capacity to minimize lost production time in the event of upset conditions and for seasonal changes. The smaller ponds are also valuable for experimentation with improved strains, operating variables, and so forth. This is particularly important when introducing a new species or product for which prior experience over a period of years is lacking.

The layout of facilities should be carefully planned to ensure that an orderly expansion of plant size can be implemented with due regard to access for construction, operation and maintenance (O & M), utilities, and other services. Buried PVC piping with separate runs for each pond is preferred, but cost considerations may dictate the grouping of pond services as the number of production units increases. This will be influenced by pond and harvesting management strategies and other factors. Piping and other buried utility runs should be restricted to utility corridors (which may also serve as access roads) as far as possible, and runs under ponds should be avoided. Gravity harvest, return, and drain piping from the ponds to centralized pumping facilities is also preferred, with only one pumping lift in the harvest and return loop if possible to minimize damage to cells. This will be influenced by topography and scale to keep depth of piping within practical limits. For large systems, multiple harvesting locations may be appropriate to avoid excessive water conveyance costs. Selection of type of pumping as well as other elements of the harvest and return loop such as screens and separation units, should take into account the sensitivity of cells to rupture or the breakup of agglomerations, setae, and other features which affect harvestability and return of debris to the ponds. Gravity supply to harvesting units, with pumping of the return water only after separation of cells, is ideal but often impractical due to excessive cost of below-grade facilities. This must be weighed against the cost of low-shear pumping equipment where conventional centrifugal pumps are unsatisfactory.

POND CONFIGURATION, HYDRAULICS, AND SIZE

The shallow mixed pond for maximizing microalgae production was introduced in the 1950s and early 1960s by Oswald[1] and is called a "high rate" pond. Most of the early ponds used a meandering configuration with relatively narrow channels and many 180° bends. Propeller pumps were used to produce a channel velocity of up to about 30 cm/sec. In the 1970s this configuration was improved by the use of paddlewheel mixers of various designs which reduced the shearing forces on the cells and mixing energy requirements through elimination of the transverse wall with small openings for pump discharge. The multiplicity of channel bends was also a source of hydraulic loss and solids deposition problems, which were minimized by a long single-loop (racetrack) configuration when suitable baffles were provided at the ends. The single-loop configuration is particularly suited to larger installations where the length of the loop makes use of the full potential of the paddlewheel mixer. Mixer design, mixing velocity, and pond operating depth have a strong influence on pond configuration and size because these set practical limits on channel length and width.

Pond Configuration

As a general rule, only one mixer should be used per pond to avoid interference between mixers, access, and other operational problems. Considerations of access, arrangement of piping and other utilities, construction, and operation and maintenance activities, all point toward the centralization of special facilities at one end of the pond. A back-to-back arrangement with single-lane access road between pairs of ponds is preferred, as this results in economies of a common wall and utilities, yet allows access for maintenance of pond walls and lining along the full pond length. This arrangement is illustrated in Figure 1. Special facilities incorporated in this design include the paddlewheel, deepened section with travelling solids remover for settleable solids control, CO_2 dissolution facilities, and inlet-outlet provisions. These, as well as pond lining and wall construction, will be discussed in subsequent sections.

Pond Hydraulics and Size

Ponds of the "high rate" type depend upon sufficient mixing velocity to keep the cells in suspension and periodically exposed to light for rapid growth and avoidance of deposited solids. The velocity needed to avoid deposition of cells varies with the settling rate of the cell or aggregation of the cells where flocculated. This rate can range from a negative value (buoyant cells) to several meters per hour for flocculated cells. A minimum mixing velocity at any point of about 10 cm/sec is generally sufficient to avoid deposition of cells, but a minimum design value of about 20 cm/sec based on an average over the channel cross-section is often used because of unavoidable variations in velocity, particularly near the pond ends. Due to the energy cost dependence on velocity, increasing approximately as a cubic function, most ponds have been operated at velocities from 10 to 30 cm/sec.

Many ponds operating in the above velocity range have encountered difficulty with solids deposition in stagnant areas. Detail A of Figure 1 shows a typical deposition pattern where a simple curved end without baffles is used. This problem is only partially overcome with the use of a few concentrically placed curved baffles, because a low velocity zone continues

269

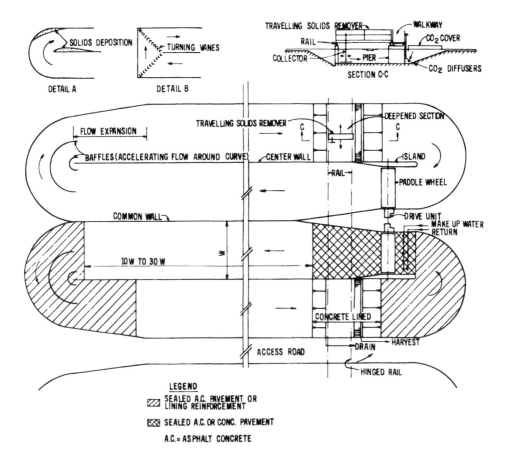

FIGURE 1. Back-to-back pond configuration.

to exist on the back side of the center wall and baffles. The preferred solution for the pond end without paddlewheel is shown on the left end of the main view of Figure 1. This consists of an eccentrically placed curved wall and baffles to create a curved zone of accelerating flow followed by a flow expansion zone after the directional change has been made. The rate of contraction of the curved zone is sufficient to avoid eddies or velocities below that causing deposition on the back side of the center wall and baffles. This concept was developed and first tested with the two 1230 m² ponds constructed in Singapore.[3] A similar approach is taken on the paddlewheel end, except that the baffles were found to be unnecessary due to the head drop created by the paddlewheel. The narrower channel resulting from the eccentric wall curvature is advantageous from the standpoint of paddlewheel design, and provides space for locating and servicing the drive units in the back-to-back configuration.

An alternative to the above end configuration is shown in Detail B of Figure 1. This is a proprietary design developed by Dainippon Ink & Chemical, Inc., using two right angle changes in direction caused by turning vanes along the diagonals from the center wall to the corners.[4] The vanes are closely spaced (less than 50 cm apart) and are supported by members along the diagonal line.

The above discussion has been concerned with pond hydraulics from a plan view, particularly with regard to maintaining a nearly uniform velocity distribution to avoid solids deposition with a minimum energy input. From this perspective, pond size is relatively unrestricted, except for such factors as wind effects on velocity distribution or wave action with very wide or long channels, and practical considerations such as effectiveness of baffles and feasible width of the paddlewheel. When viewed in profile, however, hydraulics have a much greater impact on pond size. Most small ponds are constructed with flat bottom grades, and the depth is allowed to vary to match the hydraulic profile around the loop. In recent years, there has been a trend toward a shallower depth and higher mixing velocities, both in the area of commercial production for the health food market and in theoretical research attempting to increase productivity through the "flashing light" effect.[5] Depths of 10 cm or less and velocities of 50 cm/sec and more are being proposed.[5,6] To maintain this shallow depth within a reasonable tolerance over a modest-size pond, say less than 2000 m^2 in area, a sloping pond bottom becomes necessary even at the lower range of velocities. As the velocity and pond size increase, the head loss around the loop soon exceeds the culture depth. Energy and other impacts resulting from higher mixing velocities need to be weighed against increased productivity attributed to faster mixing. What little work that has been done in this area has been at rather small scale. For example, the study by Argaman and Spivak[7] gave experimental data for the hydraulics of a high rate pond of about 280 m^2, but used an average depth of 35 cm and maximum velocity of only about 17 cm/sec. The raceways tested by Richmond[6] at a depth of 13 cm and a velocity of 50 cm/sec were 200 m^2. The largest raceway tested by Laws[5] was 48 m^2.

The cost and technical difficulty of holding construction tolerances on very flat grades to a few centimeters, consistent with very shallow depth, over many thousands of square meters must be considered when proposing large ponds. Practical O & M problems can also arise. For example, when the head loss and corresponding difference in bottom elevation around the pond loop (assuming the bottom is sloped to match the hydraulic slope) exceeds twice the operating culture depth, shutdown of the paddlewheel for any reason causes the pond water to collect in the low area to the point that the high area of the pond begins to dry out. This is undesirable because of contamination of the culture with dead biomass resulting from the drying out of inevitable puddles in the exposed area, as well as the deleterious effect on the lining due to the heat and UV exposure. If the rule were adopted that the head loss around the loop should not exceed twice the operating culture depth, a family of curves for approximate limiting pond area vs. velocity and depth could be developed, similar to those shown in Figure 2. The curves would vary for each condition of bottom roughness, and channel length-to-width ratio. A smooth bottom lining (Manning's "n" = 0.013) and channel length-to-width ratio of 40 are assumed in Figure 2. Using the Manning formula for friction loss and ignoring bend and other local losses:

$$V = \frac{R^{2/3}}{n} \left[\frac{H_L}{L} \right]^{1/2}$$

where V = velocity, m/s; R = hydraulic radius, m; n = Manning coefficient, sm$^{-1/3}$; H = head loss, m; and L = channel length, m. Substituting the depth, D, for R, which holds for very wide, shallow channels, and 2D for H$_L$ based on the above assumption, the formula may be solved for L as follows:

$$L = \frac{2\,D^{7/3}}{n^2\,V^2}$$

If the channel width is assumed equal to L/40, the above equation can be rewritten in terms of pond area, A, ha:

$$A = \frac{L^2}{40 \times 10^4} = \frac{D^{14/3}}{10^5\,n^4\,V^4}$$

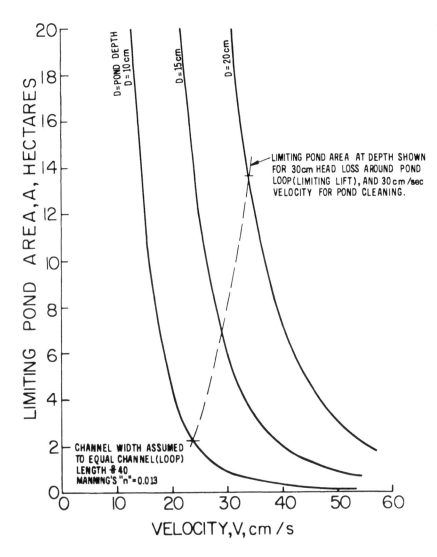

FIGURE 2. Limiting pond area, depth, and velocity relationships.

As seen from Figure 2, the limiting area for a depth of 10 cm and a mixing velocity of 50 cm/sec is only about 1/8 ha, which imposes a rather severe cost penalty based on the economy of scale applicable to pond construction and operation. Note that by increasing the depth to 15 cm at 50 cm/sec or by reducing the velocity to 30 cm/sec at 10 cm depth, the limiting area rises more than sixfold. As the velocity is reduced and depth increased, a point is reached where the limiting area becomes very large, and constraints on pond size other than the above mixing hydraulics criterion will probably govern. Note that the limiting pond area varies with Manning's "n" in a way similar to velocity, i.e., a small increase in "n" results in a rapid decrease in area. This must be considered when roughness is purposely increased to increase turbulence.

A practical limitation on pond size is imposed by the feasible lift which can be accomplished with a single paddlewheel. As discussed in the section of paddlewheel design, this lift will depend on the paddlewheel diameter and number of blades, and limitations on available facilities for installing and servicing the equipment. A total lift of about 30 cm is probably a reasonable upper limit for a paddlewheel mixer at the present time, although greater lifts might be justified for large installations which are equipped for servicing heavy equipment. A further consideration is the desirability of having the capability of periodically achieving a mixing velocity of about 30 cm/sec to assist in pond cleaning. This is relevant for ponds which are designed for slower mixing velocities where heavier solids such as dirt, chemical precipitates, and flocculated organics may accumulate on the pond bottom. Cleaning is greatly facilitated by a combination of velocity adequate to move settleable solids and a deepened section capable of capturing these solids on-stream, as discussed in a later section. Brooming may still be required in some cases to dislodge compacted heavy solids, but the above provisions significantly reduce the labor requirements and down-time for pond cleaning.

The dashed line on Figure 2 shows the limiting pond area based on a 30 cm head loss around the pond loop at 30 cm/sec velocity to meet the above paddlewheel lift limitation and cleaning velocity, respectively. The limiting pond area to the left of the dashed line will be controlled by this criterion, whereas to the right of the dashed line the previous mixing hydraulics criterion will govern. It should be emphasized that the curves of limiting pond area shown in Figure 2 are not hard and fast rules, but help to show the relationships involved and some of the implications of depth and velocity on pond size.

PADDLEWHEEL DESIGN AND CONSTRUCTION

Paddlewheels have been used for decades to mix small culture tanks and ponds, using either simple partial-depth blades, or high speed "cage" rotors similar to those used for oxidation ditch aeration. These designs are of limited usefulness for large microalgae production ponds where the main objective is to move water efficiently rather than aeration. This is due to backflow around and under the paddle or rotor blades which prevent the creation of adequate head. Where beating of the water with a high speed rotor is used to create more head, this causes damage to fragile cells and excessive energy costs. An improved paddlewheel design was introduced about a decade ago,[8] and was reported on in the Bureau of Environmental Studies (Australia) report[9] and by Dodd and Anderson[10] (see particularly the author's reply to the discussion). The design was first implemented at the Primary Production Department's high rate ponds at the Pig and Poultry Research and Training Institute in Singapore, as shown in Figure 3 and in the International Development Research Centre (Canada)[3] report on the workshop on high rate algae ponds held in Singapore in February 1980.

The operating principle of the improved paddlewheel design is that of a positive displacement vane pump, as shown in Figure 4. The paddle tips have a limited clearance with the side walls and a curved bottom section, the arc length of which is sufficient to prevent the water from escaping from between adjacent blades. Therefore, significant backflow around and under the blades is prevented and the paddle tip speed is only slightly higher than the channel velocity. This design serves to minimize energy requirements and cell damage, as well as to provide positive control of channel velocity. The limited clearance needed to control backflow is achieved by attention to details of design and construction which allow for main shaft deflection, adjustment of blade tips, and use of infill concrete in the curved bottom portion placed after paddlewheel installation. Flexible tip seals may be provided where there is danger of coarse material (e.g., loose rock bottom lining) becoming wedged between the blade and the curved bottom.

Main Shaft and Drive Unit

As the channel width increases, a choice must be made regarding the number of supports and the means for driving the paddlewheel. Where foundation conditions are such that relative movement between supports is possible, e.g., where clay soils are subject to expansion and contraction at varying moisture contents, intermediate support for the main shaft should be avoided. Unpredictable bearing loads and short bearing life can result due to shaft

FIGURE 3. High rate ponds in Singapore for treatment of piggery wastewater and production of high protein feed. Courtesy of the International Development Research Centre, Canada, and The Primary Production Department, Government of Singapore.

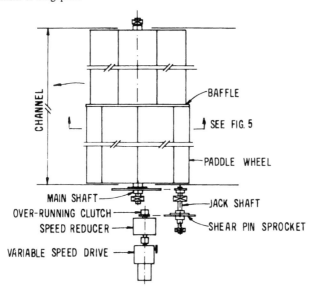

FIGURE 4. Paddlewheel drive train.

stresses induced by misalignment due either to installation conditions or subsequent support movement. A heavy duty carbon steel pipe shaft with welded hubs, sprocket flange, and stainless steel stub shafts spanning between two sealed spherical roller bearings with split housings is preferred. The pipe size and stiffness should be chosen to control deflection at midspan compatible with the tip clearance provided, which can be more critical than combined torsional and bending stresses imposed by paddle and drive chain loads. An important main shaft load which is sometimes overlooked is the wet weight of attached biological growth on the paddlewheel, which can exceed the dead load of the paddles themselves. This is particularly a problem with wastewater ponds, where periodic paddlewheel cleaning should be a part of the O & M routine. As shown in Figure 4, a chain and sprocket drive with the final sprocket bolted to a pipe flange inboard of the main shaft pillow block is preferred. This provides for simplicity of installation and avoidance of a large and costly coupling and pillow block needed to transmit the full drive torque through the stub shaft as in a direct drive design. Sufficient main shaft length should be provided for an enclosure around the chain and sprocket drive as well as the remainder of the drive equipment to isolate them from the adverse pond environment.

The welded steel plate hubs which transfer the paddle spoke loads to the main shaft should be large enough to allow a simple two-bolt spoke connection for easy installation and replacement of the paddle units, as well as adjustment of tip clearance. This facilitates installation of a large paddlewheel where only the main shaft need be lifted into place with long reach heavy equipment. Careful control of tolerances is warranted to allow inter-changeability and replacement of paddle units. Specifications, workmanship, and inspection of welding, surface preparation, and protective coatings for the main shaft should be carefully attended to, particularly where saline pond water is involved.

The selection of paddlewheel drive equipment involves the consideration of required capacity including normal mixing power and peak loads for pond cleaning, speed variation and control, overload protection, free-wheeling capability at shutoff, reliability, efficiency, and cost. Due to the slow speed of paddlewheel rotation, as low as a few revolutions per minute, and high torque requirements, enclosed speed reducers can be quite costly for a direct drive approach. Worn gears are capable of large reductions but have low efficiency and are self-locking which rules out free-wheeling through the gearing to an over-running clutch. The preferred arrangement is a double chain drive low-speed reduction through a jack shaft with shear pin for overload protection, to an over-running (cam) clutch on the speed reducer output shaft as shown in Figure 4. The over-running clutch allows the pad-dlewheel to free-wheel when the drive motor stops, which avoids the flowing water backing up against a stopped paddlewheel and possible overtopping of pond walls. This arrangement also facilitates the mounting of the final drive sprocket in line with the paddlewheel sprocket using the jack shaft to avoid the high overhung load and proximity to the pond environment that would occur if the speed reducer were used with single chain drive only. The cost savings for the smaller clutch and speed reducer greatly outweigh the cost of the extra chain reduction, particularly for larger paddlewheels.

Variable speed drive units are preferred for paddlewheels because of the flexibility provided to meet varying operational conditions such as reduced speed startup, normal mixing, and cleaning requirements. A turndown ratio of about 3:1 is usually sufficient, so a wide variety of mechanical and electrical variable speed drive types are available. Types which operate at significantly reduced efficiency at the normal operating speed for pond mixing should be avoided. Suitability for the severe environmental exposure prevalent at most pond sites and availability of qualified service technicians are important considerations, which may favor the selection of mechanical over electronic types of drive.

Power requirements for paddlewheels depend on the flow rate and differential head created by the paddlewheel, efficiency of the paddlewheel itself and the various components of the drive train, and desired service factor. Requirements should be determined both the normal mixing and cleaning velocities, using service factors appropriate for each. The difficult aspect in determining power requirements is selection of paddlewheel efficiency. This is strongly influenced by paddlewheel design, including such factors as number of paddlewheel blades, paddle geometry (including submerged support members), blade tip radius and drag

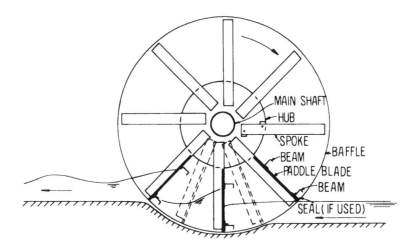

FIGURE 5. Paddlewheel transverse section.

along the curved bottom, hydrodynamic losses entering and leaving the paddlewheel, and so forth. Argaman and Spivak[7] analyzed the pumping efficiency of a partial depth cage rotor in a small high rate pond and found it to be on the order of 4%. This can be improved by using larger diameter low speed design with full depth blades. As discussed in the section on paddle design, the larger the blade tip radius, the lower the energy expended in unnecessary lifting of the water. However, as the differential head or lift across the paddlewheel increases, the greater the need for supporting members to keep paddle weight and cost within reasonable limits. Such members tend to reduce the efficiency due to internal hydrodynamic losses in the space between the blades. For these reasons, it is difficult to present a generalized equation for paddlewheel efficiency, which must be determined empirically on a case-by-case basis.

Paddle Design

In this context, the paddle includes the spokes, beams, blades, and seals (if used), forming an assembly which is bolted to the main shaft hubs. The arrangement of these various elements is shown on Figure 5 for a moderately large paddlewheel where the required lift exceeds 15 cm.

Eight paddles are used in the typical paddlewheel section (Figure 5), although 12 paddles may be appropriate for large ponds requiring lifts of 30 cm or more. The paddles are displaced at mid-channel by $22\frac{1}{2}°$ to reduce cyclic loads on the drive train as the blades enter and leave the water. This gives the equivalent of a 16-blade paddlewheel with lower power and maintenance requirements. The mid-channel circular baffle prevents short circuiting flow around the blades and also blocks the tendency for currents formed by pond end curvature to travel axially along the blades.

The blade width must be sufficient to avoid overtopping by the water wave created by paddle rotation. As may be seen for the lowermost blade in Figure 5, the blade width may be roughly estimated by the sum of the depth of the bottom curvature depression, pond water depth, lift, and freeboard allowance. The minimum blade tip radius is approximately the sum of the blade width and the hub radius needed to transfer the spoke loads. From these considerations, the minimum tip radius will be about 1.0 to 1.2 m (2.0 to 2.4 m paddlewheel diameter) for an 8-blade paddlewheel. Diameters up to 3 m using 12 blades may be necessary for large ponds.

Materials of construction are an important consideration in paddle design. Particularly where brackish water is involved, fiberglass is preferred due to its strength, durability, light weight, and availability in suitable structural shapes. For the paddle blades, sheet fiberglass with rectangular corrugations as used for building siding and roofing (opaque type) is excellent and though more expensive is much more durable than plywood which is subject to water absorption and warpage problems. Fiberglass spokes and beams are used to support the fiberglass blades (corrugations perpendicular to beams), and are of channel cross-section for maximum bending resistance without creating a water trap as the blade leaves the downstream surface. Although the horizontal lower beam causes some loss of hydraulic efficiency, it allows better use of lightweight materials and wider hub and spoke spacing than with a simple flat blade spanning between spokes. The latter construction is appropriate for smaller ponds with lifts of about 5 cm or less.

SETTLEABLE SOLIDS CONTROL

High rate ponds operated on a continuous basis over a period of months, or even weeks in the case of heavily loaded wastewater ponds, tend to accumulate settleable solids which interfere with performance. The settleable solids can originate from biological sources such as bacterial floc, old or dead algae cells which tend to settle rapidly, or from inert materials such as wind-blown dirt and chemical precipitates. Removal of such material is difficult and labor-intensive unless special provisions are made in the pond design. Pond down-time is also an important factor if the pond must be shut down, culture medium transferred, and the bottom broomed and flushed to remove solids. The very flat bottom slopes and inevitable depressions or "bird baths" make thorough cleaning, e.g., where a food grade product is involved, particularly difficult and costly.

A method of on-stream settleable solids control for high rate ponds was introduced at about the same time as the improved paddlewheel design[10,11] and also was first implemented in the Singapore high rate ponds.[3] The design incorporates a deepened or depressed section acting as an in-line rectangular sedimentation basin as shown in Section C-C of Figure 1. Accumulated solids are periodically removed from the bottom of the deepened section by a traveling suction device discharging to a trough or through a hose draped between masts. A pivoting tee collector mounted on a bridge supported by rails across the ponds is coupled to a positive displacement pump mounted on the bridge. The bridge is moved by a hand controlled drive to clean a number of ponds with the same unit, as shown in Figure 6. Where desired, the solids pumped from the deepened section may be thickened in a separate settler and the supernatant returned to the pond to minimize loss of culture medium.

The frequency of cleaning the deepened section will depend on the rate of accumulation of solids and the time for the solids to become anaerobic with possible rising sludge or other problems. For heavily loaded wastewater ponds, the interval between cleaning may be less than 1 week, whereas for clean culture ponds with saline media it may be several months.

The depth of the deepened section should not exceed about 1 m, to avoid risk of uplift and flotation when the pond is dewatered. This depth is generally adequate to achieve the desired settling provided the transition zones and bottom length are not too short. Reinforced concrete construction with a slab thickness of not less than 10 cm is recommended. Where danger of a high water table exists, provision of relief drains or added thickness should be considered.

POND BOTTOM LINING AND WALL CONSTRUCTION

The selection of type of bottom lining and wall construction is very important to the success of high rate ponds, particularly as the size of the pond increases. Failure of linings has been a source of difficulty, particularly with thin membrane linings, causing serious O & M problems or the need for premature replacement. Wall construction and lining materials

FIGURE 6.   Traveling bridge at deepened section of Singapore ponds for on-stream control of settleable solids. Courtesy of the International Development Research Centre, Canada, and the Primary Production Department, Government of Singapore.

need to be considered together as they are strongly interdependent. Construction and economic factors play a large part in the selection process, as do the purpose of algae cultivation, soil characteristics, and availability of materials, expertise, and equipment for their installation. As discussed in the section on site selection, climate can also influence selection of wall construction, particularly at sites susceptible to flooding and high winds during severe storm events.

Bottom Lining: Materials and Methods

Some form of bottom lining is usually required for high rate ponds to meet the requirements for seepage control as well as erosion and turbidity control at maximum mixing velocities. Native soils that are sufficiently fine grained to meet the seepage control requirement generally give rise to excessive turbidity at mixing velocities sufficient to keep cells in suspension and ensure good productivity. As discussed in the section on pond hydraulics, productivity may be enhanced by increased turbulence which improves transport of nutrients to the cells and exposure of the cells to light including possible "flashing light" effect. Turbulence may be induced by bottom roughness and higher mixing velocities, or by suspended vortex inducing devices such as the "foils" proposed by Laws.[5] Such approaches need to take into consideration the O & M impacts on liner maintenance and control of contaminants, as well as the limitations on pond size due to increased head loss.

Pond lining materials may be broadly classified into nonmembrane and membrane types. Nonmembrane types range from a thin blanket of coarse granular material to control erosion of impervious native soil, to various types of paving such as asphalt concrete (A.C.), pneumatically applied mortar ("gunite"), or conventional concrete. Membrane linings include the very wide range of commercially available sheet plastic and rubber materials which generally require field joining at the seams, and spray applied materials without seams. All of the above lining materials have advantages and disadvantages which affect their suitability for particular applications and site conditions. A combination of materials will usually provide the most satifactory solution to meet the varying needs in different areas of the pond.

Blanket Materials

The most common blanket materials used in pond construction are clay blankets to control seepage in the case of pervious native soils, and coarse granular blankets to control erosion and turbidity in the case of fine grained native soils or where a clay blanket is used. The clay blanket can be used as an alternative to a membrane lining to control seepage at a cost saving where suitable clay is locally available. This also avoids puncturing and other problems associated with thin membrane linings.

Coarse granular blankets provide an economical and efficient lining, and are recommended for other than food grade product applications where the native soil is suitable for seepage control or where a clay blanket is used over pervious soils. The selection of type of coarse granular material will depend on local availability and cost, but at least a major fraction of crushed rock is preferred due to the improved rolling and binding characteristics. A gradation of sizes up to about 2 cm, with sufficient fines to give good binding is recommended. with a blanket thickness of about 4 to 5 cm. Coarser material may also be used, particularly where it is desired to increase bottom roughness for greater turbulence.

Paving Materials

Although conventional paving materials are generally too costly for use throughout the pond bottom, they are appropriate for particular areas of the pond where wave action would tend to disturb other types of lining. These areas include the approach and discharge aprons for the paddlewheel and the end curvature sections as shown by hatching in Figure 1. The selection of paving materials will depend on local conditions. A.C. paving has the advantage of flexibility where expansive soils would tend to crack thin concrete or gunite slabs. Conventional A.C. paving as used for highways is pervious, and a modified mix design with higher asphalt content and/or seal coat should be used for greater durability and water tightness. Concrete paving may be less costly than A.C. where the quantities are too small to bring in special A.C. paving and rolling equipment. Control of tolerances and roughness can be a problem with gunite paving, but it can be useful particularly where special shapes or access problems are involved.

Sheet Membrane Materials

Sheet membrane liners of various plastic and rubber materials have become widely used for seepage control in water reservoirs, waste ponds, and disposal sites where the liner is typically protected by an overlying blanket of sand or soil. Their use in high rate ponds has had mixed success because of the absence of a protective blanket, leaving the thin liner exposed to puncture damage and displacement due to hydraulic and wind forces. These risks were dramatically illustrated by the recent loss in a matter of minutes of the sheet plastic liner from a 1 ha pond at a commercial algae farm in California due to high winds and insufficient water cover. Liner movement and wrinkling due to hydraulic forces, particularly at bends and near mixers, and bubbles or "whales" due to gas or water accumulation under the liner are other common problems. Claims made for liner durability and expected life based on other applications are seldom realized in high rate ponds due to the high risk of mechanical damage from the above causes, as well as necessary pond O & M activities such as pond cleaning. However, membrane liners will probably continue to be used in many applications, particularly where cleanliness and contamination control are a major factor as with food grade products.

The risks of damage to membrane liners can be substantially reduced by careful attention to pond design details. The objective should be to take advantage of the membrane as a bottom material of high uniformity and toughness which can be readily cleaned, but supplement it with other materials in areas where it is difficult to install or is subject to damage. The membrane material should be of sufficient thickness and durability to prevent wrinkling or puncture in normal service on the straight reaches of the pond. Foot traffic and the use of equipment on the liner should be minimized by the use of on-stream hydraulic cleaning as described previously. Areas of turbulence and irregular shape such as at the pond ends and transitions should be lined with other materials, either paving as previously described, or heavy duty spray applied membrane lining with fabric reinforcement as described in the following section. Vertical walls are less subject to damage from wind and wave action than sloping berms, as discussed in the section on wall construction, but create problems with sheet membrane liner installation and separation of the liner from the wall. These problems could be overcome by a combination of sheet material for the bottom lining, joined with a spray applied membrane on the vertical and transition surfaces around the pond perimeter and baffle walls. The selection of the sheet and spray applied membrane materials and joint design to ensure compatibility and soundness at the joints would be essential. Ease of installation and repair should be a major factor in the selection of membrane materials, along with such other factors as availability, cost, chemical and UV resistance, elasticity and toughness at the range of temperatures encountered, nontoxicity, and uniformity.

Spray Applied Membrane Materials

A number of commercial spray applied membrane systems have been developed which have seen limited use for high rate ponds. These include both cold applied (two component) and hot applied systems using mixtures of asphalt, rubber, and other elastomeric materials. Reinforcing fabrics may be used with either type. These fabrics improve the resistance to puncture and flow damage in areas of high traffic or turbulence. The cost of raw materials is generally less for the hot applied systems, which allows the economic use of greater thickness than with other membrane linings. This is advantageous for bridging imperfections in the surfaces being sprayed, as well as for resistance to lifting or other displacement. However, the technical skill and application equipment required limit the use of hot applied materials to areas where highly experienced contractors are available for this type of work.

The economic advantage of spray applied lining for large bottom areas lies in the use of distributor trucks capable of uniform application at high speed. In contrast with sheet membranes, no seams are involved but uniformity of thickness, pinholing, and laminating can be problems if the equipment or workmanship are deficient. The proper design, formulation, and mixing of the various components of the membrane system are also crucial, and must be adapted to the particular use, e.g., for wall or bottom service, temperature exposure, surface bonding conditions, water velocity, and so forth.

Turbulence Enhancement

Where increased bottom roughness is desired to create turbulence with spray applied membrane linings, a chip spreader similar to that used for highway seal coating could be used in conjunction with spray application. Other forms of roughness could also be used with either sheet or spray applied membranes, such as briquettes or tiles spaced uniformly over the surface. Additional liner thickness may be needed in this case to reduce the risk of puncture at corners of the roughness-causing elements. These elements or other devices such as foils used to create turbulence would have an impact on O & M activities such as pond cleaning and liner maintenance. These tradeoffs along with construction and cost impacts would need to be carefully weighed at the design stage to determine the practicality and cost effectiveness of such measures for productivity enhancement.

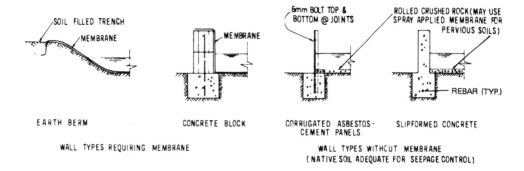

FIGURE 7.   Alternative pond wall sections.

Wall Materials and Design

Many high rate ponds have been constructed simply as "holes in the ground" with thin plastic membrane liners draped up sloping earth berms as shown in the left hand section on Figure 7. While this has the advantage of low cost and simplicity, it also can lead to operational and other difficulties and short service life. Risk of pond contamination or loss of the entire culture is inherent in the design due to the possibility of flooding and the tendency for wave run-up along the sloping banks due either to wind- or paddlewheel-induced waves. Wind blowing across the berm has a tendency to lift the liner at the edges due to the "Bernoulli effect", which can lead to the loss of the liner as previously mentioned when the water depth is reduced. Pond earthwork and liner fabrication require special methods and close tolerances, particularly at the ends and transitions to structures. Space requirements are somewhat greater than with vertical walls due to sloping sides and lack of common wall, although this is partially offset by reduced shading effect.

Concrete paved sloping banks were used in Australia,[10] but the costs of compacting and shaping the banks and placing the paving were rather high. A rolled crushed rock bottom lining was used over lime and cement stabilized clay soil. This design was revised in the case of the Singapore ponds, which used vertical free-standing walls constructed of corrugated asbestos-cement roofing panels set into a shallow concrete filled trench, as shown in Figures 7 and 8. Rolled crushed rock was used over clay without stabilization or any form of sealing other than a bead of silicone sealant between the panel joints. This proved to be a satisfactory and economical method of wall construction for both straight and curved walls. The panels with vertical corrugations had the advantage of acting as a continuous horizontal expansion joint, with sufficient vertical cantilever beam strength to resist all normal loads. No leaks were experienced in either of the two 1230 m² ponds using this construction for both external and internal walls and baffles. This wall design is recommended for use where the panel material is available. The use of asbestos-containing materials is being restricted for applications where the fibers may become an airborne contaminant either during construction or in service, and strict regulations to protect workers are in effect in many areas.[12] However, asbestos-cement pressure pipe has been one of the most common materials for potable water distribution systems and continues to be used for this purpose. Epidemiological studies[13] in both the U.S. and U.K. have concluded that this is a safe practice. Where the panels are precut to length and holes drilled at the source of supply, worksite precautions can be kept to a minimum during construction. If there are concerns about possible consumer reaction in the case of health food product applications, the use of a spray applied membrane extending up the water side of the panels will isolate them from the pond environment.

Other types of vertical wall construction illustrated in Figure 7 are concrete block and slipform concrete. Both require expansion joints at junctions with structures and at intervals along the wall length. Concrete block is pervious and requires a spray applied membrane

FIGURE 8. Crushed rock bottom lining and corrugated asbestos-cement wall at Singapore ponds. Courtesy of the International Development Research Centre, Canada, and the Primary Production Department, Government of Singapore.

for satisfactory sealing. Some vertical and horizontal reinforcement is also recommended because expansion and contraction tends to loosen mortar joints. The slipformed concrete alternative has not been tried but may be cost effective for large installations where the cost of developing the special equipment can be distributed over many thousands of meters of wall. The equipment would be similar to that used for slipformed highway barriers, and would preferably be wire or laser guided.

Construction Factors

The scheduling of construction activities should be planned to allow critical steps such as final grading and bottom liner installation to occur during favorable weather. Rough grading should be completed over the whole pond area as early as possible in the construction program, but fine grading in advance of liner installation should be deferred until most of the other work is completed to give the maximum time for soil consolidation and moisture content to stabilize. Structural concrete should preferably be in place before abutting wall construction is carried out. Wall footings should not be excavated until shortly before the footing concrete is placed, to avoid damage from rain and shrinkage cracking with expansive soils. In the slipformed concrete wall alternative, the trenching machine could be operated just ahead of the slipform machine or be integrated with it.

Fine grading is difficult with clay soils which tend to harden and laminate when dry but become sticky and unmanageable when moistened. In such cases a thin blanket of graded granular material may be needed to form a satisfactory working surface which can be accurately brought to grade and rolled smooth. Particularly in the case of spray applied membranes, defects such as exposed pebbles, cracks, loose material causing wheel rutting, segregation of coarse and fine material, wet spots, and roughness at junctions with wall footings can cause areas of liner weakness or leakage. Most of these problems can be overcome by good workmanship and inspection of surfaces prior to membrane application. Some tough-up work by hand spraying will usually be necessary following detailed inspection of the membrane prior to water testing. Special treatment of membrane edges may be necessary to assure adequate bonding to resist hydraulic forces, such as at paddlewheel aprons. A wide variety of epoxy adhesives, water resistant putty, and other bonding agents are available for this purpose.

Access for liner installation and maintenance needs to be considered at the design stage. Truck access into the pond area itself is usually necessary for all but small ponds to allow economical handling and/or application of membrane materials. This can be done by leaving out a section of wall, usually at the end without pond equipment, which is then closed after liner installation. For spray applied membranes, the practical length of hose may require vehicular access between pairs of ponds for making lining repairs and other maintenance activities, as illustrated in Figure 1. Such access roads can also serve as utility corridors for pond piping and other necessary services.

POND FEEDING, PIPING, AND OTHER SERVICES

Nutrients to support algae growth may either be added directly to the pond or from a central source to the return lines for recycled water from processing facilities. The latter approach is generally more appropriate for nutrients other than CO_2, except for batch operated or small ponds. CO_2 is a special case due to the large dissolution requirements which dictate in-pond feeding. Conventional pond management practices also generally dictate facilities for monitoring and control of pond pH and liquid level, and for recording temperature and other operating parameters. These systems and other pond facilities such as those for pond mixing and settleable solids control require piping and other services which need to be considered in pond design.

CO_2 Feeding Provisions

Except for wastewater ponds, which use CO_2 produced by bacterial respiration, most microalgae production ponds use CO_2 as the primary carbon source. Due to the slow rate of atmospheric CO_2 exchange, a means of rapidly and efficiently dissolving large quantities of CO_2 into the pond liquid is required to satisfy uptake by cells and to prevent excessive pH. Ponds growing *Spirulina* are an exception where bicarbonate may be added to elevate the pH for population control. CO_2 may be sparged through submerged diffusers either as relatively pure CO_2 or as CO_2-enriched air. Systems using relatively pure CO_2 are preferred due to lower dissolution energy and costs, except where lower CO_2 content gases from waste sources such as flue gas are available.

Dissolution efficiency increases rapidly with depth of diffuser, so it is preferable to locate the diffusers in the deepened section of the pond as shown on Figure 1. A covered section may be used for capturing CO_2 which escapes the surface, extending the transfer area and allowing redissolution via a low pressure blower and a second line of diffusers. The walkway across the deepened section provides access for servicing the diffusers as well as for the mounting of pH probe for control of CO_2 injection and other instrumentation. This system is similar in some respects to that proposed by Stengel et al.[14]

Pond Piping

Cultivation of microalgae in high rate ponds might be described as "hydraulic farming", in which the crop is moved by pumps and piping from the pond to the harvesting area. While this requires little labor in comparison with conventional farming, the piping, pumps, valving, and controls can be a significant cost item and require careful design. As indicated in the section on layout of facilities, buried PVC piping designed for gravity flow is preferred for the harvest, return, and drain lines. These are shown on Figure 1, along with the makeup water line connected to the low pressure water system. High pressure water service should also be provided to the deepened section and paddlewheel areas for washdown service. High pressure water, CO_2, and sludge piping, as well as power and instrumentation conduits, can be racked along the walkway and solids remover near side rail, supported to pass over the ponds in succession on the rail piers.

Electrical and Instrumentation

Buried PVC conduit is preferred for electrical power and instrumentation runs from the central control area to the outside end of the walkway adjacent to the end of the solids remover near the side rail. Services within the pond area are then fed by exposed conduit runs following along the solids remover near side rail in conjunction with the exposed piping runs to each pond. As with piping, buried conduit runs should not pass under the ponds. Due to the severe exposure, marine type service outlets on the walkways should be used and ground fault protection provided, particularly with saline water applications. Lighting should be provided for the walkway and paddlewheel drive areas as a minimum requirement for servicing and safety.

REFERENCES

1. **Oswald, W. J., Golueke, C. G., and Gee, H. K.,** Waste Water Reclamation Through the Production of Algae, Water Res. Center Contrib. 22. Sanitary Eng. Res. Lab., University of California, Berkeley, 1959.
2. **Becker, E. W.,** Comparative toxicological studies with algae in India, Thailand and Peru, in *Algae Biomass,* Shelef, G. and Soeder, C. J., Eds., Elsevier/North Holland, Amsterdam, 1980, 767.
3. International Development Research Centre, Wastewater Treatment and Resource Recovery: Report of a Workshop on High-Rate Algae Ponds, Singapore, 27—29 February 1980, IDRC-154 e, IDRC, Ottawa, 1980.
4. **Shimamatsu, H., et al.,** Apparatus for Cultivating Algae, U.S. Patent 4,217,728, Assigned to Dainippon Ink & Chemical, Inc., Tokyo, August 19, 1980.
5. **Laws, E. A.,** Research, Development, and Demonstration of Algae Production Raceway (APR) Systems for the Production of Hydrocarbon Resources. Prepared under Solar Energy Research Institute Subcontract No. XE-0-9013-01, University of Hawaii and Hawaii Institute of Marine Biology, 1983.
6. **Richmond, A.,** Phototrophic microalgae, *Biotechnology,* 3, 109, 1983.
7. **Argaman, Y. and Spivak, E.,** Engineering aspects of wastewater treatment in aerated ring-shaped channels, *Water Res.,* 8, 317, 1974.
8. **Dodd, J. C.,** Mixer for Algae Ponds, U.S. Patent 3,855,370, December 17, 1974.
9. Bureau of Environmental Studies, Algae Harvesting from Sewage, Environ. Study Rep. 1, Australian Government Publishing Service, Canberra, 1976.
10. **Dodd, J. C. and Anderson, J. L.,** An integrated high rate pond-algae harvesting system, *Prog. Water Technol.,* 9, 713 (C-55 to C-59 — discussion), 1977.
11. **Dodd, J. C.,** Solids Remover for High Rate Algae Ponds, U.S. Patent 3,969,249, July 13, 1976.
12. **Anon.,** CAL/OSHA Resource List and Standard for Asbestos, S-620, Department of Industrial Relations, State of California, Sacramento, 1982.
13. **Anon.,** No relation found between A/C pipe and cancer, *Civil Eng.,* 53(10), 22, 1983.
14. **Stengel, E. et al.,** Method and Arrangement for Optimally Supplying Autotrophic Organisms with CO_2 Nutrient, Assigned to Gesellschaft fur Strahlen und Umweltforschung mbH, Munich, U.S. Patent 4,084,346. April 18, 1978.

OUTDOOR MASS CULTURES OF MICROALGAE

A. Richmond

Part I

BIOLOGICAL PRINCIPLES

PRODUCTION SYSTEMS

Two basic systems for the production of microalgae are readily recognized. One is based on the use of wastewater, which is high in dissolved organic carbon (DOC). The growth medium is not precisely defined since the types of wastes in the water change as well as their concentration. Wastewater systems carry mixed populations of microalgae and bacteria, and species of the former usually change, sometimes drastically, throughout the seasons. As pointed out by Soeder,[1] the biomass from DOC-loaded systems should be referred to as algal-bacterial matter ("albazoid") and not simply as algae. This is because the harvested mass from such systems is a mixture of algae, organic residues at different stages of decomposition, and bacteria, which may amount to about 25% of the harvested product. Thus, the biology of wastewater systems, whatever the origin of the waste, differs in many basic aspects from the other system for producing microalgae which is known as the "clean method". A "clean" culture is usually grown in a well-defined and well-mixed mineral medium in shallow raceways or in narrow, transparent tubes. The bacterial population is quantitatively insignificant and the culture is meant to be monoalgal.[1] This method for growing microalgae is dealt with exclusively in this chapter.

INTERRELATIONS BETWEEN SOLAR IRRADIANCE, POPULATION DENSITY, AND MIXING

Kinetics of Light-Limited Growth

When a culture of algae is started with a sufficiently small inoculum so that only a small part of the incident light is absorbed and the algae are allowed to grow under constant incident light and constant temperature in a culture medium that is likewise kept unchanged, growth may be illustrated by plotting the packed cell volume per liter (V) (which represents the population density) against time in culture (Figure 1). Growth as depicted in Figure 1 is light limited, since at the higher light intensity (5000 lux) it is greatly stimulated. The dependence of growth on light is also illustrated in Figure 1b, in which the change in the rate of increase in biomass (V/t) is plotted against culture days. The rate of change, which decreases with time, actually decreases due to the increase in the overall population density (Figure 1a); after some 5 days V/t becomes constant at both high and low light intensities. At the higher light, however, this constant has a higher value because the higher light intensity facilitates the establishment and maintenance of a higher population density.

The same results may be illustrated by plotting the logarithm of the population density (V) as a function of culture days (Figure 2, a,b). Clearly, $\log^{10}$ V/t is constant in the early stage of growth in the culture, whereas V/t becomes constant at a later stage (from day 5 on). Thus, in a culture in which growth is limited by light, two distinct phases are evident. The earlier growth phase, in which log V/t is constant, is regarded as the "exponential phase". The following growth phase, in which V/t becomes constant, is the "linear phase".[2] The curves presented in Figures 1 and 2 show that the transition from the exponential phase

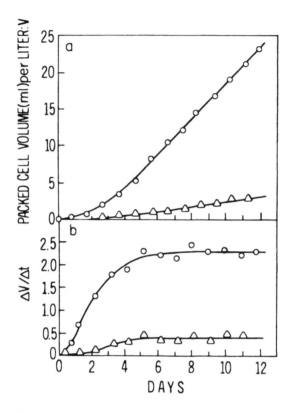

FIGURE 1.   Effect of light intensity along the growth curve.
(O—O) 5000 lux; (△—△) 800 lux. (From Tamiya, H. et al.,
Algal Culture: From Laboratory to Pilot Plant, The Carnegie
Institution, Washington, D.C., 1953. With permission.)

to the linear phase occurred at different population densities according to the light intensities
applied. In addition, the higher the light intensity, the higher the population density at which
the exponential phase ceased and the linear phase began. When, however, the culture was
exposed to very hight light intensities, approaching the intensity of solar irradiance, $\log^{10}$
V during the exponential phase was independent of light intensity. This indicated the cells
were light saturated, exhibiting a growth rate independent of the light treatment (Figure 3).

At the point when the population density greatly increases, any further rise in the quantity
of algal mass becomes increasingly dependent on light intensity and a linear phase of growth
sets in. This indeed is observed in outdoor cultures, when nutrients and temperature do not
limit growth, or at least not severely. Under these circumstances, the daily increase in
biomass is distinctly dependent on the available quantity of solar irradiance, and when the
population density is maintained sufficiently high, this dependence will be manifested without
any signs of light saturation even at the highest light flux measured outdoors[2] (Figure 4).

The strict dependence of growth on light in the linear phase is due to the phenomenon
termed "mutual shading", which results from the absorbance of the incident light by the
algal cells closest to the illuminated surface, thereby decreasing the amount of irradiance
available for the cells below. The greater the population density, the higher the extent of
mutual shading and the greater the fraction of cells that are not illuminated at a given instant.
Thus, in dense photoautotrophic algal cultures, in which growth is strictly light limited, the
meaningful quantity of light is the light available to the average algal cell in the culture.
The probability of mutual shading as a function of population density, cell size, and depth
of the culture along the direction of the light beam has been formualted by Tamiya et al.[2]

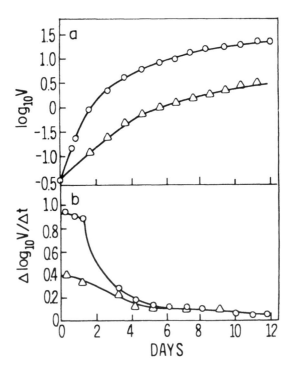

FIGURE 2. The same data as in Figure 1, presented in terms of logarithms of cell concentration ($\log_{10}V$) and their daily increase ($\log_{10}V/\triangle+$). (O—O) 5000 lux; ($\triangle$—$\triangle$) 800 lux. (From Tamiya, H. et al., *Algal Culture: From Laboratory to Pilot Plant,* The Carnegie Institution, Washington, D.C., 1953. With permission.)

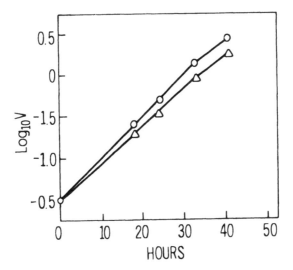

FIGURE 3. Growth curves obtained at saturating light intensities at optimal temperature (25°C). (O—O) 50 klx; ($\triangle$—$\triangle$) 10 klx.(From Tamiya, H. et al., *Algal Culture: From Laboratory to Pilot Plant,* The Carnegie Institute, Washington, D.C., 1953. With permission.)

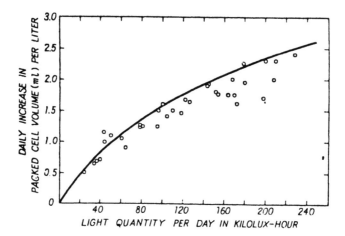

FIGURE 4.   Daily increase of algal concentration in the outdoor culture as a function of quantity of sunlight in each day. Temperature, 17 to 28°C. (From Tamiya, H. et al., *Algal Culture: From Laboratory to Pilot Plant*, The Carnegie Institution, Washington, D.C., 1953. With permission.)

The exponential phase of growth occurs when the effect of mutual shading is negligibly small. Under these conditions, $dX/dt = EX$, where X is the concentration of cells and E is a constant. The growth process becomes linear when mutual shading is so strong that all the light impinging on the surface is totally absorbed. Then, $dx/dt = L.A$, where A is the area of the illuminated surface of the culture and L is a constant. Both E and L are functions of temperature and the intensity (I) of incident light. At very low light intensities, they are both linear functions of I and are independent of temperature (light-limited growth rates: E_o, L_o); at sufficiently high light intensities, E becomes independent of I (light saturated: E_s), whereas L continues to increase, although the rate of increase gradually tapers off with the increase in light intensity. Unlike E_o and L_o, E_s and L are distinctly temperature-dependent at higher light intensities. Therefore, as the population density increases, the extent of mutual shading, no less than the amount of incident radiation, determines the irradiance available for the average cell in the culture and thus its rate of growth when other factors are not limiting.

Mutual shading, which causes each average cell in the algal culture to receive its illumination intermittently, presents a major issue in the biotechnology of algaculture. Kok[3] made a quantitative study of the effect of light intermittence on photosynthesis. In flashing light, the rate of photosynthesis depended on the ratio of the light to dark period and their absolute times as well as on the total amount of light. The method employed in that study was based on a thin layer (about 3 mm) of algal suspension exposed to light having the intensity of sunlight which was provided in light-dark cycles varying both in length and in the ratio of the dark/light periods. This was obtained with a rotating sector consisting of two blades, each with two 90°C cutouts, the relative positions of which determined the angular opening and hence the ratio of light to dark periods, which ranged from 1:1 to 1:00. The frequency of illumination was determined by the speed of revolution of the disk, which varied from 30 to 2300 rpm. Results indicated that the faster the speed of the disk, the smaller the effect of light intermittence on reducing the rate of photosynthesis. This was demonstrated in an experiment using a disk opening $(t_D + t_F)/t_F = 5.5$, where t_D represents the "dark time" and t_F the "flash time". The results are given in Figure 5 in which, for convenience, the flash time (which is inversely proportional to the disk speed) is plotted on a logarithmic scale. Clearly, the shorter the flash time, the lesser the effect of light intermittence on reducing the photosynthetic rate, compared to continuous light. The flash pattern

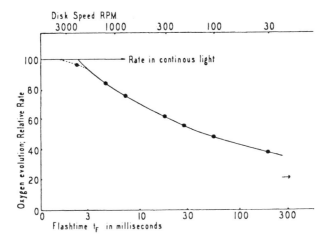

FIGURE 5. Relative rates of photosynthesis in flashing light of different flash times, the ratio of flash time to total period being constant $(t_d + t_F)/t_F = 5.5$. (From Kok, B., *Algal Culture: From Laboratory to Pilot Plant*, The Carnegie Institute, Washington, D.C., 1953. With permission.)

required for a high yield of biomass appears to put rather extreme demands on the turbulence obtainable in a flowing culture. There are indications that for maximal increase of light utilization efficiency, the cells would have to receive all their light in a layer of the suspension equal to approximately less than 10% of the total depth, then go into complete darkness in the other 90% of the suspension before receiving another portion of solar irradiance at the upper layer.

The Effect of Mixing on Light Distribution

Mixing of the culture aims primarily to induce a fast movement of algal cells from the illuminated upper layer or the photic zone to the lower, unilluminated strata in the pond and back again to the photic layer. Mixing, then, results in a dynamic light-dark pattern for the single alga and represents one of the most basic requirements for high productivity in mass cultures (see "Technological Aspects of Mass Cultivation — A General Outline").

The effect of mixing should therefore be most pronounced as the population density increases and light limitation of growth becomes more severe. Märkl[4] demonstrated this relationship in well-controlled laboratory experiments. The photosynthetic activity of algal cultures of different densities was measured as a function of the stirring speed. The results of five experiments performed with increasing density of algae are shown in Figure 6. Experiments 1 through 3 are carried out with the illumination of fluorescent tubes (55 Wm^{-2}), whereas in experiments 4 and 5 a Xenon lamp (530 Wm^{-2}) is used. The CO_2 concentration in the gas bubble supplying CO_2 was established at 1% barring the possibility for carbon limitation of the photosynthetic rate.

At a population density of 0.17 g/ℓ almost no light gradients in the culture existed. As a consequence, mixing had no influence on the photosynthic reaction rate. At higher densities of algae increasing the stirring speed resulted in an enhancement of photosynthetic activity. As expected, the influence of the "flash effect" increased with higher light gradients within the culture and with increasing light intensity. Thus, at the highest population density (2.33 g/ℓ), which would produce the highest light gradient in the culture vessel, stirring increased the protosynthetic rate by close to 50%.

The marked effect that the intensity of stirring exerted on the productivity per area or the areal output rate was demonstrated by Richmond and Vonshak[5] in experiments with *Spirulina*

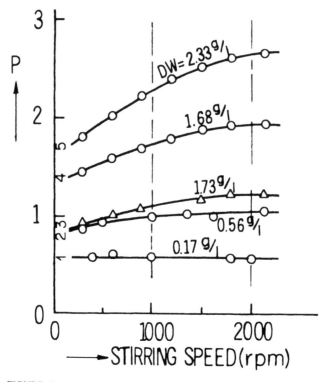

FIGURE 6. Net photosynthetic reaction rate in *Chlorella vulgaris* as influenced by mixing. T = 27°C, pH = 6.5. In experiments 1 to 3, illumination was provided by fluorescent tube lamps, Xenon lamps providing 530 W/m² being used to illuminate the cultures in experiments 4 and 5. (From Markel, H., *Algae Biomass*, Elsevier/North-Holland, Amsterdam, 1980. With permission.)

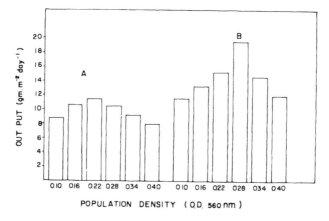

FIGURE 7. Output rate as effected by cell density and turbulence. (A) Paddle speed 15 rpm; (B) paddle speed 30 rpm. (From Richmond A. and Vonshak. A., *Arch. Hydrobiol. Brih.*, 11, 274, 1978. With permission.)

platensis grown outdoors. Two points were evident: the population density exerted a clear effect on the output rate and doubling the speed of flow resulted in an increase of over 50% in output at the optimum cell density (Figure 7). There are several reasons for the distinct effect of turbulence on the output rate as detailed in the chapter entitled "Technological Aspects of Mass Cultivation — A General Outline".

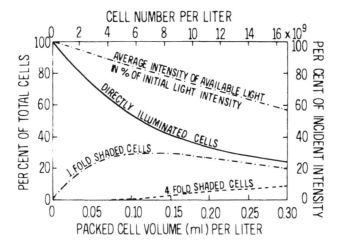

FIGURE 8. Percentages of directly illuminated cells and of cells which are shaded by an increasing number of other cells as functions of population density of cell suspension. The curve at the top represents the average intensity of light (in percentage of the intensity of incident light) available to each cell in the suspension. All the curves were calculated by equations developed by Tamiya.[2] (From Tamiya, H. et al., *Algal Culture: From Laboratory to Pilot Plant*, The Carnegie Institute, Washington, D.C., 1953. With permission.)

The economic establishment of turbulent flow that will submit individual cells to suitable flash patterns, rather than to a random distribution of intensity variations, represents a major engineering problem. Recently, an innovative approach was suggested by Laws et al.[6] to induce turbulence in algal ponds. Arrays of foils similar in design to airplane wings were placed in an algal culture flume to create systemic mixing. Vortices are produced in the culture due to the pressure differential created as water flows over and under the foils. In a flume having a flow rate of 30 cm/sec, the foil arrays produced vortices with rotation of approximately 0.5 to 1.0 Hz. This rotation rate is satisfactory to take advantage of the flashing light effect if the culture is sufficiently dense. Solar energy conversion efficiencies in an experimental culture of *P. tricornutum* averaged 3.7% over a 3-month period, representing an increase of 2.2- to 2.4-fold with the foil arrays in place vs. controls with no foil arrays. Five-day running means of solar energy conversion efficiencies reached as much as 10% during the 3-month period. The authors concluded that the use of foil arrays appeared to be an effective and inexpensive way to utilize the flashing light effect in a dense algal culture system.[6] A salient feature of the effect of mixing is that the higher the light intensity and the population density, the greater the positive effects that turbulence may yield. As long as optimal intermittence is not realized, however, only part of the possible increase in yield due to mixing may be obtained.

The relationships between the incident light, the algal cell concentration and the extinction coefficient, and the shape of the flask, which determines the length of the path of the light beam, were thoroughly investigated by Tamiya et al.[2] In Figure 8, the concentration of cells is given not only in terms of cell number per liter, but also in terms of packed cell volume (mℓ) per liter (5.6×10^{10} cells correspond to 1 mℓ packed cell volume). At a population concentration of about 6×10^9 cells of *Chlorella* per liter, only 50% of the cells are directly illuminated, yet the average intensity of light available to all cells in the suspension is as much as 85% of the incident light. This is because the rest of the cells are not heavily shaded due to the small thickness (2.8 cm) of the culture suspension. Evidently, the studies of Tamiya et al.[2] and Kok[3] point to the advantage of having the algae grown in a thin suspension, where the light/dark cycle and the available light per cell may be easier to maintain at a level close to the optimal for reaping maximal production per area.

On a large scale, however, shallow culture depth involves many practical difficulties and the compromise called for between population density, depth of culture, and rate of mixing to obtain the highest production per area must be carefully weighed, special emphasis being given to changing environmental conditions. In practice, these principles translate into a pond depth of around 15 cm. A significant increase in growth rate, however, would be obtained if at given population densities, the depth were reduced to 10 cm or even less.

Grobbelaar[7] presented a mathematical model (described later in this chapter) that facilitates the calculation of biomass production at various depths in an 18 m² outdoor algal pond as related to the population density and mutual shading. The results of such calculations (Figure 9) demonstrate the importance of population density in affecting productivity. Clearly, an optimum population density is that which yields the greatest yield of biomass per pond area under given environmental conditions. When the population density is below optimum (Figure 9, top), the light available per cell is high, but the production per area is low because the overall concentration of biomass is low. At the other extreme, when the population density is above the optimum (Figure 9, bottom), the light available for photosynthesis penetrates only a very narrow photic zone at the very top layer of the culture, most of the cells at any given instant remaining unilluminated. Under such circumstances, the net production of biomass per unit area falls significantly, due to various processes such as respiration and cell death.

INTERRELATIONS BETWEEN SOLAR IRRADIANCE AND TEMPERATURE

A major theme in the production of algal biomass concerns the effect of these two major environmental factors which, being the least controllable, represent the major practical limitations to production. The algal biomass in a pond responds continuously to variations in irradiance and temperature as they occur throughout the day and year. In an outdoor system, growth limitations by either of these factors alone is rare. Usually, the growth rate and productivity of photosynthetic microalgae are governed by both temperature and irradiance.

Tamiya et al.[2] carried out detailed studies on the effect of temperature on exponential growth rates, i.e., when mutual shading is negligible, at different light intensities. The results (Figure 10) are analyzed as follows: at all temperatures studied, and at lower light intensities, the growth rate increases linearly with light intensity (I). With a further increase in light intensity, the increase in rate gradually slows until the curves become flat and the growth rate is practically independent of light intensity. The functional relations observed may be approximately expressed by the formula for a rectangular hyperbola, as follows:

$$\frac{\Delta \log V}{\Delta t} = \frac{\alpha k_G I}{k_G + \alpha I}$$

where k_G is the maximum value of $\Delta \log V / \Delta t$ (at saturating light), and α is the initial slope of $\Delta \log V / \Delta t$ as a function of I namely:

$$\alpha = \left[\frac{d(\Delta \log V / \Delta t)}{dI} \right]_{I \to 0}$$

At all temperatures studied and especially at lower temperatures, growth was depressed at extremely high light intensities. At 7°C, growth was retarded even when light intensity was about 5000 lux and at 50,000 lux the cells ceased to grow, becoming colorless within a few days. In all cases, cell suspensions grown under weaker light were dark green, the

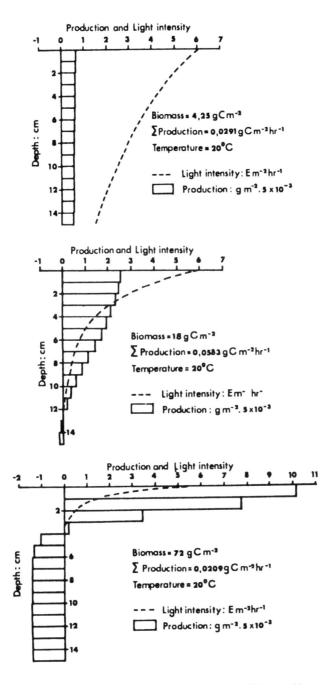

FIGURE 9. The calculated production in relation to light intensities at 10 mm intervals in a 0.15 m deep culture at three biomass concentrations. (From Grobbellar, J. U., Infections: Experiences in miniponds, U.O.F.S. Publ. Series C, No. 3, 116, 1981. With permission.)

color changing to light green with increase of light intensity, and then to brownish pale green. This fading, which stem from photoinhibition and photooxidation was also dependent on temperature; for example, the light intensity at which light green cells were formed increased with temperature as follows: 2000 lux at 6°C, 25,000 lux at 15°C, and 50,000 lux at 25°C.[2] Richmond et al.[8] and Vonshak et al.[9] investigated the interplay between seasonal variations

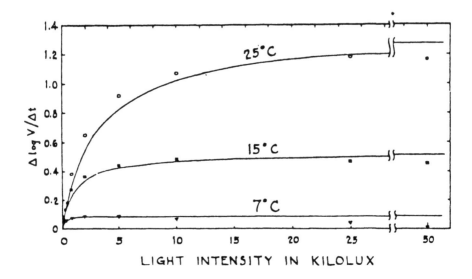

FIGURE 10.   The exponential growth rate as a function of light intensity at different temperatures. Values of $\log_{10} V/t$ are averages of several measurements. (From Tamiya, H. et al., *Algal Culture: From Laboratory to Pilot Plant*, The Carnegie Institute, Washington, D.C., 1953. With permission.)

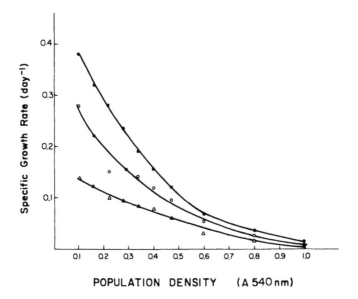

FIGURE 11.   Specific growth rate of *Spirulina* as affected by population density and the seasons of the year (A 540 nm). ●-●-● Summer (June to September). ■-■-■ autumn and spring (October to November and April to May), ▲-▲ winter (December to February). (From Vonshak, A. et al., *Biomass*, 2, 175, 1982. With permission.)

of temperature, the incident irradiance, and the light available for the individual cell as determined by the population density. The relationship between the population density and the specific growth rate throughout the seasons is shown in Figure 11. The lower the population density, the higher the specific growth rate, as is to be expected in a system which is primarily light limited. The relative effect of decreasing the population density and thus, of increasing the rate of light availability per cell, is much less pronounced in the

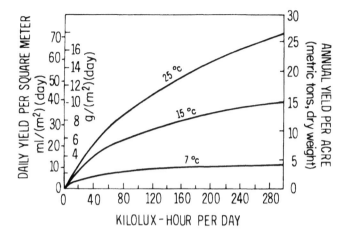

FIGURE 12. The calculated yields of algae per unit time and per unit area of illuminated surface as a function of temperature and daily quantity of light. (From Tamiya, H. et al., *Algal Culture: From Laboratory to Pilot Plant*, the Carnegie Institute, Washington, D.C., 1953. With permission.)

winter when the major environmental factor limiting the rates of growth and of output is temperature rather than light. As already explained, in light-limited growth, a close relationship must exist between the specific growth rate (μ) and cell density (X). Since μ depends strongly on temperature, this relationship varies greatly throughout the year; the more severe the temperature limitation on μ, the smaller would be the dependence of μ on X, becoming hardly noticeable in winter. At the same time, the relative response of the culture to "summer conditions" declines as the population density increases, until at very high densities (1.5 g/ℓ dry weight), no seasonal effect on the specific growth rate was observed. Clearly, in such dense cultures, light limitation due to mutual shading was so extreme that most of the growth rate potential was limited by light, thereby blocking any potential enhancing effect of temperature on algal growth.

Tamiya et al.[2] calculated the yield of *Chlorella* as a function of temperature and light. The results are shown in Figure 12 and are similar in essence to the results obtained for the factors affecting the specific growth rate (Figure 10).

In Peru, Castillo et al.[10] correlated the yield of *Scenedesmus* biomass with temperature and solar irradiation (Figure 13). While no light saturation of the growth of *Scenedesmus* was observed, the yield correlated with temperatures up to 32°C only, and decreased at higher temperatures. In *Spirulina platensis* the output rate changed markedly throughout the seasons. Nevertheless, as the population density was increased the seasonal difference in output rate diminished (see second major second of Part II).

The interrelated effects of solar irradiance and temperature on the photosynthetic activity of *Spirulina platensis*, as evidenced from the oxygen tension in the pond, were investigated.[9] At any given solar intensity, an increase in the culture temperature above 12°C resulted in an increased concentration of dissolved oxygen, and similarly, for each temperature range, an increase in irradiance elevated the concentration of oxygen (Figure 14).

It must be stressed, however, that these results were obtained in small (1 m²) ponds that were partially shaded during some parts of the day and were very well agitated; thus, no ill effects of photoinhibition or of excessive O_2 concentration were manifested in these ponds. Such ill effects may be observed on very hot, bright summer days in large (e.g., 1000 m² ponds) that are not sufficiently agitated.

By following the daily course of temperature and radiation, and correlating them with the dissolved oxygen concentration in a *Spirulina* pond Vonshak et al.[9] attempted to sort out

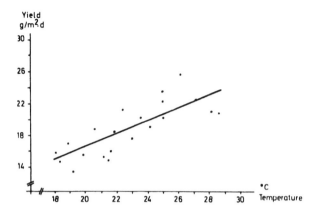

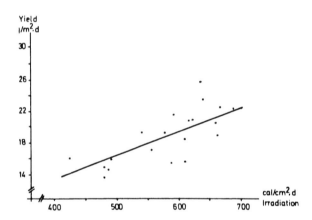

FIGURE 13.   Correlation of temperature and irradiation with yield of *Scenedesmus* in Casa Grande. (From Castilo, S. S. et al., *Algae Biomass — Production and Use*, Elsevier/North-Holland, Amsterdam, 1980.

the environmental limitations imposed on the culture (Figure 15). In winter, the peak in O_2 concentration in the pond coincided with the maximum daily temperature, and during a typical day in winter (Figure 15C) the rate of increase in O_2 evolution in the morning up to about 9 a.m. was clearly very low until a rise in temperature became evident in the late morning. In contrast, during the summer, the peak O_2 concentration coincided with the peak of irradiance. Moreover, when the temperature in the pond was maintained relatively high during the night (Figure 15A), the rates of increase in O_2 concentration and in light intensity were nearly parallel. Clearly, O_2 evolution and biomass production under the temperature prevailing in the pond in summer was mainly controlled by the rate of light irradiation.

As expected, the relationship between O_2 evolution, temperature, and irradiance during early spring and autumn (Figure 15B) was such that the peak in pond O_2 content did not coincide with the peak of either temperature or irradiance, but occurred at a point when the temperature had not yet reached its maximum while light intensity had already begun to decline. In these seasons, therefore, the photosynthetic system in *Spirulina* was apparently limited by both light and temperature throughout the day.

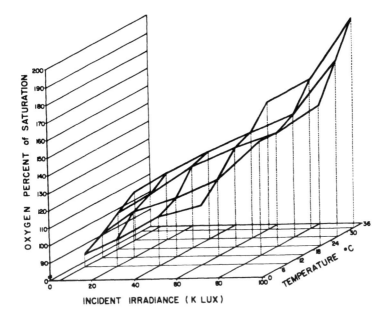

FIGURE 14.   Interrelation between maximum daily temperatures, irradiance, and photosynthetic activity throughout the seasons in culture of *Spirulina platensis*. The data represent averages of scores of observations collected during 2 years of experimentation with *Spirulina platensis* cultured in 1.0 m² ponds. The pH in the ponds was optimal for growth, ranging from 9.5 to 10.0, and no nutrient limitation could be observed. (From Vonshak, A. et al., *Biomass*, 2, 175, 1982. With permission.)

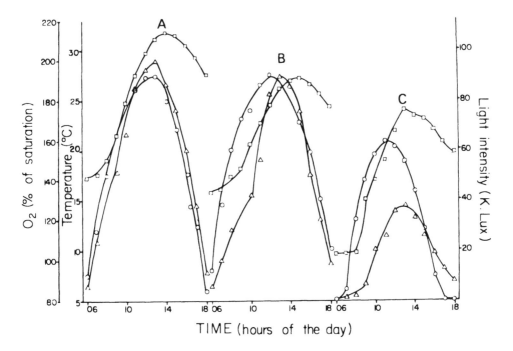

FIGURE 15.   Relationship between O₂ concentration, temperature, and light intensity in pods of *Spirulina platensis* throughout the seasons. (A) Summer, (B) spring and autumn, (C) winter. □-□-□ Temperature, O-O-O irradiance, Δ-Δ-Δ O₂ saturation. (From Vonshak, A. et al., *Biomass*, 1, 175, 1982. With permission.)

Table 1
O_2 SOLUBILITY
IN FRESHWATER
AS A FUNCTION
OF
TEMPERATURE[12]

Temperature (°C)	DO (mg/ℓ)
5	12.8
10	11.3
15	10.2
20	9.2
25	8.4
30	7.5
35	7.1
40	6.6

THE COMBINED EFFECTS OF LIGHT INTENSITY AND DISSOLVED OXYGEN

Dissolved Oxygen Concentration

Oxygen is a fundamental parameter of algal ponds, as well as of natural bodies of water, aside from water itself. Dissolved oxygen is obviously essential to the metabolism of all aquatic organisms that possess aerobic respiratory metabolism. Hence, oxygen solubility and especially the dynamics of its distribution in bodies of water are basic to an understanding of the distribution, behavior, and physiological growth of all aquatic organisms.

Air contains about 20.95% oxygen by volume, the remainder being nitrogen, except for very small percentages of other gasses. Because oxygen is more soluble in water than is nitrogen, the amount of oxygen dissolved in water from air is approximately 35% and the remainder is largely nitrogen.[11] As explained by Ben-Yaakov,[12] an aqueous solution in contact with a gas phase of a given partial pressure of O_2 (pO_2') will attain, at equilibrium, a partial pressure (pO_2) equal to the partial pressure in the gas phase. This relationship pO_2' = pO_2 is independent of the temperature and the salinity of the solution and expresses the equivalence of the O_2 phase.

The oxgyen content of water may be measured in units of weight of dissolved oxygen (DO) per volume of solution (mg/ℓ), or as percent of saturation, which is a measure of pO_2 and hence of O_2 activity in the water. The partial pressure pO_2' of a solution 100% saturated with air is, therefore, equal to the partial pressure of oxygen in air. Since air contains about 21% O_2 by volume, the partial pressure of O_2 in air (and of a 100% air saturated solution) is about 0.21 atm. This is true, provided the total atmospheric pressure is 1 atm (760 mmHg), and corrections must be made for variations in barometric pressure. The total amount of oxygen that will dissolve in solution depends on oxygen solubility, which is a function of both temperature and the ionic composition of the solution (type and quantity of dissolved salts). The total amount of oxygen dissolved in a solution 100% saturated with air as a function of temperature is presented in Table 1. The temperature effect is rather large and amounts to approximately 2%/°C at 25°C. The salinity effect is relatively smaller. It amounts to about 6%/g/ℓ.[12,13]

Supersaturation with Oxygen

Unlike wastewater systems, in which the O_2 produced by photosynthesizing biomass is taken up by heterotrophic microorganisms, very high concentrations of DO may occur in the "clean" culture medium, due to very intensive photosynthetic activity of rather dense (e.g., 500 mg dry wt/ℓ) algal populations in a well-balanced nutrient solution. Concentrations of over 35 mg O_2/ℓ and 400% saturation have been recorded.[11] In the author's field installation of *Spirulina*, the O_2 concentration often rises to over 300% saturation on bright, hot days. Were it not for the rather vigorous stirring (about 50 cm/sec) the DO would have risen to over 400% saturation. Such high concentrations of DO in the algal culture can be toxic to the cell and diminish its rate of photosynthesis.

Oxygen Toxicity and Photooxidation

According to Fridovich,[14] the toxicity of oxygen may not be easily comprehended. The apparent ease with which aerobic microorganisms, animals, and plants survive at atmospheric levels of oxygen is due to elaborate defenses against its considerable toxicity. These defenses are adequate under ordinary conditions, but can be overwhelmed at high oxygen concentrations. The defenses that allow aerobes to prosper in its presence relate to the fact that although the reactions of oxygen with organic substances are exothermic, they do not occur at perceptible rates at ordinary temperatures. The explanation for the low reactivity of molecular oxygen lies in the special electronic arrangement of its atoms, causing in effect a "spin restriction" that arrests the reactivity of oxygen under most circumstances. The spin restriction applies to divalent reduction of O_2. It can be circumvented by combining the O_2 with a metal, a course followed by many oxidases which contain Cu(II) or Fe(II) or other metals. Another way to annul the spin restriction is to add electrons to O_2 one at a time. Such univalent pathways for the reduction of oxygen are quite common and therein lies an explanation for the toxicity of oxygen.

The reduction of O_2 to $2H_2O$ requires four electrons and the univalent pathway involves intermediates that are much more reactive than O_2 itself. Thus, the superoxide anion radical (O_2^-), hydrogen peroxide, and the hydroxyl radical are all involved in the reduction of oxygen and may play havoc on the algal cell when its internal defense mechanisms weaken. Active defense against harmful radicals is provided by three classes of enzymes: catalases, peroxidases, and superoxide dismutases. The first two catalyze the divalent reduction H_2O_2, whereas superoxide dismutase catalyzes the following reaction:

$$O_2^- + O_2^- + 2H^+ \rightarrow H_2O_2 + O_2 \quad \text{(Reference 14)}$$

Numerous observations of a sudden and spontaneous disintegration of algal blooms in lakes and ponds, particularly of cyanobacteria, have been reported. The phenomenon is referred to as "photoxidative death" and results from the lethal effect to cells of exposure to light in the presence of oxygen. Boyd et al.[15] gave a detailed account of a sudden death of a dense bloom of the cyanobacterium *Anabena variabilis* as follows: abundant growth of *A. varibilis* was first observed in late March 1974 and by April 18 it amounted to more than 99% of the phytoplankton at all depths. The depth distribution of *A. variabilis* was relatively uniform on the first sampling dates, and no surface scums were seen. The depth to which 1% of the incident light penetrated varied between 0.5 and 1.0 m during the period April 18 to April 26. A slight surface scum was first noted on April 27 and a thick scum was present on the morning of April 29. The density of filaments in the upper 10 cm was 37,300/ mℓ. Upward migration of *A. variabilis* was obviously the source of the scum, since the average density of filaments in the 0.1 to 2.5 m water column decreased. During that period, *A. variabilis* in the surface scum quickly began to deteriorate and blotches of blue-green substances were seen in a few places on the afternoon of April 29. Algae in the scum had a bleached appearance and in some areas the water was brown with dead algae. On the next day, the entire pond was brown and turbid with dead *A. variabilis*, the brown turbidity resulting primarily from dead algae which did not float to the surface. The few intact filaments

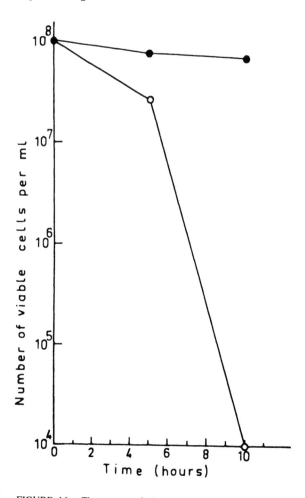

FIGURE 16.   The course of photooxidative death. *A. nidulans*
cells were suspended in 0.05 *M* potassium phosphate buffer (pH
7.0) and incubated in the light at 35° in an atmosphere of 99% O_2
+ 1% CO_2 (●) or in 100% O_2 (○). (From Abeliovich, A and
Shilo, M., *J. Bacteriol.*, 682, 1972. With permission.)

of *A. variabilis* present between April 30 and May 4 were in poor condition and no filaments
were recognizable on May 5.[15]

Abelovich and Shilo[16] investigated the frequent occurrences of sudden and spontaneous
disintegration of cyanobacterial algal blooms which often takes place in the summer in Israeli
fish ponds. They tested whether a photooxidative effect was involved in this phenomenon,
since oxygen supersaturation and high light intensities prevailed at that time.[17] A characteristic feature preceding massive die-off of these algal blooms was a low CO_2 concentration
(expressed as elevated pH during the light period) in the highly alkaline waters. Simulating
photooxidative condtions in the laboratory, they found that wild-type *Anacystis nidulans* and
Synechococcus cedrorum die out rapidly when incubated in light under 100% oxygen in the
absence of CO_2 at temperatures of 4 to 14°C. Also quick death of *A. nidulans* occurred
under these conditions at 35°C and of *C. cedrorum* at 26°C.

The role of CO_2 in either inhibiting or accelerating photooxidative death was shown by
adding DCMU (3-(3,4-dichlorophenyl-1,1,dimethyl urea), an established inhibitor of photosynthetic oxygen evolution and CO_2 assimilation (Figure 16), which inhibited photosynthetic activity in *A. nidulans* within 10 sec. In the presence of DCMU 99% of *A. nidulans*

cells lost viability in a 100% O_2 atmosphere at 35°C, even in a medium that had sufficient carbon supply. In a complete medium containing Na_2CO_3 without DCMU, cell viability was not impaired under the same conditions. The conclusion was that in the absence of CO_2, photooxidative death at 35°C under an atmosphere of pure oxygen proceeds in spite of the well-established protective function of the carotenoids in these organisms, which under these conditions did not afford sufficient protection against photooxidation. The role of CO_2 in preventing photooxidative death in *A. nidulans* and *S. cedrorum* suggested that the protective mechanisms were associated with the photosynthetic activity of the cell. Indeed, addition of DCMU, which blocks photosystem II (PS II), leads to photooxidative death even in complete (CO_2-containing) medium, hence, the hypothesis that photooxidation at physiological temperatures in the absence of CO_2 involves a peroxide or a superoxide radical produced by direct reduction of oxygen by some reduced electron carrier that accumulates when photosynthesis is inhibited.[16]

This hypothesis, however, fails to explain lethal photooxidation at low temperatures which, according to Abeliovich and Shilo,[16] probably proceeds by direct photosensitization of the photosynthetic pigments. This may be the reason that photooxidative death of carotenoid-containing *A. nidulans* and *S. cedrorum* occurred at a much more rapid rate at the lower temperature. The considerable lag (approximately 8 hr) before the onset of mortality at 35°C and the even longer lag (48 hr) when cells were preincubated in an atmosphere of 5% CO_2 indicated that the protective mechanism may be sustained by metabolic reserves possibly supplying endogenous CO_2 by respiration, even after photoassimilation of exogenous CO_2 has ceased. From a practical standpoint it is well to remember that photooxidative damage may be detected before it becomes irreversible, since under photooxidative conditions, progressive damage to the photosynthetic capacity of the cells takes place during the period preceding the onset of death. One indication of early photooxidative damage to *A. nidulans* cells is the initiation of light-dependent oxygen consumption.

An enzyme possibly taking part in the prevention of photooxidative death is superoxide dismutase, which seems to play a key role in the protection of various aerobic organisms against oxygen toxicity. This is revealed when a culture of *A. nidulans* is transferred to conditions that cause photooxidative death after a lag period of 6 to 8 hr. Superoxide dismutase activity begins to decrease early in this period and is gradually depleted to 10% of its initial level by the 6th hour. The lag preceding photooxidative death can be prolonged to 40 to 48 hr by growing the cells under an atmosphere of air enriched with 5% CO_2 prior to the shift to photooxidative conditions.[16] Under such conditions superoxide dismutase activity was found to be depleted to 10% of its normal value only after 40 hr. If, on the other hand, cells devoid of superoxide dismutase (grown under pure nitrogen) were shifted to photooxidative condtions, death would commence after a brief lag period of 2 to 4 hr (Figure 17). In all cases photooxidative death at physiological temperatures (35°C) occurred when superoxide dismutase activity remained below or was depleted to 10% of the normative value. The level of superoxide dismutase depended on the oxygen concentration in the growth medium. Abeliovich and Shilo[16] concluded that induction of its synthesis by oxygen may constitute an important mechanism of protection against photooxidation.

Accordingly, when the extent of supersaturation with oxygen, which readily occurs in mass cultures, surpasses a certain level, the protective activity of superoxide dismutase and other scavengers of oxygen radicals (as well as by carotenoids) may simply not be sufficient. The combination of high dissolved oxygen, temperature, light intensity, population density, and pH, which deplete the CO_2 supply, increase the likelihood of photooxidative damage.

Before the onset of an irreversible stage, photooxidative damage may be readily corrected under laboratory conditions; recovery of superoxide dismutase activity occurs when cells that were subjected to photooxidative conditions for 6 hr are shifted back to normal aerobic growth conditions. Within 30 min the cells recover all their superoxide dismutase activity.

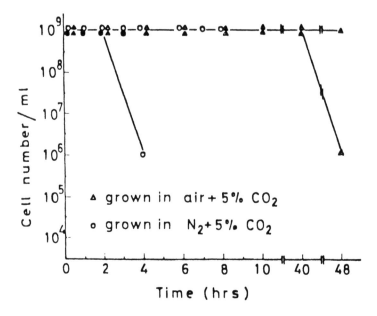

FIGURE 17.   Viability of *Anacystis nidulans* cells, preincubated in nitrogen atmosphere or in air, when grown in photooxidative conditions: 35°C, atmosphere of pure oxygen, light (○, △) or dark (●, ▲). Preincubation: N_2 atmosphere or air containing CO_2. (From Abeliovich, A. and Shilo, M., *J. Bacteriol.*, 682, 1972. With permission.)

Both photosynthesis and protein synthesis seem to be necessary for this process, since both chloramphenicol or darkness prevent this recovery.

In mass algal cultures, photooxidative death constitutes a major risk, particularly with cyanobacteria. This is because outdoors, a combination of factors may readily exist that may endanger certain sensitive species, such as *Spirulina platensis*. The course of developments that culminated in the death of the mass culture of *Spirulina* in the author's laboratory at Sede Boqer, Israel is described below.

During the summer months, the environmental conditions for growth of *Spirulina* are optimal. The maximum temperature in the culture approaches the optimum for this algae and the radiation flux is at maximum. Under these conditions, photosynthetic activity in *Spirulina* is at its peak and the concentration of O_2 may reach well over 350% of saturation in ponds that are not vigorously stirred. If in addition, the concentration of biomass is allowed to increase over the optimum for maximum output, photooxidative damage ensues, which in extreme cases results in the death of the entire culture. This is shown in Figure 18, which depicts growth and O_2 concentration in a *Spirulina* pond. Vigorous growth was evident from day 0 (in mid-July) to day 4, when the turbidity reading was 650 Klet units and the chlorophyll concentration was 15 mg/ℓ. Thereafter, growth continued at a somewhat lower rate. The concentration of O_2, which was approximately 350% of saturation up to day 3, began to decline, becoming steady at 250%. This decline, which is the earliest indication for some possible cellular damage, may have several possible explanations. It may be due to a decrease in photosynthetic activity and perhaps to increased photorespiration. On day 8, the pond was harvested by removing some 50% of the biomass. Following a lag after the heavy harvest satisfactory growth occurred for 2 days and the chlorophyll concentration reached 9.5 mg/ℓ on day 12. Nevertheless, while this increase in biomass took place, DO continued to decline and it became clear that although growth seemed satisfactory, cellular metabolism was modified. Between days 12 and 16, growth continued, albeit at a somewhat slower pace. This warning signal was followed by an accelerated decline in dissolved O_2.

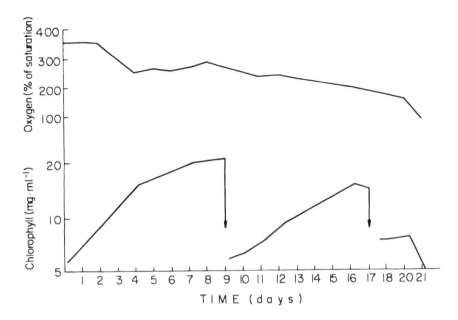

FIGURE 18.   The pattern of growth and O_2 concentration in a 100 m² *Spirulina* pond, revealing the course of photooxidation and eventual death of the culture. Measurements of chlorophyll and O_2 concentrations were taken daily at midnight.

On day 17 the performance of the culture gave obvious reasons for alarm: instead of a continued rise in biomass there occurred a clear, although slight, decrease or halt in the accumulation of algal mass. More seriously, the concentration of O_2 fell drastically. The culture was judged to be in danger of deterioration and approximately 50% was removed and replaced with fresh medium. This treatment, which on similar occasions previously helped the culture to recover, had little effect and on day 18 only a negligible growth response was observed. On day 19 the O_2 concentration did not change. The next day a sharp decline in chlorophyll concentration took place and concomitantly, the concentration of oxygen dipped to below saturation. Analysis of the course of deterioration (Figure 18) as well as of the conditions prevailing in the pond indicated that it was due to photooxidation. The slight halt in the increase in biomass evident on day 13, which was associated with a persistent decline in pond O_2, indicated that the sudden death occurring on day 20 was preceded by a period of general decrease in cellular performance. It is possible, however, that the initial damage to the culture may have already taken place on day 4, when the first signs of what turned out to be a continuous decline in O_2 concentration were first observed. Clearly, a better understanding of the process of photooxidative death in an algal pond requires further research.

Oxygen Inhibition

In addition to the direct damage that high oxygen concentration may inflict upon cell, excess oxygen in the growth medium decreases photosynthetic activity. Warburg and Negelein reported that the rate of photosynthesis in *Chlorella* decreased when the oxygen pressure was increased from 1/50 to 1 atm. Spoeher and Milner[19] found that high concentrations of oxygen resulted in an appreciable reduction in yield as compared with cultures grown with a gas mixture of CO_2 in air or in nitrogen (Table 2). Doubling the light intensity under air or nitrogen more than doubled the yield of *Chlorella*. Under high O_2 tensions, however, this increase in yield did not occur, reflecting the severe inhibition by O_2 of photosynthesis. Radmer and Kok[20] investigated the details of oxygen effects on CO_2 assim-

Table 2
THE EFFECT OF THE PARTIAL
PRESSURE OF OXYGEN ON THE
YIELD OF *CHLORELLA*[19]

Gas mixture 5%	Light (W)	Days	O (%)	Yield (g/2 ℓ)
Air	100	11	20	0.7098
Nitrogen	100	11	Trace	0.6912
Oxygen	100	17	95	0.5034
Air	200	15	20	1.4740
Nitrogen	200	15	Trace	1.8980
Oxygen	200	17	95	0.4155

ilation. Using a special mass spectrometer to directly monitor the gaseous components in a suspension of algae, they concluded that CO_2 and O_2 are in direct competition for photosynthetically generated reductants.

Photorespiration

Oxygen in the culture medium may have yet another negative effect, i.e., stimulation of photorespiration (see "Photosynthesis and Ultrastructure in Microalgae"). Photorespiration, which may be defined as the oxygen-sensitive loss of CO_2 during photosynthesis, is often faster than the previously recognized, conventional, "dark" respiration. Photorespiration is characteristic of a group of plants (C3), which includes most if not all algae. The great differences in net photosynthesis between efficient and less efficient plant species may be largely attributed to differences in photorespiration. It seems apparent that decreasing the efflux of photorespiratory CO_2 should increase net CO_2 assimilation during photosynthesis in the less efficient species and this has indeed been verified experimentally.

From the standpoint of production of algal biomass, a pertinent point concerning photorespiration is that it may be greatly stimulated by increases in temperature and O_2 concentration. Another aspect to consider relates to the association that has been established in the past few years between photorespiration and the nitrogen economy of plants. Cullimore and Sims[22] demonstrated the operation of photorespiratory N cycle in *Chylamydomonoas*. NH_3 release by this process is light-dependent and sensitive to changes in the partial pressures of O_2 and CO_2. Much of the released NH_3 must ultimately have been derived from catabolism of amino acids released from protein.

The rate of photorespiration and the magnitude of loss of organic carbon from the algal cell is still not clear because of difficulties in measuring photorespiration at alkaline pH, where CO_2 represents only a small portion of the total dissolved inorganic carbon (DIC). Thus, much of the CO_2 evolved by the cells in light would immediately combine with OH^- to form HCO_3^- and would not be detected by the techniques used. Birmingham et al.[23] circumvented this difficulty by appropriate techniques and demonstrated the existence and magnitude of O_2-sensitive CO_2 release in the light with the algae tested. Table 3 shows that the loss of CO_2 in the light is greatly stimulated when the O_2 concentration was elevated from 2 to 21%, which is also characteristic of photorespiration in higher plants.

It is worth noting that photorespiration is detectable only in algae exposed to subsaturating levels of DIC, i.e., <50 mM DIC. This is in agreement with reports that high CO_2 concentrations inhibit photorespiration in C3 plants and suggests that under natural conditions O_2-sensitive CO_2 release would not occur in these algae.

According to Merret and Armitage,[24] decreasing the oxygen concentration from 21 to 2% (v/v) at a constant CO_2 concentration of 0.03% (v/v) gave a twofold increase in dry weight yield for *Eugleana gracilis*, a result consistent with the operation of a functional glycolate

Table 3
THE EFFECT OF THE TEMPERATURE AND OXYGEN CONCENTRATION ON PHOTOSYNTHESIS AND PHOTORESPIRATION ON SOME SPECIES OF FRESHWATER ALGAE

Temp. (°C)	Algae	2% O_2 (mol CO_2 mg/Chl/hr)			21% O_2 (mol CO_2 mg/Chl/hr)		
		Photosynthesis	Photorespiration	% TPS[a]	Photosynthesis	Photorespiration	% TPS
20	*Chlamydomonas reinhardii*	44.1	3.2	8.7	36.4	10.4	28.2
	Chlorella vulgaris	54.3	1.8	3.2	57.3	11.4	19.8
	Navicula pelliculosa	64.6	0.8	1.4	67.1	10.6	16.3
25	*Anabaena flos-aquae*	114.7	7.8	6.7	104.0	18.4	17.7
	Anacystis nidulans	62.3	3.0	4.9	72.8	7.7	10.3
	Coccochloris peniocystis	58.0	1.3	2.7	59.9	1.4	2.4
	Pormidium molle	97.3	3.9	3.8	93.9	4.8	6.2

[a] Photorespiration expressed as a percentage of true photosynthesis.

From Birmingham, C. et al., *Plant Physiol.*, 69, 259, 1982. With permission.

pathway in this alga. Nevertheless, a similar effect of oxygen concentration on dry weight yield was not observed with *Clamydomonas reinhardii*.

It thus remains to be clarified whether photorespiration is significant in mass algal cultures grown at population densities two to four orders of magnitude higher than those that ordinarily occur in natural habitats. At least in some species or strains, the extremely high values of DO that may occur under such conditions, together with the high cell density (which may decrease the availability of dissolved carbon for each cell), may be conducive to elevated levels of photorespiration.

PRODUCTIVITY OF MASS CULTURES

The efficiency with which solar irradiation is converted to chemical energy by the algal cell is critical in biomass production. The photosynthetic efficiency (PE) of biomass growth is defined as the energy stored in biomass per unit of light energy absorbed. This parameter is of fundamental importance in biomass production, because the biotechnology for algae production should naturally aim to achieve the maximum attainable photosynthetic efficiency so as to produce the most biomass possible per unit area. To date, the actual biomass yields from algal ponds have been very low relative to the theoretical maximum, the determination of which has rarely been attempted. In general, the theoretical maximum has been estimated from the classical representation of photosynthesis as

$$CO_2 + H_2O + rhv = 1/6 \ C_6H_{12}O_6 + O_2 \qquad (1)$$

Reaction 1 is of course oversimplified since it ignores nitrogen assimilation and the synthesis of proteins, nucleic acids, and lipids. Another difficulty is that PE has been based on measurements of the light absorbed and the O_2 evolved in resting (nongrowing) algal cells. As pointed out by Pirt et al.,[26] estimates of the PE from resting cell O_2 evolution have been highly variable, giving values for the quantum yield (n) which have ranged from about 4 to 12.[27-29] In general, the value n = 8 has been selected as the minimum largely because it accords with the current model of photosynthesis;[27] chemically, the oxidation of two water molecules requires the transfer of four electrons to ferredoxin (Fd) molecules. Both electrons must be transferred through a number of steps, two of which are photochemical reactions. In each of these two light reactions one photon of light is used to transfer one electron with a quantum efficiency of 1.0. The light requirement for the transfer of four electrons is eight photons:

$$2H_2O + 4Fd^{+3} + 8 \ photons = 4H^+ + O_2 + 4fd^{+2} \qquad (2)$$

Accordingly, the utilization of 8 Ei (moles of photons) by the thylakoid photochemical apparatus produces 4 mol of reduced ferredoxin and about 3 mol of ATP, which are needed to bring about the reduction of 1 mol of CO_2 to sugar in the "dark" reactions that follow.

Pirt et al.[26] pointed out that the use of resting cells to determine the bioenergetics of growing cells is not really valid. This is because PE is an expression of the growth yield and maximum growth can be expected only when the energy source is growth limiting. They concluded that the maximum PE deduced from the classical representation of photosynthesis (Reaction 2) may be wrong, and that it was possible to obtain efficiencies some 50% higher. As pointed out by Pirt,[30] the central issue is whether the minimum quantum yield is less than or exceeds 8 hv/O_2. Exclusion of values less than 8 seems highly arbitrary, for such values may be interpreted to mean that PS I and II may, under certain conditions, work in parallel[26] (see "Photosynthesis and Ultrastructure in Microalgae").

Today's accepted explanation for the calculation of PE is as follows: green plants use only light with wavelengths from 400 to 700 nm. This photosynthetically active radiation (PAR) constitutes about 43% of the total solar radiation at the earth's surface, and has an energy input equivalent to that of monochromatic light at 575 nm. One einstein of 575 nm light contains 49.74 kcal of energy, and assuming that the free energy stored in photosynthesis is 114 kcal/mol of reduced CO_2, the theoretical maximum energy efficiency for the photosynthetic reduction of CO_2 to glucose with white light is $114/(8 \times 49.74) = 0.286$, or approximately 29%.[31] This value, then, was in effect challenged by Pirt et al.[26] who used light-limited chemostat cultures of *Chlorella* (as well as a mixed culture of *Chlorella* sp.) and three heterotrophic bacterial species. The maximal photosynthetic efficiencies obtained were 34.7 and 46.8% for the *Chlorella* and the mixed culture, respectively.

The maximum percent conversion of solar energy to the free energy of biomass is obtained from the formula:

$$[PE(\%) \times (PAR) \times (MC)] \qquad (3)$$

Where PE is the fraction of the light absorbed which is stored in the biomass, PAR ($=$ 0.43) is the fraction of total solar energy available for photosynthesis, and MC is the fraction of energy available after correction for maintenance energy and photorespiration. Loss of energy due to reflectance is negligible in an algal pond, but loss of energy due to maintenance is estimated at about 10%. Accordingly, the maximum possible conversion of total solar energy to biomass free energy with the mixed culture was estimated by Pirt et al.[25] to be $46.8 \times 0.43 \times 0.9 = 18.1\%$. This value is about 50% more than the conventional maximum which is obtained by multiplying 28.6% (the conventional maximal efficiency for photosynthesis) by 0.43 to yield 12.3% (the maximal conventional figure for PE in microalgae), assuming total light absorption and no dark respiration. In comparison, the photosynthetic efficiency in producing the U.S. maize crop is given as 1.26%. An efficiency of 18.1% would correspond to a dry biomass output of about 500 t/ha/year in the sunny parts of the world.[26] This figure is so remote from the actual output obtained in practice that it will probably never be obtained. It is not necessary, however, to adopt these maximum PE values to realize the great potential of a microalgal agriculture, for even the commonly accepted net PE maxima (8 to 10%) would yield a huge output of approximately 250 t/ha/year. Thus, the theoretical limits for photosynthetic efficiencies suggest that photosynthetic microalgae have the potential for becoming the most efficient means for photobiological utilization of solar energy, comparable in efficiency with photovoltaic cells and solar thermal energy.

It should be stressed that under field conditions, much lower photosynthetic efficiencies have so far been obtained. Therefore, the challenge to the algal industry may be defined as producing biomass on a large scale under outdoor conditions with an efficiency similar to that obtained under controlled laboratory conditions. These include a short (a few centimeters) path of the light beam, vigorous stirring, and low light intensities.

Goldman[32] summarized information on algal yields reported over the past 30 years in some 30 major studies of mass cultivation of microalgae. These studies represent a great variety of culture systems, algal species, geographical locations, and pond sizes. Algal yields have risen dramatically during the past decade of research (Table 4). Although the early efforts in mass algal culturing resulted in yields no better than 10 to 15 g dry wt/m²/day for short periods, today it is not uncommon for sustained yields to reach 20 to 25 g dry wt/m²/day (Figure 19).

The economic potential of algal production depends to a large extent on the yield or the output rate. The maximum yield potential of algae may be represented by models that integrate the factors limiting production. Goldman[33] presented a thorough study of the maximum yield potential of algae based on the thesis that algal production under the proper conditions is limited only by light. Goldman's model was based on earlier models by Van

Table 4

SUMMARY OF WORLD PROGRAMS FOR MASS CULTIVATION OF MICROALGAE[32]

Year	Location	Species	Culture systems Size (m²) Unit	Total	Operation[a]	Best yields (duration) g dry wt. $^{-2}$/day Maximum	Average
1954—55	Tokyo, Japan	*Chlorella*	2.8—5.5	13.8	N, F, SC	28 (3)	16 (27)
1960—77	Dortmund, Germany	*Scenedesmus*	80	320	N, F, SC	28 (?)	10 (?)
1967	Trebon, Czechoslovakia	*Scenedesmus*	50	50	N, F, SC	25 (10)	16 (65)
1968	Rupite, Romania	*Scenedesmus*	50	50	N, F, SC	30 (10)	23 (63)
1969	Firebaugh, California	*Scenedesmus*	1000	1000	TD, F, C	35 (10)	10 (70)
1974	Fort Pierce, Florida	*Diatoms*	4	8	WSW, M, C	25 (15)	
1976—77	Mexico City, Mexico	*Spriulina*	?	200,000	B, F, SC	20 (?)	10 (?)
1976	Haifa, Israel	*Green algae*	120	270	W, F, C	35 (30)	15 (365)
1974	Bangkok, Thailand	*Scenedesmus*	87	609	N, F, SC	35 (?)	15 (?)
1974	Bangkok, Thailand	*Spirulina*	87	609	N, F, SC	18 (?)	15 (?)
1977	Taiwan	*Chlorella*	250—500	180,000	N, F, SC	35 (7)	18 (365)[c]
1977	Richmond, California	*Micractinium*	12	48	W, F, SC	32 (30)	
1980	Sede Boger, Israel	*Spirulina*	100	6	N, F, SC	25 (30)	15 (300)

[a]   N = Artificial nutrients; W = wastewater; WSW = wastewater-seawater mixture; B = brine; TD = agricultural tile drainage; F = freshwater algae; M = marine algae; SC = semi-continuous harvest; C = continuous harvest.

[b]   Included nonalgal solids from wastewater.

[c]   Algae grown on combined CO_2 and organic substrates.

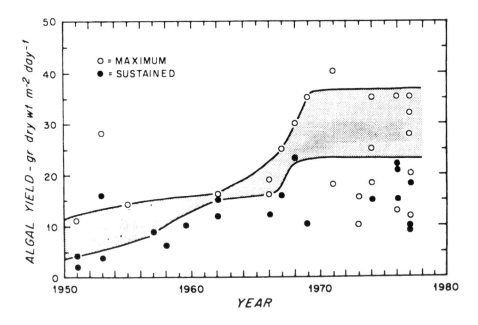

FIGURE 19. Increase in reported algal yields over the past 30 years. Shaded area represents difference between short-term maximum and longer term sustained yields. Each datum point represents an individual study from over 30 world-wide studies. (From Goldman, J. C., *Water Res.*, 13, 1, 1979a. With permission.)

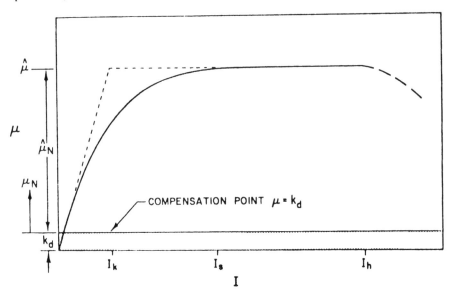

FIGURE 20. Detailed relationship between algal growth rate (μ) and irradiance (I). (From Goldman, J. C., *Water Res.*, 13, 119, 1979b. With permission.)

Oorschot[34] and Shelef et al.,[35] and was primarily related to the current understanding of photosynthetic efficiency in microalgae and the kinetic response of algal growth to light intensity. All these models are based on an analysis of the generalized shape of the curve relating algal growth to light intensity (Figure 20). According to Goldman,[33] the five main features of this curve are (1) at some low growth rates, growth is balanced by decay, i.e., cell death, respiration, and excretion (k_d is compensation point), and the net growth rate is zero; (2) the initial slope of the curve represents the maximum efficiency of growth in response to light; (3) the maximum possible growth rate ($\hat{\mu}$) is a function of the saturation light intensity (I_K). I_K is the light intensity at which the extrapolated initial slope of the curve intersects with $\hat{\mu}$; (4) there is a saturating light intensity (I_s) for which $\mu = \hat{\mu}$ (the relationship between I_s and I_K is a function of the shape of the light curve; (5) at some light intensity I_A $>I_s$ the growth rate is inhibited and $\mu < \hat{\mu}$. The overall shape of the curve is the main determinant in estimating the maximum potential yield of a given light intensity when light is the sole limiting factor.[33]

The generalized equation has the form

$$P = \frac{E_q E_s E_R E_I E_V E_D I'_o}{J} - D_r.$$ (4)

in which P = algal yield, g dry wt/m²/day; J = heat of combustion algae, kcal/g dry wt; E_q = thermodynamic efficiency; E_s = light utilization efficiency; E_R = light transmission efficiency at air-liquid interface; E_I = sunlight visible region efficiency; E_V = visible fraction of total sunlight; E_O = light dissipation efficiency; I_o = total incident light energy at culture surface, cal/cm²/day; D_r = overall decay factor, g dry wt/m²/day. By using the available information on each of the above factors Goldman[33] reduced Equation 4 to:

$$P = 0.28 \ I_s \left(\ln \frac{0.45 \ I_o}{I_s} + 1 \right).$$ (5)

in which I_s = is in units of cal/cm²/day.

Since the maximum available intensity of sunlight striking the earth's surface is <800 cal/cm²/min, Equation 5 predicts that, for a range of I_K values between 0.02 and 0.06 cal/cm²/min as is typical for microalgae, the upper limit for algal yield is 30 to 60 g dry wt/m²/day.

According to this model, the factor that most strongly influences algal yield (Equation 5) is I_K. A threefold increase in I_K from 0.02 to 0.06 cal/cm²/min results in an almost doubled algal yield at high solar irradiance.[33]

It should, however, be remembered that in reality outdoors, light is rarely the sole limiting factor for algal growth and production of biomass. As mentioned in the second major section of Part I (this chapter), temperature usually significantly affects the output of biomass, even when irradiance constitutes the major growth limitation. In addition, the population density may have a far greater effect on the light actually available for the cells than the intensity of the incident light. The rate of stirring also has a marked effect on the light-dark regime of the average cell and thusly on its utilization of light. These considerations cast some doubt on whether the pattern of growth response to light can provide in itself enough of a basis for estimating the potential of mass algal cultures outdoors. The highest average outputs reported to date were 53 g/m²/day was obtained with *Scenedesmus obliquus* and a maximum production rate of 54 g/m²/day was achieved with *Chlorella* sp.

Grobbelaar[37] suggested a model for describing algal production in mass cultures based on the work of Tiwari et al.,[38] but modified to accept only temperature and light as input variables, since all nutrients were provided in excess. Respiration and organic excretions were treated as a single loss and the equation derived was as follows:

$$P = K_1 X_1 K_2^T [l/(1 + l_s k_3^T)] - X_1 (k_4^T/100)$$ (6)

where P = production (mg/ℓ/ton), X_1 = algal biomass (mg/ℓ), l = light intensity (E/m²/hr), Is = light intensity at half-maximal growth, k_1 = maximum light utilization efficiency, k_2 = Q_{10} for photosynthesis, k_3 = Q_{10} for light half saturation constant, k_4 = respiration

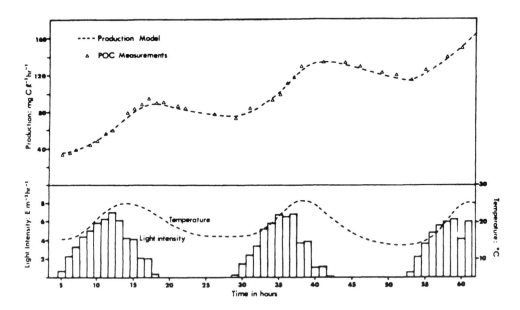

FIGURE 21. Temperature, light intensity, measured POC, and calculated POC values against time of culturing. (From Grobbelaar, J. U., Open semi-defined systems for outdoor mass culture of algae, U.O.F.S., Publ. Series C, No. 3, 24, 1981. With permission.)

and excretion, T = (temperature − reference temperature/10), E = Einsteins (4,6 E/m²/ sec ~ W/m² ~ 1433 × 10⁻³ cal/cm²/min). The following parameter values were used:

$$I_s = 0.36 \ Em^2/hr \quad \text{(Reference 39)}$$

$k_1 = 6.6\%$, $k_2 = 2.0\%$, $k_3 = 1.5\%$, $k_4 = 1.5\%$ (reference temperature = 10°C).

The model was tested by measuring the total organic carbon (TOC), dissolved organic carbon (DOC), and particulate organic carbon (POC) at hourly intervals on samples from an 18 m² open outdoor pond. The alga (*Ankistrodesmus* sp.) was grown as a batch culture, with excess N and P, and Co_2 as a pure gas on demand from a pH-controller to maintain a pH of 7.5. A culture depth of 0.15 m was maintained by adding tapwater daily to compensate for evaporation losses. Culture temperature, radiation, measured POC, and calculated POC were recorded continuously (Figure 21). The cumulative pattern of production of algal mass is typified by a daily increase in biomass (POC) with a maximum of about 1800 hr followed by a decrease to a minimum at 0600 hr. The maximum temperature measured was 26°C and the minimum was 14°C. The initial value of biomass measured was used as the input biomass concentration in the model and the measured light intensities and culture temperatures served as input variables for calculating hourly changes in biomass concentration.

Data from the first 24-hr cycle were used to calibrate the model. Respiration and extracellular excretion were found to be responsible for a loss of about 16% during the dark period, which amounted to approximately 36% over a 24-hr period. Hence, a value of 1.5/ hr was used for k_4 in Equation 1.

As respiration rates are influenced by various factors,[41] and since extracellular organic excretion can amount to 70% of assimilated carbon,[42] the total loss of 36% was judged not to be excessively high. Nevertheless, it is doubtful whether this high figure represents the norm in algal mass cultures.

In summing up, the challenge confronting the grower of outdoor mass cultures is to harvest the highest yields of biomass per pond area. This may be accomplished by consideration of the following:

1. Selection of algal strains that have a high light saturation constant. This is an obvious requirement for a more efficient utilization of solar light and hence for higher yield potential.
2. Selections of species with an overall low rate of decay processes, i.e., respiration and excretion, as well as a high resistance to photooxidative damage.
3. Conditions in the pond chosen so that irradiance should be, as much as possible, the sole limitation to productivity.
4. High turbulence, which is required for high yields of biomass, particularly when the incident radiation is high.

In general, the goal perhaps unreachable in outdoor mass culture is to achieve, under outdoor conditions, the high photosynthetic efficiency obtainable in the laboratory with low light intensities.

Part II

PRINCIPLES OF POND MAINTENANCE

EVALUATING CULTURE PERFORMANCE

General

For successful maintenance of algal cultures so as to obtain maximum output under given environmental limitations, one must monitor the culture constantly and correctly assess its relative performance. The importance of obtaining detailed information to facilitate reliable, quick evaluation of the physiological state of the culture cannot be overemphasized. Warning signals must be recognized early to prevent situations which, within a day or two, could culminate in the loss of the entire culture. Thus, a basic requisite for correct pond maintenance entails consistent evaluation of the degree to which the photosynthetic and other metabolic systems of the algal cells function at the maximum potential obtainable under the given environmental conditions.

Clearly, the optimal course for pond management is to ensure that no nutritional limitations exist and that algae growth is limited only by climatic factors, i.e., light and temperature. Ideally, solar irradiance should be the sole limitation to growth, and even this limitation must be regulated throughout the seasons so that the population density (and mutual shading) is adjusted in relation to the intensity of solar irradiance.

Maintaining Nutritional Status

Tests must be made continuously to ensure that no deficiency in mineral nutrients is developing. The population density must be monitored daily, together with microscopic examinations to identify possible contamination by foreign plankton as well as possible morphological changes in the cultured species. Perhaps most importantly, the dissolved oxygen concentration in the pond should be monitored continuously.

Maintenance of the nutrient status in the pond relates to the fact that in a continuous culture, the levels of nutrient elements in the medium continuously decrease as biomass is removed from the pond. In addition, some minerals may gradually precipitate, nitrogen and NH_3 gas may be lost due to high pH or to denitrification, and anaerobic pockets may occur when mixing is inadequate. The concentration of the major elements, nitrogen and phosphorous, should be analyzed routinely, preferably by standard colorimetric procedures that are quick and reliable. In the author's laboratory, the level of nitrogen in the pond serves as a guideline for adding, in equivalent amounts, the entire formula of the growth medium (except for carbon and phosphorus) assuming that the relative depletion of nitrogen is roughly equivalent to that of the rest of the nutrient elements in the growth medium. Phosphorus is added separately, as is carbon in the form of bicarbonate or CO_2.

From the economic standpoint, the concentrations of the various nutrients should be maintained at their minimum values, i.e., at slightly more than the concentration at which inhibition of growth occurs. Relatively low nutrient concentrations decrease losses due to precipitation of minerals and volatilization of nitrogen. Also, the price of the growth medium is reduced, a factor of particular importance in case deterioration of the culture, a harmful increase in salinity, or contamination by other species mandate a change of the medium. Using nutrients at low concentrations requires, however, much caution to ascertain that their levels do not decrease below the threshold concentrations which limit growth. To prevent this, input of nutrients, balanced so as to replace the nutrients depleted by the harvested biomass, should be continuous. Exceptions to this practice will be made in cases when a particularly high concentration of some mineral nutrients provides an advantage to the cultured algae over potential competitors. A good example is the concentration of bicarbonate in *Spirulina* culture. Four grams per liter of Na bicarbonate are sufficient to sustain maximum growth in mass cultures of *Spirulina*. At this concentration of bicarbonate, however, infestation with *Chlorella* presents a serious problem. Elevating the concentration of bicarbonate in the *Spirulina* medium arrests the take-over of the culture by *Chlorella*, as will be further discussed in this chapter.

Monitoring the pH

A rise in pH reflects depletion of CO_2 or HCO_3 through photosynthesis. The pH may be maintained by a suitable inflow of CO_2 or by the addition of mineral acid to a medium of high alkalinity. In the latter case, the alkalinity must be brought back to its original levels by adding carbonates in stoichiometric amounts. Sodium bicarbonate may be used in addition of CO_2 to adjust the pH in growth media for algae such as *Spirulina* which have a pH optimum of 8.5 to 10.5 and thrive at a high concentration, i.e., $0.2 M$ or more of $NaHCO_3$. It is noteworthy that carbon nutrition ususally represents one of the major components in the operating costs of an algal culture.

In a *Spirulina* culture grown at a pH of 10.0 to 10.5, a failure of the pH to rise may not necessarily indicate the absence of photosynthetic acitivity. This is because at such an elevated pH, a substantial amount of CO_2 is absorbed from the air to compensate for the depletion of carbon through photosynthesis. Observations in the author's field laboratory indicated that at pH 10.5, CO_2 from the air could provide the carbon needed for a net production of approximately 7 g dry wt/m²/day. Outputs greater than this require continuous addition of carbon; in a pond of *Spirulina* growing vigorously in midsummer, the pH would rise by 0.10 to 0.15 pH units daily.

Measurement of Algal Population

The population density of a culture should be checked at least twice daily. The major parameters to follow routinely in this respect are the overall turbidity, the concentration of Chlorophyll, and the dry weight per liter (for details see "Algal Nutrition"). In a continuous culture maintained in what may be regarded, for practical purposes, as steady state, there exists a good correlation between these parameters. Thus, measurement of turbidity, which is the most rapid way to determine changes in the population density, may be used at the standard method to obtain a quantitative measure of the algal population in the pond.

Measurements of turbidity or dry weight give only a quantitative indication of the variations in biomass that take place in a culture. These measurements should always be accompanied by a detailed microscopic evaluation, because environmental changes may cause morphological modifications in some species which should be carefully recorded and analyzed.

However, the most important reason for microscopic examination is to monitor any development of "weed", i.e., foreign algal species as well as protozoa or fungi that may threaten to take over the cultivated species or cause severe damage to the culture. When the invading organisms are detected early, they should be easier to control. No less important, the mere appearance of significant populations of other organisms in the culture should be regarded as a warning signal that the cultured species have come under some pressure. It may be that the temperature decreased or increased too much, giving a relative advantage to another species. It may also signal that the concentration of a particular nutrient had declined below the concentration threshold for the cultured species. Indeed, any weakness of the cultured species provides an opportunity for foreign organisms to rapidly increase their population in the pond.

A useful practice is to take photographs of any modification in the culture's population for the record. With time, accumulating experience permits ready diagnosis of the irregularities that occasionally develop in the culture.

Measurement of Oxygen

The relative concentration of dissolved O_2, DO, represents a most useful parameter for evaluation of the photosynthetic activity of the culture. When photosynthesis is rapid, DO rises quickly to values which represent 200 to 400% saturation. Vigorous stirring of the culture decreases the steady-state level of super-saturated O_2 and this in fact is one advantage of intense stirring.

Experience in this author's laboratory, as well as reports by other workers,[43] indicated that DO, which is readily measured with an oxygen electrode, serves as a reliable and sensitive indicator of conditions that impede growth and productivity. The following illustrates this point. In a rapidly growing *Spirulina* culture at the author's laboratory the oxygen concentration on a typical summer day in 1979 was approximately 18.0 mg/ℓ. The next day, the O_2 level dropped suddenly to 14.5 mg/ℓ. Microscopic examination revealed that the decline in O_2 was associated with the initiation of familiar features of deterioration in *Spirulina*, i.e., decreased filament length and the appearance of fragments. This became more prevalent the following day, when the concentration of O_2 reached the dangerously low level of 8.5 mg/ℓ. At this point it was decided to replace 50% of the pond volume with fresh culture medium. Within 1 day, a substantial improvement was evident in both the O_2 concentration, which reached 18.5 mg/ℓ, and in the appearance of the *Spirulina* filaments. Thus, observation of the marked decline in pond O_2 was instrumental in reaching the costly decision to dispose of 50% of the culture. Experience indicated that a delay in detecting the deterioration in the pond could have resulted in complete loss of the culture.

The course of photooxidative death in a *Spirulina* pond was described in Part I of this chapter. The continued decrease in DO, in spite of apparently normal growth as evidenced by the daily increase in turbidity and chlorophyll, emphasized the importance of a careful monitoring of the dissolved O_2 in an actively photosynthesizing algal pond. In such a case, the O_2 concentration is related to the activity of the photoautotrophic biomass and a continuous monitoring of the relationship between these two major parameters would be of great assistance for correct maintenance of the culture. Gutterman and Ben Yaakov[44] developed a microcomputer-controlled system for monitoring and control of environmental factors in an algal culture. The system measures the major parameters from which the performance and the state of the culture may be deduced: pH, DO, turbidity, light intensity, and water and air temperatures. In this system, DO is related to cell density, making it possible to estimate on line with the aid of the computer, the actual rate of growth at any instant. When fully developed, such systems will permit the alga grower to evaluate the culture's performance in terms of productivity. Essentially, the task is to ascertain that the given inputs at the time of measurement, i.e., solar irradiance, temperature, nutrients, cell population, and mixing energy, are yielding the highest possible productivity per area. Moreover, on-line computerized monitoring and control would finally permit a push-button, automatic maintenance of the pond which should result in greater accuracy and reduced labor costs.

ESTABLISHING AND MAINTAINING THE OPTIMUM POPULATION DENSITY

The population density is a major factor in the production of biomass, exerting far-reaching effects on the general performance and productivity of the culture. This is so because mutual shading, a parameter greatly affected by the population density, reduces the quantity of light available to the average algal cell. For example, even when the cell density of *Spirulina* is very low, approximately 100 mg dry wt/ℓ sufficient radiation (i.e., above the compensation point) is available to no more than one half the cells in the culture at any given instant. At cell densities of 300 to 500 mg dry wt/ℓ, which were found optimal for maximum photosynthesis efficiency in pond cultures of *Spirulina,* solar irradiance is almost completely absorbed in the upper 2 to 6 cm of the pond, leaving some 70 to 80% of the cells practically in complete darkness at any given moment. This illustrates the importance of distributing light to the individual cells, which is affected by turbulence.

The relationship between the population density and the growth rate varies throughout the seasons, as shown in Figure 11. Clearly, the lower the population density, the higher the specific growth rate, as is to be expected in a system which is primarily light limited. As already explained (Figure 11), the relative effect of decreasing the population density, and of increasing the availability of light for each cell, is most pronounced in the summer and much less so in winter.

Extensive experimentation to elucidate the relationships between the specific growth and output rates and the optimum population density (defined as that cell concentration which yields the highest output rate under given environmental conditions) indicated that the optimal density varied according to the season. Plotting the output rate of the pond against the population density during the course of the year (Figure 22) revealed that the higher the temperature of the available irradiance per cell, the more pronounced the dependence of the output rate on population density. Thus, in the summer, the maximum output rate reached in a culture maintained at optimum density was almost three times as high as that obtained in a culture maintained at the highest population density used in our experiments. In the winter, when the total output was a fraction of that obtained in summer, the maximum output from the culture maintained at optimum density was only double that obtained from the culture with the highest density.

A decline in output rate was always associated with a high population density, i.e., over 600 to 700 mg/ℓ. Since the output is a product of the specific growth rate and the population density, this decline cannot be explained merely by the effect of increased mutual shading, since the increase in population density should compensate for the decrease in the specific growth rate. Thus, an additional factor involved in the phenomenon of decreased output at high densities must stem from the increase in metabolic maintenance energy, which was more pronounced due to the reduction in photosynthetic activity.

A decrease in population density below the optimal point also resulted in a significant decrease in the output rate. This apparently indicates that a high intensity of solar irradiance per unit pond area cannot be exploited at peak efficiency when the population density is relatively low. Thus, although the specific growth rate (μ) increased progressively as the population density (x) decreased and the limitation by light was relieved (Figure 11), this was not sufficient to compensate for the decline in output, which affected the decrease in the overall density. Indeed, the maximum output rate was obtained at a relatively high population density, exhibiting specific growth rates some 50% lower than the maximal growth

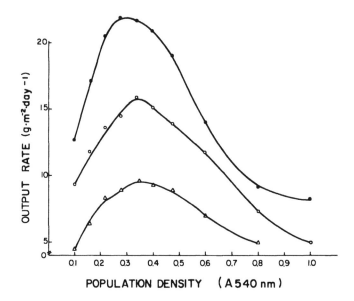

FIGURE 22.   The output rate of *Spirulina* as affected by population density and the seasons of the year. O-O-O Summer (June to September), □-□ autumn and spring (October to November and April to May), △-△-△ winter (December to February). The data represents averages obtained from several experiments carried out throughout the year. (From Vonshak, A. et al., *Biomass,* 2, 175, 1982. With permission.)

rates manifested by algae maintained at as low a population density as was practically possible.

It seems most advisable to maintain the cell density in the pond at its optimum and to determine the harvesting regime accordingly. The experience with mass *Spirulina* cultures in the author's laboratory indicates that a good procedure is to harvest each time the population density reaches a concentration of about 500 mg dry wt/ℓ, removing some 25% of the pond biomass. When productivity is at its peak, the culture must be harvested every 2 days so as to maintain it at the optimum cell density.

As a rule, harvesting should be performed at the uppermost limit of the optimum cell concentration. This facilitates more efficient harvesting and permits the removal of larger amounts of biomass in each harvesting operation.

A highly turbulent flow, e.g., 50 cm/sec, creates some foam, which may be disregarded as long as it disperses quickly. High amounts of foaming, however, which create a more permanent layer of foam over the culture and decrease the light availability are harmful. Usually, foaming presents a problem only when the organic solutes in the pond, particularly protein, increase. In a photoautotrophic culture, the organic solutes have no nutritional advantage and their mere presence suggests cell lysis or enhanced cell excretion, both indications of stress in the culture. Foam that results from cell lysis will have a definite green or brown tinge.

Excessive foaming, then, should be regarded as a signal of deterioration in the culture, requiring special precautions. One way to reduce foaming and improve the overall conditions in the culture is to discharge the spent medium during harvest, replacing this effluent with an equivalent volume of new medium.

MAINTAINING MONOALGAL CULTURES

Maintaining a monoalgal culture is perhaps the most formidable challenge facing the producer of algal biomass. Naturally, the cultivation of algal monocultures outdoors is hampered by contamination by other algae, bacteria, zooplankton, and fungi. Contaminants reduce the uniformity of the product and may greatly reduce the yield. In extreme cases, or if left unattended, another algal species may take over and become the dominant species.

Tactics to combat contaminants are based on two principles. First, conditions in the pond should be maintained so as not to diverge too far from the optimum for the cultivated algae. The second principle is based on maintaining appropriate conditions in the pond to selectively promote the growth of the alga of interest and/or inhibit that of alien organisms.

Various environmental factors have been suggested as selective forces in algal production systems. Nutrients were shown by Schanz and co-workers[45] to affect the relative growth of two strains of *Anabaena*. In a culture containing *Anabaena* and *Chlorella*, however, light was shown to control their relative growth: light optima for the two algal species were distinctly different, the former becoming saturated at one half the light intensity required to saturate the latter. The temperature and pH optima for the two species were also significantly different. *Anabaena* grew best at a temperature range of 28 to 35°C and *Chlorella* between 23 and 28°C. *Chlorella* grew well at a pH of 5 to 9, with an optimum at pH 7 to 8. In contrast, *Anabaena* grew well between pH 6 and 11, with an optimum at pH 9 to 10.[46]

Mur and co-workers[47] also reported light to be the selective factor in the competition between green and blue-green algae. They suggested that mutual shading provided conditions more favorable for a number of blue-green than the green algae. Dominance relations between various marine phytoplankton species was found by Goldman[33] to be strongly affected by temperature. Below 19°C the diatom *Phaeodactylum tricornutum* was dominant, whereas above 27°C the cyanobacterium *Oscillatoria* became progressively dominant (see "Cell Response to Environmental Factors"). The initial abundance of the different species in a culture seemed also to affect the outcome of the competition. In a mixed culture, *Anabaena* dominated *Chlorella* except when the initial concentration of *Chlorella* was high relative to that of *Anabaena*.

Cyanobacteria were reported to be more efficient than green algae in utilizing low concentrations of CO_2. Thus, when the pH rises, as in enriched lakes, cyanobacteria predominate.[48] Accordingly, addition of nutrients and CO_2 resulted in a spectacular population changeover from blue-green to green algae.[49]

Richmond et al.[50] and Vonshak et al.[9] elucidated the conditions that affect the competition between *Spirulina* and *Chlorella* in outdoor ponds. Under field conditions, a bicarbonate concentration as low as 4 g/ℓ, i.e., 25% of the concentration in the standard Zarouk[51] formula, was sufficient to support the maximum growth rate in *Spirulina*. At this bicarbonate concentration, however, a significant increase in contamination by *Chlorella* species was always observed. Also, a factor that intensified the contamination of a *Spirulina* pond with *Chlorella* was the mode of harvesting of *Spirulina*, i.e., harvesting by filtration on a 375 mesh vibrating screen (Sweco, Los Angeles, Calif.), which did not retain the *Chlorella* cells and which were returned to the pond with the effluent. Continuous enrichment of the *Spirulina* culture with *Chlorella* sp. thus followed, and after a few such harvests, over 50% of the algal biomass in the pond belonged to species other than *Spirulina*. It was, in fact, impossible to maintain a monoalgal culture of *Spirulina platensis* outdoors even with 8 g/ℓ of bicarbonate (Figure 23) and the dominance of *Spirulina* could be maintained only with 16 g/ℓ bicarbonate (Figure 23) with this mode of harvesting.

The shift in the pond population under the low bicarbonate regime to dominance by *Chlorella* sp. resulted in the development of a food chain consisting of several *Chlorella* grazers belonging to a variety of ciliates as well as their predators (e.g., copepodes, termatodes), some of which are shown in Figures 23 A and B.

Thus, in ponds in which 4 g/ℓ bicarbonate was maintained and in which contamination by *Chlorella* was heavy, total chlorophyll, expressed as percent of the total organic dry

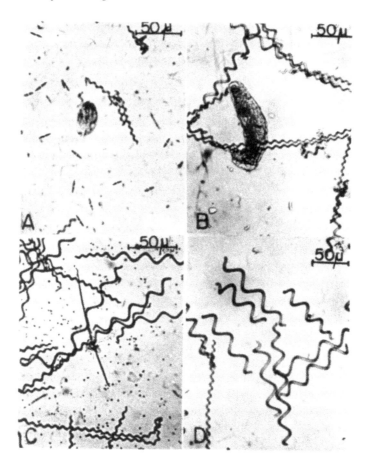

FIGURE 23.   Effect of bicarbonate concentration on the purity of *Spirulina* outdoor cultures grown in 100-m² ponds: (A) Zarouk medium containing 2 g/ℓ NaHCO₃, (B) Zarouk medium containing 4 g/ℓ NaHCO₃, (C) Zarouk medium containing 8 g/ℓ NaHCO₃, (D) full Zarouk medium. (From Vonshak, A. et al., *Biomass*, 2, 175, 1982. With permission.)

weight fluctuated drastically, reflected rapid shifts in the populations of *Chlorella* and its predators (Figure 24). In contrast, in a pond containing a monoculture of *Spirulina*, the proportion of chlorophyll remained constant at about 1.5%.

In addition to bicarbonate, gaseous CO_2 exerted a decisive influence on the competition between the two algae, which was clearly evident when a mixed culture of *Spriulina* and *Chlorella* was grown in a turbidostat.[50] When CO_2 was withheld by replacing the stream of air enriched with 1.5% CO_2 with CO_2-free air, an immediate, sharp decline in the population of *Chlorella* occurred (Figure 25). When the supply of CO_2 was resumed just before *Chlorella* was washed out, a complete recovery in the population of *Chlorella* took place, and within 70 hr it reached the same population density as before the supply of CO_2 was stopped. Richmond et al.[50] considered the possibility that it was the rise in pH, which occurred when the CO_2 was withheld, rather than the CO_2 itself, that caused the sharp decline in the *Chlorella* population. When the CO_2 supply was withheld but the pH was kept constant by continuous titration with 6 N HCl, the *Chlorella* still disappeared, indicating that its population declined in response to the lack of dissolved CO_2 rather than to the rise in pH.

For the maintenance of monoalgal cultures, it is imperative that the pond temperature not deviate too much from the optimum for the cultivated species. Vonshak et al.[52] studied the effect of temperature on the competition of *Spirulina* with *Chlorella* in outdoor cultures.

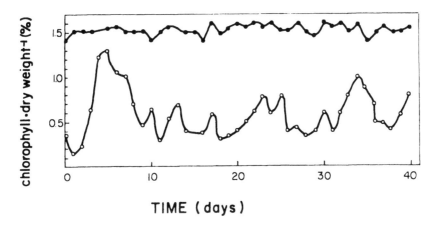

FIGURE 24. Effect of bicarbonate concentration on the chlorophyll dry weight ratio in a *Spirulina* outdoor culture: (●-●) full Zarouk medium, (○-○) Zarouk medium with 4 g/ℓ NaHCO₃. (From Vonshak, A. et al., *Biomass*, 2, 175, 1982. With permission.)

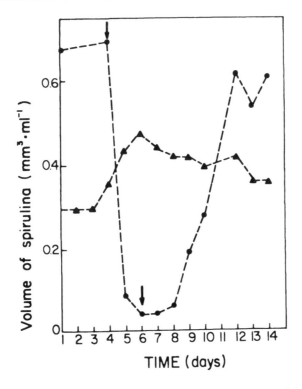

FIGURE 25. The effect of withholding CO_2 in a continuous mixed culture of *Spirulina* (▲) and *Chlorella* (●), grown in a Zarouk medium (see the fifth chapter) at 35°C. At the beginning of the experiment, the populations of the two algal species were at steady state, as indicated by the steady cell volume of each species in the mixed culture observed for 3 consecutive days to be 0.069 hr. On day 3, CO_2 was withheld (upper arrow), by passing CO_2-free air through the turbidostat. On day 5 (lower arrow) CO_2 supply was resumed by passing 1.5% CO_2 enriched air. The pH was 8.7; during the period that CO_2 was not supplied, the pH rose to 9.8. (From Richmond et al., *Plant Cell*, 23 (8), 1411, 1982. With permission.)

With the advent of winter, as the daily temperature became increasingly lower, net growth of *S. platensis* (optimum temperature 37°C) ceases altogether and the overall biomass in the pond declined steadily. A steady take-over of the culture by a local strain of *C. vulgaris*, the optimal temperature for which was approximately 10°C lower than that of *Spirulina*, became evident. To raise the temperature, the pond was covered with 0.2 mm-thick polyethylene sheeting of the kind used for agricultural greenhouses. This raised the maximum daily temperature of the culture by up to 8°C and markedly affected the culture; the growth of *Spirulina* was resumed, and concomitantly, the population of *Chlorella* steadily decreased.

PEST CONTROL

An organically rich suspension, such as an open outdoor mass algal culture, is subject to infections caused by bacteria, fungi, viruses, and protozoa. In addition, grazing by rotifers could quickly decimate a large population of microalgae. Many reports have shown that contaminants reduce the quality as well as the overall yield of the algal product.

Information on contamination of mass algal cultures is generally very meager. One reason is that study of infection depends on cultures becoming infected and cannot usually be repeated exactly under controlled conditions. Infections occur erratically, and are reported on as casual observations made during the normal operation of outdoor cultures.[7]

Viruses

The occurrence and properties of cyanophages, i.e., viruses attacking cyanobacteria, have been well documented. The first virus capable of lysing a filamentous cyanobacterium, *Plectonema* sp., was isolated from an oxidation pond. Viruses may be the explanation for the observation that blue-green algae seldom reach large numbers in such ponds. The salinity range of cyanophages is of interest; some phages have been isolated from brackish fish ponds with salinities reaching 2000 mg of Cl^{-2}/ℓ.[53]

Padan and Shilo[53] reported that the interactions of the cyanophage with its host are markedly influenced by environmental conditions. The tolerance of the cyanophages to alkalinity accords with the alkaline range of *Plectonema* (pH 7 to 11) which shows little or no growth below neutrality. Most other cyanobacteria also prefer the alkaline range. Survival of cyanophages at high pH must have ecological significance as a selective factor affecting cyanobacterial blooms.

Temperature affects the survival of free phage and the development of cyanophage in hosts. Some common cyanophages (LPP and SM-1) show similar temperature sensitivity: 85% of the free phage particles remain infective at temperatures up to 40°C, whereas only 55% infect at 45°C, and less than 0.001% at 50°C. Two LPP isolates from nature are temperature sensitive, multiplying normally at 26 to 29°C and producing only early symptoms of infection, i.e., invagination of the photosynthetic lamellae and cessation of CO_2 photoassimilation, above 31°C.[53]

Provided that the adenosine triphosphate supply is ensured, certain cyanophages (LPP) can grow and replicate in the dark and under conditions inhibiting host growth (such as the presence of DCMU or the complete absence of external carbon sources), whereas typical bacteriophages require actively multiplying hosts. This wide tolerance of the LPP cyanophages makes them quite suited for success under natural conditions.[53]

As pointed out by Padan and Shilo,[53] the high degree of host specificity, the selection of resistant host mutants, and the dependence on environmental factors indicate the complexity of the interaction between alga and cyanophage, whose outcome depends not only on the inherent properties of both phage and host, but also on fluctuations in external conditions. It may be well for growers of mass algal cultures to remember this point. Nevertheless, it should be stressed that in the only cyanobacteria grown commercially to date, *Spirulina* sp., damages inflicted by a cyanophage have not yet been reported.

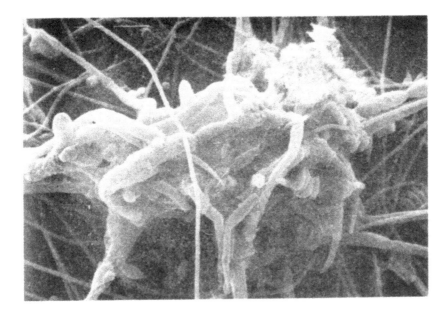

FIGURE 26. Fungal mycelia entangled with algae taken from a mass outdoor algal culture (marker = 10 μm). (From Grobbellar, J. U., Deterministic production model for describing algal growth, U.O.F.S. Publ., Series C, No. 3, 1981. With permission.)

Likewise, viruses attacking eukaryotic green algae are very poorly documented. Such viruses definitely exist, but most of the evidence for their presence comes from electron microscopy, where they have been observed incidentally as part of other studies.[54] Heussler et al.[36] tested for pathogenic viruses some 50 samples from healthy and diseased outdoor cultures of *Scenedesmus* and from various bodies of water in Peru. Seven test plates with *Scenedesmus* showed a positive reaction, but beyond this there was no basis for assuming that the outdoor cultures of *S. acutus* were contaminated by a virus. Clearly, direct unequivocal evidence for eradication of a population of green algae by a virus is lacking.

Fungi

Fungal parasites of fresh water algae are commonly called chytrides. The chytrides (Chytridiales) resemble a group of primitive fungi that were discovered as parasites on various algae such as *Diatomeae* and *Euglenophyta*. They may occur as epidemics, thereby destroying whole generations of host cells (Figure 26).

Morphological characters such as the size and shape of sporangia and rhizoids and the types of zoospores and resting spores, have been used to identify individual species of aquatic fungi. Muller and Sengbusch[55] used induced fluorescence for visualization of aquatic fungi parasitizing algae. They showed that a decline of an *Anabaena flos-aquae* bloom was accelerated or even caused by a fungal attack. Several algal species from the same plankton sample exhibited epidemic infection, each host species being infected by a different parasite. A narrow host-range of fungi was revealed, explaining their coexistence in the same ecosystem. for at least two host-parasite pairs, parasitism was associated with direct cell-to-cell contact and not to secretion of a lytic agent by the fungi.

Canter and Jaworski[56] studied parasitism of the diatom *Fragilaria crotonensis* by chytridiaceous fungi. They discovered strong specificity in host-parasite relationships. In clones of two morphological types of *F. crotonensis,* states of resistance or susceptibility existed with regards to parasitism by two Chytrid species. Also, slight differences in susceptibility toward a particular fungus could occur within various *Fragilaria* clones of the same morphological type. Clearly, physiological races with respect to the response to pathogenic fungi seem to exist within a single morphological form of *F. crotonensis*.

According to Heussler et al.,[36] the fungus *Chytridium* sp.[57] appeared often in their outdoor cultures as parasites of *Scenedesmus* cells, usually together with the flagellate *Aphelidium*. The causative organism has not yet spread to such an extent that it has become a problem.[36]

As summarized by Muller and Sengbusch,[55] fungal growth may be affected by several external factors. For a number of host-parasite interrelationships temperature seems to be the most important factor controlling chytrid parasitism.[58,59] Also, an optimum oxygen concentration, low concentration of potassium and magnesium, and sufficient light were essential for infection of *Scenedesmus*.[60] Zoospores can remain alive in darkness, but the rates of adhesion to host cells are low. Upon illumination, adhesion to host cells increased greatly.[58,61]

Bacteria

According to Scott and Chutter,[54] numerous species of myxobacteria capable of lysing fresh water algae are known.[62] These organisms are generally aerobic, flagellated, Gram-negative rods. The myxobacteria are capable of attacking a large number of cyanobacteria, as well as green algae such as *Chlamydomonas* and *Spirogyra*, but cell-free filtrates did not cause algal lysis. In addition, a non-myxobacterial *Spirillum*-like bacterium attacking *Scenedesmus* has been described by Schnepf et al.[63] Scott and Chutter[54] suggested that since the myxobacteria are less specific than viruses in their host range, they are likely to be more important in affecting algal populations and merit further study. In *Spirulina* ponds at the author's laboratory, the bacterial count in healthy, rapidly growing cultures of algae outdoors is usually relatively small, ranging between 1 and 5×10^3 bacteria per milliliter. The small size of this population, which may be quantitatively detected by plating aliquots from the pond on enriched agar medium, may be due to the UV radiation outdoors or to algal-bacteria interactions. When deterioration begins in the pond and the organic load is thus greatly increased, the bacterial count may quickly increase by an order of magnitude or even more. In mass cultures of *Spirulina*, very long filamentous bacteria (up to 100-μm long) proliferated in cultures exhibiting relatively slow growth rates where cell breakdown was evident.

Heussler et al.[36] reported on a project in which 100 strains of bacteria were isolated trom healthy and diseased algal cultures. Approximately 75 of these strains were harmful to *Scenedesmus acutus* if inoculated individually into old, axenic laboratory cultures of the alga (stationary growth stage). No damage was caused to young, vigorous algal cultures, and moreover, only relatively few bacteria could be found in healthy cultures in the laboratory or outdoors.

Thus, there is no evidence that bacteria may directly control the size of the algal population in well-managed, rapidly growing mass cultures of algae.

Invertebrates

Protozoa and rotifers are almost always present in outdoor algal cultures. Their numbers are usually small, but occasionally they increase and seriously influence yields. Infections by *Philodina* sp., *Amoeba* sp., Ciliata, *Vorticella* sp., *Stylonichia* sp., and *Strobilidium* sp. have been identified. Counts of up to 3.09×10 invertebrate organisms per liter have been made in heavily infected cultures.[7]

Protozoa

Infections by protozoa are known to have caused shifts of algal dominance in outdoor cultures. Grobbelaar[7] described a case in which a culture containing mainly *Chlorella* sp. was infected by a *Stylonichia* sp. Within 5 days *Scenedesmus* sp. dominated the algal population, since its colonies were too large to be taken in by the *Stylonichia*. Selective grazing by the *Stylonicha* effected a change in species dominance.

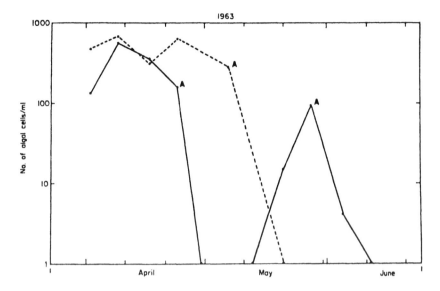

FIGURE 27. Decreases in algal populations in relation to grazing by protozoa. Continuous left-hand line *Eudorina elegans;* continuous right-hand line, *Gemellicystic imperfecta,* broken line, *Synura* sp., all in Esthwaite Water during 1963. (A) date when protozoans were first seen. (From Canter, H. M. and Lund, J. W. G., *Proc. Linn. Soc. Lond.,* 179 (2), 203, 1968. With permission.)

Canter and Lund[64] investigated the role of protozoa in controlling the abundance of planktonic algae in lakes. Species of the protozoa *Pseudospora* were found to feed on colonial algae and can reduce the algal population by nearly 100% in a week or two (Figure 27).

According to Goulder,[65] the ciliate *Loxodes magnus* can consume 4 to 12 *Scendesmus* cells per hour. The grazing rate of the ciliate, however, was estimated to be slower than the growth rate of *Scenedesmus,* and it was concluded that the ciliate thus would not significantly influence the size of the *Scenedesmus* population.[54] In contrast, when the ciliate *Colpoda ateinii* was introduced into a culture of the cyanobacterium *Anacystic nidulans,* the ciliates could, under optimal conditions, double every 3 to 4 hr and approximately 1000 algal cells were needed to produce one ciliate. The algal population was soon exhausted and the ciliates formed cysts.[66]

A few specialized protozoans consume filamentous cyanobacteria, among which species of the genus *Nassula* seem most important. Brabrand et al.,[67] studied the effect of grazing by gymnostomid ciliate *Nassula ornata* Ehrenb. on populations of *Oscillatoria agardhii* Gom. var., in laboratory and field experiments. Ingestion of *Oscillatoria* by *Nassula* increased exponentially over the temperature range of 5 to 20°C and the grazer could depress the population of *Oscillatoria* at all temperatures (Figure 28).

Huang and Wu[68] described amoebae grazing on filamentous cyanobacteria in rice paddy fields. A lobose and a filose amoebae were isolated and have been maintained in axenic culture by feeding with *Anabaena* as the sole prey. Their grazing behavior, observed by simplified microculture technique, revealed that the amoeba approached any part of the filamentous algae and engulfed a cell or a segment of the filament. In the presence of these amoebae, the population of *Anabaena* was reduced to a very low level within a few days.

Ho and Alexander[69] attempted to assess the influence of two major environmental variables, temperature and pH, on the rate and extent of protozoan replication at the expense of selected algae. Raising the temperature from 20 to 37°C increased both the protozoan growth rate and the maximum cell yield, and no replication of the amoeba was evident at 4°C (Figure 29). The predators were very sensitive to the pH. *Amoeba discoides* grew well

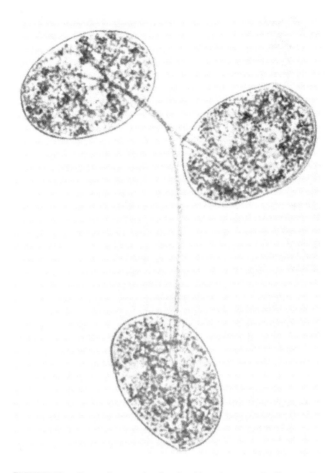

FIGURE 28. Photomicrographs showing *Nassula ornata* feeding on *Oscillatoria agardhii*. When feeding, *Nassula* adheres to a filament of *Oscillaria*, bends it, and draws the double filament into the cell through a cytopharyngeal basket. The algal filament is coiled inside the ciliate before it is broken into pieces and digested. The food vacuoles contain pigments from the ingested algal making the whole animal appear red or green depending on pigmentation of the algae. (From Brabrand, A., et al., *Biological Control of Cyanobacteria Oecologia.* With permission.)

at pH 8.0 but not at all at pH 6.0. In contrast, *Hartmannella castellanii* proliferated at the lower, but not at the higher pH. At neutrality, both these amoeba species grew well. When the suitability of several of the algae to *A. radiosa, A. discoides,* and *H. castellanii* at pH 6.0 and 8.0 was evaluated, the growth rates and protozoan abundance were markedly affected by the food source. A good nutrient source for one predator failed to sustain much replication of a second. The study of *Amoeba radiosa* demonstrated that the protozoan abundance was determined by the pH even when the same algal species was the prey, and also that the pH governed the relative suitability of two prey species.

Zooplankton

Grazing by zooplankton could, under favorable conditions, completely suppress algal growth in sewage ponds and hyperfertilized fish ponds.[70] Scott and Chutter[54] computed that to produce one mass of *Daphnia* about five masses of algae were required. When the algal concentrations were high, a large proportion of algae would pass through the gut of a *Daphnia* without being digested.

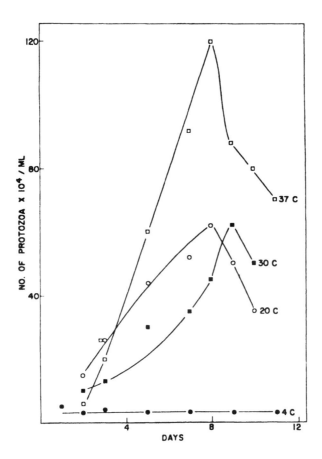

FIGURE 29.   Effect of temperature on the feeding of *Pandorina morum* by *Amoeba radiosa*. (From Ho, T. S. and Alexander, M., *J. Phycol.*, 10, 95, 1974. With permission.)

Loosanoff et al.[71] stated that the most common difficulty experienced in growing *Chlorella* on a large scale was the invasion of the cultures by various forms of zooplankton, the common offenders being crustaceans, especially members of the subclass *Copedoda*. Upon entering cultures of *Chlorella*, these pests rapidly multiply to such an extent that they consume most of the plant cells. Likewise infections of outdoor cultures of *Scenedesmus* by the zooflagellate *Aphelidium* sp. caused serious difficulties, frequently culminating in the complete loss of cultures so that economic production was not possible without measure to combat the parasite.[36] For this purpose, Heussler et al.[36] developed a model which was based on the following equation:

$$A_t = \frac{A_{max} \times A_o \times e^{kAt}}{(A_{max} - A_o) + A_o \times e^{kAt}} - t \times P_o \times e^{kpt}$$

The equation permits the calculation of the number of algal cells (A_t) produced in time (t) on the basis of several parameters, some of which, such as the initial number of algal cells (A_o), the maximum possible number of algal cells (A_{max}) and the initial number of parasite cells (P_o), are determined directly; other parameters, such as the logarithmic growth factors K_A (algae) and K_P (parasite), may be computed in the course of the growth of the culture.

The average growth factor for *Scenedesmus acutus* was calculated by Heussler et al.[36] to be k_A = 1.06 (log$_e$ 2.9), but under extremely favorable conditions values of up to k_A = 1.31 (log$_e$ 3.7) were obtained. The multiplication factor of *Aphelidium* (k_p) lies in the range of 0.69 to 0.92 (log$_e$ 2.0 and log$_e$ 2.5). By introducing the actual growth parameters of the host and parasite, this model allows prediction of the course of the infection and makes possible the selective use of control measures and an estimation in advance of their effect.

The model makes it clear that efficient parasite control is possible only under favorable conditions for growth of the algae. When growth is light limited a key role is played by the population density, since the growth rate is reduced as the density increases. As soon as the algal growth rate is lower than that of the parasite, the infection becomes troublesome. The conclusion was that the culture density must thus be maintained in the optimum (linear) region by daily harvest. For treatment of a diseased culture, however, reduction of the population density into the exponential region of the growth curve may become necessary. This should also be the practice when harmful environmental conditions (shortage of light, cool weather) occur. In general, the culture density should not exceed 70% of the maximum algae concentration attainable under the prevailing environmental conditions.[36] According to Heussler et al.,[36] the danger of poor harvests as a result of parasite attack may be almost eliminated if the above recommendations are followed. During 12 months of growing outdoor cultures of *Scenedesmus* with the tactics mandated by the model, only one loss of a culture due to *Aphelidium* was recorded.

Although the cyanobacteria have long been regarded as inadequate food for fresh water zooplankton because of their filamentous or clumped morphology and their resistance to digestion, there is convincing evidence that cyanobacteria may be a useful diet for certain planktonic rotifers. Starkweather[72] concluded that *Brachionus calyciflorus* is fully capable of ingesting a broad size range of *Anabaena flos-aquae* filaments, the alga readily supporting normal maturation and reproduction of this zooplankton. Filamentous cyanobacteria grown in mass culture may thus also be attacked by rotifers, which currently are serious pests mainly in species belonging to the Chlorophytae.

Control of Infections

Controlling infection of mass algal cultures as well as maintaining a monoalgal culture by preventing alien algal species from proliferating is a critical issue in commercial alga-culture. Grobbelaar[7] summarized the tactics for controlling and combating infections as follows:

1. Maintain optimum conditions for algal growth. This is by far the most important approach. Numerous observations indicate that healthy, rapidly growing populations of algae do not generally suffer from infections. One reason may be related to the extracellular substances that many algae liberate to the medium. These substances may stimulate or inhibit the growth of the cultured species as well as of other organisms. In fact, they strongly affect species-specific blooms and ecological succession. A prerequisite for maintaining monoalgal cultures for months on end is that the cultures be maintained under conditions close to the optima in terms of nutrients, pH, population density, rate of mixing, and temperature.

2. Chemical applications. At present, there are no chemicals that have been produced specifically to control infections and contamination of algal cultures. Many potentially suitable products, however, exist among the pesticides that are being continuously developed for crop protection. Soeder and Maiweg[57] used the fungicide "Orthocid 50" (*N*-trichloro-methyl-thio-tehrahydrophtalimid) while Abeliovich and Dikbuck[60] made use of the fungicide "Benomyl" [methyl-1-(butylcarbomoyl benzimidazolecarbamate)], Heussler et al.[36] tested a number of biocides to combat growth of *Aphelidium* and arrived at the conclusion that *p*-toluenesulfonamide was the most attractive because of its relatively low cost. Lincoln et al.[73] reported that infestations of rotifers (*Brachionus rubens*) and cladocerans (*Diaphano-*

soma brachyurum) in a 0.1 ha high rate algal pond were eliminated by temporarily raising the nonionized (free) ammonia concentration to approximately 20 mg/ℓ (as N) by addition of ammonium hydroxide solution. The 24-hr LD_{100} for *Branchionus* was 17 mg/ℓ free ammonia-N and that for *Diaphanosoma* just under 20 mg/ℓ. Ammonium hydroxide treatment did not inhibit microalgal populations but rather promoted rapid recovery of algal densities to pre-infestation levels. The conclusion was that manipulation of free ammonia concentrations provided an economical and effective means of controlling zooplankton in mass algal cultures. Grobbelaar[7] used the organophosphate mercaptothione to control infections by *Philodina*. Concentrations >2.0 mg/ℓ killed most rotifers within 48 hr, but *Amoeba* and *Stylonichia* remained. Mercaptothione had no adverse effect on the algae.

3. Physical means for removing the unwanted organisms. Good results have been obtained by placing mesh screens of about 50 μm porosity in the channels for some time each day. These not only removed most of the parasites, but other detrital materials as well. This measure is obviously suitable only when the infecting or contaminating organism is significantly larger than the cultured species, which is not often true.

4. Subject the culture temporarily to an extreme change in the environment without significantly damaging the cultured species. Heating or cooling the culture represents one such treatment. Grobbelaar[7] reported that most parasites could be killed in the heated cultures during winter by allowing the cultures to cool below 5°C for a few hours, a treatment which had no adverse effect on the algae. Another method of shock treatment has been to add an organic acid to reduce the pH to 3.5 for a few hours and then neutralize the acid with an inorganic salt. Most rotifers and protozoa were killed, without adverse effects on the algae. The procedure was repeated for a few days until all animal parasites were killed.[7] In contrast, O'Brien and de Noyelles[74] observed that there was a close agreement between high pH values and the disappearance of crustacean zooplankton, whose population decreased as pH levels of 10.4 and 10.8 were reached. Oxygen concentration could also be manipulated to extremes. In dense algal cultures which are mixed only gently during the night, O_2 tension in the culture could decline considerably, effecting a reduction in zooplankton numbers.[54]

REFERENCES

1. **Soeder, C. J.,** *Productivity of Microalgal Systems,* Ser. C, No. 3, Institut fur Biotechnolgie, Kernforschungsanlage Julich GmbH, Julich, Germany, 1981, 9.
2. **Tamiya, H., Shibata, K., Sasa, T., Iwamura, T., and Morimura, Y.,** Effect of diurnally intermittent illumination on the growth and some cellular characteristics of chlorella, in *Algal Culture: From Laboratory to Pilot Plant,* Publ. no. 600, Burlew, J. S., Ed., The Carnegie Institution, Washington, D.C., 1953.
3. **Kok, B.,** Experiments on photosynthesis by chlorella in flashing light, in *Algal Culture: From Laboratory to Pilot Plant,* Publ. no. 600, Burlew, J. S., Ed., The Carnegie Institution, Washington, D.C., 1953.
4. **Märkl, H.,** Modelling of algal production systems, in *Algae Biomass,* Shelef, G. and Soeder, C. J., Eds., Elsevier/North-Holland, Amsterdam, 1980.
5. **Richmond, A. and Vonshak, A.,** *Spirulina* culture in Israel, *Arch. Hydrobiol. Beih. Erg. Limnol.,* 11, 274, 1978.
6. **Laws, E. A., Terry, K. L., Wickman, J., and Chalup, M. S.,** A simple algal production system designed to utilize the flashing light effect, *Biotechnol. Bioeng.,* 25, 2319, 1983.
7. **Grobbelaar, J. U.,** Open simi-defined systems for outdoor mass culture of algae, Unversity of the Orange Free State, Publ., Series C, No. 3, -24-30, 1981.
8. **Richmond, A., Vonshak, A., and Arad, S.,** Environmental limitation in outdoor production of algal biomass, in *Algae Biomass,* Shelef, G. and Soder, C. J., Eds., Elsevier/North-Holland, Amsterdam, 1980.
9. **Vonshak, A., Abeliovich, A., Boussiba, S., and Richmond, A.,** Production of *Spirulina* biomass: effects of environmental factors and population density, *Biomass,* 2, 175, 1982.
10. **Castillo, S. J., Merino, M. F., and Heussler, P.,** Production and ecological implications of algae mass culture under peruvian conditions, in *Alage Biomass,* Shelef, G. and Soeder, C. J., Eds., Elsevier/North-Holland, Amsterdam, 1980.

11. **Wetzel, R. G.,** *Limnology,* W. B. Saunders, Philadelphia, 1975.

12. **Ben-Yaakov, S.,** A portable dissolved oxygen analyzer for the fish farming industry, *Bamidgeh,* 31(3), 69, 1979.

13. **Weiss, R. F.,** The solubility of nitrogen, oxygen and argon in water and seawater, *Deep Sea Res.,* 17, 721, 1970.

14. **Fridovich, I.,** Oxygen in toxic, *BioScience,* 27(7), 462, 1977.

15. **Boyd, C. E., Prather, E. E., and Parks, R. W.,** Sudden mortality of a massive phytoplankton bloom, *Weed Sci.,* 23, 61, 1975.

16. **Abeliovich, A. and Shilo, M.,** Photooxidative death in blue-green algae, *J. Bacteriol.,* 111, 682, 1972.

16a. **Abeliovich, A., Kellenberg, D., and Shilo, M.,** Effect of photooxidative conditions on levels of superoxide dismutase in *Anacystis nidulans, Photochem. Photobiol.,* 19, 379, 1974.

17. **Abeliovich, A.,** Water blooms of blue green algae and oxygen regime in fish ponds, *Verh. Int. Verein. Limnol.,* 17, 594, 1969.

18. **Warburg, O. and Negelein, E.,** Uber den Energiensatz bei der Kohlensaire-assimilation, *Z. Physik. Chem.,* 102, 235, 1922.

19. **Spoeher, H. A. and Milner, H. W.,** The chemical composition of chlorella: effect of environmental conditions, *Plant Physiol.,* 24, 120, 1948.

20. **Radmer, R. J. and Kok, B.,** Photoreduction of O-2 Primes, *Plant Physiol.,* 58, 336, 1976.

21. **Zelitch, I.,** The biochemistry of photorespiration, *Comment. Plant Sci.,* 6, 44, 1973.

22. **Cullimore, J. V. and Sims, A. P.,** An association between photorespiration and protein cat studies with *Chlamydomonas, Planta,* 150, 392, 1980.

23. **Birmingham, B. C., Coleman, J. R., and Colman, B.,** Measurement of photorespiration in algae, *Plant Physiol.,* 69, 259, 1982.

24. **Merrett, M. J. and Armitage, T. L.,** The effect of oxygen concentration on photosynthetic biomass production by algae, *Planta,* 155, 95, 1982.

25. **Emerson, R.,** *Ann. Rev. Plant Physiol.,* 9, 1, 1958.

26. **Pirt, S. J., Lee, Y.-K., Richmond, A., and Pirt, M. W.,** The photosynthetic efficiency of *Chlorella* biomass growth with reference to solar energy utilization, *J. Chem. Technol. Biotechnol.,* 20, 225, 1980.

27. **Hill, R.,** in *Essays in Biochemistry I,* Academic Press, London, 1965, 121.

28. **Warburg, O. and Burk, D.,** *Arch. Biochem.,* 25, 410, 1950.

29. **Brackett, F. S., Olsen, R. A., and Crickard, R. G.,** *Gen. Physiol.,* 36, 563, 1953.

30. **Pirt, S. J.,** Maximum photosynthetic efficiency: a problem to be resolved, *Biotechnol. Bioeng.,* 25, 1915, 1983.

31. **Bassham, J. A.,** Synthesis of organic compounds from carbon dioxide in land plants, in *Biological Solar Energy Conversion,* Mitsui, A., Miyashi, S., Pietro, A. S., and Tamura, S., Eds., Academic Press, New York, 1971, 151.

32. **Goldman, J. C.,** Outdoor algal mass cultures. I. Applications, *Water Res.,* 13, 1, 1979a.

33. **Goldman, J. C.,** Outdoor algal mass cultures. II. Photosynthetic yield limitations, *Water Res.,* 13, 119, 1979b.

34. **van Oorschot, J. L. P.,** *Conversion of Light Energy in Algal Culture,* Doctoral thesis, University of Wageningen, the Netherlands, 1955.

35. **Shelef, F., Oswald, W. J., and Golueke, C. G.,** Kinetics of algal systems in waste treatment: light intensity and nitrogen concentration as growth-limiting factors, SERL REP. 68-n, San Engr. Res. Lab. Univ. California, Berkeley, 1968.

36. **Huessler, P., Castillo, S. J., and Merino, M. F.,** Parasite problems in the outdoor cultivation of *Scenedesmus, Arch. Hydrobiol. Beih. Erg. Limnol.,* 11, 223, 1978.

37. **Grobbelaar, J. U.,** Deterministic production model for describing algal growth in large outdoor mass algal cultures, *U.O.F.S. Publ.,* Series C, No. 3, 1981, 173—181.

38. **Tiwari, J. L., Hobbie, J. E., Reed, J. P., Stanley, D. W., and Miller, M. C.,** Some stochastic differential equation models of an aquatic ecosystem, *Ecol. Modelling,* 4, 3, 1978.

39. **Talling, J. F.,** Photosynthetic characteristics of some freshwater plankton diatoms in relation to underwater radiation, *New Phytol.,* 56, 29, 1957.

40. **Goldman, J. C. and Carpenter, E. J.,** A kinetic approach to the effect of temperature on algal growth, *Limnol. Oceanogr.,* 19, 756, 1974.

41. **Harris, G. P.,** Photosynthesis, productivity and growth. The physiological ecology of phytoplankton, *Arch. Hydrobiol. Beih. Arg. Limnol.,* 10(I), 1, 1978.

42. **Margue, T. H., Friberg, E., Hughes, D. J., and Morris, I.,** Extracellular release of carbon by marine phytoplankton; a physiological approach, *Limnol. Oceanogr.,* 25, 262, 1980.

43. **Stengel, E. amd Reckerman, H.,** *Arch. Hydrobiol.,* 82, 263, 1978.

44. **Gutterman, H. S. and Ben Yaakov, S.,** Department of Electrical Engineering, Ben-Gurion University, personal communication, 1983.

45. **Schanz, F., Allen, E. D., and Gorham, P. R.,** *Can. J. Bot.,* 57, 2443, 1979.

46. **Vincent, W. F. and Silvester, W. B.,** Growth of blue-green algae in the Manukau (New Zealand) oxidation ponds. I. Growth potential of oxidation pond water and comparative optima for blue-green and green algal growth, *Water Res.,* 13, 717, 1979.
47. **Mur, L. R., Gons, H. J., and Van Liere, L.,** Some experiments on the competition between green alga and blue-green bacteria in light-limited environments, *Microbiol. Lett.,* 1, 335, 1975.
48. **King, D. L.,** The role of carbon in eutrofication, *J. Water Pollut. Contr. Fed.,* 42, 2035, 1970.
49. **Shapiro, A.,** Blue-green alga: why they become dominant, *Science,* 179, 382, 1972.
50. **Richmond, A., Karg, S., and Boussiba, A.,** Effect of bicarbonate and carbon dioxide on the competition between *Chlorella vulgaris* and *Spirulina pltensis, Plant Cell Physiol.,* 23(8), 1411, 1982.
51. **Zarouk, C.,** Contribution a l'Etude d'une Cyanophycee. Influence de Divers Facteurs Physiques et Chimiques sur la Crossiance et al Photosynthese de *Spriulina maxima* Ph.D. thesis, University of Paris, Paris, 1966.
52. **Vonshak, A., Boussiba, S., Abeliovich, A., and Richmond, A.,** Production of *Spirulina* biomass: maintenance of monoalgal culture outdoors, *Biotechnol. Bioeng.,* 25, 341, 1983.
53. **Padan. E. and Shilo, M.,** Cyanophages — viruses attacking blue algae, *Bacteriol. Rev.,* 37(3), 343, 1973.
54. **Scott, W. E. and Chutter, F. M.,** Introduction: infections and predators, *U.O.F.S. Publ.,* Ser. C, No. 3, 103—109, 1981.
55. **Muller, U. V. and Sengbusch, P.,** Visualilzation of aquatic fungi (*Chytridiales*) parasitizing on algae by mean of induced fluorescence, *Arch. Hydrobiol.,* 97(4), 471, 1983.
56. **Canter, H. M. and Jaworski, G. H. M.,** A further study of parasitism of the diatom *Fragilaria crotonensis* kitton by chytridiaceaous fungi in culture, *Ann. Bot.,* 52, 549, 1983.
57. **Soeder, C. J. and Maiweg, D.,** Eiflub pilzreicher parasiten auf unsteril massenkulturen von *Scenedesmus, Arch. Hydrobiol.,* 66, 48, 1969.
58. **Muller, U. V., Sengbusch, P., Barr, D. J. S., and Hickman, C. J.,** Chytrids and algae. II. Factors influencing parasitism of *Rhizophydium spaherocarpum* on *Spirogyra, Can. J. Bot.,* 45, 431, 1967.
59. **Blinn, D. W. and Button, K. S.,** The effect of temperature on parasitism of *Pandorina* sp. by *Dangeardia mammilata* B. Schroder in the Arizona mountain lake, *J. Phycol.,* 9, 323, 1973.
60. **Abeliovich, A. and Oickbuch, S.,** Factors affecting infection of *Scenedesmus obliquus* by a *Chytridium* sp. in sewage oxidation ponds, *Appl. Environ. Microbiol.,* 34, 832, 1977.
61. **Canter, H. M. and Jaworski, G. H. M.,** The effect of light and darkness upon infection of *Asterionella formosa* Hassal by the chytrid *Rhizophydium planktonicum* Canter emend, *Ann. Bot.,* 47, 13, 1981.
62. **Daft, M. J., McCord, S. B., and Stewart, W. D. P.,** Ecological studies on algae lysing bacteria in fresh waters, *Freshwater Biol.,* 5, 577, 1975.
63. **Schnepf, E., Hegewald, E., and Soeder, C. J.,** Elektronenmikroskopische beobachtungen an parasiten a us *Scenedesmus*-massenkulturen. IV. Bakterian, *Arch. Microbiol.,* 98, 133, 1974.
64. **Canter, H. M. and Lund, J. W. G.,** The importance of protozoa in controlling the abundance of planktonic algae in lakes, *Proc. Linn. Soc. London,* 179, 203, 1968.
65. **Goulder, R.,** Grazing of the ciliated protozoon *Loxodes magnus* on the alga *Scenedesmus* in a eutrophic pond, *Oikes,* 23, 109, 1972.
66. **Bader, F. G., Tsuchiya, H. M., and Fredrickson, A. G.,** Grazing of ciliates on blue-green algae: effects of ciliate encystment and related phenomena, *Biotechnol. Bioeng.,* 18, 311, 1976a.
67. **Brabrand, A., Faafeng, B. A., Kallqvist, T., and Nilssen, J. P.,** Biological control of undesirable cyanobacteria in culturally eutrophic lakes, *Oecologia,* 60, 1, 1983.
68. **Huang, T. and Wu., H.,** Predation of amoebae on the filamentous blue-green algae, *Bot. Bull. Acad. Sinica,* 23, 63, 1982.
69. **Ho, T. S. S., Alexander, M., and Theresa, S.-S.,** The feeding of amebae on algae in culture, *J. Phycol.,* 10, 95, 1974.
70. **Uhlmann, D.,** Influence of dilution, sinking and grazing rate on phytoplankton populations of hyperfertilized ponds and micro-ecosystems, *Mitt. Int. Ver. Limnol.,* 19, 100, 1971.
71. **Loosanoff, V. L., Hanks, J. E., and Ganaros, A. E.,** Control of certain forms of zooplankton in mass algal cultures, *Science,* 125, 1092, 1957.
72. **Starkweather, P. L.,** Trophic relationships between the rotifer *Brachionus calyciflorusz* and the blue-green alga *Anabaena flos-aguae, Verh. Int. Ver. Limnol.,* 21, 1507, 1981.
73. **Lincoln, E. P., Hall, T. W., and Koopman, B.,** Zooplankton control in mass algal cultures, *Aquaculture,* 32, 331, 1983.
74. **O'Brien, W. J. amd de Noyelles, F., Jr.,** Photosynthetically elevated pH as a factor in zooplankton mortality in nutrient enriched ponds, *Ecology,* 53(4), 605, 1972.
75. **Grobbelaar, J. U.,** Open semi-defined systems for outdoor mass culture of algae, in *Wastewater for Aquaculture,* Grobbelaar, J. U., Soeder, C. J., and Toerien, D. F., Eds., UOFS Publ., Series C, No. 3, 24—30, 1981.
76. **Goldman, J. C.,** Physiological aspects in algal mass cultures, in *Algae Biomass,* Shelef, G. and Soeder, C. J., Eds., Elsevier/North-Holland, Amsterdam, 1980.

ALGAE IN WASTEWATER OXIDATION PONDS

A. Abeliovich

Combined systems for intensive mass algae culture coupled to wastewater treatment in high rate oxidation ponds (HROP) are at present, at best, only at an experimental stage. Thus, HROP cannot be considered as an established technique with standard operational criteria. In addition, local conditions (water quality and composition as well as the climate) play a cardinal role in selecting a specific management regime. Therefore, it is possible to present only general guidelines for the design and construction of an HROP. As shall be described later in some detail, adaptations must be made for each specific case.

GENERAL CONSIDERATIONS

Combining oxidation ponds for wastewater treatment with algal biomass production requires special adaptations, such as changes in the design of the system and its operational regime. Such sacrifices deviate from the optimal and are justifiable only if the product can pay for the difference in price between a conventional oxidation pond system and one which is adapted for algae production.

Conventional oxidation ponds are shallow bodies of water (1 to 2 m deep) with water detention time and organic load designed for achieving effluents of maximal purity at minimal cost. Thus, detention times for reducing approximately 99% of biological oxygen demand (BOD) of domestic wastewater might vary between 10 to 30 days, depending on temperature and the initial organic load.

This process requires practically no energy, since oxygen for decomposing the organic wastes is provided by algal photosynthesis. However, production of algal biomass in these ponds is very slow, and their standing crop is low, about 10 to 100 mg/ℓ of dry weight of algae. Since one of the most energy consuming stages in algal biomass production is the separation of the algae from water, obtaining maximal concentration of algae per volume of pond water is of prime economic importance. Growth of dense cultures of algae can be achieved best in HROP. These are shallow (30 to 40 cm) ponds constructed in the form of a raceway, whereby various means the water is being stirred constantly at a speed sufficient to mix all the algae that tend to settle at the bottom. Depending on the algal species dominating the pond, this can be accomplished at water movement rates between 1 to 30 cm/min. Species dominance depends primarily on the organic load.[1] Indeed, oxidation ponds follow this rule in general, and highly polluted anaerobic water usually permit the development of *Euglena* and *Chlamydomonas* species only. Then, in order of decreasing organic load, appear *Scenedesmus, Chlorella,* and *Micractinium* sp. Development of blooms of other green algae, phytoflagellates, cyanobacteria, diatoms, etc., are generally indicative of low organic load. In general, algal standing crop can reach 1.0 to 1.5 g dry wt/ℓ in the HROP.[2]

In a wastewater treatment plant, based on oxidation ponds, the HROP replaces only one stage, that of the faculatative stabilization pond. There still is a need for an anaerobic pretreatment, and a series of maturation or polishing ponds. The reason for the need for these latter ponds is that algal growth rate is directly proportional to the availability of nutrients and therefore, maintainance of optimal nutrient concentrations is essential for maintenance of maximal growth rate. The polishing ponds are needed to deplete the remaining nutrients from the effluents before discharge. This is true not only for inorganic nutrients such as N or P, but also for organic ones, as it has been estimated that about 50% of the nutrition of the algae in the HROP is heterotrophic.[3]

In the oxidation pond, light for photosynthesis (400 to 700 nm) penetrates the water only to a very limited extent. Usually, about 90% of the light that penetrates the air/water boundary is being absorbed at 10 to 20% of the water depth due to the high density of the algal suspension. The actual depth of this zone is 10 to 50 cm in the standard oxidation pond, being 3 to 8 cm in the HROP. The light that reaches deeper in the water column is practically below the compensation point, particularly as the respiration rates in these ecosystems are relatively high. As daylight is available for only 50% of the time on an average yearly basis, this means that on the average, in a totally mixed reactor, each algal cell receives light for only 5 to 10% of its lifetime, i.e., 90 to 95% of the time it remains in the dark.

Nevertheless, if all other factors are maintained at optimal levels, algal growth rate can be maintained in this system at a doubling time of 3 to 4 days, for very long periods of time.[2] Algal growth under these conditions depends, therefore, to a large extent upon the supply of organic nutrients available in the untreated wastewater and upon an external supply of oxygen which is introduced into the deep layers of the water by mixing. The algae must compete with bacteria on available nutrients. Thus apparently, only those algal species capable of heterotrophic growth survive in an oxidation pond. Cyanobacteria, which have difficulties adapting to dark heterotrophic nutrition can be found and cultivated only after most, or all BOD has been depleted or completely removed.[4]

BIOLOGICAL STEADY STATE

An HROP is expected to perform two tasks, which are in some respects contradictory: production of maximal algal biomass coupled with maximal wastewater purification. It is expected to perform as a stable and reliable system year round, irrespective of chemical changes that result from fluctuations in the composition of wastes. The pond should operate, if possible, without changes in the algal species which dominate the pond, to which the harvesting technique was tailored and the potential use of the product specified. Obviously, frequent changes in species dominance will introduce further complications in pond management.

Maintaining stable biological equilibrium in such a complicated system may seem to be next to impossible. In fact due to the extreme conditions prevailing in the HROP (high organic load, extreme diurnal fluctuations in O_2 and pH, and short detention times), only a very limited number of algal species survive. Thus, if pond management is carried out correctly, a stable population can be maintained without difficulty. The specific alga that will dominate the HROP is a function of local conditions. It can hardly be predicted in advance and no existing technique is available for maintaining a specific predetermined composition of algal species.

The HROP can be regarded as a large chemostat, and the first prerequisite for establishing a stable algal community in it is the need for an equilibrium between the algal growth rate and the rate of algal dilution by raw wastewater. Optimal algal growth rate means optimal rate of photosynthesis at the upper water layer and oxygenation of the total water volume. This is accomplished with algae growing either photosynthetically or heterotrophically, utilizing organic matter (see below) and molecular oxygen, which is available in this ecosystem primarily through photosynthesis and partly by mechanical aeration. Thus, maintainence of optimal photosynthesis and optimal O_2 concentrations in the pond are the key to a successful and stable pond operation. Two factors that are important in this respect are ammonia and BOD concentrations. Ammonia is known to be an uncoupler of photosynthesis in isolated chloroplasts. In whole algal cells, nonionized ammonia was also reported to inhibit photosynthesis.[5,6] The exact intracellular concentration causing this inhibition is very difficult to establish, but external concentrations of 2 mM at pH 8.1 inhibit approximately 50% of photosynthetic O_2 evolution in any of the algal species tested so far.[5] Therefore, a main concern in deciding on the detention time of water in the HROP should be to avoid

RETENTION TIME (hrs)

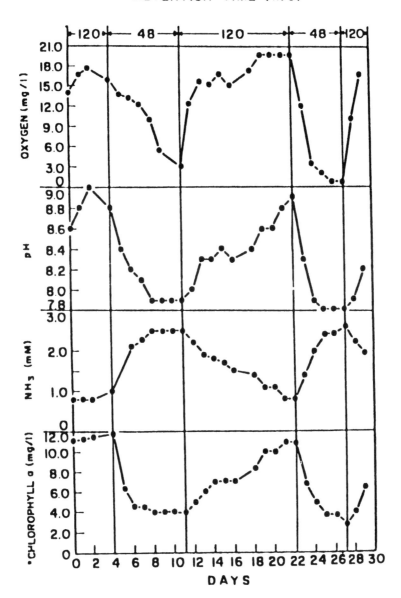

FIGURE 1. Chlorophyll-a, ammonia, and oxygen concentrations, and pH values in a high-rate sewage oxidation pond during operation at short and long detention times (48 and 120 hr). pH and oxygen concentrations were measured at noon. Samples for the determination of ammonia and chlorophyll-a were taken into the laboratory and analyzed within 30 min. (From Abeliovich, A. and Azov, Y., *Appl. Environ. Microbiol.*, 31, 801, 1976. With permission.)

this critical concentration of ammonia as shown in Figure 1. As the concentration of ammonia in wastewater differs from one place to another, the proper detention time has to be determined locally.

Localities in Israel with wastes of low ammonia concentrations have established detention times as low as 48 hr. In contrast, in localities where wastewater contains high ammonia, any attempt to operate a pond at too short detention times (below 100 hr) resulted in an increase of the ammonia to a concentration that caused a washout of the algae from the pond.[5] Maintenance of optimal BOD is crucial. As already mentioned, organic load is a primary factor in determining the dominating alga.[1,7] The presence of *Euglena* or *Chlamydomonas* means an overloaded system, while the appearance of diatoms or cyanobacteria usually means reduced BOD as a consequence of too long detention time (the effect of which will be considered later). Between these extremes, usually, any of the following species of *Chlorella, Scenedesmus,* or *Micractinium* will establish itself as the dominating species in the great majority of cases.[1,7] Photosynthesis elevates the pH in the pond and if there exists a high concentration of ammonia, i.e., >2.0 mM, it leads to inhibition of photosynthesis when the pH reaches 8.1 to 8.3 (Figure 1). The only mechanism available for lowering the pH in this biotope is respiration, which in turn enables photosynthesis to proceed undisturbed, as long as the pH is kept below 8.1. Under optimal conditions, the rate of photosynthesis is 5 to 6 times faster than respiration. The two processes may be balanced and maintained at the same rate, as photosynthesis is restricted to the surface, while respiration takes place throughout the water volume. Therefore, when ammonia concentration is high, a careful balance should be maintained between its concentration and the concentration of BOD, to enable the continued lowering of the pH via respiration, so that photosynthesis can proceed. Paradoxically, therefore, too low BOD in the HROP means if ammonia concentration is high, development of anoxic conditions and washout of the algal population, due to inhibition of photosynthesis.[6,8]

If a stable steady state is to be maintained in the HROP, it is very important to avoid development of a complicated food chain in the pond. Rotifers, copepods, larvae of mosquitos and flies, and various waterbugs mean an unstable algal population which is at the mercy of population dynamics of herbivorous grazing zooplankters and their predators.[9] The result is usually sudden and frequent collapses and disappearances of the algal populations. To avoid this type of disturbances, the oxygen regime should be established in such a way as to produce diurnally anaerobic conditions in the pond for a short period. This will prevent development of any significant food chain based on aerobic organisms. In addition, we have noticed that infections of the algae by parasitic fungi (mainly various species of chytridia) are very sensitive to oxygen levels in the water.[8] Once constant aerobic conditions prevail, massive infections occur, destroying all susceptible algae. In this case too, maintenance of anaerobic or microaerophilic conditions for 1 hr every night prevents these infections. This also can be achieved by carefully controlling detention time vs. BOD.

DESIGN PARAMETERS

Pond Construction

In general, ponds are constructed as a raceway of variable size. However, two parameters must be chosen carefully. Depth of the pond should be such that under optimal growth condition, i.e., maximal output of algal biomas, at steady state, the compensation point should be at approximately 10 to 20% of water depth. This will ensure proper oxygenation of the pond water without too much oversaturation, which frequently exerts harmful effects both in terms of algal viability[10] or retardation of growth. In determining the ratio between the photic and a-photic zones in the pond, the prime consideration should be given to production of microaerophilic conditions in the pond for a short duration toward the end of the night. This, as already mentioned, is essential for the prevention of development of a food chain based on obligate aerobes. The pond should therefore be constructed so that water level can be regulated between 30 to 50 cm. The second critical parameter is designing a HROP is the flow rate of the water in the raceway. The aim of this is to provide total mixing of the algal population, as well as aeration if this is needed. As this is an operation requiring high energy input, every attempt should be made to minimize the flow rate to the lowest possible level, which in turn depends primarily on the buoyancy of the dominating algal species. The author has operated very stable HROP dominated by *Scenedesmus obliquus* with a flow rate of 1 to 2 cm/sec, yet, other algae may require much higher flow rates, up to 15 cm/sec.[11,12] The simplest and apparently the most efficient way for moving and mixing the water is by a paddlewheel,[13] although other techniques, airlift or conventional pumps, have been described.[14,15]

Another point that should be taken into consideration when determining the pond size is the difference in performance of the pond in winter and summer seasons, with respect both to the algal growth rate and the water treatment efficiency. These can slow down by a factor of 4 during the cold season.[6]

Extra ponds should be constructed for expanding the overall pond area during the cold season. The exact ratio between winter and summer pond area has to be determined locally, based on experimentation. Another possibility is to maintain the ponds much deeper during winter. This is possible if the organic load per unit area is not too high. Once again, exact figures cannot be given as these must be determined locally by optimizing pond depth to the maximal organic load and the lowest temperatures prevailing in the pond.

Nutritional Considerations

In general, the place of HROP in a layout of wastewater treatment system should be such as to avoid any nutrient limitations for algal growth, which would cause reduced productivity. The most expensive stage in the whole operation is separation of the algae from the water. Anything less than optimal algal growth and maximal possible standing crop means a sharp increase in operational costs.

After separation of algae, therefore, the effluents from the HROP are still rich in nutrients and should pass through further treatment stages for stripping of the remaining nutrients. It has been shown[3] that in HROP fed by raw sewage, there are no nutritional limitations with respect to carbon, nitrogen, and phosphorus. When fed by diluted wastewater or by secondary effluents, however, carbon limitations might develop.[12] Up to 50% of the carbon assimilated by the algae in HROP is by direct heterotrophic nutrition.[3] Domestic raw sewage might contain, among other things, about 80 to 100 mg of total carbohydrate as glucose equivalents per liter, and the relatively small number of bacteria present in it (about $1 \times 10^6/m\ell$) cannot account for its mineralization in the HROP.[3] Also, direct utilization of solar energy cannot account solely for the extremely high productivity rates which have been reported in several studies of HROP (as high as 40 g $C/m^2/day$).[3,5,17]

Therefore, for reasons outlined earlier, it may be safely assumed that a very significant part of algal nutrition in the HROP is heterotrophic and also that there is a selective pressure in the pond in favor of algae capable of efficient heterotrophic growth. In addition, this estimate is supported by the following considerations. Suppose the total BOD of domestic sewage is 600 to 700 and that approximately 60% is in soluble form, and 90 to 99% of this quantity disappears within a 4 day retention time, being equivalent to around 50 mg $C/\ell/$ day. Given an average 40 cm depth for HROP, this figure corresponds to approximately 20 to 25 g C/m/day. Furthermore, *Scenedesmus obliquus*, for example, can utilize at least four amino acids instead of inorganic N supply: glycine, arginine, leucine, and phenylalanine.[26] For these reasons, the HROP should always be optimized primarily for algae production and not for effluent quality, as it has to operate at optimal nutrient supply.

Physical Parameters

Obviously, the operation of a photosynthetic high rate oxidation pond depends on optimal solar radiation and temperature, and much literature is available on the relative importance of both parameters. Different algal species, however, will respond differently to different parameters that affect growth — temperature and light, organic load, availability of inorganic nutrients (N, P), salinity, and water hardness. These factors affect the selection of species that will dominate the HROP. As various algae react differently to specific combinations of light-temperature regimes, the outcome is contradictory data which are presented in the literature on the relative importance of light and temperature. One finds articles stating that temperature has much greater effect than light on performance of oxidation ponds,[8] while others[19,20] claim that at lower organic loads, light plays the major role while temperature does not change algal growth rates at a range of 5 to 33°C, but there is a change in species dominance. These are just two of many extreme examples, reflecting the simple fact that specific local conditions play the dominant role in determining the relative effect of light and temperature on pond performance. If the organic load is such that it will permit growth of several algal species, each with different temperatures optimum for growth, temperature will not affect pond performance, while if the organic load is such that it inhibits the growth of most algal species so that only one species dominates the pond, its peak performance will coincide with this alga's optimal growth temperature (as light is, under normal operational conditions, a limiting factor in this ecosystem).

There is no reliable way, therefore, for predicting the relative importance of light and temperature upon a specific set up, as well as optimizing, in advance, light and temperature for maximal algal growth and pond performance, except in broad and general terms.

pH and O$_2$

The interrelationships between pH, ammonia, and photosynthesis have already been described in detail.[5] As previously mentioned, if ammonia concentration in the pond exceeds 2 mM, photosynthesis depends on the initial pH of the raw wastes entering the HROP and their buffer capacity. High initial pH (>8.0), coupled with high concentrations of ammonia and low BOD load force long detention times, while low initial pH (<7.5) and low concentrations of ammonia enable a stable operation of the HROP at short detention times. The difference between the two extremes is very significant, ranging between 2[9] to 4 or even 5 days.[5] When photosynthesis is optimal, pH values in the HROP might reach values well above 10 at the surface of the pond without having any adverse effects on the stability of the algal community or biomass production.[5]

Oxygen concentration in the pond is, of course, intimately coupled to the pH. High rates of photosynthesis mean not only high oxygen concentrations (>30 mg/ℓ), but also a high pH. This combination is potentially dangerous to the algae because of the possibility of photooxidative damage to the photosynthetic apparatus and other vital functions in the algal cell.[10] Here, proper mixing of the water serves to control this problem in several ways. It increases transfer of oxygen from the supersaturated medium to the atmosphere, as well as transfer of oxygen for algal and bacterial respiration below the photic zone. Proper mixing is also important for the supply of oxygen during the night and the minimal mixing rate should be determined and used for each set of conditions separately. A general outline of optimal operation parameters is presented in Table 1.

HARVESTING THE BIOMASS

There is by no means a generally accepted harvesting technique to suit all situations. At present, the available means and devices for microalgae harvesting include centrifugation, coagulation, filtration, and microstraining. The method of choice depends on several factors such as the suitability of a particular alga to a specific technique and the price per unit product.

Centrifugation

A self-cleaning plate separator was found to effectively concentrate all conceivable types

Table 1
COMPILED PARAMETERS FOR A STABLE OPERATION OF HIGH RATE OXIDATION PONDS (DOMESTIC SEWAGE ONLY)[2,3,5,6,15-17,25]

Raw sewage		Detention time (days)			Effluent	
Ammonia (mg/ℓ)	Total BOD (mg/ℓ)	8—14°C	22—35°C	Algal conc. g/ℓ dry wt.	Ammonia (% of initial conc.)	BOD (% removed; less removed algal biomass)
80—90	400—500	10—12		0.25—0.5	70—80	90—99
40—50	300—400	6—7		0.1—0.2	40—50	90—95
80—90	400—500		4	0.5—1.0	80—90	90—99
40—50	300—400		2—3	0.2—0.3	40—50	90—95

of microalgae.[21] The average total soluble solids (TSS) of the concentrate is about 12%, but the energy demand of the process, 1 kWh/m³, is high. Other centrifuge types are also available, such as nozzle centrifuge or screw centrifuge. These means require a preconcentration stage.[21]

Microstraining

Microstrainers consist of a rotary drum covered by a straining fabric. The algal suspension enters axially and the algae are trapped on the inner surface of the drum. The system operates with a very low pressure difference between the inner and outer sides of the drum and a backwash collects the algae from the screen.[22] The operational costs are low, as energy consumption is only 0.2 kWh/m³, but the algal concentration is also low: 1.5% TSS.[21] In addition, several algae, among them strains of *Scenedesmus* and *Chlorella,* could not be harvested by microstraining, easily passing through the filter.[23]

Vibrating Screens

Whenever the size of algal cell or colony permit, this is a relatively low cost process with good reliability.[21] For example, it is possible to concentrate *Coelastrum* sp. to 7 to 8% TSS on a vibrating screen.[21] The use of small pore size >1000 mesh filters, when available, will enable using this apparatus for concentrating unicellular microalgae.

Chemical Flocculation

A comprehensive description of the process was given by Moraine et al.[24] According to Benemann et al.,[22] chemical coagulation with lime or alum followed by sedimentation or dissolved air flotation is the most reliable and cost effective method for harvesting algae, costing approximately $100/1000 m³. The slurry obtained by this method contains typically 7 to 8% dry matter. According to Berk,[23] acidification of the slurry to pH 3.5 and boiling it at 100°C enables further easy filtration, resulting in a cake of 14 to 19% dry matter. This procedure might be particularly useful, since no matter what drying procedure is chosen, the slurry has to be sterilized because of health considerations. Usually, sterilization and drying are carried out on a drum dryer in an expensive process. Autoclaving the slurry with subsequent simple filtration, however, produces a cake containing about 80% water. This can be used directly as a protein additive for production of animal feed when mixed with other dry ingredients, without having to be totally dried. Such a procedure should significantly improve the economic attractiveness of algae production in HROP.

REFERENCES

1. **Palmer, C. M.**, A composite rating of algae tolerating organic loading, *J. Phycol.*, 5, 78, 1969.
2. **Abeliovich, A.**, Factors limiting algae growth in high rate oxidation ponds in *Algal Biomass*, Shelef, G. and Soeder, C. J., Eds., Elsevier/North Holland, Amsterdam, 1980.
3. **Abeliovich, A. and Weisman, D.**, Role of heterotrophic nutrition in growth of the alga *Scenedesmus obliquus* in high rate oxidation ponds, *Appl. Environ. Microbiol.*, 35, 32, 1978.
4. **Shelef, G., Azov, Y., Moraine, R., and Oron, G.**, Algal biomass production as an integral part of a wastewater treatment and reclamation system, in *Algal Biomass Production and Use*, Shelef, G. and Soeder, C. J., Eds., Elsevier/North Holland, Amsterdam, 1980, 163.
5. **Abeliovich, A. and Azov, Y.**, Toxicity of ammonia to algae in sewage oxidation ponds, *Appl. Environ. Microbiol.*, 31, 801, 1976.
6. **Abeliovich, A.**, The effects of unbalanced ammonia and BOD concentrations on oxidation ponds, *Water Res.*, 17, 299, 1983.
7. **Azov, Y., Shelef, G., Moraine, R., and Levy, A.**, Controlling algal genera in high rate wastewater oxidation ponds, in *Algal Biomass Production and Use*, Shelef, G. and Soeder, C. J., Eds., Elsevier/North Holland, Amsterdam, 1980, 245.
8. **Abeliovich, A. and Dickbuck, S.**, Factors affecting infection of *Scenedesmus obliquus* by a *chytridium sp.* in sewage oxidation ponds, *Appl. Environ. Microbiol.*, 34, 832, 1977.
9. **Groeneweg, J., Klein, B., Madin, F. H., Runkel, K. H., and Stengel, E.**, First results of outdoor treatment of pig manure with algal bacterial systems, in *Algal Biomass Production and Use*, Shelef, G. and Soeder, C. J., Eds., Elsevier/North Holland, Amsterdam, 1980.
10. **Abeliovich. A. and Shilo, M.**, Photooxidative death in blue-green algae, *J. Bacteriol.*, 111, 682, 1972.
11. **Oswald, W. J.**, Complete waste treatment in ponds, *6th Int. Conf. Water Pollut. Res., Jerusalem*, Pergamon Press, Oxford, 1972, B361.
12. **Azov, Y., Shelef, G., and Moraine, R.**, Carbon limitation of biomass production in high rate oxidation ponds, *Biotechnol. Bioeng.*, 24, 579, 1982.
13. **Soong, P.**, Production and development of *Chlorella* and *Spirulina* in Taiwan, in *Algae Biomass Production and Use*, Shelef, G. and Soeder, C. J., Eds., Elsevier/North Holland, Amsterdam, 1980.
14. **Persoone, G., Movales, J., Verlet, H., and De Paum, N.**, Air lift pumps and the effect of mixing on algal growth, in *Algal Biomass Production and Use*, Shelef, G. and Soeder, C. J., Eds., Elseveir/North Holland, Amsterdam, 1980.
15. **Abeliovich, A.**, Operation of a deep well mixed high rate photosynthetic oxidation pond, *Water Res.*, 13, 281, 1979.
16. **Azov, Y., Shelef, G., Moraine, R., and Oron, G.**, Alternative operating strategies for high rate sewage oxidation ponds, in *Algal Biomass Production and Use*, Shelef, G. and Soeder, C. J., Eds., Elsevier/North Holland, Amsterdam, 1980.
17. **Shelef, G., Schwartz, M., and Schechter, H.**, Prediction of photosynthetic biomass production in accelerated algal bacterial wastewater treatment systems, in *Proc. 6th Int. Conf. Water Pollut. Res. Jerusalem*, Pergamon Press, Oxford, 1972, PA591.
18. **Arthur, J. P.**, The development of design equations for facultative waste stabilization ponds in semi-arid areas, *Proc. Inst. Civ. Eng.*, 71(2), 197, 1981.
19. **Goldman, J. C. and Ryhter, J. H.**, Temperature influenced species competition in mass cultures of marine phytoplankton, *Biotechnol. Bioeng.*, 18, 1125, 1976.
20. **Goldman, J. C.**, Biomass production in mass cultures of marine phytoplankton at varying temperatures, *J. Exp. Mar. Biol. Ecol.*, 27, 161, 1977.
21. **Mohn, F. H.**, Experiences and strategies in the recovery of biomass from mass cultures of microalgae, in *Algae Biomass Production and Use*, Shelef, G. and Soeder, C. J., Eds., Elsevier/North Holland, Amsterdam, 1980.
22. **Benemann, J., Koopman, B., Weissman, J., Eisenberg, D., and Goebel, R.**, Development of microalgae harvesting and high rate pond technologies in California, in *Algae Biomass Production and Use*, Shelef, G. and Soeder, C. J., Eds., Elsevier/North Holland, Amsterdam, 1980, 457.
23. **Berk, Z.**, Thermal dewatering of algal slurries, in *Algae Biomass Production and Use*, Shelef, G. and Soeder, C. J., Eds., Elsevier/North Holland, Amsterdam, 1980, 571.
24. **Moraine, R., Shelef, G., Sandbank, E., Bar-Moshe, Z., and Shvartzburg, L.**, Recovery of sewage borne algae: flocculation, flotation and centrifugation techniques, in *Algal Biomass Production and Use*, Shelef, G. and Soeder, Eds., Elsevier/North Holland, Amsterdam, 1980, 531.
25. **Azov, Y. and Shelef, G.**, Operation of high rate oxidation ponds. Theory and experiments, *Water Res.*, 16, 1153, 1982.
26. **Abeliovich, A. and Weisman, D.**, Unpublished data.

NUTRITIONAL PROPERTIES OF MICROALGAE: POTENTIALS AND CONSTRAINTS

E. W. Becker

INTRODUCTION

Intense efforts have been made to develop new, alternative, and unconventional protein sources in anticipation of increasing world population and insufficient protein supplies. The idea of utilizing proteins from bacteria, yeast, fungi, and algae was considered as a distinct possibility for at least partially meeting the global protein shortages. The term single cell protein (SCP) was coined in 1966 to designate the protein obtained from microbial sources. More recently, the term biomass protein (BMP) has been introduced to describe the protein production of biomass from different microbial sources. Both terms are not quite correct, because SCP from algae is definitely more than protein, the algal biomass consisting of a whole spectrum of compounds typical for living organisms, viz., protein, peptides, free amino acids, amines, nucleic acids, various carbohydrates, lipids, vitamins, minerals, etc.

The concept of utilizing SCP is not completely new, as such proteins are already being used in foods and feed in different regions of the world at varying levels: the commerical interest in these proteins, however, is new. Among the different sources of unconventional proteins, algae probably have the longest history, especially as food.

The cyanobacterium *Spirulina* was already eaten by the Aztecs in ancient Mexico, who called it "tecuitlatl".[1,249] The same algae forms part of the food of the Kanembou tribe north of Lake Chad in Central Africa, who make the dried algae ("dihe") into a sauce called "biri". Besides *Spirulina,* there are several other algal species, mainly marine forms, which are regularly incorporated into various food preparations, especially in Asian countries.

Although the utilization of algae and also other forms of SCP has some tradition, it was only in the last decade that the fast-growing market for SCP prompted international organizations such as the International Union of Pure and Applied Chemistry (IUPAC) and the Protein-Calorie Advisory Group (PAG) of the United Nations to publish recommendations and guidelines for the utilization of these novel proteins. The guidelines stipulate various quality criteria which should be fulfilled before the particular product can be declared safe and suitable for use as food and feed (Table 1).

These guidelines are intended to serve as a general recommendation rather than as a series of mandatory procedures. In particular, the extent of animal testing considered necessary prior to undertaking trials in humans will depend on the protein product. Products intended for use in animal feeds may not require as extensive testing as stipulated for human foods; foods derived from animal sources, however, must be examined for the possible presence of residues transmitted from animal feeds.

In the following sections, the criteria compiled in Table 1 are discussed in relation to the nutritional quality and other prerequisites for algal biomass to be utilized as animal feed or as human food. These include the following points:

- Proximate chemical composition (nitrogen, amino acid profile, fat, crude fiber, carbohydrates, etc.)
- Biogenic toxic substances (phycotoxins, nucleic acid, hemagglutinin, other toxicants)
- Nonbiogenic toxic substances (heavy metals, pesticides, residues from harvesting and processing)
- Protein quality studies (Protein Efficiency Ration, PER; Net Protein Utilization, NPU; Biological Value, BV; Digestibility Coefficient, DC)

Table 1
RECOMMENDED EVALUATION FOR ESTABLISHING SAFETY OF SCP

1. General characteristics
   a. Nature, biological properties, and harmlessness of the microorganisms, including description of tests to assure constancy and purity of the strain culture
   b. Characteristics of the substrate and sources of principal nutrient supplement and of other special agents used in the process
   c. Conditions for concentration and drying
   d. Constancy and sanitary quality of the product
2. Product characteristics
   a. Microscopic morphology
   b. Physical properties: solubility, wettability
   c. Detailed chemical composition: dry matter, total lipids, nonsaponifiable fat compounds and percentage of fatty acids, phospholipids, fat-soluble pigments; composition of hydrocarbons; composition of carbohydrates; composition of nitrogen fraction (total N, protein N, amino acid profile, nucleic acid N); composition of inorganic macro- and micro-elements; fat and water-soluble vitamins; any other organic substances present
3. Nutritional studies in rodents
   a. Digestibility of the nutrients, protein efficiency ratio
   b. Biological value, net protein utilization
   c. Digestible and metabolizable energy
   d. Supplementation properties in conventional feeding stuffs
4. Feeding trials in target animals
   a. Tests for maximum levels of incorporation in normal rations
   b. Acceptability studies
   c. Investigation of possible depressant effects
5. Safety studies
   a. Studies on bacteriological and mycological purity and harmlessness of any microorganisms in the product
   b. Analyses for contaminations with organic and inorganic pollutants
   c. Short- and medium-term safety studies with rodents, pigs, fowls (egg production, hatchability), and other target species
   d. Long-term studies including investigations of possible carcinogenicity with at least two species
   e. Reproduction studies
   f. Multigeneration studies
   g. Teratogenicity studies
   h. Mutagenicity studies
6. Clinical studies with human subjects

- Biochemical nutritional studies (in vitro tests on protein and carbohydrate digestibility)
- Nutritional studies by protein-depletion and -repletion experiments
- Supplemental value of algae to conventional food sources
- Sanitary analyses (microbial examinations for contaminations)
- Safety evaluations (short- and long-term feeding trials with experimental animals)
- Clinical studies (tests for safety and suitability of the product for human consumption)
- Acceptability studies (organoleptic factors, recipes, etc.)

It should be stressed that, in general, algal biomass is not intended as the sole source of protein for humans or animals, but as a supplement to the basic diet.

As shall be discussed in detail later, the nutritive value of the algae depends to a considerable degree on the method of postharvest processing. With the exception of the blue-green algae *Spirulina*, most of the other types of algae have a relatively rigid cell wall which makes the untreated algae indigestible to nonruminants. If the biomass is to be used as human food or feed for monogastric animals, the algal cells have to be ruptured, either by chemical methods such as butanol-induced autolysis, breaking of hydrogen bonds with phenol, formic acid, or urea, or by physical methods of disruption such as boiling, spray-drying, drum-drying, sun-drying, etc. The major problem encountered in chemical treatments is the necessity of recovering the solvent and ensuring that the final product has not been rendered toxic.

CHEMICAL COMPOSITION

The chemical composition of algae gives some basic information on the nutritive potential of the algal biomass. In many algae the chemical composition can be modified relatively easily by chemical means, i.e., nitrogen or phosphorus depletion in the medium, or by physical changes such as osmotic pressure, radiation intensity, temperature, and physiological parameters like age of culture, population density, light or dark growth, etc.

Spoehr and Milner[2] were among the first to give detailed information on the effect of environmental conditions on the chemical composition of *Chlorella*. Changes in the composition pattern of basic proteins and the amino acid profile of the same algae are described by Kanazawa and Kanazawa.[3,4] Variation in carbohydrates and fatty acids during the growth cycle of *Chlorella* has been reported by Matsuka et al.[5] In addition, Altmann et al.[6] reported variations in soluble protein during the growth cycle. Manipulations of the protein in mass cultured algae are described by Mostert and Grobbelaar.[7] Changes in carbohydrate, protein, and lipid content in green algae have been studied using the techniques of synchronous culture. Tamiya[8] described two distinct forms of *Chlorella* possessing widely different characterisitics. A young form developed during the dark phase, is smaller in size and richer in chlorophyll and protein than the other form, which grows during the light phase. Environmental changes that modify algal cell wall chemistry have also been described by Aaronson et al.,[9] Durant and Jolly,[10] and others.[11-14,239-241]

The effect of transferring exponentially growing cells to a medium lacking a source of assimilable nitrogen are of interest from both the physiological and the ecological points of view. In cells treated this way the flow of carbon fixed in photosynthesis is switched from the path of protein synthesis to that leading to formation of carbohydrates. Cell division does not continue after cell nitrogen has fallen below a limited value. On prolonged nitrogen starvation, lipids begin to accumulate in the cells, probably as a result of the lipid-synthesizing enzyme systems which are less susceptible to disorganization than the carbohydrate-synthesizing system. Thus, the major proportion of the fixed carbon is bound in lipids.

One always should keep in mind that algal cultivation basically represents a special form of agriculture and is exposed to various environmental influences. The chemical composition of cells within one algal species may vary to some extent depending on the growth conditions, therefore, it is not suprising that different chemical analyses for the same algae are found in the literature.

The high protein content of microalgae was one of the reasons for selection of these organisms as an unconventional protein source. The data on the gross composition of selected algae, summarized in Table 2, confirm that the bulk of these algae is crude protein, varying from about 30 to 65% of the dry matter.

Besides protein, there are other cellular consitituents which are equally important in determining the value of the algal biomass as food, feed, or as a source of chemicals such as carbohydrates, lipids, or vitamins.

Various analyses on algal constituents have been published in the literature.[9,15] A compilation of recent data on gross chemical composition of different algae is given in Table 2.

Protein

The use of the term "crude protein", i.e., N × 6.25 is often criticized, but it is a useful criterion commonly employed in evaluating food and feed. Since substantial parts of the total nitrogen in algae, as well as in other forms of SCP, consist of nonprotein nitrogen

Table 2
GROSS CHEMICAL COMPOSITION OF DIFFERENT ALGAE
(% OF DRY MATTER)

Alga	Protein	Lipids	Carbohydrates	Nucleic acid	Ref.
Spirulina platensis	46—50	4—9	8—14	2—5	16
	62.5	3.0	8.5	3.9	17
Spirulina maxima	65.0	2.0	20.0		18
	60—71	6—7	13—16	2.9—4.5	19
Chlorella vulgaris	51—58	14—22	12—17	4—5	20
Chlorella pyrenoidosa	57.0	2.0	26.0		9
Scenedesmus obliquus	50—56	12—14	10—17	3—6	21
	52.0	9.0	12.5	6.0	22
Scenedesmus quadricauda	47.0	1.9			23
Dunaliella salina	57.0	6.0	32.0		24
Dunaliella bioculata	49.0	8.0	4.0		25
Synechococcus sp.	63.0	11.0	15.0	5.0	20
Euglena gracilis	39—61	14—20	14—18		26
Prymnesium parvum	28—45	22—38	25—33	1—2	27
Hormidium sp.	41.0	3.8			23
Ulothrix sp.	45.0	1.1			23
Uronema gigas	58.0	1.7			23
Stigeoclonoiuim sp.	51	1.2			23

(NPN) — which to a large extent originates from nucleic acids — the calculation of crude protein by multiplying total nitrogen by 6.25 produces an overestimate of the proportion of nucleic acid relative to protein. Thus, for correct calculations, purine nitrogen must be determined separately and the nucleic acid content calculated by appropriate methods. Since the nitrogen content of pyrimidines is about 40% that of purines and both are present in equimolar amounts in most nucleic acids, the purine nitrogen should be multiplied by the factor 1.4 in order to obtain nucleic acid nitrogen. A "corrected protein nitrogen" is obtained by subtracting 1.4 times purin nitrogen from crude protein nitrogen. Multiplying by the proper factors (6.25 for protein, 9.0 for nucleic acids from purine nitrogen), the values for "corrected protein" and "nucleic acids" are obtained.[28] Subbulaksmi et al. found a nonprotein nitrogen contents of approximately 12% of the total nitrogen in *Scenedesmus*.[29] Gibbs and Duffus[30] reportes that in *Dunaliella* 6% of the crude nitrogen is NPN. According to Durand-Chastel,[31] 11.5% of the total nitrogen in *Spirulina maxima* derives from nucleic acid nitrogen. As an average figure, it can be assumed that in algae about 10% of the crude protein consist of NPN.

In *Spirulina,* about 20% of the protein consists of C-phycocyanin, which serves as a nitrogen storage compound in the cell. C-phycocyanin differs from other storage materials since its synthesis occurs largely during the logarithmic growth phase of the algae and ceases during the stationary phase, while other storage products are usually accumulated at the end of the growth phase or in the final stage of the life cycle.

The role of C-phycocyanin as storage substance for nitrogen can be demonstrated by transferring *Spirulina* into a nitrogen-free medium. During nitrogen starvation the algal cells continue to grow at the initial growth rate for one generation. The amount of total protein does not change, while the concentration of C-phycocyanin decreases at a rate (30 to 50%) sufficient to maintain the concentration of nonphycocyanin protein at a constant level.[32,33]

Amino Acid Composition

The nutritive quality of proteins is determined by the content, proportion, and availability of its amino acids. While plants are capable of synthesizing all amino acids, animals are limited to the biosynthesis of certain amino acids only (nonessential amino acids). The remaining (essential) amino acids, i.e., isoleucine, leucine, lysine, methionine, phenylalanine, threonine, tryptophan, and valine have to be supplied through food. Arginine and histidine are considered semi-essential, because they have to be provided exogenously during growth or upon the appearance of signs of deficiency. Special cases are cystine and tyrosine, which can be synthesized by the organism from the essential amino acids methionine and phenylalanine if the intake through food is insufficient. Different animal species differ quantitatively and in part also qualitatively in their amino acid requirements.

Amino acid profiles, as determined either microbiologically or chromatographically on protein hydrolizates fail to differentiate between the total amount and the degree of availability of the amino acids. This is of special importance for methionine and lysine. During prolonged storage or heat treatment of biomass, the free ε-amino group of lysine tends to form compounds with reducing carbohydrates (Maillard reaction), making the lysine unavailable for digestion. This aspect has to be considered in connection with the different drying steps applied during the processing of algal biomass. The effects of heat treatment, pH, and reducing sugars on a product from *Spirulina* as well as on the availability of lysine and methionine was studied by Adrian et al.,[34,35] who found a considerable reduction of the digestibility of amino acids due to these treatments. A chemical method commonly employed for the determination of available lysine has been reported by Carpenter.[36]

The amino acid profile of a given protein allows certain conclusions as to its value.[37] The calculation of the nutritional quality of a protein based on amino acid analyses presupposes detailed information on the required amount of (essential) amino acids in the target organisms in comparison with a reference composition as a standard. The problem is that the requirements differ according to strain, age, sex, etc. In addition, nearly all published amino acid profiles lack statistical analysis making comparisons between various reports difficult. Two different calculations are common for estimating the value of a protein from its amino acid composition, i.e., chemical score and essential amino acid index (EAA). The chemical score is based on the ratio of the essential amino acids in greatest deficit in the sample compared with its content is a reference protein.[38]

$$\text{Chemical Score} = \frac{\text{amount of limiting amino acid in sample}}{\text{amount of amino acid in reference}} \times 100$$

The EAA is predicted on the hypothesis that the biological value of a protein is a function of the levels of all the essential amino acids in relation to their content in the reference protein:[39]

$$\text{EAA} = \sqrt[n]{\frac{100a}{a_r} \times \frac{100b}{b_r} \times ,..., \times \frac{100n}{n_r}}$$

$a,b,...,n$ = Essential amino acid concentration in the sample
$a_r,b_r,...,n_r$ = Essential amino acid concentration in the reference

It was found that the chemical score, based as it is on a single limiting amino acid, tends to underestimate the biological value of the protein, whereas the EAA index gives figures more closely correlated with the biological value determined in feeding tests. Depending on the consumer (animal, human), different amino acids are regarded as essential. This fact is considered by the latter calculation, because only those amino acids are written under the root which are essential in the particular case. Animal proteins generally have a higher protein score than plant proteins, which tend to be low in the essential amino acids tryptophan, methionine, and lysine.

Table 3
AMINO ACID PATTERN OF DIFFERENT ALGAE (g/16 g N)

Amino acid	FAO	Egg	Spirulina maxima	Spirulina maxima	Spirulina platensis	Spirulina sp.	Scenedesmus obliquus	Scenedesmus obliquus	Scenedesmus obliquus	Chlorella ellipsoidea	Chlorella pyrenoidosa	Chlorella vulgaris	Dunaliella primalecta	Dunaliella bardawil
Ile	4.0	6.6	6.8	6.0	6.7	4.8	3.6	3.7	3.4	4.5	3.4	3.2	5.5	4.2
Leu	7.0	8.8	10.9	8.0	9.8	8.4	7.3	8.1	9.6	9.3	4.0	9.5	11.1	11.0
Val	5.0	7.2	7.5	6.5	7.1	5.4	6.0	5.5	5.4	7.9	5.1	7.0	5.6	5.8
Lys	5.5	7.0	5.3	4.6	4.8	4.7	5.6	5.6	6.4	5.9	7.9	6.4	5.3	7.0
Phe	6.0	5.8	5.7	4.9	5.3	4.0	4.8	5.0	5.8	4.2	4.5	5.5	5.4	5.8
Tyr		4.2	5.9	3.9	5.3	—	3.2	—	2.8	1.7	2.7	2.8	3.7	3.7
Met	3.5	3.2	2.3	1.4	2.5	2.3	1.5	2.1	1.4	0.6	1.8	1.3	1.9	2.3
Cys		2.3	0.7	0.4	0.9	1.0	0.6	1.4	—	0.7	—	—	0.8	1.2
Try	1.0	1.7	1.5	1.4	0.3	1.5	0.3	1.4	—	—	1.4	—	—	0.7
Thr	4.0	5.0	5.6	4.6	6.2	4.6	5.1	4.8	6.7	4.9	3.2	5.3	5.5	5.4
Ala		—	9.0	6.8	9.5	6.9	9.0	7.8	10.3	12.2	5.9	9.4	7.5	7.3
Arg		6.2	7.2	6.5	7.3	6.6	7.1	5.3	6.3	5.8	5.6	6.9	6.1	7.3
Asp		11.0	12.2	8.6	11.8	9.1	8.4	9.0	10.2	8.8	5.9	9.3	11.3	10.4
Glu		12.6	17.4	12.6	10.3	12.2	10.7	10.0	12.4	10.5	9.3	13.7	11.7	12.7
Gly		4.2	6.6	4.8	5.7	4.9	7.1	5.5	6.8	10.4	4.8	6.3	5.8	5.5
His		2.4	2.0	1.8	2.2	1.6	2.1	1.6	1.7	1.7	1.4	2.0	0.5	1.8
Pro		4.2	4.1	3.9	4.2	3.8	3.9	4.6	5.3	5.0	4.0	5.0	—	3.3
Ser		6.9	4.9	4.2	5.1	4.9	3.8	4.3	5.9	5.2	2.2	5.8	4.7	4.6
Ref.	42	41	43	44	17	45	22	45	46	47	48	46	30	49

Algal protein is often deficient especially in the sulfur-containing amino acids methionine and cystine.[40] Selected data from the extensive literature on amino acid profiles of different algae are summarized in Table 3. The amino acid composition of egg protein and of a reference pattern, recommended by FAO/WHO are given for comparison.[41,42]

Deficiencies in essential amino acids of algal proteins can be compensated for by supplementing the algal protein with the particular amino acid or protein from other sources rich in these amino acids. Another method to improve the quality of algal protein could be breeding of selected strains with positive SCP characteristics. The breeding of green algae of the genus *Chlamydomonas* is well advanced, although not yet used for practical purposes.[50] Successful isolations of high-methionine mutants of *Chlorella* and *Spirulina* are mentioned by Soeder.[21]

Lipids

Algal lipids are typically esters of glycerol and fatty acids having carbon numbers in the range of C_{12} to C_{20}.[58,59] Nearly all of the fatty acids found in algal lipids are straight chain molecules with an even number of carbon atoms due to their biosynthesis from acetate by α-addition. Freshwater green algae in general seem to contain few fatty acids with more than three double bonds or 18 carbon atoms. However, analyses performed with *Scenedesmus* showed considerable amounts of a 16:4 acid and small amounts of fatty acids with more than 18 carbon atoms (Table 4). Similar findings have been reported by Chuecas and Riley[60]

Table 4
FATTY ACID COMPOSITION OF LIPIDS FROM
DIFFERENT ALGAE (% OF TOTAL LIPIDS)

Fatty acid	Spirulina platensis	Spirulina platensis	Spirulina maxima	Spirulina maxima	Spirulina sp.	Scenedesmus obliquus	Scenedesmus obliquus	Scenedesmus obliquus	Chlorella vulgaris	Dunaliella bardawil[a]
12:0	0.4	—	tr.[b]	0.4	—	—	1.8	0.3	—	—
14:0	0.7	—	0.3	1.1	tr.	1.3	2.4	0.6	0.9	—
14:1	0.2	—	0.1	—	—	—	—	0.1	2.0	—
15:0	tr.	—	tr.	—	—	—	—	—	1.6	—
16:0	45.5	43.4	45.1	33.7	33.0	12.1	18.0	16.0	20.4	41.7
16:1	9.6	9.7	6.8	3.0	2.0	9.5	4.1	8.0	5.8	7.3
16:2	1.2	tr.	tr.	3.6	—	—	3.1	1.0	1.7	—
16:4	—	—	—	—	—	13.1	—	26.0	—	3.7
17:0	0.3	—	0.2	0.2	—	—	1.1	—	2.5	—
18:0	1.3	2.9	1.4	tr.	4.0	0.9	33.7	0.3	15.3	2.9
18:1	3.8	5.0	1.9	4.0	7.0	15.3	10.5	8.0	6.6	8.8
18:2	14.5	12.4	14.6	22.3	12.0	9.0	5.3	6.0	1.5	15.1
α18:3	0.3	tr.	0.3	0.3	27.0	30.7	—	38.0	20.5	
γ18:3	21.3	21.4	20.3	17.9	—	—	—	—	—	—
20:2	—	—	—	—	—	2.7	—	—	1.5	—
20:3	0.4	—	0.8	—	—	—	13.1	20.8	—	
Others	—	—	—	14.3	—	1.6	6.9	2.5	19.6	—
Ref.	51	52	51	19	53	54	55	56	55	57

[a] % of polar lipids.
[b] tr. = traces.

for *Dunaliella tertiolecta*. The lipids of this algae contained 7% C 16:4 and 10% 20:5 fatty acids. The major components of algal lipids are triglyceride, sulphoquinovosyl diglyceride, monogalactosyl diglyceride, digalactosyl diglyceride, lecithin, phosphatidyl glycerol, and phosphatidyl inositol.[61] Different types of lipids are found in the various taxonomic groupings of algae. Diatoms, which produce lipids as a storage product, synthesize only small amounts of linolenic acid, whereas this fatty acid is common in Chlorophyceae. Cyanobacteria tend to contain large amounts of polyunsaturated lipids (35 to 60% of total), while in the eukaryotic algae saturated and monounsaturated fatty acids predominate. Triglycerides are common storage products in algae and may contribute up to 80% of the total lipid fraction.

The total lipid fractions in algae vary between 1% and more than 40% of dry weight.[53,62-64] Environmental factors can affect both the relative proportions of fatty acids as well as the total amount.[65] Spoehr and Milner[66] noted in 1949 that nitrogen starvation is most influential on lipid storage and reported that the lipid fractions may be as high as 70 to 80% of dry weight in extreme cases. Other algal species, however, produce carbohydrates rather than lipids under such conditions.[64]

Poly-β-hydroxybutyrate (PHB) has been identified by Campbell et al.[67] in *Spirulina platensis* as lipid reserve, which accumulated during exponential growth to 6% of the total dry weight. The authors suggest that other cyanobacteria might also accumulate PHB if grown under conditions of high CO_2 concentrations (5%), as was the case in their experiments with *Spirulina*.

Other factors have also been noted to influence lipid production in algae. A shift in lipids to triglycerides with increasing light intensity has been observed by Orcutt and Patterson for *Nitzschia closterium*. Temperature also appears to have an influence on lipid production and composition, however, the data published are not conclusive.[68,69]

Modifications of the culture conditions not only change the relative amount of the gross chemical constituents of the algae (i.e., protein, lipids, carbohydrates, etc.), but also influence their chemical composition. The effect of varying CO_2 concentrations on the fatty acid composition in *Chlorella fusca* has been studied by Dickson et al.[70] Autotrophically grown cells supplied with 1% CO_2 in air have 16:0, 16:3, 16:4, 18:1, 18:2, and 18:3 as the major fatty acids with a considerable amount of 16:4. In cells grown in the presence of glucose and air containing 1% CO_2, the 16:4 acid disappears, the linolenic acid (18:3) content is reduced and the other acids increase complementarily.

While in the chlorophyceae large amounts of α-linolenic acid and only traces of γ-linolenic acid can be found, in *Spirulina* there are no or only little amounts of α-linolenic acid, but high quantities of γ-linolenic acid. Nevertheless, *Spirulina* isolated and identified by Kenyon et al. contained α-18:3 as the only linolenate species.[51-53] In this context Wood[58] raised the question as to the validity of the present classification of *Spriulina*. The presence of the essential fatty acids α-linolenic and γ-linolenic acid is important in relation to the use of a algae as a source of human food or animal feed.

The unsaponifiable matter in the lipids of various algal species has been studied by Paoletti et al.[64] Its general composition and the components of the hydrocarbon fraction are given in Table 5. The unsaponifiable content in chlorophyceae is higher than that of cyanobacteria, with the exception of *Chlorella*. The major constituents are sterols and hydrocarbons. Although each algal species has a specific pattern of hydrocarbons, n-C_{17} is always one of the most abundant components. In general, the cyanobacteria show simpler hydrocarbon pattern than the chlorophyceae, which contain larger amounts hydrocarbons of high molecular weight and a higher degree of unsaturation.

Ash

The ash content in algae varies quantitatively and qualitatively depending upon culture conditions in general, but especially upon the composition of the culture medium. While marine algae generally contain large amounts (up to 50% of the dry matter) of ash, microalgae usually contain less than 10% of their dry weight as ash, with the exception of some halophile algae and diatoms, which due to their siliceous skeleton, have higher ash contents. Since the mineral composition and amount of the ash of microalgae is comparable to that found in higher plants, normally their effect on the nutritional quality of the algal biomass is only marginal. However, caution is advisable if elevated concentrations of heavy metals (see section on toxicology/heavy metals) are present or if the ash content considerably exceeds 10% of the dry weight caused by unused minerals from the culture medium. In the latter case the proportion of the other major cell constituents would be unfavorably changed. Results of chemical analyses on the mineral content of different algae are summarized in Table 6.

Another point which might be of importance in connection with the utilization of algae as feed is the Ca:P ratio in the algae. Due to imbalances in the culture medium there is a chance that an unfavorable ratio may lead to teeth anomalies in rodents[74] (see also the section ''Toxicology/Toxicological Studies with Animals'').

Nucleic Acids

Of particular importance for the nutritional suitability of SCP for humans is the nucleic acid content of the microbial biomass. Algal biomass contains higher amounts of nucleic acids than most conventional foods, but considerably less than other sources of SCP.

The reported concentrations of nucleic acids (RNA and DNA) in the algal species, which

Table 5
RELATIVE COMPOSITION OF ALGA UNSAPONIFIABLES AND QUANTITATIVE COMPOSITION OF ALGE HYDROCARBONS (% of TOTAL HYDROCARBON)[63]

Compound	Spirulina platensis (Chad)	Spirulina platensis (Mexico).	Calothrix sp.	Nostoc commune	Scenedesmus quadricauda	Chlorella sp.	Uronema gigas	Uronema terrestre	Selenastrum gracile
Alcohols	52.2	40.0	39.6	40.4	43.3	44.4	43.6	37.1	35.5
Sterols	5.3	8.0	7.8	14.5	23.0	21.8	17.3	21.7	28.1
Hydrocarbons	32.7	40.6	48.0	41.8	16.8	27.3	31.7	36.0	23.1
n-C-14	0.2	—	tr.[a]	0.1	1.0	0.2	0.4	0.5	0.4
N-C-15	3.7	2.9	0.3	1.5	2.3	2.0	1.2	0.9	0.9
n-C-16	3.8	2.8	0.7	1.6	2.5	1.0	1.2	1.4	1.1
Δ-C-17	3.1	5.2	tr.		tr.		tr.	0.3	tr.
				3.2		35.8			
iso-C-17		tr.	0.5		1.9		26.7	2.5	0.8
n-C-17	84.0	71.7	47.4	50.4	41.1	25.4	13.3	24.6	61.1
iso-C-18	tr.	tr.	1.4	24.4	tr.	0.9	0.5	tr.	tr.
n-C-18	0.5	0.7	30.9	1.0	2.0	tr.	1.2	1.7	1.8
iso-C-19	1.3	2.9	0.9	tr.	tr.	3.8	3.6	10.1	15.1
n-C-19			10.4	0.6	2.5	—			—
Δ-C-20	tr.	—	tr.	tr.	2.3	tr.	0.4	1.4	0.9
iso-C-22	tr.	tr.	0.1	tr.	0.7	tr.	1.1	1.7	0.5
n-C-22	0.5	1.7	0.2	1.4	2.0	2.8	0.4	1.4	1.1
Δ-C-23	tr.	tr.	0.1	0.4	1.1	0.3	0.2	0.3	0.5
iso-C-25	—	tr.	0.1	0.2	1.7	0.5	0.7	0.5	0.6
n-C-25	0.3	1.2	0.3	1.4	11.4	5.0	6.1	3.2	1.4
n-C-26	0.2	0.6	0.1	0.5	1.4	1.2	1.0	0.8	0.5
Δ-C-27	—	—	—	—	—	0.4	9.7	—	—
Squalene	0.9	1.3	0.8	2.2	5.7	3.9	3.7	24.7	4.7
Unidentified	0.4	1.0	0.8	4.5	9.4	3.3	24.1	12.7	2.3

Note: tr. = traces.

are considered as SCP, vary between 4 and 6% of dry weight with the ratio of RNA:DNA of about 3:1.[31,56,75-77] Since increased nucleic acid concentrations in food preparations might be a health hazard to the consumer, eventual toxicological effects of nucleic acids in SCP are discussed in more detail in the section "Toxicology/Nucleic Acids".

Vitamins

Microalgal biomass represents a valuable source of nearly all important vitamins which improve the nutritional property of this unconventional protein. Unfortunately, information on the vitamin content of different algae is rare and scattered in the literature, probably because the determinations of various vitamins in a particulate material such as microalgae are rather difficult.[78-81]

The vitamin content in algae, as in higher plants, depends on the growth conditions.

Table 6
MINERAL COMPOSITION OF
SOME ALGAE

Component	*Spirulina maxima*	*Spirulina platensis*	*Scenedesmus obliquus*
Ash (%)	8.9	—	8.7
Na (mg/100 g)	1380	450	120
K	1800	1420	710
P	1220	1450	1070
Ca	103	700	1110
Mg	334	900	300
Fe	72	115	320
Mn	4.8	—	39.1
Zn	18.5	—	11.3
Cu	1.6	—	2.9
Cl	463	—	38.5
S	656	—	—
Co (μg/kg)	1.9	—	0.9
Ni	2.5	—	4.9
Cr	2.7	—	3.5
F	—	9.5	19.0
J	—	—	2.2
Pb	5.1	3.9	7
Cd	1.5	0.6	0.4
As	—	0.9	4
Hg	—	0.07	0.6
Ref.	71	17	22

Besides fluctuations caused by environmental factors, the methods applied for drying and processing algae reduce the concentrations of several vitamins. This is true especially for the heat-unstable vitamins B_1, B_2, C, and nicotinic acid, which may decrease by 40, 25, 60, and 25%, respectively, compared to the concentration in fresh algal material.[82]

A comparison of the amounts of vitamins in algae with those in soya, spinach, and beef liver is given in Table 7. Plant foods are not generally considered a source of vitamin B_{12}, thus its presence in green algae is rather surprising. It is assumed that these algae do not have the ability to synthesize this vitamin and that the vitamin actually found was synthesized by closely associated bacteria and then absorbed and concentrated by the algae.[85] The vitamin B_{12} concentration in *Spriulina* is about 10 times higher than in the green algae, indicating a close phylogenetic link of this algae to bacteria which are able to synthesize this vitamin.

Cell Wall

Cell walls represent about 10% of the algal dry matter and consist of a variety of macro-molecules. The amounts and chemical composition of these compounds in the algal cell wall are group-, species-, and even strain-specific. In general, cell walls are composite materials and at least two components, one fibrillar and the other mucilaginous were identified. The latter is usually regarded as forming a non- or paracrystalline matrix in which the former is embedded. There is evidence that these components may occur in part as alternating layers. The microfibrils form the most inert and resistant part of the cell wall, the most common of the skeletal components being cellulose. (For details on cell wall composition see Mackie and Preston[87] and Percival.[88])

Table 7
VITAMIN CONTENT OF DIFFERENT ALGAE IN COMPARISON TO
CONVENTIONAL FOOD SOURCES (mg/kg DRY MATTER)

Vitamin	*Spirulina platensis*	*Spirulina maxima*	*Spirulina* sp.	*Scenedesmus obliquus*	*Scenedesmus quadricauda*	*Chlorella pyrenoidosa*	Spinach (dry matter)	Beef liver (fresh)
Provitamin A	840	1400	—	230	554	480	—	—
Vitamin E	120	140	190	—	—	—	122	10
Thiamin	44	54	55	8.2	11.5	9.9	4.9	3
Riboflavin	37	41	40	36.6	26.9	35.9	9.7	290
Pyridoxine	3	3	3	2.5	—	22.9	9.7	7
Cobalamine	7	6	2	0.4	1.1	0.02	—	0.65
Vitamin C	80	70	—	20	396		2480	310
Biotin	0.3	0.4	0.4	0.2	—	0.15	0.34	1
Inositol	380	340	350	—	—	—	—	—
Folic acid	0.4	0.6	0.5	0.7	729	—	3.65	2.9
d-Ca-Pantothenate	13	11	11	16.5	46	20	14.8	73
Nicotinate		—	118	120	108	240	29.2	136
Ref.	51	51	19	22	85	86	41	41

The rigid cell wall, one of the characteristics of most chlorophyceae, poses serious problems in processing and utilizing these algae, expecially with regard to digestibility and biological value. Since the cellulosic cell wall is not digestible for nonruminants, expensive treatments are necessary to disrupt the cell wall, making the algal protein acessible for proteolytic enzymes. Among the various processing methods tested, drum-drying proved to be most effective.[89-96] It is assumed that the poor results obtained in earlier nutritional studies with green algae were due mainly to unsuitable processing.

The cell walls of the cyanobacteria are different from those of the chlorophyceae. Chemical analyses of wall fractions revealed the absence of cellulosic material and a close relationship to the structure of Gram-negative bacteria. This is verifiable by the demonstration of an outer membrane containing lipopolysaccharide and a peptidoglycan layer, which seems to be true for both the unicellular and the filamentous forms.[87,88,97,98] The cell wall of the cyanobacterium *Spirulina* does not represent a barrier to proteolytic enzymes as demonstrated by the fact that this algae can be digested by monogastric vertebrates without rupturing the cell walls (see the section "Nutritional Studies").

Polysaccharide

The aqueous extract of *Spirulina*, after removal of contaminant proteins, contained a polysaccharide which comprised approximately 15% of the biomass, the acid hydrolysate of which contained only glucose. The physicochemical characteristics of this glucan showed that more than 85% of the glucose residues are linked α-1, 4, the remaining residues being linked α-1, 6. The polysaccharide was degraded by β-amylase, yielding maltose (50 to 55%) and a dextrin as major products.[100] According to Durand-Chastel,[19] the amount of carbohydrates in *Spirulina maxima* is about 16.0%, composed of 9% rhamnose, 1.5% glucan, as well as glucosamine and glycogen. Recent analyses, however, could not confirm the high content of rhamnose.

Pure starch consisting of amylose and an amylopectin was isolated from *Chlorella pyrenoidosa*. The reserve polysaccharide paramylum of the euglenoids consists solely of β-1, 3 linked D-glucose residues and occurs in water insoluble, membrane-bound cellular inclusions. In addition, maltose, sucrose, and trehalose have been identified in *Euglena*.[102]

Pigments

Lipophile pigments such as chlorophylls and carotenoids constitute 3 to 5% of the dry algal biomass. With a few exceptions, the pigment pattern of the chlorophyceae is similar to that found in higher plants, as follows: chlorophyll-a — β-carotene, lutein, antheraxanthin, and chlorophyll-b — violaxanthin, neoxanthin. In addition to these major carotenoids, small amounts of other carotenoids have been detected in green algae,[103] e.g., loroxanthin, siphoaxanthin, pyrenoxanthin, siphonein, and α-carotene.

The pigment composition of the procaryotic cyanobacteria differs from that of eucaryotic chlorophyceae. Besides the fact that the prokaryotes do not contain chlorophyll-b, their characteristic carotenoids are echinenone and zeaxanthin. Cyanobacteria synthesize carotenoid glycosides; the predominant pigments of this type are myxoxanthophyll, oscillaxanthin, and aphanizophyll. The pigments of *Spirulina*, the only cyanobacterium of commercial importance, have been determined as chlorophyll-a — β-carotene, echineneone, α-cryptoxanthin, zeaxanthin, myxoxanthophyll or myxoxanthophyll-like myxol-glycoside, oscillaxanthin or oscillaxanthin-like oscillol-glycoside. The reported occurence of lutein in *Spirulina*[31] is no doubt an error. Besides these lipophile pigments, cyanobacteria contain water-soluble pigments, such as c-phycocyanin and allophycocyanin in *Spirulina*.

Some of these pigments are of commercial value. The β-carotene content of certain algae can be very high; in *Dunaliella bardawil*, the concentration of this pigment can be as high as 8% of the dry weight.[49] β-Carotene acts as pro-vitamin A and can serve as natural food color. If algae are used as feed, these carotenoids cause darker pigmentation of the yolk of eggs and impart a reddish color to the feathers of certain birds, as well as in the skin of fish or in the meat of chicks, in which it may reach unacceptable levels.[104]

According to recent reports from Japan, chlorophyll degradation compounds such as phaeophorbide-a caused photosensitized skin irritations in humans who consumed *Chlorella*.[107,108,235,236] Phaeophorbides are formed from chlorophylls by the removal of magnesium through the action of dilute mineral acids (formation of phaeophytin) followed by enzymatic removal of the phytol ester on the C-7 propionate group by chlorophyllase. The enzyme chlorophyllase is still active in cells dried at moderate temperatures, its action being enhanced by light and humidity. Intense heat treatment (100°C, 3 min) is required to inactivate this enzyme.

The irritation seems to be due to a light-sensitization of the skin caused by peroxide-induced formation of fatty acids (arachidonic acid) in the cell membrane. The minimum effective dose was reported to be an intake of 25 mg phaeophorbide per day. Rats fed phaeophorbide-a were tested on photosensitization, the LD_{50} was 45.5 mg/100 g body weight, the MLD was 12 mg/100 g body weight. These findings prompted the Japanese Ministry of Health in 1981 to instruct that the concentration of phaeophorbides in *Chlorella* products sold commerically should not exceed 100 mg/100 g material.

PROCESSING OF ALGAL BIOMASS AND THE PROBLEM OF DIGESTIBILITY

The fact that algal proteins (except those from cyanobacteria) are poorly utilized when the intact cells are fed to monogastric animals or humans has led to investigations on the effect of different treatments on increasing algal digestibility.

To increase the availability of cell-bound protein in *Scenedesmus*, mechanical (ball mill), enzymatic (cellulolytic enzyme), and chemical (hydrogen peroxide) methods of degrading the algal cell wall were investigated by Hedenskog et al.[89] The amount of nitrogen liberated

Table 8
DATA ON IN VITRO DIGESTIBILITY OF DIFFERENT ALGAE

| Alga | Treatment | Digestibility (%) | | Ref. |
		Trypsin	Pepsin	
Chlorella sp.	Fresh	46.2		110
Chlorella sp.	Dried	27.4; 43.9		89
Chlorella sp.	Lyophilized	62.5—65.5		110
Chlorella sp.	Protein-isolate	92.7		110
Scenedesmus quadricauda	Fresh		31.1	89
Scenedesmus quadricauda	Spray-dried		29.5—32.2	89
Scenedesmus quadricauda	Sand-grounded		59.5	89
Scenedesmus quadricauda	Boiled (5 min)		33.8	89
Scenedesmus quadricauda	Fresh	54.0	62.0	23
Scenedesmus quadricauda	Fresh, disintegr.		78.0	89
Scenedesmus quadricauda	Spray-dried		32.0	89
Scenedesmus quadricauda	Spray-dried, disintegr.		51.0	89
Scenedesmus obliquus	Spray-dried, Meicelase, 2.5%		54.0	89
Scenedesmus sp.	Protein-isolate	93.1		110
Hormidium sp.	Fresh	60.0	84.0	23
Ulothrix sp.	Fresh	88.0	81.0	23
Uronema gigas	Fresh	66.0	88.0	23
Uronema sp.	Fresh	85.0	92.0	23
Stigeoclonium sp.	Fresh	63.0	86.0	23
Spirulina sp.	Protein-isolate oven-dried		45.0	109
Spirulina sp.	Protein-isolate acetone-washed		70.0	109
Spirulina sp.	Protein-isolate cold extract		50.0	109

by pepsin has been used as a criterion to evaluate the efficiency of the different treatments.

No significant increase in the amount of pepsin-liberated nitrogen was achieved by the hydrogen peroxide treatment or by the cellulolytic enzyme Meicelase (produced by a strain of *Trichoderma*) (Table 8). The mechanical method yielding a disintegration of 70 to 90% of the cells offered a very effective means for increasing the algal digestibility. Hindak and Pribil[23] investigated several filamentous algae for their digestibility (Table 8). A common feature of all the filamentous algae studied was that their proteins were readily digestible in vitro, using either pepsin or trypsin or a combination of the two enzymes. It was suggested that the lamellar structure of the cell wall in the filamentous algal probably makes it possible for digestive enzymes to enter the cell interior.

Al'bitskaya et al.[109] studied the effect of isolation techniques on the digestibility of *Spirulina* proteins. The methods applied were (1) protein extraction from disintegrated cells by 0.4% NaOH at 25°C for 30 min and subsequent drying at 105°C for 12 hr; (2) same extraction conditions, dehydration by threefold washing with acetone followed by centrifugation and air-drying; (3) extraction at 5°C for 22 hr and dehydration with acetone. The best results were obtained with method 2, followed by methods 3 and 1 (Table 8).

Becker and Venkataraman[54] reported on studies of in vitro protein and carbohydrate digestibility of differently processes *Scenedesmus obliquus* and *Spirulina platensis*. The protein digestibility (after 24 hr) was examined by simulating the intestinal enzyme system pepsin-pancreatin.[111] The data, compiled in Table 9, demonstrate that the method of processing had a considerable influence on the digestibility of *Scenedesmus*, drum-drying being significantly superior to all the other methods. Similar results have been reported for *Chlorella*, where digestibility did not increase after prolonged boiling.[85] Tamiya[112] showed that

Table 9
IN VITRO DIGESTIBILITY OF
SCENEDESMUS AND *SPIRULINA*
PROCESSED BY DIFFERENT
METHODS[111]

Alga	Processing	Digestibility (%)
Scenedesmus	Drum-dried	75
	Cooked, sun-dried	50
	Sun-dried	43
	Fresh	30
Spirulina	Fresh	82
	Freeze-dried	70
	Sun-dried	65

Table 10
EFFECT OF PROCESSING ON THE
DIGESTIBILITY OF *SPIRULINA*[113]

Treatment	Digestibility (%)
Fresh	76.3
2N HCl	77.6
Boiling with Ca(OH)$_2$	74.6
Autoclaved (105°C, 15 min)	74.9
Autoclaved (105°C, 30 min)	76.1
Autoclaved (120°C, 15 min)	76.2
Autoclaved (120°C, 30 min)	75.2
Sonification (20 min)	76.0
Boiling with water	76.6

the in vitro digestibility of decolorized *Chlorella* was 66.2% while it was 55.3% only in freeze-dried samples.

In the case of *Spirulina*, however, only marginal differences could be observed between the various treatments. The fresh algal sample had the highest digestibility of 82%, while the sun- and freeze-dried algae had the lowest digestibility, 65 and 70%, respectively.

Similar results were obtained by de Hernandez and Shimada,[113] who subjected this algae to several physical or chemical treatments (Table 10). All these methods did not alter the digestibility significantly, confirming the observation that *Spirulina* per se has a high digestibility and does not require further processing. A certain improvement of the in vitro digestibility of *Spirulina* due to processing was reported by Baranowski et al.[133]. The four methods of drying applied (freeze, drum, cabinet and solar) increased the amount of available lysine and in vitro enzymatic protein digestibility in all samples over that of an undried control.

Other components of the algal cell such as carbohydrates will affect the overall digestibility. Carbohydrate digestibility of *Scenedesmus obliquus* was reported by Becker and Venkataraman.[54] The algal material was incubated with α-amylase and the amount of maltose released was employed as a measure of digestibility. For "gastric simulation" prior to amylolysis, a few samples were preincubated with HCl or HCl-pepsin in order to release starch that may be bound and thus be acted on by α-amylase.

The result of the different treatments on the carbohydrate content of *Scenedesmus* are summarized in Table 11.

Table 11
IN VITRO CARBOHYDRATE DIGESTIBILITY OF
SCENEDESMUS PROCESSED BY DIFFERENT
METHODS[56]

Treatment	Digestibility
Fresh algae, extracted with hot alcohol, cooked for 30 min, HCl-pepsin pretreatment	100
As above without cooking	86
Extracted with hot alcohol, disintegrated with glass-beads, cooked for 30 min	72
As with first treatment, with HCl treatment instead of HCl-pepsin	65
Extracted with hot alcohol, autoclaved 120°C, 15 min	56
Drum-dried, autoclaved 120°C, 15 min	56
Drum-dried, cooked 30 min	51
Extracted with hot alcohol, pretreated with HCl	47
Extracted with hot alcohol, cooked 30 min	47
Drum-dried	46
Extracted with hot alcohol	38

Note: Amount of maltose released per 100 mg algae after 4 hr: 5.00 mg maltose = 100%.

Uncooked algae had the lowest digestibility. Cooking increased the amylolysis considerably, while autoclaving or mechanical disintegration further improved the digestibility. The amylolysis was higher in the samples pretreated with HCl or HCl-pepsin than in the untreated samples. The increase in the digestibility due to the different methods of processing could be attributed to swelling and rupturing of starch granules and to better gelatinization of starch.

The digestibility of carbohydrates of *Scenedesmus* seems to be satisfactory, but whether they are likely to cause gastrointestinal disturbances, fluid retention, etc., can be only established by in vivo experiments.

Amylolysis of sun-dried *Spirulina* was tried under conditions similar to those described earlier.[54] As determined by the extent of maltose release there was no amylolysis in this algae, even after 4 hr of incubation. The possible reason may be the low starch content (1%) in the algal material or that α-amylase was ineffective against the carbohydrates in this algae.

NUTRITIONAL STUDIES

Although the chemical composition of an algae holds valuable information concerning the prospective nutritional quality, the chemical composition alone cannot be considered as a substitute for bioassays because the nutritional value of the proteinaceous algal biomass is determined by the availability and digestibility of the nutritive constituents to the consumer. To study these important parameters, several animal feeding experiments have been performed by many investigators. The first systematic animal experiments for the nutritional evaluation of microalgal protein were carried out with rats and chickens in the 1950s and 1960s and led to contradictory results. The obvious reason for the divergent observations were the different methods which have been applied by the researchers to process the algae. To illustrate this point, a survey on data concerning the apparent digestibility of algae which were differently processed are given in Table 12. These data stress the importance of the drying method for the utilization of microalgae.

Table 12
APPARENT DIGESTIBILITY OF ALGAE
PROCESSED BY DIFFERENT METHODS

Alga	Processing	Digestibility (%)	Ref.
Casein		90	93
Scenedesmus	Fresh, untreated	24	96
	Disintegrated with glass beads	74	95
	Spray-drying	26—44	114
	Vacuum-drying	57—68	96
	Boiling (6—8 min)	73	96
	Microwave cooking	77	115
	Drum drying	77	93
Chlorella	Freeze-drying	51	115
	Boiling (6—8 min)	70—73	85
Air-drying	73	116	
Oocystis	Drum-drying	81	117
	Deamination	81	117
	Irradiation	80	117
	Autoclaving	78	117
Spirulina	Drum-drying	77	246

In this context it was reported that in the feces of rats fed with oven-dried *Chlorella,* intact cell walls were abundant and that the dietetic quality of *Scenedesmus* was altered favorably by drying processes.[118,119] Similar observations were made by McDowell and Leveille[120] who concluded that the poor digestibility of microalgae was due to the tough cell wall.

Animal Feeding Tests

The most common and simplest method to evaluate proteins by animal feeding tests is the determination of the Protein Efficiency Ratio (PER). The PER is based on short-term (3 to 4 weeks) feeding trials with weanling rats. The response to the diets fed is expressed in terms of weight gain per unit of protein (N × 6.25) consumed by the animal:

$$PER = \frac{\text{Weight gain (g)}}{\text{Consumed protein (g)}}$$

However, it is impossible by this method to distinguish between digestibility and the true quality of a given protein. The PER evaluation also ignores differences in dose response at suboptimal levels. The method can be improved by a multilevel assay in which the growth responses at several levels of the tested protein are compared with suboptimal levels of the reference protein. In order to obtain reliable data on the protein quality by determining the PER it is absolutely necessary that the material tested is fully digestibile. This probably was not the case in some of the earlier studies reported.

The PER value obtained is compared normally with a reference protein such as casein. Because of differences in response to the same standard casein even in the same animal house, the PER values for casein are customarily adjusted to an assumed value of 2.5, which requires a corresponding correction of the experimental values.[121]

Since the estimation of the PER reveals the quality of the *protein* tested and not that of the total algal *biomass,* all other nutrients (vitamins, minerals, etc.) must be given in sufficient amounts.[122] The composition may differ slightly from laboratory to laboratory, however, such a variation will not affect the results as long as there is no deficiency in any of the listed components.

This method was employed by several authors to demonstrate the influence of post-harvesting treatments on the digestibility of algae. Becker et al.[54,91] reported on the effect of different methods of drying on the PER of *Scenedesmus obliquus* and *Spirulina platensis*. For drum-drying, the algal slurry was sprayed on a drum drier and heated at 120°C for 12 sec. For sun-drying, the centrifuged algal slurry was further concentrated, thinly spread over plastic sheets, and dried in the sun. For the cooked sun-dried samples the concentrated algal slurry was cooked for 20 min at 100°C and sun-dried subsequently. These algal samples were incorporated in the experimental diets at 10 or 20% protein levels. In one diet, drum-dried *Scenedesmus* and sun-dried *Spirulina* were supplemented with methionine. Drum-dried *Scenedesmus* yielded significantly higher PER values than the samples dried by the other methods. PER of drum-dried *Scenedesmus* was lower at 20% protein level than at 10%, indicating that 20% represent supraoptimal levels of protein. In contrast, the sun-dried algae, both uncooked and cooked, showed higher PER values at the 20% protein level than at 10%.

When marginal deficiencies in sulfur amino acids were made up by supplementation with methionine, drum-dried *Scenedesmus* gave PER values quite high compared to vegetables, cereals, and soy protein and was close to that of casein.[123] The effect of methionine supplementation seems to vary with different algae. Omstedt et al.[94] have also reported increased nitrogen efficiency ratios with drum-dried *Scenedesmus*, but no such effect was noticeable in drum dried *Spirulina platensis* when the diets were supplemented with methionine. Thus, it seemed that in the case of *Spirulina*, methionine was not a limiting amino acid. It was evident that sun-dried *Spriulina* also contained a protein of fairly good quality because the values obtained with sun-dried *Spirulina* were higher than those found for sun-dried *Scenedesmus*. The PER values for *Spirulina*, however, were lower than those of drum-dried *Scenedesmus*, demonstrating the effect of this method of drying on the digestibility of these algae. Omsted et al.[94] compared the value of lyophilized with drum-dried *Spirulina* and reported that the drum-dried sample showed a higher nutritive value. Bourges et al.[124] compared *Spirulina* samples from two different regions with casein at 10% protein level. One algal sample was a commercial product from Mexico, the other algal sample originated from Lake Chad and was cultivated in France. For the first sample the authors found the PER to be 2.20 while for the second sample a lower value of 1.86 was obtained.

To study the apparent digestibility in in vitro tests, de Hernandez and Shimade[113] subjected *Spirulina* to several physical or chemical treatments. The treatments did not significantly alter the digestibility which varied between 73 and 78%.

Pabst et al.[125] investigated the wholesomeness of very high concentrations (up to 80% of dry weight) of algal biomass in feed, and determined the utilization of carbohydrates, fats, fiber, phosphorus, and nitrogen from drum-dried *Scenedesmus* by balance studies in adult male rats. Utilization of the algal carbohydrates was 60%, whereas added glucose was completely absorbed. About 90% of the fat was absorbed, but since added edible oil outweighed the lipids from the algae, definite conclusions were not possible. As a result of the bulk of the algae, the diet contained large amounts of raw protein and phosphates so that these substances were in positive balance. The animals showed normal weights and no organ abnormalities could be detected after the termination of the investigation.

A study by Erchul and Isenberg[115] evaluated the protein quality of seven algal biomasses produced by a water reclamation experimental plant. The PER obtained ranged from 0.68 to 1.98; the biomass with the highest protein quality compared favorably with soybean meal. Nitrogen balance experiments suggested that the variations observed were due largely to differences in digestibility, and to a lesser extent to differences in nitrogen retention. The reasons responsible for these differences were not determined by the authors, but from the data reported it could be assumed that several factors were involved, i.e., insufficient drying (several days at room temperature), high range of protein content in the samples (27 to 50%), and extensive storage times of the material before the first test (up to 120 weeks).

In nearly all nutritional studies rats were employed as experimental animals. One of the very few studies on mice was performed by Bowman et al.[26] who fed 6-week-old mice with a diet containing 70% air-dried (35°C) *Chlorella* for 16 days. Although average feed consumption was comparable between the experimental diet and a control diet (3 g/day per mouse), the weight-gain pattern was significantly different. While the control animals gained 6.3 g (males) and 3.0 g (females), the increment in weight for the animals of the experimental group was 2.3 and 2.1 g, respectively. When the experiment was repeated with air-dried *Anacystis* fed to 4-week-old mice, weight gain was irregular because of inadequate amounts supplied and probably because the blue-green algae was less digestible. However, all animals remained active and healthy and no diarrhea and other abnormalities were noted.

As drum-drying is the most expensive stage in commercial production of algae, simpler methods of processing were also tried by Cook,[85] who tested algae processed by the other three methods, viz., sun-drying, cooking and sun-drying, and freeze-drying and came to the same conclusion: algae processed by these methods gave much lower PER values compared to drum-dried material. The reason, obviously, is the incomplete breakage of the cell wall and the inability of the digestive enzymes of monogastriers to hydrolyze the cellulosic algal cell wall. Cooking would be expected to improve the PER values. This treatment, however, is rather delicate; if too short it does not break the cell wall effectively and if extended too long, it impairs the quality of the algal material.

Several other authors have reported on PER studies with different algae. The published data confirm the results described above. To provide an overall view of the major findings, comparative PER values are summarized in Table 13. The data represent only a part of the reported figures relating to the nutritive value and give a very general idea about the protein efficiency ratio of various algae, the experimental conditions, and the diet compositions used.

If the algal biomass consists of a mixed population of different algal species grown on sewage, the nutritive quality may differ greatly depending on the ratio of the algal species present in the biomass as well as on the presence of other microorganisms, rotifers, insect larvae, residues from nutrients, etc.

Summing up the available information it can be stated that particularly those algae, which are considered generally as the most potential sources of unconventional protein (*Chlorella, Scenedesmus,* and *Spirulina*) are of relatively good nutritive quality as judged by the PER studies. As a matter of fact this rating presupposed that, especially in the case of Chlorophyceae, the biomass is processed by proper treatments and is fully digestible. The importance of the processing step on the digestibility of algae is stressed graphically in Figure 1. The protein value of these algae is high compared with other plant proteins and reaches about 80% of the value of casein. Supplementation with methionine improves the PER of most of the algae tested, indicating that this essential amino acid is limiting the quality of the algal protein.

METABOLIC STUDIES

Although the estimation of PER is the method most commonly used to evaluate the quality of proteins, it has certain limitations, especially for Chlorophyceae, as already pointed out. Therefore, more specific methods have to be used to evaluate the nutritive quality of algal protein. By applying nitrogen balance studies it is possible to distinguish between the digestibility of the proteinacous matter and the quantity of nitrogen retained for storage and/ or anabolism.

The following methods are used for the metabolic studies; the test animals for these investigations are usually rats. The diets are prepared in the same way as for the PER studies, however, the animals are housed in individual metabolic cages in which urine and feces can be collected separately.

Table 13
COMPARATIVE PER VALUES OF DIFFERENT ALGAE PROCESSED
BY DIFFERENT METHODS

Alga	Protein level (%)	Processing[a]	PER	Ref.
Casein	10		2.50	
Scenedesmus	10	DD	1.99	127
Scenedesmus (sew)	10	DD	2.22	128
Scenedesmus + 0.3% met	10	DD	2.00	129
Scenedesmus	15	SD	0.87	115
	10	DD	1.86	93
	10	Cooked 8 min	1.78	93
Chlorella	10	DD	1.89	127
	10	Raw	0.84	131
	10	Autoclaved	1.31	131
	15.3		1.38	40
	10	FD	1.66	132, 133
Chlorella + 0.2% met	10	FD	2.20	132
Chlorella	15	SD	0.68	115
	11	Oven-dried 70°C	1.49	134
	10	Air-dried	1.72	116
	20	Air-dried	1.52	116
Chlorella + 0.3% met	10	Air-dried	2.60	116
Chlorella (sew) centrifuged	12	Oven-dried 80°C	1.57	135
Chlorella (sew) alum flocculated	12	Oven-dried 80°C	0.99	135
Coelastrum	10	DD	1.84	93
	10	DD	1.68	130
Uronema	10	DD	1.35	127
	10	DD	1.43	93
Oocystis (sew)	10	DD	1.39	136
Oocystis (sew) + met	10	DD	1.79	136
Oocystis (sew) + lys	10	DD	1.62	136
Oocystis (sew) + met + lys	10	DD	2.31	136
Spongiococcum	15.3		0.94	40
Spongiococcum + 0.5% met	14.8		0.34	40
Micractinium (sew)	10	DD	2.00	128
Spirulina	10	SD	2.20	124
	10	SD	1.86	124
Spirulina	20.5	SD	2.10	138
Spirulina + 0.3% met	10	SD	1.74	129
Spirulina	11	Oven-dried 70°C	1.82	134
Chlorella/Euglena (sew)	10	DD	2.22	128
Chlorella/Euglena (sew) + soya (1:1)	10	DD	2.13	128
Micractinium (sew) + soya (1:)	10	DD	2.23	128
Scenedesmus/Chlorella	12	Raw	1.61	139
Scenedesmus/Chlorella (sew) (10:1)	12	SD	1.31	85
	12	Autoclaved 30 min	0.70	85
	12	Cooked 30 min	1.60	85
	12	Cooked 120 min	1.52	85
Scenedesmus/Chlorella	15.1		1.22	40
Scenedesmus/Chlorella (3:1)	15	SD	1.62	115
Chlorella/Scenedesmus/Ankistrodesmus (15:2:1)	15	SD	1.98	115

[a] Sew: sewage-grown, DD: drum-dried, SD-sun-dried, FD: freeze-dried.

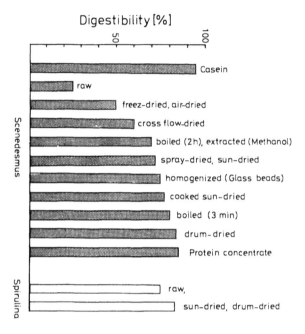

FIGURE 1. Influence of processing on the digestibility of algal proteins.[84]

Biological Value (BV)

The biological value of a protein is measured as the ratio of the absorbed nitrogen to the total nitrogen intake. The absorbed nitrogen is defined as the difference between the ingested and the intestinally excreted nitrogen.

The BV can be calculated by using the following equation:

$$BV = \frac{I - (F - F_k) - (U - U_k)}{I - (F - F_k)} \quad \text{or} \quad \frac{B - B_k}{I - (F - F_k)}$$

where, I = total nitrogen intake, F = fecal (excreted) nitrogen, F_k = endogenous fecal nitrogen loss, U = urinary (excreted) nitrogen, U_k = endogenous urinary nitrogen loss, B = body nitrogen (measured at the end of the test period on animals fed the test diet), B_k = body nitrogen at zero nitrogen intake (measured at the end of the test period on animals fed a nonprotein diet).

Digestibility Coefficient (DC)

The digestibility of the tested protein, i.e., the proportion of food nitrogen that is absorbed by the animal, can be calculated by the following equation, using the experimental data already obtained by the aforementioned experiment on BV.

$$DC = \frac{I - (F - F_k)}{I}$$

Net Protein Utilization (NPU)

A further method to establish the quality of a given protein is the estimation of the net protein utilization value by rats. This actually requires the hydrolysis of the total carcasses of the test animals and the estimation of the total animal nitrogen. If this time-consuming

Table 14
NITROGEN BALANCE DATA ON RATS FED *SCENEDESMUS* AND
SPIRULINA[92]

Alga	Protein (%)	Protein intake (g/day)	Nitrogen excreted (mg/day)			Nitrogen retained (%)
			Urinary	Fecal	Total	
Casein	10	1.65	42.9	18.9	61.8	76.5
Scenedesmus						
Drum-dried	10	1.52	50.7	50.2	100.9	58.4
Drum-dried	20	2.73	126.6	104.2	230.8	47.0
Sun-dried	10	0.92	42.0	45.2	87.2	40.7
Cooked, sun-dried	10	1.18	53.4	47.6	91.0	51.9
Sun-dried	10	0.89	14.0	36.3	50.3	64.8
Sun-dried + 0.3% met	10	0.92	27.3	22.3	50.0	65.9

procedure is applied, the following formula represents the product of the digestibility coefficient of the nitrogenous component of the tested biomass and its biological value:

$$NPU = BV \times DC = \frac{B - B_k}{I}$$

If a hydrolysis of the carcasses of the test animals cannot be performed, the following formula can be applied, using the parameters obtained from the BV estimation:

$$NPU = \frac{I - (F - F_k) - (U - U_k)}{I}$$

Since the results of digestibility testing and PER studies with *Spirulina* and *Scenedesmus* showed the importance of the drying step on the nutritive value of these algae, experiments were performed by Becker et al.[92] to evaluate DC, BV, and NPU of sun-dried *Spirulina* and *Scenedesmus* which were differently processed.

The nitrogen balance for *Scenedesmus* and *Spirulina* are given in Table 14. Food consumption and consequently nitrogen intake by the rats were comparable for casein and drum-dried algae at the 10% level; the intake of nitrogen by rats fed on 20% drum-dried algae was nearly doubled. Excretion of fecal nitrogen was greater in rats fed with algal diets than with casein, while urinary nitrogen excretion was not very much different between all the diets containing 10% protein. The percentage of nitrogen retained was higher with drum-dried algae than with algae subjected to the other drying processes used in the study. The values obtained by BV, DC, and NPU are summarized in Table 15.

In general, the drum-dried algae had a nutritional quality which was about 85% that of casein. At 10% protein level, all parameters of the drum-dried *Scenedesmus* were better than those of the samples dried by the other methods. The BV of sun-dried algae was low and cooking did not lead to any significant differences. At higher protein level (20%), the values were lowered in the drum-dried algae. NPU was significantly lower for drum-dried algae than for casein, indicating that the unfortified algal protein is limited by at least one of the essential amino acids (probably methionine).

The data on the nitrogen balance studies for *Spirulina* confirm that this alga with its thin and fragile cell wall does not present serious problems in protein utilization. Thus, even simple sun-drying is sufficient to obtain acceptable values in metabolic studies from *Spirulina*.

In a similar approach Lee et al.[117] investigated the effects of different processing methods

Table 15
EFFECT OF PROCESSING ON BIOLOGICAL VALUE
(BV), DIGESTIBILITY COEFFICIENT (DC), AND NET
PROTEIN UTILIZATION (NPU) OF *SCENEDESMUS*,
SPIRULINA, AND *CHLORELLA*

Alga	Protein (%)	BV	DC	NPU	Ref.
Casein	10	87.8	95.1	83.4	92
Scenedesmus					
Drum-dried	10	80.8	81.4	65.8	
Drum-dried	20	67.1	77.4	52.0	
Sun-dried	10	72.1	72.5	52.0	
Spirulina					
Sun-dried	10	77.6	83.9	65.0	
Sun-dried + 0.3% met	10	79.5	91.9	73.0	
Chlorella					
Protein extract	10	79.9	83.4	66.2	116
Protein extract	20	78.6	84.3	66.3	
Protein extract + 0.37% met	10	91.1	86.1	78.4	
Chlorella ^{15}N-method Protein extract	10	83.5	79.0	65.9	
Egg	10	94.7	94.2	89.1	

on the nutritive value and digestibility of *Oocystis* in rats. The algae were processed by drum-drying, deamidation (800 g algae were treated with 400 mg L-amino acid oxidase for 48 hr at 4°C), irradiation (20% algae suspension was irradiated with Cobalt-60 cell at a dose of 5 Mrad), and autoclaving (30 min at 121°C). The nutritional quality of the biomass was assayed by the determination of the Net Protein Ratio (NPR) as:

$$NPR = \frac{\text{Gain of test animal} + \text{Loss of control animal (N-free diet)}}{\text{Protein consumed by test animal}}$$

This estimation is an improved modification of the PER since it measures the protein used for growth and the protein used for maintenance. The results are independent of food intake.[247]

It was found that *Oocystis* proteins were superior to soybean protein, but inferior to casein. With the exception of autoclaved algae, there were no significant differences in the protein of digestibility (Table 12) and NPR values (Table 16) among the processing methods.

Metabolic studies on rats were performed by Clement et al.,[44] who compared the nutritive value of fresh and stewed (1 hr) *Spirulina* with the reference protein casein. The authors observed that the values for NPU, DC, and BV were higher for the fresh, unprocessed algal samples (48, 76, and 63, respectively) than for the diet containing stewed algae, which gave values of 38, 74, and 51, respectively.

NPU of *Spirulina* samples from Lake Texcoco and Lake Chad (cultivated in France) were examined by Bourges et al.[124] For the Mexican *Spirulina* the NPU was 56.6, while the African strain resulted in a slightly lower value of 52.6 (casein was 61.5). The values are higher than those obtained by Clement et al.,[44] however, no explanation is given by the authors for these differences.

Narasimha et al.[140] evaluated DC, BV, and NPU of *Spirulina platensis* with and without methionine supplementation (0.2%). While the digestibility remained similar between both the samples (75.5 and 75.7) BV and NPU were improved significantly by the addition of methionine to the algal diet. BV increased from 68.0 to 82.4 and NPU from 52.7 to 62.4.

Table 16
NET PROTEIN RATIO (NPR)
OF *OOCYSTIS* PROCESSED
BY DIFFERENT METHODS IN
COMPARISON TO CASEIN
AND SOYBEAN MEAL[117]

Protein source	NPR
Casein	5.43
Soybean meal	3.40
Alga, drum-dried	3.72
Alga, drum-dried + 0.5% met	3.86
Alga, deaminated[a]	3.96
Alga, irradiated[a]	3.63
Alga, autoclaved	3.06

[a] For details, see text.

The results disagree to a certain extent with other reports,[113,129] where supplementation with methionine did not improve the nutritional quality of *Spirulina*.

In view of the fact that earlier nutritional studies resulted in reports that large amounts of metabolic fecal nitrogen were excreted by animals given algal preparations and an overall poor digestion of the algae because of cellulose-containing substances in the cell wall, Yamaguchi et al.[116] tested the nutritive value of extracted *Chlorella* protein in comparison with whole-egg protein and casein in rat feeding experiments (Table 15). The nutritive values determined with ^{15}N gave slightly lower digestibility and similarly higher retention after absorption compared with the corresponding values for TD and BV determined by the conventional method. The improvement in the values of the protein with a methionine supplement was clearly shown in the results obtained by ^{15}N as well as by the conventional method. This improvement was due to increased utilization of absorbed N and not to higher digestibility of the protein. This effect was particularly evident through an increase caused in total N content of liver and spleen, probably owing to the reutilization of body protein. Although the values for the algal protein were somewhat inferior to those of whole-egg protein or casein, they are very high compared to values reported by others for algal protein (Table 17).

Almost contradictory are the findings of an early study reported by Bock and Wünsche[141] who tested the nutritive value of *Chlorella vulgaris* and *Scenedesmus obliquus* by the N-balance method. The utilization of the protein was unsatisfactory with both algae, the biological value amounted to 52.9 and 47.0, and the digestibility of nitrogen to only 44.1 and 26.1, respectively. Since it had to be assumed that the algal material used in this study was not processed sufficiently, the authors repeated the study with drum-dried *Scenedesmus* from another source.[142] The data obtained in this second evaluation are in good agreement with the findings of other similar investigations. The authors reported BV of 73.1, DC of 79%, and NPU of 70.5, and attributed the algae a high nutritional quality.

In a series of nutritional experiments lasting about 3 years, Fink and Herold[118,143-145,248] evaluated the quality of 12 different batches of *Scenedesmus obliquus* in rat feeding experiments. The algal material was dried in thin layers under infrared lamps (60 to 70°C). The authors found a very high biological value of the algal protein which was equal or superior to that of egg or milk protein. It was also superior to the proteins of green leaves of higher plants such as spinach or alfalfa. In contrast, feeding experiments with fresh algal biomass showed that with this material no growth could be produced, the animals decreased in weight, and some of them died of dietetic liver necrosis. The results of these experiments

Table 17
COMPARABLE DATA ON BIOLOGICAL VALUE (BV), DIGESTIBILITY COEFFICIENT (DC), AND NET PROTEIN UTILIZATION (NPU) OF DIFFERENT ALGAE

Alga	Method of processing[a]	BV	DC	NPU	Ref.
Algae grown in synthetic media					
Scenedesmus	Air-dried	60.0	51.0	31.0	115
	DD	81.3	82.8	67.3	93
Chlorella	Air-dried	52.9	—	31.4	141
	DD	71.6	79.9	57.1	127
Coelastrum	DD	75.3	77.8	58.6	93
Uronema	DD	54.9	81.8	44.9	93
Spirulina	Air-dried			56.5	124
	Sun-dried	75.0	83.0	62.0	146
	Stewed	51.0	74.0	38.0	44
	Raw	63.0	76.0	48.0	44
	DD	68.0	75.5	52.7	140
Spirulina + 0.2% met	DD	82.4	75.7	62.4	140
Algae grown on sewage					
Scenedesmus/Chlorella (9:1)	Air-dried	54.3	65.4	35.5	85
	Autoclaved	54.5	65.5	35.6	85
	Cooked 30 min	56.0	73.0	40.9	85
	Cooked 120 min	48.7	69.8	44.0	85
Chlorella/Scenedesmus (75:20)	Air-dried	76.0	75.0	57.0	115

[a]   DD: Drum-dried.

must be treated with skepticism. In spite of the good nutritional quality of the algal biomass, it is quite improbable that the algal protein would be superior to egg or milk protein.

In general, the results of the N-balance experiments; confirm the findings of the PER studies. Although the values differ depending on the algal strain or species tested, it is evident that the algal material processed by drum-drying (except for *Spirulina* which can be dried by sun-drying) is superior to material dried by other methods. Neglecting extreme values it can be stated that after proper processing the average nutritional quality of most of the algae examined is about 80% of that found for high-quality reference proteins. This was confirmed by comparative studies, performed by Pabst[327] who determined the nutritional value of different algae and some conventional protein, (the results are compiled in Figure 2). Here the average crude protein intake of the animals is taken as 100% for each protein carrier. The figure proves methionine-enriched casein to be an easily digestible protein of very high quality. The results for the three algal species are within a very narrow range in all criteria, approximating the values for fish meal and closely followed by soya protein. Distinctly lower is the digestibility and the protein value of yeast (*Saccharomyces cerevisae*).

The long series of different and independent investigations which analyzed the various metabolic parameters in different animal species (rats, chicken, pigs, ruminants, fish) has repeatedly and unequivocally confirmed the high nutritional quality of algal protein, which was found equal or even superior to conventional plant proteins.

PROTEIN REGENERATION STUDIES

Many of the problems associated with growth of new tissue in young animals are met

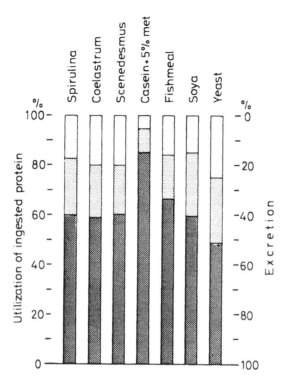

FIGURE 2. The utilization of protein of different sources. White section: protein not absorbed; dotted section: protein excreted in urine; dark section: protein retained and assimilated.[327]

again in regeneration of adult tissues. Thus, a further possibility to evaluate the nutritional quality of proteins is regeneration studies on protein-depleted animals. This procedure is a reliable method for determining the nutritive value of protein by first depleting and then repleting the protein reserves of adult animals. The advantages of this method are that the time for the assay is relatively short, the samples for the testing are small, and the fact that the same animal can be used several times for different tests.

This method uses a basic principle common to many bioassays, viz., the production of a deficit followed by a measurement of the replacement of that deficit. Depletion can be accomplished by feeding a protein-free diet until the rats have lost about 25% of their initial body weight. The animals are then fed nitrogen in the test diet and the rate of repletion is measured. For estimations of the nutritive values seven days of repletion are usually sufficient. Significant correlations exist between gain in weight during repletion with the regeneration of blood, liver, or carcass proteins, making weight recovery alone a good measure of nutritive value.

The extent of regeneration in protein-depleted rats has been used as an index for evaluating the quality of conventional protein,[18,147,242] but there are very few reports on algal protein quality evaluations based on this method.

Anusuya Devi et al.[148,326] reported on protein regeneration studies with *Scenedesmus* and *Spirulina*, evaluated in independent experiments with rats. The animals were divided into five groups with eight animals per group. One group was fed on stock diet and was used as positive control without repletion or depletion; the other four groups were fed on a protein-free diet for a period of 12 days. After this depletion period the rats of one group were examined. The other groups of depleted rats were repleted by feeding diets containing either

Table 18
DATA OF DEPLETION-REPLETION STUDIES WITH RATS FED DIETS
CONTAINING DRUM-DRIED *SCENEDESMUS*[148]

Diet	Feed consumed (g/day)	Weight gained (+) or lost (−)	(g/100 g Body weight)			Liver protein (g)	Serum protein (g/100 mℓ)		
			Liver	Kidney	Heart		Total	Albumin	Globulin
Stock	21	+ 85	2.87	0.63	0.29	22.20	7.4	4.8	2.6
Depletion[a]	15	− 36	2.37	0.61	0.36	16.45	4.7	2.8	1.9
Repletion[b]									
Casein	18	+ 83	2.79	0.57	0.33	21.25	7.2	4.7	2.5
Algae (10%)	20	+ 45	2.79	0.66	0.35	18.60	6.2	3.4	2.8
Algae (15%)	22	+ 60	2.72	0.62	0.31	21.10	6.5	3.6	2.9

[a] Data after depletion (12 days).
[b] Data after repletion (15 days).

casein protein (10%) or algal protein (10% and 15%). The period of repletion was 15 days.

After this experimental period the rats were sacrificed, the livers were collected and analyzed for nitrogen, and the activities of the enzymes succinic dehydrogenase (SDH), alanine amino transferase (AAT), and alkaline phosphatase (AP) were determined.

During the depletion period of 12 days, the rats lost about 20 to 25% of their weight. During the following repletion period fed on *Scenedesmus,* the food intake by the animals was slightly higher compared to the other diets, while weight gain and final body weight of the rats fed on diets containing 10% casein protein was found to be higher than in rats fed with algal diets (Table 18). Higher algal contents in diets resulted in increased body weight gains, as the weight of the kidney, heart, and liver were found to be comparable in all the repleted groups.

The regeneration of liver protein by the 10% casein protein diet was comparable to that of the algal diet containing 10% protein. The serum albumin level was low (2.8%) in depleted animals when compared with the value of 4.8% in animals fed on stock diets. In rats repleted on 10% casein protein, the albumin level increased markedly to 4.7, which is nearly the same level as in rats fed on stock diet. The increase in serum albumin in rats fed on 10% algal protein diet was 3.4, and 3.6 on a 15% algal protein diet. It is evident that algal protein was less effective than casein for the regeneration of serum albumin.

The extent of regeneration of liver enzymes and serum proteins following depletion may be attributed to the quality and level of dietary proteins. Liver enzymes have been shown to be sensitive to protein levels in the diet. Low activity of SDH on a protein-restricted diet may be due to decreased availability of -SH groups for activation, or to a decrease in the absolute amount of the enzyme. Increased activity of alkaline phosphatase during protein depletion has been reported by others and attributed to the accumulation of plasma phosphatase in the liver, due to impaired elimination.[149,243,244]

Rats fed on stock diets showed higher levels of SDH and AAT and lower activity of AP in the liver as compared to the animals kept on the protein-free diet (Table 19).

Among the different groups of repleted rats, regeneration of enzyme activity was more pronounced with the casein diet compared to diets containing 10 and 15% protein from algae. The protein quality of sun-dried *Spirulina* was evaluated by the same method. Diets of 10% protein level were used for this experiments.[326]

The data on body weight and food uptake of the different groups are given in Table 20. Weight gain and final body weight of the rats fed on a 10% casein protein diet were found to be higher than of the rats fed on algal diets, indicating the nutritive value of the algae was less than that of casein.

Table 19
LEVELS OF HEPATIC ENZYMES
IN RATS FED DRUM-DRIED
SCENEDESMUS AND SUN-DRIED
SPIRULINA IN DEPLETION-
REPLETION STUDIES[148]

Diet	SDH	AAT	AP
Scenedesmus			
Stock	1.29	87.84	28.34
Depletion[a]	0.49	57.44	48.41
Repletion[b]			
Casein (10%)	0.78	83.07	26.22
Algae (10%)	0.69	78.52	24.75
Algae (15%)	0.73	82.75	25.37
Spirulina			
Stock	1.23	84.23	28.20
Depletion[a]	0.47	51.62	50.60
Repletion[b]			
Casein (10%)	0.78	82.42	26.98
Algae (10%)	0.71	77.75	23.44
Algae (10%) + 0.3 met	0.74	79.43	24.62

SDH: Succinic dehydrogenase, μg TTC reduced/mg
fresh liver/10 min.
AAT: Alanine amino transferase, μM pyruvate lib-
erated/g fresh liver/10 min.
AP: Alkaline phosphatase, μM phosphorus liber-
ated/g fresh liver/hr.

[a] Data after depletion (12 days).
[b] Data after repletion (15 days).

Rats fed on stock diet showed higher levels of SDH and AAT and lower activity of AP than rats fed the experimental diets. On depletion the levels of SDH and AAT decreased significantly while there was a sharp increase in AP activity (Table 19). Among the different groups of repleted rats, regeneration of enzyme activity was more pronounced in the casein diet as compared to algal diets, although the group fed methionine-fortified *Spirulina* reached nearly the same level as the casein group.

In summary, algal protein represents a vegetable protein of relatively high nutritional qualities, as could be demonstrated by many nutritional studies. Algal protein, however, is less effective than casein. This difference seems to be more pronounced in the depletion-repletion method than in the other methods employed, which might be evidence for the sensitivity of the first method. Since no other conventional protein sources were included in the experiments, the results obtained in depletion-repletion studies were not sufficiently substantiated.

SUPPLEMENTATION STUDIES

Although certain algae have been considered as a food by some population groups, it is without a doubt a new type of food for most human populations so its immediate utilization should not be expected. The nutritional studies performed with algal biomass have clearly demonstrated that it represents a valuable source of protein, minerals, and vitamins with a good potential to complement and enrich other conventional foodstuffs.

Table 20
DATA OF DEPLETION-REPLETION STUDIES WITH RATS FED DIETS CONTAINING SUN-DRIED *SPIRULINA*[148]

Diet	Feed consumed (g/day)	Weight gained (+) or lost (−)	Liver	Kidney	Heart	Liver protein (g)	Total	Albumin	Globulin
			g/100 g Body weight				Serum protein (g/100 mℓ)		
Stock	21	+84	2.84	0.65	0.30	22.10	7.4	4.8	2.6
Depletion[a]	15	−34	2.36	0.60	0.38	16.13	4.7	2.7	2.0
Repletion[b]									
Casein	17	+82	2.78	0.64	0.34	21.16	7.1	4.6	2.5
Algae (10%)	20	+56	2.77	0.63	0.22	18.45	6.3	3.4	2.9
Algae (10%) + 0.3% met	21	+58	2.79	0.65	0.33	18.98	6.6	3.6	3.0

[a] Data after depletion (12 days).
[b] Data after repletion (15 days).

Table 21
SUPPLEMENTARY EFFECT OF ALGAL PROTEIN ON FLOUR AND BREAD[150]

	Weight gain (g/27 days)	FCE[a]
Flour	6.4	—
Flour + lys (0.75%)	25.5	—
Flour + thr (1.20%)	7.3	—
Flour + *Scenedesmus* (4.0%)	20.3	—
Flour + lys (0.75%) + *Scenedesmus* (4.0%)	53.2	—
Flour + thr (1.20%) + *Scenedesmus* (4.0%)	18.6	—
Bread	14.8	11.7
Bread + lys (0.75%)	96.3	3.2
Bread + lys (0.75%) + thr (1.20%)	139.0	2.6
Bread + lys (0.75%) + soy protein (2.1%)	99.1	3.5
Bread + *Scenedesmus* (4.0%)	40.8	—
Bread + *Chlorella*	133.0	2.9

[a] FCE: Feed conversion efficiency = feed consumed (g)/weight gained (g).

Rice, wheat, corn, and ragi (*Eleusine coracana*) form the major sources of protein in the diets of many poorly nourished populations. These cereals are often deficient in lysine and threonine. Several studies aimed at improving the nutritive value through supplementation either with the deficient amino acids directly or with legumes or unconventional protein sources such as leaf-protein or yeast have been reported. Over 25 years ago, rat feeding experiments were carried out by Hundley and Ing[150] to study the feasibility of supplementing wheat and bread with algal preparations (Table 21).

Scenedesmus improves growth significantly with both flour and bread diets. These effects can be attributed almost certainly to amino acid supplementation. *Scenedesmus* contributed considerable lysine, although the lysine content of this algae was not sufficient for a full growth response.

The experiment with *Chlorella* indicates that this algae is an effective source of threonine in bread as judged by both growth and feed conversion efficiency. According to these results, this algae is a better source of threonine than purified soy protein.

Table 22
PROTEIN EFFICIENCY RATIO (PER) AND NET
PROTEIN UTILIZATION (NPU) OF PROTEIN
MIXTURES BASED ON *SPIRULINA* AND CEREALS[124]

	Protein content (%)	PER	NPU
Casein	10.0	2.50	54.4
Corn	7.0	1.23	30.5
Wheat	9.0	1.43	32.5
Spirulina + corn (1:1)	10.0	1.72	34.7
Spirulina + corn (3:1)	10.1	1.80	37.2
Spirulina + corn (3:1)	20.1	1.76	37.9
Spirulina + wheat (1:1)	10.2	2.25	51.1
Spirulina + wheat (1:1)	14.5	1.96	48.4
Spirulina + wheat (1:3)	10.8	1.74	41.1
Spirulina + corn + oat (5:3:2)	10.0	1.91	35.0
Spirulina + corn + oat (5:3:2)	15.5	1.85	41.4
Spirulina + corn + rice (2:2:1)	10.0	1.95	45.1

Bourges et al.[124] have performed a similar study to evaluate the supplementation of some cereals with *Spirulina*. The PER and NPU of the different protein mixtures are given in Table 22.

In general, the PER and NPU values are in close parallel when they are examined together. The mixture *Spirulina*/corn oats, however, has a relatively high PER but a relatively low NPU value, and the mixture *Spirulina*/wheat (1:3) has a low PER with a comparatively not-so-low NPU value.

Both mixtures of algae with corn had a protein quality higher than corn alone, the effect being more apparent in the mixture with higher algal content. At a protein concentration of 10%, the utilization seems to be more efficient than at higher concentrations (14.5%). This is probably due to the fact that the essential amino acid pattern of the 10% mixture is already close to the pattern of well-balanced proteins.

To study the supplementary effect of algae to the conventional protein sources in India, several combinations of algae and food ingredients have been tested by carrying out PER experiments.[151,330] The results of the studies on the supplementation of rice, wheat, and ragi with drum-dried *Scenedesmus* and sun-dried *Spirulina* are given in Table 23.

As judged by rat growth experiments, *Scenedesmus* and *Spirulina* have a promising supplementary value to common cereals. Since all diets used in the study were fortified with minerals and vitamins, the effect of algae in improving the growth rate in cereal diets which were devoid of these nutrients could not be demonstrated. Though the yeast contained about 45% protein, its PER was considerably lower than that of algae. The reasons for the low value may be complex and due to several factors such as unbalanced amino acid pattern, insufficient processing of the algae, or poor consumption by the animals. As can be seen, addition of *Spirulina* to yeast-based diets improved the PER.

Cook et al.[139] tested the protein quality of the wastewater-grown algae, *Scenedesmus* and *Chlorella*. Different mixtures of algae with cereal and nonfat milk powder were fed to rats for 28 days. Nonfat dry milk was used as reference (Table 24). When this standard was given as the sole source of protein in the diets, the PER was significantly higher than that of uncooked or cooked algae or cooked oatmeal. Algae boiled for 30 min showed a higher PER than the unboiled due to the improved digestibility. The PER of cooked oatmeal was better than that of cooked algae. Since in raw or cooked algae the proteins are only partially digestible, low PER values may be expected. Nevertheless, addition of algae to the combinations of the other protein sources still improved the PER values.

Table 23
SUPPLEMENTARY VALUE OF *SCENEDESMUS*
AND *SPIRULINA* TO DIETS BASED ON RICE,
RAGI, WHEAT, AND YEAST[151,330]

Diet	Protein (10%)	PER
Scenedesmus	10.0	2.39
Rice	5.8	2.46
Rice + *Scenedesmus* (3:1)	8.5	2.65
Rice + *Scenedesmus* (1:1)	9.3	2.95
Ragi	6.0	1.38
Ragi + *Scenedesmus* (3:1)	8.6	2.10
Ragi + *Scenedesmus* (1:1)	8.8	2.45
Wheat	10.0	1.36
Wheat + *Scenedesmus* (3:1)	10.3	1.82
Wheat + *Scenedesmus* (1:1)	10.0	2.28
Wheat + rice + *Scenedesmus* (1:1:1)	9.5	2.64
Spirulina	10.0	2.01
Rice + *Spirulina* (3:1)	8.0	2.49
Rice + *Spirulina* (1:1)	9.0	2.56
Wheat + *Spirulina* (3:1)	10.0	1.51
Wheat + *Spirulina* (1:1)	10.0	1.96
Yeast	10.0	1.05
Yeast + *Spirulina* (1:1)	10.0	1.50
Yeast + *Spirulina* + *Scenedesmus* (1:1:1)	10.0	1.43

Table 24
SUPPLEMENTARY VALUE
OF BOILED ALGAE
(*CHLORELLA/SCENEDESMUS*)
TO CEREAL AND DRY MILK[129]

Diet	Protein (%)	PER
Fresh algae	12	1.61
Boiled algae	12	1.85
Nonfat dry milk	13	2.64
Boiled oatmeal	12	2.35
Boiled algae	5.3	
+ Boiled oatmeal	1.4	
		2.55
+ Cracked wheat	1.4	
+ Nonfat dry milk	3.8	

In order to evaluate the possibilities of substituting conventional protein sources in animal feed with algae, Cheeke et al.[131] performed nutritional studies with rats on *Chlorella*, grown on swine manure. The authors estimated the feed conversion efficients (FCE) of autoclaved (30 min at 120°C) *Chlorella* in comparison with other protein sources. At 13% protein level, fish meal and soybean meal supported higher growth rates and better food conversion efficiency than autoclaved *Chlorella* when used as the sole protein supplement to corn. At 18% protein, level, the performance of algae was equal to that of fish meal and soybean meal at 13% protein concentration. The results suggest that in a practical feeding system, algae could give satisfactory performance as protein supplements to grain, if used at a higher than traditional supplements.

Table 25
EFFECT OF AMINO ACID FORTIFICATION ON
CHLORELLA PROTEIN[131]

Diet	Daily gain (g)	FCE[a]
Chlorella (autoclaved)	3.8	5.5
Chlorella + lys (0.3%)	5.5	3.8
Chlorella + met (0.3%)	4.1	4.9
Chlorella + met (0.3%) + lys (0.3%)	6.1	3.6
Soybean meal	5.0	3.8

[a] FCE: Feed conversion efficiency = feed consumed (g)/weight gained (g).

It has been reported repeatedly that the fortification of algal diets with methionine improves the quality of the protein resulting in PER values close to that of casein. This indicates that methionine is a limiting amino acid.[54,91,94,123] This assumption was supported by experiments performed by fortifying *Chlorella* grown on swine manure with lysine and methionine, as can be seen from Table 25.

Supplementation with lysine alone resulted in growth equal to that obtained with soybean meal, indicating that in practical feeding situations where corn is supplemented with algae, methionine should be adequate. Thus, while by amino acid analysis or PER determination methionine appears to be the limiting amino acid, in practical feeding lysine is first limiting. Since algae are high in lysine content, it is necessary to identify why lysine availability is low; inactivation during heat processing and/or lysine "tie-up" as part of the indigestible cell wall are two possibilities to consider.

Contradicting observations are reported by Gross et al.[152] on the supplementation of *Scenedesmus* with amino acids. The authors found a significant continuous decrease of the PER with increased quantities of methionine, starting at 0.1% in the diet. A possible explanation for the negative effect caused by the addition of methionine might be related to the fact that feed preparations with higher concentrations of methionine can get a bitter taste, resulting in poorer palatability and subsequently lower consumption of the diet. In addition, the authors studied possible interactions of methionine and isoleucine supplementation (Table 26). No significant differences between the groups with regard to the PER and food intake could be detected. In the case of isoleucine the addition of 0.05% may perhaps be too low to observe a statistically significant improvement of the PER.

The urea content of the plasma is another parameter for the estimation of protein quality. In this case also, the methionine and/or isoleucine supplementation did not show any positive variation, and the uric level in the plasma did not vary significantly.

A detailed study on the nutritive value of *Spirulina* has been published recently by de Hernandez and Shimada,[113] based on the in vivo experiments in rats to determine limiting amino acids and digestibility. None of the amino acids studied (lysine, methionine, and histidine) added alone or in combination to diets containing 10% protein provided exclusively by *Spirulina* seems to be limiting; however, the results could be masked by the low palatability and acceptability of the products by the rats. As suggested by the authors, the PER were low and ranged between 1.21 and 1.65, compared to a soybean reference diet (Table 27). Another possibility might be that amino acid imbalances have been caused by adding pure amino acids, which are absorbed much more quickly than those bound in proteins.

Supplementation studies with *Spirulina* plus barley have been reported by Narasimha et al.[140] A diet containing 10% protein, provided in equal amounts from algae and barley, gave DC, BV, and NPU values of 81.1, 75.5, and 61.2 compared to the values obtained for *Spirulina* alone which were 75.5, 68.0, and 52.7, respectively. This result demonstrated

Table 26
EFFECT OF AMINO ACID FORTIFICATION ON
***SCENEDESMUS* PROTEIN[152]**

Diet	Digestibility in vivo	Uric acid (mg/100 mℓ)	PER
Casein	89.40	1.57	3.00
Scenedesmus	78.72	1.79	2.63
Scenedesmus + met (0,05%)	77.74	1.87	2.63
Scenedesmus + ile (0.05%)	78.87	2.04	2.70
Scenedesmus + ile (0;05%) + met (0.05%)	79.45	2.11	2.60

Table 27
EFFECT OF AMINO ACID FORTIFICATION ON
***SPIRULINA* PROTEIN[113]**

Diet	PER
Soya (23.5%) + met (0.27%) + lys (0.07%)	2.50
Spriulina (19.6%) + met (0.29%)	1.35
Spirulina (19.6%) + met (0.29%) + lys (0.27%)	1.21
Spirulina (19.6%) + met (0.29%) + lys (0.79%)	1.25
Spirulina (22.0%) + met (0.13%) + his (0.13%)	1.65
Spirulina (22.0%) + met (0.13%) + his (0.13%) + lys (0.29%)	1.40
Spirulina (19.6%) + met (0.30%) + (0.15%) + lys (0.35%)	1.21

that the good nutritive quality of *Spirulina* can be further improved by the addition of barley, which by itself gave BC, BV, and NPU values of 82.0, 71.2, and 58.0, respectively.

In conclusion, algal proteins are marginally deficient in sulfur-containing amino acids while lysine content often is higher than the nutritionally required amount. Combinations of algae and cereals would serve to improve the nutritional value of cereals and at the same time could prevent acceptability and tolerance problems, thus providing a sound base of industrialized products.

TOXICOLOGICAL ASPECTS

Whether or not a particular algal biomass is suitable for a utilization as animal feed or human food is determined by several factors mostly related either to the nutritional quality or the toxicological safety of the product. Algal products, therefore, should not be recommended for use unless both criteria have been answered satisfactorily. What follows is a description of the factors which should be considered in determining the toxicological safety of algae.

Nucleic Acids

A major limitation in the use of microorganisms as a food source may be a high nucleic acid (RNA and DNA) content. This is approximately 4 to 6% of the dry weight for algae, 8 to 12% for yeasts, and up to 20% for bacteria. Since man is lacking the enzyme uricase, the purines in nucleic acid are metabolized and excreted as uric acid.[153,154] Elevated serum levels of uric acid increase the risk of gout. Increased urinary concentration of uric acid may result in the formation of uric acid stones in the kidney, nephropathy, or other health hazards.

RNA, the predominant source of purines in single cell products, causes an increase in urinary uric acid of 100 to 150 mg/g RNA, indicating that only part of the purines in this nucleic acid are absorbed and eventually excreted as uric acid by the kidney. The effect of DNA on uric acid level in both urine and plasma is less than half that of RNA. Plasma uric acid concentration of normal men is 5.1 ± 0.9 mg/100 m/ℓ as measured by the enzymatic uricase method.[155] The plasma uric acid concentration of normal women is about 1 mg lower than that of the male. Standards for children are not so well established, but appear to fall in the lower range of the values for adults. Most authorities accept 6.0 mg uric acid per 100 mℓ plasma as the lower limit of the high-risk population and most of the cases with secondary gout had levels of 7.0 mg/100 mℓ or more.

Due to the higher nucleic acid and purine content in SCP, the Protein Advisory Group (PAG) recommended that the daily nucleic acid intake from any unconventional source should not exceed 2.6 g (with total nucleic acid from all sources not to exceed 4.0 g/day, thus setting a safety limit.[28]

According to Zöllner and Griebsch, feeding 1.0 g daily of RNA per 70 kg body weight increased serum uric acid concentration by 0.9 mg/100 mℓ in persons taking an otherwise purine-free diet; the intake of 1 g of DNA raised the concentration by 0.4 g/100 mℓ.

Edozien et al.[157] found that in young men the serum uric acid level of 4.5 mg/100 mℓ on a purine-free diet increased to approximately 7.0 mg/100 mℓ after taking 45 g of yeast (= 2.9 g total nucleic acids) daily. Doubling of the yeast quantity increased uric acid levels to about 8.5 mg/100 mℓ.

Data on the effect of algae-containing diets on the uric acid concentration in humans are scarce. Experiments conducted in Germany and Bangkok in clinical studies with volunteers did not reveal an alarming increase of uric acid concentration due to the consumption of algae-containing diets. This was confirmed by Felheim,[158] who reported the results of experiments performed in Thailand. In this study, 20 adult persons received a normal diet with the addition of 21 g of dry algae (*Scenedesmus*, 7 g/meal) and another 20 persons received the same diet without algae. The actual intake was 12 to 14 g algae per day per person. The duration of the experiment was 20 days; on day 11 the two groups were interchanged. During this evaluation the concentration of uric acid in plasma was estimated five times in all experimental individuals. It was found that the addition of algae to the diet did not change the concentration of uric acid normally present in the tested individuals (4.9 to 5.5 mg/100 mℓ). Higher algal concentrations were used by Waslien et al.[153] who determined uric acid levels in men fed algae as the sole protein source. In their experiments 7 healthy men were given casein- or algae-containing diets at two dietary levels to provide 25 and 50 g protein per day; pure RNA was added to the casein. At a protein/purine-free diet the mean uric acid concentration in the plasma was 5.4 mg/100 mℓ. After a daily intake of 25 and 50 g of protein from casein, providing 1.8 and 3.7 g of RNA, the uric acid concentration increased to 6.9 and 8.7 g/100 mℓ, respectively. The intake of equal amounts of algal protein (1.7 and 3.6 g of nucleic acid), elevated the plasma uric acid concentration to 7.4 and 9.7 mg/ 100 mℓ, respectively, reaching the lower limit of the abnormal range.

Such elevated plasma uric acid levels could not be substantiated by Griebsch and Zöllner[159] who examined the effect of drum-dried *Scenedesmus* on the uric acid metabolism. They administered high algal concentrations, which provided the minimum requirement to maintain protein balance, i.e., 0.62 g/kg body weight as well as one half and double this amount (0.31 and 1.24 g/kg body weight). They found that at a daily intake of 0.31 g of algal protein per kilogram of body weight, uric acid levels, even in "latent" hyperuricemic persons, remained within the norm (6.5 mg/100 mℓ), while the level increased to 13 mg/ 100 mℓ when 1.24 g algal protein per kilogram body weight were consumed daily. Almost contradictory and deviating from all other reports are the data of Müller-Wecker and Kofrany,[154] who tested the correlation between uric acid production in the bodies and alga intake.

In their study, the test person was kept at 100% whole-egg protein for 9 days during which the average uric acid concentration in the serum was 5.07 mg/100 mℓ. Therefore, 22% of the egg protein was replaced by *Scenedesmus* crude protein (20 g of algal material). At this diet the average serum concentration of uric acid dropped to a lower level of 4.31 mg/100 mℓ.

Based on these experiments and by assuming a maximum nucleic acid content of 6% in algae, the uppermost tolerable limit for an addition of unicellular algae to normal diets would be 30 g/day for adult persons. By considering the large number of endangered latent hyperuricemic persons, this level would better be reduced to about 20 g/day or 0.3g/kg body weight.

Swine have a very low excretion capacity for uric acid (ten times less than man). This characteristic, however, is well compensated for by the ability of this species to build high amounts of allantoic acid, which is harmless in respect to health disturbance. Therefore, it can be expected that pigs may tolerate higher amounts of algae in their diets than man. Also, broiler production on algal mass presents no problem from the standpoint of nucleic acids because poultry have a high excretion capacity of uric acid as related to body weight. Even lesser problems can be expected with fish which transform most compounds of the the the nitrogen to urea and ammonia.

Toxins

As part of the chemical characterization of the algal material, the samples have to be analyzed for the presence of toxic compounds. For better understanding, these toxins have been divided into two groups: biogenic and nonbiogenic toxins. According to this grouping, biogenic toxins include all those compounds which are either synthesized by mircoorganisms or formed through the decomposition of microbial metabolic products. Since outdoor algal cultures can be contaminated with several microorganisms, algal samples have to be examined not only for algal toxins (phytotoxins), but for fungal (mycotoxins) and bacterial toxins.

Nonbiogenic toxins or environmental contaminants comprise all such substances which are not produced by the algae or other microorganisms, but enter the culture from the environment and are absorbed and accumulated by the algae. Most toxins of this group are of anthropogenic origin (pollution with heavy metals, pesticides. etc.). Nutrients from the culture media or residues from processing stages represent other possible sources of contamination. While the occurrence of biogenic toxins is an intrinsic characteristic of certain microorganisms, impurities with nonbiogenic toxins can be avoided in most cases by proper site-selection for algal plants, i.e., in areas without industry, heavy traffic, etc., and also by using appropriate nutrients.

Biogenic Toxins

Phyco/Mycotoxins

One of the drawbacks to using bloom-forming freshwater cyanobacteria as a source of food or feed relates to the fact that toxic strains of several species of cyanobacteria are often morphologically indistinguishable from nontoxic strains. Coccoid forms of these toxic algae *(Microcystis)*, which are smaller than *Scenedesmus* and *Chlorella* may contaminate cultures of these chlorophyceae. More than 12 species belonging to 9 general toxic fresh water cyanobacteria have been implicated in cases of animal poisoning. The most common forms are *Microcystis aeruginosa*, *Anabaena flos-aquae*, and *Aphanizomenon flos-aquae*. The phycotoxins of these algae caused casualties in livestock and can cause allergic reactions or gastroenteritis in humans. Fortunately, no such case has ever been reported in relation to the mass cultivation of algae, due perhaps to the fact that culture conditions are not favorable for the toxic algae.[160]

Since there has been concern about the possibility that algae selected for utilization as a food source may contain toxins, detailed investigations have been performed in Germany for the presence of toxins in *Spirulina* and *Scenedesmus*.[17,22] The following compounds have been sought after: aflatoxin, ochratoxin A, sterigmatocystin, citrinin, patulin, penicillic acid, zearalenone, T_2-toxin, diacetoxyscirpenol, and thrichothecene. Dermatological toxicological tests with trichothecene, toxicity studies with mice on tremorgens (fumitremorgen and verrucolugen), and investigations on antibiotic activities (penicillium toxins) have also been conducted. None of the above-listed toxins could be detected in the algae samples and all biological tests gave negative results. Also, various feeding trials performed with quite different algal samples have not revealed any pathological symptoms due to biogenic toxins. It could be concluded that the algal species examined were free from phycotoxins or any other toxins. It seems quite unlikely that these toxins occur in clean cultures and the presently cultivated species can be assumed to be safe as food or feed.

A possible antinutritive effect of hemagglutinin was studied by Gross et al.[152] Tests with different human blood types and with cow and rabbit blood showed no hemagglutinating activity in *Scenedesmus obliquus*.

ENVIRONMENTAL CONTAMINANTS

Heavy Metals

For quite some time, the contamination of algae with heavy metals was one of the major problems in algal cultivation because algae, like many of the microorganisms, are capable of accumulating heavy metals at concentrations which are several orders of magnitude higher than the concentrations in the surrounding media.[161-163] This fact deserved serious consideration, since it has been repeatedly reported that the metal concentration in algae may reach levels which are toxic for the potential consumer. Among the different metals found in algae, lead, cadium, mercury, and arsenic, are those which require special attention because of their health hazard.[35,84,164-166]

The only recommendation regarding the permitted intake of heavy metals through food is published by WHO/FAO.[167] These figures state that an adult person of 60 kg body weight should not consume more than 3 mg of lead, 0.5 mg of cadmium, 20 mg of arsenic, and 0.3 mg of mercury per week. However, one must take into consideration that (especially in the case of lead) children react much more sensitively than adults and therefore, the tolerable amounts for children have to be much lower than the calculations indicate.

As long as no legislative directions stipulate the heavy metal content in algae or other unconventional protein sources, the only guidelines in this matter are the recommendations of PAG and IUPAC. To illustrate the situation, a few recommendations and indications of heavy metal concentrations in microalgae are presented in Table 28. These data, obtained by analyses of samples from various algal cultivation units, reveal that the culture techniques as well as the locations of some of the present algal cultivation plants have to be modified. The main reasons for high values of heavy metals concern the fertilizer and the water used as well as with air pollutants. How these contaminants can be avoided must be discussed with respect to the local conditions. After all, it should be possible (by choosing the ideal location and production process) to obtain an algal product which meets the strict regulations on heavy metal contaminations in food products.

It is known that the metal accumulation in microorganisms is a relatively rapid process. Bacteria can attain the equilibrium distribution of heavy metals in the cells and the liquid phase within 15 min. In the case of algae, the equilibrium time is certainly longer, but a saturation of the algae with metals will be reached within 24 hr; any exposure longer than this time will not increase the accumulation capacity of the algal cells.[162,171-173] Thus, the accumulation of heavy metals, if they are present in the medium, cannot be avoided. The only possibility to control the amount of metals in the algae is to monitor the metal con-

Table 28
DATA ON HEAVY METAL CONTAMINATIONS (PPM)

Source	Pb	Cd	Hg	As	Ref.
Upper limits in drinking water (WHO)	0.1	0.01	0.001	0.05	168
Upper limits in SCP (IUPAC)	5.0	1.0	0.1	2.0	169
Maximum weekly intake (mg/adult)	3.0	0.5	0.3	20.0	167
Scenedesmus (Thailand) without fertilizer	—	0.35	—	0.06	165
Scenedesmus (Thailand) with fertilizer	6.03	1.67	0.07	2.36	165
Scenedesmus (Germany)	34.8	2.46	0.09	2.36	165
Scenedesmus (Peru)	0.58	0.30	0.43	0.91	54
Spirulina (India)	3.95	0.62	0.07	0.97	54
Spirulina (Mexico)	0.4	0.1	0.24	1.1	31
Spirulina (Mexico)	5.1	0.5	0.5	2.9	170
Spirulina (Chad)	3.7	—	0.5	1.8	170

centration in the culture medium, taking care to use fertilizers and water with the lowest possible metal content.

Besides the amount of heavy metals, accumulated within the algae during cultivation, there can be another source for these impurities in the harvested biomass. Under alkaline conditions and in the presence of phosphate or sulfate ions, cadmium and lead form slightly soluble compounds which precipitate or float adhering to small particles in the culture medium. In harvesting, most of these metal salts will be harvested together with the algae and subsequently will remain in the algal powder after drying.

With the present technology it is not possible to separate the algae from these compounds by introducing additional separation steps or by a simple water rinse. It seems that the reported high values of accumulated metals in algae are often not primarily due to the metals contained in the algae itself, heavy metal ions often do not penetrate into the cell, but remain absorbed in or on the other cell wall, and can be removed by treating the algae with complexing agents.[174-176]

Yannai et al.[177] evaluated the wholesomeness of algae grown on wastewater by feeding them to poultry and carp. This investigation is of special importance with regard to heavy metal contamination since the algae used in their investigation contained relatively high levels of contaminants. The study was divided into two parts. The first part dealt with the safety of test animals raised on algal-containing ratios. The second part included analyses of the feed used and of tissues taken from the test animals for the presence of toxic metals.

The heavy metal contents in the algae used for this investigation are given in Table 29. The chickens used in the experiments were raised on ration containing 15% of dried algae, the carp on ration containing 25% of dried algae; controls were grown on commercial diets.

For the toxicity test, called "secondary toxicity test" by the authors, weanling rats were fed on meat of the chickens which were raised on algae. At the age of 3 months the rats were mated and the second generation was raised on the same diets. No significant difference in growth performance, general appearance, behavior, survival, fertility, and lactation could be detected between rats fed with diets containing up to 30% of dehydrated meat from chicken raised on the algae-containing ration and rats fed with control diets. However, all the diets containing the large amount of dehydrated meat were apparently not very palatable to the rats.

The histological examination carried out on 18 different tissues did not reveal any noticeable abnormalities in the rats maintained on algae-fed chickens. The meat of the chickens fed by algae-containing ratios must have been wholesome, judged by all the parameters employed in this study.

No appreciable accumulation of any of the measured elements in the chicken or carp

Table 29
METAL CONTAMINATIONS IN SEWAGE-GROWN ALGAE USED FOR CHICKEN AND CARP FEEDING TRIALS (mg/kg)[177]

Metal	Scenedesmus	Micractinium	Chlorella
Mercury	0.30	0.64	0.26
Copper	45.30	33.10	24.20
Cadmium	1.6	1.3	1.4
Lead	3.8	2.9	8.1
Aluminum	33.90	7.4	0.33
Arsenic	1.1	1.3	3.6

a   Harvested by flocculation with alum.
b   Harvested by flocculation.

Table 30
METAL CONTENT (MG/KG) IN TISSUES TAKEN FROM CHICKENS FED DIFFERENT WASTEWATER-GROWN ALGAE[177]

Ration	Metal	Liver	Leg muscle	Tibia
Control	Hg	0.09	0.045	—
	Cu	3.00	0.83	4.23
	Cd	0.36	0.10	2.05
	Pb	0.45	0.40	3.70
Scenedesmus	Hg	0.04	0.04	—
	Cu	3.18	1.12	1.89
	Cd	0.24	0.12	0.60
	Pb	0.39	0.83	0.96
	Al	—	—	1.40
	As	0.81	0.11	—
Micractinimum	Hg	0.06	0.047	—
	Cu	3.20	0.83	2.81
	Cd	0.16	0.066	1.49
	Pb	0.37	0.27	0.44
	Al	—	—	6.40
	As	1.46	0.18	—
Chlorella	Hg	0.07	0.06	—
	Cu	3.20	0.84	2.39
	Cd	0.16	0.08	1.40
	Pb	0.23	0.34	4.70
	Al	—	—	5.10
	As	1.07	0.21	—

tissues could be detected (Tables 30 and 31), in spite of their fairly high concentrations in the animals' ration.

The fact that the relatively high concentration of heavy metals failed to bring about corresponding high levels in the animal tissues can perhaps be explained by the unusually large amount of phosphorus (apparently in the form of phosphate) present in all the algae samples assayed. All the metals mentioned above form water-insoluble phosphates and it is conceivable therefore, that these metals are rendered unabsorbable by the gastrointestinal tract of the animals. These observations were supported by another study conducted by Yannai and Mokady,[178] in which sewage-grown Micractinium with elevated levels of heavy metals was fed to Japanese quails. As can be seen from Table 32 the percentages of metal

Table 31
HEAVY METAL CONTENT (mg/kg) IN
MUSCLE TISSUE AND LIVER OF CARP
FED *CHLORELLA*[177]

Tissue	Ration	Hg	Cu	Cd	Pb	As
Muscle	Control	41	890	50	50	100
	Alga	16	940	40	40	70
Liver	Control	41				
			Not estimated			
	Alga	44				

Table 32
METAL CONTENTS IN ALGAE AND RATIONS AND THEIR
ABSORBABILITY IN JAPANESE QUAILS[178]

Metal	Algae	Control ration		Ration with 20% algae	
		Metal content	% Absorbed	Metal content	% Absorbed
Al (mg/g)	50.9	1.01	25.7	11.1	3.0
Cd (mg/kg)	1.34	0.57	36.0	0.73	38.7
Cu (mg/kg)	74.0	10.8	27.0	29.4	30.0
Pb (mg/kg)	2.0	0.12	—	0.49	6.4
Hg (mg/kg)	0.26	0.05	—	0.11	—

absorption were quite low. Also, the residue of aluminum in algae, resulting from aluminum-containing flocculants, does not seem to be harmful when administered to animals.[179]

Potential toxic effects of heavy metals accumulated in algae were also studied by Boudene et al.[170] The authors fed *Spirulina* contaminated with fluorine and heavy metals (F: 112, As: 1.3, Cd: 0.09, Hg: 0.10, and Pb: 0.52 ppm) to rats over a period of 75 weeks. The diets contained 14% protein, provided either by algae or casein (control). The animals submitted to the test diet showed no difference from the control animals. The increase in weight was comparable between the two groups and no evident toxicity related to the sample could be detected. An elevated level of 3 ppm arsenic was found in the tissue of the test animals after 5 weeks which leveled off to 1.3 ppm after the 12th week, while the level in the control groups was 0.2 ppm during the entire experiment.

Organic Compounds

Another group of environmental contaminants which might be accumulated in cultivated algae are organic compounds such as polychlorinated biphenyls or polycyclic aromatic hydrocarbons. Several substances of these groups are highly toxic and known carcinogens. Very few reports on the accumulation of these chemicals in algae can be found in the literature, probably due to the very costly and extensive analytical methods required. The available data are summarized in Table 33.

Analyses for dialkylnitrosamines of dimethyl and dipentyl (symmetrical and as metrical) nitrosopiperidine, nitrosomorpholine, and nitrosopyrrolidine in samples of *Scenedesmus* from Peru and *Spirulina* from India were negative (detection limit 0.1 μg/kg).[17] A tolerable maximal concentration for several of these compounds cannot be given since the health risks increase with increasing concentrations and because of the probability of accumulation in animal or human tissue.

Regulations about concentration limits for many pesticides or their residues in food products are in force in many countries. WHO has published values for an acceptable daily intake

Table 33
ANALYTICAL DATA OF ALGAL SAMPLES FOR ORGANIC TOXIC COMPOUNDS (µg/kg)

	Scenedesmus (Mexico)	Scenedesmus (Germany)	Scenedesmus (Thailand)	Scenedesmus (Israel)	Chlorella (Israel)	Micractinium (Israel)	Spirulina (India)	Spirulina (Mexico)	Spirulina (Mexico)
Fluoranthen	3.0	444.0	91.2						
Benzo (b) fluoranthene	4.1	85.4	5.9						
Benzo (k) fluoranthene	1.2	36.3	2.4						
Benzo (a) pyrene	2.7	39.5	1.4					2.0—4.3	2.6—3.6
Benzo (ghi) perylene	1.4	52.1	3.9						
Indeno (1,2,3-cd) pyrene	1.3	50.9	2.4						
α-HCH	0.01—0.03						18.0		
β-HCH	—						3.0		
γ-HCH	0.07—0.1						5.0		
HCB	0.007—0.009						0.04		
Dieldrin	0.03—0.08						4.0		
DDE	0.05			50.0[a]	trace	50.0[a]	0.6		
DDD	0.04			40.0[a]		60.0[a]	0.7		
o'p'-DDT	—			40.0[a]					
DDT	—			80.0[a]		30.0[a]	6.0		
PCB	—			700.0	600.0		0.5		
Ref.	54	165	165	179	179	179	17	180	19

[a] Part of the concentration is due to the presence of PCB.

(ADI) of various pesticides in its Technical Report Series 525, 454, and 574. As long as no specific regulations are available for SCP, these stipulations will be applied as guidelines for the utilization of algal biomass. According to the recommendations of IUPAC, the level of 3,4-benzopyrene (benzo(a)pyrene) in SCP should not exceed 1 ppb.

BIOLOGICAL CONTAMINATIONS

Certain kinds of contaminations in outdoor algal cultures are inevitable in view of the nonaseptic conditions of cultivation, where neither the medium nor the surroundings are sterile. However, it is necessary to monitor and to control these contaminants in order to obtain an algal biomass without harmful impurities and to keep the always-present contaminants within tolerable limits. This aspect has not been given adequate attention so far, but will become important as soon as algae are marketed for use in foods and feeds. The major types of contaminants in clean algal cultures (besides bacteria) are other algal forms, zooplankton, virus, fungi, and insects. The types and the extent of contamination may vary with geographical location, climatic conditions, and the particular cultivation system.

Nonbacterial Contaminations
Alien Algae

Contaminations of outdoor algal cultures with other algal forms, even in freshwater cultures, cannot be avoided completely since the culture conditions are fairly suitable for several different species of algae. However, in general it seems that within certain limits, healthy algae establish monocultures that suppress the spreading of other algal species, perhaps by releasing growth suppressing substances. Since none of the algal contaminants is considered harmful. the occurrence of alien chlorophyceae in cultures of green algae does not represent a serious problem. Benthic algae can be found on paddle wheels and on the walls of the algal tanks in the photic zone. The contaminations of *Spirulina* with other algae is limited to very few forms (*Chlorella*, diatoms). This is mainly due to the very specific culture conditions of this algae (high salinity and pH). which are not favorable for most other algae.

A special problem is presented by contaminants of wastewater treatment schemes. Since in these cases the algal population is usually a natural mixture of different algal species, changes in the dominant algae are quite frequent. It has been described that within a few days one algal species would be replaced by another or that algal cultures are completely wiped out by predators.[181,182]

Fungi

According to the former German algal project at Dortmund. the common contaminants in the *Scenedesmus* cultures were the parasitic fungi *Aphelidium* and *Chytridium*.[238] One of the major problems encountered in *Scenedesmus* cultivation in Peru was contamination with *Aphelidium* and similarly heavy contaminations with the same organism have also been reported from Bangkok.[185,237] This type of infection in *Scenedesmus* cultures has also been reported from projects in Israel and South Africa.[183,184] Since the zoospores and zoosporangium of this fungus are very small. it is quite likely that this contamination has been overlooked in several cases. The identification of this fungus can be done by special staining techniques using methylene blue.[185] *Aphelidium* can be controlled to some extent by using suitable fungicides; however, these chemical should not be applied repeatedly, especially in algal samples destined as food, in order to avoid fungicide residues.[209]

Detailed chemical and biochemical analyses of *Scenedesmus* culture infested by *Aphelidium* have not revealed any mycotoxin which could have been synthesized by this fungus. Based on available information it can be assumed that this parasite reduces the yield of algal cultures but does not represent any health hazard, if the infected algal biomass is used as feed or food.

Insects

Among the insect contaminants, *Anopheles* and *Chironomus* in *Scenedesmus* and *Ephydra* in *Spirulina* cultures have been reported. They commonly occur in outdoor cultures, especially in tropical/subtropical regions.[186]

The eggs of *Chironomus* hatch into worm-like. reddish larvae, which can be found in large numbers during certain seasons on the bottom of the algal ponds. Once the larvae have entered the algal culture, they can grow unhindered because of abundant supply of food (algae) and the lack of natural enemies or predators in this artificial biotope. The easiest method to keep this infection within limits is to remove the eggs of this insect from the culture before the larvae hatch. This can be done by inserting fine meshwire sieves into the current of the agitated algae and filtering the mucous eggs threads of the medium.

Protozoa and rotifers are almost always present in outdoor algal cultures. Grobbelaar[184] described infections of *Chlorella* cultures by *Philidina* sp., *Ciliata*, *Vorticella* sp., *Stylonichia* sp., and *Strobilidium* sp. Counts of up to 3×10^6 invertebrates per liter of medium have been made in dense, heavily infected cultures. He also reported on a shift of algal dominance from *Chlorella* to *Scenedesmus* caused by an infection with *Stylonichia*. While the protozoa could graze on *Chlorella* cells, the *Scenedesmus* colonies were too large to be taken in. Venkataraman and Sindhu Kanya[186] reported on occasional infections of *Scenedesmus* cultures by the rotatoria *Brachionus*, which in extreme cases completely spoiled the cultures. The most effective way of controlling this has been to lower the pH of the algal culture to about 3.0 by addition of acid, and to allow the culture to stand at this pH for 1 or 2 hr. Following this. the pH is readjusted to 7.5 with KOH. This treatment does not affect the

Table 34
PROBABLE SAFETY LIMITS OF
MICROBIAL CONTAMINATIONS IN SCP
AS RECOMMENDED BY IUPAC[169]

Organism	Amount
Total aerobic bacteria	100,000/g
Enterobacteriaceae	10
Staphylococcus aureus	1
Clostridia (total)	1,000
Clostridium perfringens	100
Streptococci of Lancefield's group D	10,000
Viable yeasts and moulds	100
Salmonella	1/50 g

algal cultures but very effectively eliminates the *Rotaroria*. (See "Elements of Pond Design and Construction.")

Contaminations with this zooplankton are infrequent in *Spirulina* cultures, probably due to the high pH and salt concentration in the medium for this algae. Contamination by zooplankton were reported in the high rate algal sponds at Singapore. The most serious problem encountered there was the frequent appearance of the algal predator *Moina* ssp. (Phyllopoda). Blooms of *Moina* resulted in rapid depletion of the algal population in the ponds. Elevating the pH to above 9 was effective against the predator: however, this procedure affected the algae and reduced productivity. Continuous mixing of the culture appears to be effective in controlling predation as well as in improving pond stability and productivity.[187]

Bacterial Contamination

While the contaminations described in the previous paragraphs certainly affect algal growth and yields, it can be assumed that they do not represent health risks if they are consumed by man or animals together with the processed algae.

Large-scale algal cultures are always contaminated with certain types of bacteria. Heterotrophic bacteria are even found in cultures grown autotrophically in inorganic media, because algal cells exude several organic compounds into the medium which are likely to promote bacterial growth.

Since possible public health hazards due to such contaminations in algal cultures will be a point of concern, sanitary safety is one of the important prerequisites for the utilization of algal biomass. IUPAC has published guidelines for acceptable upper limits of bacterial contamination in SCP[169] (Table 34), which are similar to the ones recommended by PAG.[188]

Analyses on the microbial flora of *Spirulina* biomass grown in open ponds have been performed by Faggi.[189] In this investigation the total bacterial count in dried samples never exceeded 10^4 cells/g, while it was 10^5 cells/mℓ in fresh samples (7% dry matter). No *Staphylococci, Streptococci, Enterobacteria,* or coliforms could be detected in the dry material. The author tested for some fungal species (*Aspergillus* and *Penicillium* spp.) yeasts (mostly *Rhodoturola*) and among the bacteria spore-forming bacilli, micrococci, and actinomycetes. All the findings were negative.

A more detailed study on the microbial load in mass culture of *Scenedesmus obliquus* and its processed powder has been reported by Mahadevaswamy and Venkataraman.[190] The authors compared the amount of infection in inorganic and organic culture media. In the organic medium sugar cane molasses was added as additional carbon source (Table 35).

As can be seen the total microbial counts increased during the cultivation period. This indicates that the algal culture favors the bacterial growth probably due to excretion of growth-promoting substances. However, this stimulation differed from case to case. There

Table 35
RANGE OF MICROBIAL CONTAMINATIONS IN OUTDOOR CULTURES OF
***SCENEDESMUS*[190]**

Culture conditions	Age of culture (day)	Algal conc. (mg/ℓ)	Total colonies $\times$ 10^5 units/mℓ	Yeast and molds $\times$ 10^2 colonies/mℓ	Coliforms MPN/mℓ[a]
Autotrophic (CO$_2$)	Initial	80	0.2—4.0	0.3—2.0	—
	3	200	8.0—25.0	3.0—8.0	11.0—75.0
	6	320	25.0—60.0	7.0—25.0	210.0—460.0
Heterotrophic (molasses)	Initial	80	0.2—4.0	0.3—2.0	—
	3	170	5.0—20.0	1.0—6.0	75.0—210.0
	6	260	25.0—80.0	4.0—10.0	120.0—240.0
Mixotrophic (CO$_2$/molasses)	Initial	80	0.2—4.0	0.3—2.0	—
	3	290	20.0—80.0	3.0—25.0	240.0—460.0
	6	480	200.0—500.0	20.0—60.0	460.0—1100.0

[a]   MPN = Most probable number.

Table 36
INFLUENCE OF DIFFERENT ALGAE ON
WEIGHT GAIN AND PROTEIN EFFICIENCY
RATIO (PER) OF GROWING CHICKENS[40]

Diet	Weight gain in 3 weeks (g)	PER
Soya + 0.54% met	237.0	3.04
Scenedesmus + *Chlorella*	87.0	1.55
Chlorella	9.0	0.31
Spongiococcum	13.0	0.43

are reports on pilot plant cultivation of algae, where no significant increase in bacterial load was observed,[181] indicating that other parameters besides organic excretion products influence the occurence of bacteria in algal cultures. Specific bacterial-algal interactions should be considered.[182]

Addition of molasses enhances the bacterial load in the cultures, which can be attributed primarily to the presence of sugars in the medium. Yeast and mold counts seem to increase almost parallel to the increase in the number of total bacterial colonies.

The different methods of drying affect the total bacterial count substantially. The lowest loads are found in algal powder obtained from drum-drying while sun-dried material resulted in the highest contamination rates. This difference can be expected since the drum-dried algae are treated by high temperature (120°C) for a short period. Which results in the destruction of most of the bacteria.

If the algal culture is raised with molasses as an additional carbon source, microbial load in the dried product is slightly higher even when the algal slurry is dried on a drum-drier. This observation might be due to the fact that the amount of spore-forming bacteria (which survive the heat-treatment) increase in the algal cultures when molasses is added as substrate.

Table 36 indicates storage studies conducted for a 3-month period using drum-dried *Scenedesmus* sealed in aluminum bags. The bacterial load did not increase during storage. No *Salmonella* or *Staphylococcus* were detected in any of the samples analyzed before or after storage. Similar tests on bacterial contaminations were performed with sun-dried *Spi-*

rulina. Outdoor cultures of this algae had an initial bacterial load of 2.0×10^3 CFU/mℓ, which increased to 9×10^4 CFU/mℓ after 15 days of open exposure. On an average, the microbial contamination in *Spirulina* cultures is more than one order of magnitude less than that in *Scenedesmus* cultures. This difference may probably be attributed to the high alkalinity of the *Spirulina* culture medium. No pathogenic forms such as *Salmonella* or *Staphylococcus aureus* were found either in the cultures or in the sun-dried *Spirulina* powder.

Regular monitoring of algal cultures, and in particular of the dried algal powder, is necessary since the microbial load is likely to affect the quality and safety of the final product. Guidelines for such a decision may serve the recommendations by IUPAC or PAG or other regional directions for sanitary safety of protein-rich foodstuffs such as milk products or infant foods.

TOXICOLOGICAL STUDIES WITH ANIMALS

Any form of SCP being used in animal feed or human food should be subjected to a strict control of the microorganism used, the raw materials, and the production technique, in order to guarantee that no toxic substances are present in the final product.[193-195] This means that carefully designed toxicological tests in several species of experimental animals have to be conducted for a sufficiently long period of time to prove the harmlessness of the biomass. Guidelines for these testing procedures were stipulated by international organizations such as the Protein Calorie Advisory Group or the UPAC.[28,155,169,196-204] National laws will probably follow the international recommendations, but until then investigations and tests should be continued according to these guidelines.

According to the recommendations of international organization, the required toxicological tests for all kinds of SCP include the following criteria:

Analyses of chemical composition	Protein, fat, carbohydrate, ash, amino acid profile, vitamins
Analyses for biogenic toxic substances	Nucleic acids, hemagglutinins, microbial toxins, and other toxicants
Analyses for nonbiogenic toxic substances	Heavy metals, pesticides, residues from cultivations and processing
Nutritional studies	PER, NPU, DC, BV
Sanitary analyses	Microbial contaminations
Safety evaluations	Short- and long-term feeding studies with rodent and other species (pig, quail, monkey), multigeneration studies, teratogenic and mutagenic studies
Clinical studies	Human testing of algae and supplementary food mixtures
Acceptability studies	Sensory evaluation, development of recipes

Toxicological testing of microalgal biomass has not reached the same level as the experimental evaluation of other types of SCP. Probably the most detailed toxicological study on algae was published by UNIDO.[205] This report comprises evaluations with *Spirulina* on subacute and chronical toxicity, reproduction, lactation, mutagenicity, and teratogenicity. In addition to a control diet and a commonly used laboratory diet, three different diets with increasing concentrations of algae (10, 20, and 30%) were employed substituting soybean meal in the rations. The respective amounts of soybean meal were 30, 14, and 0% (control 44%), and all diets contained 8% fish meal.

In the first set of experiments, weight gain in rats was monitored for a period of 13 weeks.

Table 37
RELATIVE ORGAN WEIGHTS (FRESH) OF
RATS FED SUN-DRIED *SPIRULINA* AND
CONTROL DIET FOR 12 WEEKS (g/100 g
BODY WEIGHT)[54]

Organ	Casein (10% protein)	Alga (10% protein)	Alga +0.3% met (10% protein)
Liver	2.78	3.24	3.46
Kidney	0.55	0.66	0.64
Heart	0.29	0.36	0.34
Lung	0.36	0.37	0.42
Spleen	0.16	0.17	0.17
Testes	0.70	1.17	1.27

No significant differences could be observed between control and experimental groups. The relative weights of various organs (liver, heart, kidney, lungs, brain, spleen, testes, ovaries, thymus, thyroid, adrenals, and pituitary) of all animals were comparable. Detailed histopathological examinations did not show any anomalies. Analyses of blood, urine, total serum protein, and some enzymes (GOT, GPT, AP) were within the normal range. A second similar study lasting 80 weeks came to the same conclusions.

In order to reveal any long-term effect of the algae, a 2-year reproduction and lactation study was conducted over 3 generations. During all generations, fertility, gestation, and lactation periods were recorded. The third generation was subjected to a subacute toxicity study, and the following parameters were examined: weight gain, feed consumption, blood, urine, organ weight, and histopathology. No effect was observed regarding fertility and litter size; the study on subacute toxicity failed to show any harmful effect caused by the algal diets.

For the mutagenicity study rats and mice were fed over 3 months on experimental diets containing 30% *Spirulina*. After this period of time the animals were mated, 16 days later the females were sacrificed, and the fetuses in the uterus were examined. No mutagenic effect of the algae was observed, confirming the results of the previous study. The teratogenicity study included rats, mice, and hamsters which were fed with the five different diets mentioned above. No resorption of the embryos or anomalies of the fetuses could be observed.

In summary, it may be stated that none of the parameters tested with the three algal concentrations in the diet showed any acute variations from the control values. Differences observed were not caused by dose effects, but were the results of other external factors and could not be reproduced. The same alga was examined toxicologically by Boudene et al. in rat feeding experiments lasting 75 weeks.[170] No negative results were reported which might have been caused by the algal containing (25%) diet.

In order to evaluate the effect of continuous feeding of sun-dried *Spirulina* as the sole source of protein, short-term feeding experiments on rats were carried out by Becker and Venkataraman[54] for a period of 12 weeks. *Spirulina, Spirulina* +0, 3% methionine, and casein (control) were incorporated at 10% levels in diets. The consumption of algal diets was lower compared to casein and a similar trend was found in the weight gain. The relative organ weights of the animals are listed in Table 37. The weights of the organs of the animals in the experimental groups were similar or slightly higher than those of the control group. Histological investigations of heart, liver, kidney, and spleen did not reveal any serious abnormalities in any of the animals. The absence of negative effects caused by the feeding of *Spirulina* was confirmed by Contreras et al.[138] who tested this algae as the sole source of protein in sexually maturing rats.

The liver weights of the *Spirulina*-fed rats were similar to those measured in rats fed the commercially available diet; however, both groups had significantly lower liver weights than the casein group. Although the reason is not clear, these results are similar to those reported by others.[129] There were no significant differences between the weights of testes, pituitary, or ventral prostate gland of *Spirulina*-fed rats as compared to the control. All animals had normal development of the reproductive system as judged by organ weights, testicular histology, and reproductive hormone levels. No apparent signs of organ or body toxicity were observed. Chung et al.[134] have reported on the breeding performance of female rats fed with *Spirulina* as the sole source of protein from the weaning age to the second week of lactation.[134] All algae-fed females conceived, produced litters, and lactated satisfactorily. The average litter size and weight at birth were 10.3 and 58.3 g for the *Spirulina* group, and 9.3 and 52.8 g for the casein group, respectively.

Spirulina seemed to support lactation successfully; postweaning growth rats were also normal. Histological observations of liver, kidneys, heart, lungs, spleen, stomach, intestines, testes, and lymph nodes showed no abnormalities of male rats fed with *Spirulina* for 2 weeks. The livers of female rats which had been fed *Spirulina* from weaning to the time their litters were weaned (100 days) were also histologically normal.

The tolerance of rats to *Spirulina*-rich diets (260 and 730 g algae per kilogram diet) was also tested by Bourges et al.[124] In spite of the high algal content of both diets, all animals survived with apparent health, all organs were macroscopically and microscopically normal, and no differences were found between the control and experimental groups in any of the parameters studied.

Toxicological studies with drum-dried *Micractinium*, grown in ponds containing industrial effluents and harvested by alum flocculation, were performed by Yannai and Mokady.[178] The diets were incorporated at different levels into ration for chickens, Japanese quails, and mice with a view to establish their safety for the above species as well as the safety of chicken meat for humans. After 7 weeks of feeding on diets containing 7.5 and 15% algae to chicken, no abnormalities and no difference in growth were found between any of the groups.

The growth rates of young Japanese quail reared for 4 weeks on ration containing 10% or 20% dried *Micractinium* did not differ significantly from that of the controls. The birds fed ration with the greater amount of algae produced a somewhat smaller number of eggs per female, had a slightly lower percentage of fertilized eggs, and exhibited poorer hatchability. However, the average weight of eggs and the 1-day chicken were comparable for the three groups. As was noted for the chickens and the quails, in mice (males and females) all the groups exhibited similar growth, no abnormalities were observed, and their reproduction performance appeared to be normal.

In a so-called ''secondary toxicological test'' (see also section on toxicology/heavy metals) the same authors fed chicken meat (at a level to provide 16% protein, i.e., approximately 30 to 35% of dried meat), which were raised on diets containing 15% algae, to 4 generations of rats. In none of the generations were any abnormalities observed.

Long-term feeding of rats with a control diet and a diet containing protein from *Spirulina* were carried out by Bizzi et al.[211] In their study 60% of the total amount of protein (22.2%) in the experimental diet were substituted by algal protein. Micro- and macroscopic autopsy findings revealed no difference between the experimental groups, except for an increase in the organ weight/body weight ratio of some organs of the female rats fed the experimental diet. All the blood chemical and biochemical findings were similar.

A chronic toxication test of *Spirulina* was reported from Japan.[188] The algae was fed to Wistar rats for 6 months. The diet was a 4:1 mixture of ordinary diet and dried *Spirulina* or the ordinary diet alone. Comparisons were made between experimental and control groups in growth, external appearance, hematological tests (Hb, MCHC, urea-N, serum protein, albumin/globulin ratio, uric acid, creatinine, lipids, triglyceride, phospholipids, cholesterol, bilirubin, Ca, P, Fe and enzyme activities of alkaline phosphatase, leucine amino peptidase, LDH, and SGOT) observations and histological examinations of several organs (brain, heart, stomach, liver, spleen, kidneys, testes or ovaries, hypophysis, and adrenal glands) were performed at the end of the study. No differences were observed between experimental and control groups in growth, weight, and histological findings of the organs. Although the hematological tests showed some differences, abnormalities were not detected. Thus, the authors conclude that the algae does not cause any toxic effects on rats after a feeding period of 6 months.

In 1976, a very detailed animal feeding study with *Scenedesmus* was started at Bangkok, following the recommendations of the PAG Guidelines.[74] This investigation was designed as a 2-year toxicological and multigenerational study including all important parameters (blood chemistry, urine analysis, autopsy, histopathology, number of offsprings, etc.). The experiment included 596 animals: the experimental diet had a crude protein content of 24% and contained 25% drum-dried *Scenedesmus* which partly substituted fish meal and soya protein. Two months after commencement the animals began to develop anomalies of the incisors. This anomaly started with a discoloration of the teeth followed by a deviation from the normal growing pattern. After 1 year all animals fed on algal diets had developed teeth abnormalities, while none of the other parameters analyzed showed significant differences between experimental and control groups. In an attempt to explain which factors might have caused these anomalies several additional experiments were carried out with different diet compositions. It was suspected that an unfavorable Ca:P ratio in the algae, caused by the cement plastering of the algal ponds, could have been the reason for the abnormalities. The normal Ca:P ratio of the diet was 1:0.75%, while the algal sample had a ratio of 1:0.91%. Since vitamin D metabolism is associated with unfavorable Ca:P ratios, additional examinations were planned with different vitamin D concentrations in order to test whether high Ca and P concentrations or unfavorable Ca:P ratios have any influence on teeth formation. Due to technical problems over 75% of the animals in this experiment diet so that no conclusion could be drawn from the data collected; however, offspring of parents with and without anomalies were fed with algal diets. Independent of their parents, all animals developed the same abnormality. Offspring of the same parents kept on the control diet showed normal teeth growth. These experiments confirmed that the anomalies were caused by the algae and not by environmental influences or genetic factors. In a second experiment, different feed components were tested for their influence on teeth development. Vitamin D_3, $CuSO_4$, $CaCO_3$, Fe^{++}, gluconate, methionine, NaF, and K_2HPO_4 were added individually or in combination to the diets. After 3 months all rats, except the animals kept on control diet, showed anomalies of the incisors.

Analysis of the experiments showed that all algal diets, with or without additions or modifications, resulted in pathological teeth developments. These anomalies did not occur in animals kept on algae-free diets, independent of different Ca:P ratios or other alterations in the feed.

It seemed that disorders in tooth formation due to insufficient mineralization appeared as a result of limited Ca or P concentrations in the blood. Since no other morphological deformations of the skeleton were detected. it was concluded that toxic substances in the algae were not responsible for the abnormalities.

To the possibility that the phosphorus concentration in the algae was already very low, one could add that due to unfavorable pH, other minerals from the algal media in a secondary reaction formed insoluble. nonmetabolizable complexes with phosphorus in the feed.

For various reasons the employment of mice in toxicological studies is not very common. One of the few examinations of algae performed with this animal was reported by Pabst et al.[206] who tested the toxicological safety of drum-dried *Scenedesmus* by feeding this algae to mice. In their study 7 generations of mice of both sexes were given a diet containing 20% algae for a period of 80 weeks each. Body weight, feed efficiency, reproductive capacity, life span, organ weights, hematology, and blood chemistry were recorded. Body weights increased in the algal diet group by 10% over the controls, litter size was decreased by 11%, mean birth weight of pups increased by 4%, and survival of female mice was increased by 48%. In older females, liver-to-body weight ratio increased by 15%, absolute spleen weight increased by 19%, and absolute kidney weight increased by 13%.

No major differences were detected in the weight of brain, kidney, or testes of the males. The liver weights expressed as percentage of body weight remained higher in the test group, especially in young animals. No definite reasons can be attributed to these differences. According to the authors, the addition of nearly 20% algal powder in a balanced diet involves a considerable alteration in the individual nutrients and in the crude fiber content. These factors may be responsible for the alterations in weight of those organs that are characterized by intensive metabolism.

The only detectable hematological difference between the test and control groups was the 3% reduction of hemoglobin and hematocrit in the algae-fed female mice. The females on the algal diet produced significantly smaller litters (− 11%) mean the animals of the control groups, however, there was an increase in mean birth weight (+ 4%) in the algae-fed groups. The gestation, viability, and lactation indexes were similar in both the algal and the control groups. The average duration of life for males was the same in the test and control groups, but algae-fed females displayed a highly significant increase in life expectancy. After 80 weeks the number of surviving females was nearly 50% higher than in the other groups.[207]

In a similar study,[208] differently processed *Scenedesmus* (drum-dried, sun-dried, sun-dried after previous cooking) were incorporated at 10% protein level in the experimental diets. The objective was to test whether any of the drying methods would cause toxic symptoms in the test animals during a feeding period of 12 weeks. The consumption of diets containing casein and drum-dried algae was similar and considerably higher compared to diets containing the sun-dried and cooked sun-dried algae. The weight gain followed a similar pattern with 155, 150, 38, and 30 g of casein, drum-dried, cooked sun-dried, and sun-dried algae, respectively. The ventral side of rats fed sun-dried and cooked-sun-dried algae were fairly free of hair. The loss of hair progressed from the center of the animals outward and was observed after about 8 weeks of feeding. No hair loss was seen in rats fed drum-dried algae or casein. Liver sections of rats fed on 10% casein diet showed mild to moderate centrilobular fatty infiltration. The livers of rats fed on drum-dried algae showed generalized fatty infiltration, but this tended to be more around the central vein, otherwise the histological features were almost normal. The livers of rats fed on sun-dried and cooked sun-dried algal diets showed comparatively less fat around the central vein.

Besides these effects, which are related to the physical method of processing and not the algal material itself, no adverse symptoms or indication have been observed which would limit the utilization of *Scenedesmus*.

Krishnakumari et al.[212] evaluated possible acute oral and dermal toxicity of *Scenedesmus* and *Spirulina*. Testing for allergic effects of the algae, different dosages of drum-dried *Scenedesmus* and sun-dried *Spirulina* (0.5, 1.0, and 2.0 g algae per kilogram of body weight) were applied on to the clipped skin on the dorsal side of the animals. None of the rats showed any sign of erythema or edema of the skin at the treated areas after 24 hr nor during the observation period of 2 weeks; hair growth was resumed as in controls.

In a further study the toxicity of algae was tested in a single oral application to rats and poultry. For rats, doses of 200, 400, and 800 mg algae per kilogram of body weight were administered orally to the animals by catheter. None of the rats showed any sign of discomfort or toxic symptoms, there were no significant differences in the body and organ weights, and the tissues showed normal histology.

For poultry experiments, single doses of feed containing 1.0, 2.0, and 3.5 g of algae per kilogram body weight were fed orally to adult cocks. No symptoms of intoxication or diarrhea could be observed. Autopsy after 2 weeks showed no change of histology of the organs.

In summary, it is clear that the rodents (rats, mice) employed for the toxicological investigations accepted the experimental diets very well, including those diets of high algal concentrations. This excellent acceptability, which resulted in increased feed consumption, may have been the cause for several minor variations observed in the experimental animals, i.e., increased body weight, increased liver weight, fatty liver infiltrations, reduced litter size, etc. Thus, future studies should be performed by applying the "restricted feeding" method. Also, no serious anomalies were reported, either after feeding algae for short periods of some weeks, or after periods of 2 to 3 years, except the teeth anomalies reported from Thailand which have to be attributed to some malpractice in the production of the algae.

Studies on acute or chronic toxicity failed to reveal any evidence which would restrict the utilization of algae, provided the material is processed suitably and administered in adequate amounts. Multigenerational feeding experiments demonstrated that the incorporation of algae into the diets has no significant effect on reproduction or on the performance of the offspring. Although the data available on toxicological investigations only partially cover the full spectrum of the tests recommended internationally, it is doubtful whether further studies will reveal any toxic properties of the algae in question, which would bar the incorporation of algae into feed or food preparations. Indeed, based on the existing analytical data, the Ministry of Agriculture of the Federal Republic of Germany has given a limited permission for the incorporation of *Scenedesmus* into animal feed ratios.

NUTRITIONAL STUDIES WITH HUMANS

Careful clinical tests must yet be performed to conclusively prove the safety of algae for human consumption. The point on which the controversy may arise is when and where the recommended program of these clinical tests will commence.[245]

Preclinical tests, including long-term feeding studies, multigenerational studies with different animal species, teratogenic or mutagenic studies, etc., are to be completed before tests on man can be taken up. It is not clear whether it is legally possible to begin clinical tests at present based on available information which has been summarized in the previous sections.

Despite a great deal of work on algal cultivation technology and studies on the physiology and nutritional quality of algae, relatively few human nutrition studies have been reported so that information on metabolic changes caused by the intake of algae is scarce. It is noteworthy that the majority of the nutritional studies with humans were conducted before 1970. One reason was a worldwide alarm during the 1960s concerning a global protein shortage, which prompted several researchers to develop alternative and unconventional protein sources such as microbial proteins. With the intention of collecting basic information on the pros and cons of algal diets, tests with humans were performed at a very early stage of the development of the algal cultivation technology. At that time researchers contemplated utilizing microalgae in long-lasting flights in space as food and as a method for the dual purpose of wastewater clearance and biological respiratory gas exchange. However, the great expectations for the use of algae as food, which were projected by the initial developments, have subsided. Criticism of the exploitation of unconventional protein sources has affected the aspirations relating to the immediate use of algae as supplementary protein for human consumption. Thus, there was no longer the urgent need for human trials.

The second reason for the lack of more recent studies is the increasing awareness of possible harmful side effects caused by non-nutritive factors which may be present in algae. Since a new field of research was opened concerning large-scale production of microorganisms for food purposes, it was sometime in the early 1970s that international recommendations and guidelines were published, stipulating the procedures for a meaningful and responsible testing program.[196-204] These recommendations demand detailed preclinical evaluations with animals before studies with humans shall be taken up. Since the full spectrum of these preclinical tests has not yet been completed, it is doubtful whether trials with humans will be performed in the near future. The little information on nutritional studies with humans is still too divergent to permit general conclusions. Different objectives have been aimed at by the different investigations reaching from studies with malnourished infants up to mass feeding trials in government subsidized kitchens. In nearly all the reported studies, either *Scenedesmus* or *Chlorella* were employed. As for Spirulina, no reliable clinical studies were published (disregarding exaggerated claims of algae producing companies).[200]

Even in scientific studies contradictory results about the reaction of humans to algal diets were reported. Some authors reported cases of persons living on algae alone for certain time periods without developing any negative symptoms, while other authors reported (among other symptoms) discomfort and vomiting, and poor digestibility of relatively small amounts of algae. In what follows, studies dealing with toxicological aspects and acceptability problems are reviewed. The effects of algae on the uric acid metabolism have been described separately in the section concerning toxicology/nucleic acids.

Probably the first scientific report on the consumption of phytoplankton as human food was published in 1891 by Herdman;[213] this publication, however, is more of historical curiosity than of scientific relevance. One of the first publications on a directed long-term intake of algae by humans is the report by Jøgensen and Convit.[241] "Plankton soup", consisting largely of green algae, was fed in amounts up to 35 g/day to patients in a Venezuelan leper colony for 1 to 3 years. With many patients acceptability was good and improvement in the general health was noted; toxic effects were not described. A mixture of autotrophically grown *Chlorella* and *Scenedesmus* was used in a study performed by Powell et al.[215] with 5 healthy young men, 18 to 23 years of age. After harvesting, the algae were exposed for 1 min at 100°C and then dried under vacuum. In a later review of this study it was mentioned by the same authors that the algae were boiled for 2 hr in an autoclave at 160°C,[120] both techniques not being the proper method of processing this algae. During 7 periods the subjects received diets supplemented with increasing amounts of algae, i.e., 0, 10, 20, 50, 100, 200, and 500 g per man per day. The 2nd, 3rd, and 4th periods were 6-day periods, the 5th and 6th periods were 3-day periods, and the 7th period lasted 2 days.

Only two persons completed the 500 g feeding period, but complained of nausea, abdominal cramping pain, headache, and malaise. All laboratory tests remained within normal limits and physical examinations failed to reveal any abnormalities other than the ones associated with the gastrointestinal tract. It must be assumed that these symptoms were partly due to insufficient processing of the algae, leaving a great amount of the algae indigestible.

As a matter of principle it has to be stated that the results of this investigation are of little significance besides the fact that the health of the test persons was endangered unnecessarily. Feeding of 200 or 500 g of algae per day is about 10 and 20 times the amount which commonly is accepted as the maximum daily intake. The daily consumption of 500 g of algae means a intake of approximately 30 g of nucleic acids (international organizations recommend a maximum of 2 g/day). Furthermore, the various feeding periods of this evaluation are too short to reveal toxicological symptoms due to the intake of algae.

In 1965, similar experiments were conducted by Dam et al.[216] using algae as the principal source of nitrogen for human subjects. This represented approximately 90 to 95% of the total nitrogen intake. In a first experiment, *Scenedesmus obliquus* (lyophilized and autoclaved at 121°C for 30 min) at a rate of 7.1 g N/day was fed as the whole green algae to 5 young men, 27 to 33 years of age.

During the first 4 days the persons were able to ingest the diet although they complained about the color and bitter taste of the algae, upset stomach, and of feeling bloated. On the 5th day, when algae was mixed with other foods, the complaints of nausea were accentuated.

In a second experiment, using 5 different subjects (4 males, 1 female 24 to 34 years of age), 2 levels of *Chlorella pyrenoidosa* (autoclaved at 121°C for 15 min and subsequently extracted by ethanol for 144 hr) were fed (6.0 and 10.0 g N/day) each for 10 days. High fecal excretion of nitrogen was characteristic of all levels of the algal diet, and the algae had low apparent nitrogen digestibility, 68% for dried *Scenedesmus* and 58% for ethanol-extracted *Chlorella*. This low digestibility has to be attributed to the unsuitable processing method applied. High-temperature boiling for 15 or 30 min does not improve the digestibility of the algae, and impairs the quality of the material due to the destruction of nutritive substances and an acceleration of the Maillard reaction.

The second experiment demonstrated that the test persons were capable of consuming algae as the principal source of protein in the diet for a 20-day period without ill effects. As in the case of *Scenedesmus*, it has been suggested that algae toxicity evidenced by gastrointestinal distress may be caused by bacterial contamination or by the presence of antibacterial substances in the algae.[215,217]

The effect of 3 weeks of feeding diets containing 50, 100, and 150 g of freeze-dried algae (a mixture of *Chlorella* and *Scenedesmus*) was tested on volunteers by Kondratiev et al.[218,219] The authors analyzed blood (residual nitrogen, urea, ammonia, cholesterol, phospholipids), urine (total nitrogen, urea, ammonia, creatine, creatinine), and feces (total nitrogen, fat, ash, carbohydrates) of the persons tested. For the groups receiving 50 and 100 g of algae daily, only insignificant changes in the metabolic indices studies were observed. However, inclusion of 150 g of algae in the diet led to some shifts in the state of health in the majority of the persons under examination. Based on their findings the authors conclude that the maximum daily intake of algae for an extended period of time is about 100 g.

The duration of the experimental periods employed by Powell et al.[215] and Dam et al.[216] should be evaluated critically in view of the findings obtained by Kofranyi and Jekat.[220] These authors determined the biological value of different proteins (including algae) alone and in mixtures with egg protein. They found that the experimental period for a certain mixture has to last 3 to 4 weeks, since it takes 9 to 11 days until an equilibrium has been established between N-intake and excretion. The minimum requirements to achieve nitrogen balance were, for instance, 0.5 g of egg, 0.57 g of milk, 0.62 of algae (*Scenedesmus obliquus*, drum-dried), 0.7 g of maize, or 0.8 g of wheat per kilogram of body weight. However, a mixture of 60% algae and 40% egg was superior to egg alone by about 10% (0.45 g/kg body weight). During the experimental period of 6 months neither gastrointestinal complications nor other adverse symptoms caused by the algae were observed. In a second study reported by Müller-Wecker and Kofranyi,[154] test persons were initially given a diet containing 25% protein of drum-dried algae and 75% whole egg protein over a period of 4 weeks. In the course of the study, the dosage of algae was increased in steps until it served as the only protein source. The daily intake of *Scenedesmus* at this stage was about 77 g, the volunteers being kept for 4 weeks at this level. No ill effects were observed during this period. The relative biological value of the *Scenedesmus* protein varied between 81.5% in the first series of experiments and 96% (second test series) of the whole egg standard.

The data of Kofranyi and Jekat[220] and Müller-Wecker and Kofranyi could be confirmed basically by the findings of Lee et al.[221] who evaluated the supplementary value of algal protein in human diets. The authors investigated the nutritive value of algal protein (*Chlorella pyrenoidosa*) for maintaining nitrogen balance in human adults (3 males, 3 females) fed 6 g N/day during 10 experimental periods of 5 days each. The protein sources studied were algae, fish flour, soybean flour, dried whole egg, rice, and gelatin. The diets contained algae alone or algae plus gelatin (mean values −0.06 and −0.38 g N/day, respectively).

While the intact proteins of fish, soybean, and egg fed singly resulted in comparable nitrogen retentions, only fish protein appeared to be improved slightly by supplement of 2 or 4 g N from algae. The value for the algae-rice combination was similar to those obtained for the high quality proteins (fish, soybean, and egg), demonstrating the beneficial effect of algae as an excellent source of lysine and threonine in improving the protein quality of rice. Apparent nitrogen digestibility of the algae was improved from 66 to 75% when part of the algae was replaced by other proteins. The obvious poor digestibility of the algal sample might be due to an unfavorable processing method, which was not mentioned in this study. These findings point at a close correlation between digestibility and processing of algae. It has to be assumed that the algal material used in the earlier investigations was not processed properly, because the importance of the postharvesting treatment was not fully recognized at that time. The consumption of large amounts of improperly processed algae may explain the reported complaints.

More recent studies showed that the digestibility of the algae can be improved by applying suitable methods of processing, and no gastrointestinal discomfort was reported following consumption of such algal material in reasonable quantities.

More recently, nutritional studies with humans have been reported from Peru by Gross et al.[222] Young naval cadets and school children were employed to evaluate their tolerance to drum-dried *Scenedesmus* as a food ingredient. At the beginning and end of the experiment, hemoglobin, hematocrit, serum protein, uric acid, blood pressure, urine, and body weight of the test persons were analyzed. In the 4-week test, 10 g of algal powder for the adults and 5 g for the children were added to their normal rations. All persons showed an increase in the body weight, but all other parameters remained within the normal physiological range. Additional nutritional studies were performed with slightly (group I) and seriously undernourished children (group II). For 3 weeks the children of group I (4 years old) received 10 g of algae in addition to their normal diet. A highly significant increase in weight (up to 27 g/day) was recorded for the children nourished with the algal diets compared to the unusually low weight increment of 15 g/day obtained with 256 children of the same age fed with a regular diet. All other recorded data were normal and the algal dishes were well accepted by the children and did not cause any discomfort. At the beginning of these experiments the children of group II had an average weight which corresponded to only 63% of the standard weight; the daily gain in weight was 3.8 g.

Following a daily addition of 0.87 of algae per kilogram of body weight to the normal diet, the daily weight gain increased up to 30 g and all anthropometric values shifted toward the standard values at the end of the study. Since algal protein constituted 8% of the total protein in the diet only, the positive effects of the algal diets could be attributed to other therapeutic factors. It was observed that diarrhea was cured by administering algae. It was assumed furthermore that the algae stimulated a better intake of nutrients by an activation of epitheral cell regeneration in the gut. In all these studies no symptoms occurred with the test persons which would jeopardize the utilization of algae as food.

In Thailand, nutritional studies with volunteers were performed in order to detect possible physiological changes caused by the intake of algae.[223] During the actual experimental phase of the investigation (3 weeks), the protein in the diet consisted of 50% *Scenedesmus* protein and 50% vegetable protein (soy and mung bean). Since the algal diet caused an increase of the uric acid concentration between 2 to 3 mg/100 mℓ in all persons, modification of the diet composition were considered because the uric acid concentration constantly approached the safety limit of 10 mg/100 mℓ. The digestibility of the algal protein was good, but the excretion of nitrogen was higher as compared to egg protein. The nutritional quality of the algae was slightly lower than that of other vegetable proteins.

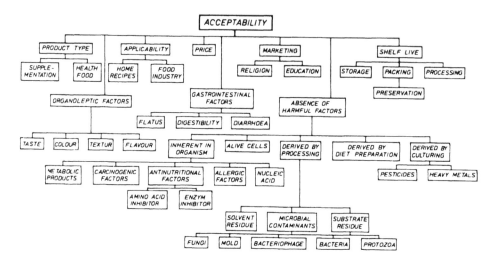

FIGURE 3. Factors affecting acceptability.[37]

ACCEPTABILITY STUDIES

Successful utilization of algae as an unconventional food product will be determined not by the nutritional quality of the algae, but primarily by the acceptability of the product by the consumer. In general, the preference for a certain product is considered as "acceptance". Acceptance, however, is a phenomenon of great complexity which consists of many inter-related components.[37] To clarify this situation, the most important factors and their inter-actions are shown schematically in Figure 3. It can be seen that the actual organoleptic criteria contribute only a small portion of all the factors and that the cooperation of different disciplines is required to produce an acceptable product.

With the intention of testing the preparation, composition, and taste of algae-containing food products, acceptability studies with volunteers were performed in Thailand and Peru.[224,225,233,234] A study of low income families in Bangkok demonstrated that different dishes with 3 to 7 g of algal protein per meal were well accepted. The initial rejection due to the green color of the dishes ceased after 4 to 8 weeks. In an acceptability study conducted with 20 nurses, the test persons received a normal Thai diet with a daily addition of 21 g of drum-dried *Scenedesmus* (7 g per meal). At the outset the green color of the dishes impaired consumption, but in the course of the study the persons got used to food color and accepted the dishes. In a more detailed study over 10 months 300 school children (6 to 11 years old) received a meal supplemented with algae powder twice a week. To find out the optimal algal concentration, three different concentration (10, 5, and 2.5 g) of algae were added to the dishes and researchers found that a preparation with 10 g of algae was not acceptable, and that all algae-containing dishes were accepted by the children to a lesser extent than the normal dishes during the first 6 to 8 weeks. Hence, a certain familiarization period has to be considered before a final statement can be given on the acceptance of the algal diets.

In Peru, in a 4-week test 15 men (21 years of age) and 15 boys (10 years of age) received daily 19 g and 5 g, respectively, of drum-dried *Scenedesmus* in their normal ration. This study confirmed the findings from Thailand that the acceptance of the algae-containing diets increased in the course of the investigation, especially with respect to the appearance of the preparations. The dishes, which were rejected at the beginning were well accepted after they were offered about 10 times. Similar trends were observed with taste preferences. Five of the tested dishes were tested by two independent sensory panels; one consisted of 25 blind persons so that possible unconscious influences of color could be controlled. Four dishes were accepted well by both panels, but the sighted group rejected one dish. In a mass test with 1745 persons of the social middle classes these five selected dishes were evaluated in a Government-subsidized kitchen. Three fourths of the persons accepted the dishes, the average daily amount of algae consumed was 13 g per person. This test was performed only once, therefore it cannot be decided whether the same results would have been obtained after offering the dishes for a longer period of time. Curiosity plays a decisive role during these tests.

To popularize algae they had to be incorporated into a commercial foodstuff. As a first step in this direction noodles were enriched with algae and tested for acceptance. In a test with school children, ordinary noodles were accepted by 77% of the children. Among the other three preparations offered (spinach, 5% and 7.5% algae), noodles containing 7.5% algae were accepted best (75%) followed by spinach noodles and 5% algal noodles. The addition of 7.5% algae to the noodles represents an increase of the protein content from 7.7 to 13%.

In developing novel food products for different target groups it has to be considered that according to age-specific experiences, preferences are developed in the consumer in relation to appearance, taste, texture, etc., of the normal food products. Unconsciously, novel and traditional products are compared and certain expectations are linked with the properties of the new product. In many cases the new product will be rejected if these expectations are not met. Various studies on color preferences of children showed that the introduction of new food items is easier with young persons since their range of expectations regarding traditional food products is still limited, thus young consumers are more willing to accept a new product. A possibility to take advantage of this fact and to utilize algae as infant feed has been reported from Mexico.[226] In this case, a beverage, formed by 50% of a suspension of *Spirulina*, was given as bottle feed to babies who accepted the ''green milk'' without rejection. In Japan, Morimur and Tamiya[227] added small amounts of *Chlorella ellipsiodea* to different food preparations, which were accepted by the test persons. The tolerated amounts of algae in the foods were tea, 20%; soup, 5 g per dish; noodles, 6%; bread, 6%; rolls, 7%; cookies, 1.5%; ice cream, 12 g/cup. The addition of these concentrations of algae caused an increase of 20% in protein and 75% in fat in the case of bread and noodles and about 30% in protein and 15% in fat in the case of ice cream. A few experiments conducted with dried powder of *Chlorella pyrenoidosa*, *Scenedesmus* sp., *Chlorococcum* sp., and *Dunaliella salina* showed that all algae tasted similar. On the other hand, Tamura et al.[228] reported that algae-containing diets (30 g) were poorly accepted by test persons because of unappealing color, taste, and flavor.

Similar discouraging results were reported by Hayami and Shino,[229] who investigated the rate of absorption of decolorized *Chlorella* in men. They found that much of the objectionable taste and odor of the algae was removed by decolorization, but the remaining smell and taste of the *Chlorella* was intolerable and impaired the appetite. To overcome these obstacles some authors emphasized that for a satisfactory utilization of algae the following must be achieved: removal of cell wall, removal of color and flavor, and utilization of components, such as vitamins and protein by separate extraction.[230] Since a large portion of proteins is not soluble in salt solutions which are generally employed as protein extractant, the authors recommend a urea soaking method, based on the disruption of hydrogen and (to a lesser extent) hydrophobic bonds in protein aggregates. It has to be seen whether this method is feasible for large-scale applications and does not cause new problems. Modifications of green algae to be used as food has been proposed by others.[231]

Clinical studies of algal nutrition under medical supervision including detailed analyses of the data have been conducted at only a very few places. Some were published in Japan concerning nutritional experiments on healthy and ill persons.[228,232] It cannot be verified, however, by which scientific methods these trials have been performed. The reported amazing effects and successful treatments of various maladies seem to be the result of a publicity campaign rather than the outcome of a careful scientific investigation.

Although it is not possible to analyze the reasons for the differing results, it seems quite likely, especially in the earlier studies, that the algae (at that time predominantly *Chlorella*) were not processed correctly and therefore remained indigestible.

While the strong uncommon taste and the intensive green color of the algae make acceptability difficult, some investigations have demonstrated that drum-dried algal powder was acceptable when added to conventional foods in small amounts. However, it cannot be ignored that earlier publications reported abdominal distentions and other negative effects caused by the consumption of even smaller quantities of algae. As long as no comprehensive clinical tests have been performed, no final statement about potential health hazards can be given.

Antinutritional or antagonistic constituents, which have been found in other protein sources, amino acid antagonists, enzyme-inhibitors or anti-vitamins are other factors which may possible limit the use of algae as food. To overcome these potential difficulties, processing and purification methods have to be standardized. The fact that no toxic compounds in the native algae have been found so far represents a hopeful starting point. Bacteriological controls, proper packaging and storage are absolute requirements for positive acceptability. In general, it seems almost impossible to guarantee the absolute safety of any given foodstuff. There will always be instances in which people complain about undesired side effects caused by the consumption of any food.

This novel protein source, whether for feed or food, requires that the basic principles of food legislation be revised. Until a few years ago food legislation was mainly concerned with the wholesomeness of food, little attention being paid to nutritional value. With the introduction of new food preparation, a revision of the basic principles of food legislation began in some countries, and the notion of nutritional requirements began to be introduced.

Although no fundamental observations have been reported which would restrict the utilization of algae as a novel protein source, it is evident that at the moment neither the general public nor the technical community are ready to accept the novelty of using algae either for animal or human feeding, except for some pilot studies in combination with experimental algal cultivation plants.*

The use of algae in human feeding must be preceded by substantial and long-term studies with animal and humans. Such testing is likely to be difficult and the unanimous acceptance of the results both by consumers and the authorities, may take a long time.

ALGAE AS ANIMAL FEED

The preliminary objective of many research projects is to test the use of algal biomass as animal feed. Detailed feeding trials with laboratory animals have proven the good nutritional quality of several microalgae, but further studies must be performed to evaluate the quality of the algae on the target animals. These studies are of commercial interest, especially with regard to the maximum amount of algae to be incorporated in the feed as substitute for conventional protein sources (fish meal, soybean meal, etc.), without causing negative effects in the animals. In order to evaluate the potential of algae as feed it is important to distinguish between the utilization for monogastric vertebrates and for ruminants. Since monogastric vertebrates are not capable of digesting cellulosic plant material, the cellulosic cell wall —

---

* Nevertheless, today there is a significant number of people who consume microalgae in the form of pills made of *Chlorella* and *Spirulina*. The annual world production of these microalgae is estimated approximately at 1500 tons, thus it seems safe to assume that at present some 1 million people the world over, consume microalgae.

Table 38
EFFECT OF DIET CONTAINING VACUUM-DRIED *CHLORELLA*
ON WEIGHT GAIN AND FEED EFFICIENCY RATIO OF CHICKEN[78]

Diet	Weight after 4 weeks (g)	Feed efficiency ratio %[a]
Basal ration	135	3.1
Basal + 10% *Chlorella*	262	2.4
Basal + 10% *Chlorella* + 0.1% met	298	2.3
Basal + 10% *Chlorella* + vitamin + 0.1% met	292	2.3
Basal + vitamin + 0.1% met	316	2.2
Complete broiler mash	342	2.2

[a]   Feed consumed/weight gain.

particularly in *Chloreophyceae* — must be ruptured to make the proteinaceous cell content digestible for proteolytic enzymes.

Ruminants such as cattle or sheep are able to utilize cellulose through the action of cellulase synthesized by symbiotic bacteria in the digestive system (the rumen). Naturally, it should be possible to feed native algal cells to ruminants; nevertheless, very few experiments on this form of application have been reported. In the following chapter data on the utilization of algae as feed for different animals are summarized.

Poultry

Most of the earlier studies of algae as poultry feed have been made with marine forms rather than with unicellular freshwater species.[250] Combs[78] was probably among the first who reported on the effect of dried *Chlorella* (21°C *in vacuo*) as a source of nutrients for chicken. In his experiments, the protein source for the basal diet was soybean, which was supplemented with *Chlorella* and amino acid as shown in Table 38.

Grau and Klein[251] studied the value of wastewater-grown algae (a mixture of mainly *Chlorella* and *Scenedesmus*) and the effect of dietary aluminum of chicken. Aluminum was investigated since algae harvested by alum flocculation approximates more closely material which could be produced on a large scale. It was found that aluminum-free algae can be tolerated at levels as high as 20% of the diet. The effect of aluminum (8% in the algae) in reducing growth was evident in that with this sample even 10% algae did not permit maximum growth. Neutralization and acid extraction were unsuccessful in removing the harmful effects of aluminum. The authors suggested that algal meals prepared with a minimum amount of alum compare favorably with meals without aluminum and indicate that chicken can tolerate at least 0.5% dietary aluminum. These observations are contradictory in part with the results of Yannai et al.[179] who found no depressing effect of aluminum-containing diets on the growth of chicken.

Arakawa et al.[252] concluded that feeding growing while Leghorn with a supplement of *Chlorella* to the mixed feed at the rate of 10%, along with soybeans, resulted in favorable growth. The nutritional value of *Chlorella* powder as chicken feed was similar to that of uncooked soybean meal. When feed including *Chlorella* was fed to laying hens, the yolk of the egg contained more pigment, especially β-carotene and xanthophylls, than the egg yolk of hens on a combined feed without algae.

Similar experiments were performed by Leveille et al.[40] who studied the protein value of several algae for the growth of chicken. Their major observations, given in Table 36, demonstrate that all the algae tested were inferior to the control diets, in which soybean

Table 39

**PERFORMANCE OF BROILER CHICKEN FED ON
DIETS CONTAINING INCREASING CONCENTRATIONS
OF ALGAE**[254]

	Algal content in diet (g/kg)				
	0[a]	75[a]	150[a]	0[b]	150[b]
Weight gain (g)	484	485	474	1878	1732
Feed consumption (g)	777	795	773	4132	3967
FCE[c]	1.60	1.64	1.63	2.20	2.29

[a] 1 to 3 weeks.
[b] 1 to 8 weeks.
[c] FCE: Feed conversion efficiency = feed consumed/weight gain.

meal supplemented with methionine served as the protein source. Out of the three algal species tested, a mixture of *Chlorella* and *Scenedesmus* gave better results than obtained when each species was fed separately.

The generally poor PER that was typical to all the algae-containing diets indicated that the algae for this experiment were not properly processed. It has been pointed out repeatedly that algal proteins are deficient in some essential amino acids. This was confirmed in experiments showing that all of the algae tested are deficient in methionine for growing rat and chicken, the mixture of algae being also deficient for chicken in glycine. With the aim of replacing conventional proteins by algae, poultry feeding trials have been conducted in Israel using sewage-grown algae.[253] Since the algal population in high-rate algal ponds usually consists of several different species, the authors examined the most common species by feeding them to young chicken as a replacement for soya protein.

All the algal species tested successfully replaced 25% of soy protein, equivalent to 7.5% of algae in the rations. Also, algae could not replace fish meal as a source of growth factors. Other investigations on sewage-grown algae were conducted by Lipstein and Hurwitz[254] who evaluated the nutritive value of *Chlorella* for broiler chickens. The performance of the birds fed with increasing algal contents is shown in Table 39.

In similar experiments conducted with hens up to the age of 8 weeks, the food gain ratio was found to vary between 2.89 and 2.66. Despite the high proportion of algal meal in the diets during the entire growth period, there was no growth retardation or mortality of the birds, but a marked increase in the intensity of pigmentation of the skin was observed. The trials suggested that there was no deleterious effect up to 100 g algae per kilogram of diet. However, higher concentrations, up to 150 g algae per kilogram diet, resulted in significant depression of growth and an increase in food/gain ratio.

In a second study, the same authors also tested the effect of dried *Chlorella* in layer diet.[225] The results of this experiment, obtained with algae at levels up to 120 g/kg diet, are summarized in Table 40.

No difference was found in egg production rate, egg weight, food conversion, etc., between controls and birds receiving any of the algal diets. The results suggest that for practical layer diets, algae can serve as almost the sole protein source, in contrast to the finding with broilers.

The beneficial effect of these high algal concentrations in poultry rations could not be confirmed by Walz et al.[104] and Brune and Walz[256] who studied the effect of *Scenedesmus* in combination with other conventional protein sources in broiler feed. High amounts of algae added to some groups caused a significant growth depression, corresponding to the percentage of algae in the diet.

Table 40
PERFORMANCE OF LAYING HENS FED ON DIETS
CONTAINING INCREASING CONCENTRATIONS OF
CHLORELLA FOR 2 MONTHS[255]

Parameter	Algae (g/kg diet)				
	0	30	60	90	120
Eggs/100 birds	84.5	82.1	80.0	84.1	80.6
Average egg weight (g)	59.9	59.8	59.3	59.8	59.5
Egg output (g/bird)	50.6	49.1	47.7	50.3	47.9
Food intake/bird (g)	110	109	110	110	110
Food consumed (g/egg)	131	133	139	131	137

The authors reported that the feces were thin and mucous but no animal died during the experiment. Therefore, only limited amounts of *Scenedesmus* (up to 5%) were included in the other diets. Whereas group 7 (2.9% algal protein) gave a growth performance as good as the control group, the other groups with higher algae contents reached the limit of practical application. At higher protein levels, severe growth depressions occurred. Nevertheless, histological anomalies of the liver, which might have caused the nutritional disorder, were not detected. Since the natural pigments of the algae were incorporated into the broiler's tissue, the authors recommended that the algal powder content should be limited in feed mixtures up to about 3 to 5%.

At the Primary Production Department of the Republic of Singapore a project is in progress that is demonstrating the functional use of high-rate algal ponds for treating piggery wastes.[216] The algal biomass is being harvested, dried, and used for livestock feeding trials. In poultry feeding experiments the normally fed soybean meal was gradually replaced (33, 66, and 100%) by a mixture of sewage-grown algae, which were boiled with hot steam for 5 min. This replacement led to decreasing gains in weight (control, 1390 g; 33% algae, 1224 g; 66% algae, 1097 g; and 100% algae, 969 g) and poorer feed conversion efficiencies of 2.32, 2.29, 2.43, and 2.62, respectively. These results are in agreement with other observations showing that only limited amounts of algae can be incorporated in poultry feed without negative effects.

While most of these investigations were aimed at supplementing soya by algae, Reddy et al.[258] tested the performance of chickens from the 1st to the 5th week which were fed *Scenedesmus obliquus* (5% protein level) in place of fish meal with and without supplemental methionine, commercial feed being used as reference. The feed efficiency was 4.68 for the reference, 4.50 for the algal diet without methionine, and 4.10 for the diet with methionine. The results indicated that algal protein could be used in poultry rations in place of fish meal.

In view of replacing conventional proteins in poultry feed, Hennig et al.[263] estimated N-digestibility and N-utilization of *Scenedesmus* by comparing one reference group and one experimental group over a period of 10 days. In the test group 8% peanut meal and 4% fish meal were replaced on the basis of crude protein equivalence by 14% of algal meal. True digestibility of the raw protein of the algal ration was 7% lower than in the reference group. The N-balance of the algae group was 23% lower than in the group without algae; the biological value was also slightly lower by 4.4%. The supplementing effect of drum-dried *Spirulina* was tested in another study.[210] Algae at different concentrations were incorporated in the diets of starters and layers, replacing fish meal partially or completely. As can be seen in Table 41, weight gain of the starters in all algal groups was significantly less compared to the control, indicating that algae can be recommended in the diet only up to 5%. The effect of algae on the layers was smaller, since weight and number of eggs produced did not vary much within the groups.

Table 41
PERFORMANCE OF POULTRY TO DIETS
CONTAINING DIFFERENT CONCENTRATIONS OF
ALGAE IN THE RATIONS[210]

Starter chicken

Algae in diet (%)		Weight gain (g)	Feed efficiency ratio[a]
0 (Control)		667	3.27
Scenedesmus			
5.0		461	4.17
7.5		385	4.95
10.0		374	5.07
Spirulina			
5.0		636	4.53
10.0		466	5.08
0 (Control)	62.5	57.5	3.09
Scenedesmus			
5.0	58.4	59.6	3.30
10.0	59.0	59.5	3.26
0 (Control)	37.5	48.2	5.5
Spirulina			
5.0	39.8	48.0	5.2
10.0	31.3	46.6	6.9

[a] Feed consumed/weight gain.

Compared to the large amount of information available on the suitability of *Chlorella* and *Scenedesmus* for poultry feed information on the value of *Spirulina* is limited. This is probably due to the fact that *Spirulina* is not a common species in high-rate algal ponds and most of the supply on the world market is sold as "health food".

In 1975, Blum and Calet[257] reported on broiler experiments with *Spirulina*. From the first day, weight increase was always depressed when *Spirulina* replaced traditional protein sources (soybean meal, fish meal) in a complete diet. The reduction in growth was less with up to 5% algae in the diet. The gain in weight was reduced by 16 and 26% for *Spirulina* levels of 20 and 30%, respectively.

The influence of a partial substitution of *Spirulina* for soybean meal in hen rations on egg quality traits (odor, flavor, free amino acid content of the yolk, egg weight, shell breaking strength, Haugh units, frequency of blood spots, yolk color, foaming power of albumin, and emulsifying capacity of the yolk) were tested by Sauveur et al.[264] and Colas et al.[265]

The addition of *Spirulina* at a concentration of 11.5% to the diet (total protein content 15%) spoiled the sensory characteristics of the egg ("chemical flavor") but did not considerably modify the free amino acid content of the yolk. All other parameters remained unchanged except for a slight increase in the Haugh unit score and the very high coloring effect of the algae on the yolk.

Since most of the poultry feeding trials have been performed with *Chlorella*, *Scenedesmus*, and *Spirulina*, there are only limited data available on the nutritional value of other algal species which could be used as animal feed.

One of the few studies on uncommon algal species was reported by Lincoln and Hill.[260] who used *Euglena* and *Synechocystis*, grown on diluted pig manure, as poultry feed. Increasing amounts (5 to 20%) of these algae, dried in a solar drier, were incorporated into the broiler feed. Diets with 5 and 10% gave variable but occasionally superior results, while inclusion of 15 and 25% algae reduced the growth of the animals. No specific toxicity was noted in the experiment. However, the authors indicated that *Synechocystis* may be toxic under certain culture conditions.

Broilers were used to compare alum flocculated and ferric sulfate flocculated algae (*Chlorella vulgaris*, *Synechocystis aquatilis*, and *Euglena viridis*).[266] The algae samples were grown on effluent from an anaerobic lagoon containing swine wastes. The flocculated algae were removed by autoflocculation, concentrated with a gravity filter, and then dried by solar radiation. The chemical composition of the algal meal varied slightly but showed very high ash (average 39.3%) in all samples and considerably lower protein (average 23%) content. Chickens were fed 0, 10, or 20% *Chlorella* flocculated with either alum or ferric sulfate and fortified with methionine and lysine. Similar experiments were conducted with *Synechocystis* and with *Euglena*. The conclusion was that in the case of *Chlorella* and *Euglena*, the addition of up to 10% algae in the rations resulted in growth performances comparable to the control diet. Higher algal concentrations caused decreasing growth rates. In the case of *Synechocystis*, concentrations of more than 5% in the diet resulted in significant growth retardations.

The nutritive of a new type of *Chlorella* (*C. vulgaris* Al-25) was evaluated by Yoshida and Hoshii[259] in poultry feeding experiments. This algae has a very thin cell wall and it was suspected, therefore, that protein and other nutrients would be easily digestible. The algae contained less protein (24%) but more carbohydrate (53% nitrogen-free extract) than the usual type of *Chlorella*. Air-dried samples were incorporated at 16.3 and 20% levels in the rations. The energy available from the algal portion of the diet was 3.34 kcal/g with availability of 73%. Gross protein value with supplemental methionine (2%) was 71%. Palatability was excellent and no signs for acute toxicity could be detected. True digestibility of gross energy and protein in adult rats were 80 and 90%, respectively.

It remains to be seen whether isolation and breeding of cell wall-free chlorophyceae or at least the selection of strains with thin cell walls offer possibilities to overcome the problems related to processing and digestibility of algae such as *Chlorella*.

In most of the nutritional studies the whole algal biomass was incorporated in the diets. To evaluate the value of specific cell fractions, Brune[262] tested the performance of starter broilers fed on *Scenedesmus* and *Spirulina*, with and without extraction of the lipids by methylene chloride, as the sole source of protein in the diet. To all the rations methionine, vegetable oil, minerals, and vitamins were added. *Spirulina* was superior to *Scenedesmus* and even to the conventional starter ration, lipid extraction reducing the value only marginally. While the quality of *Scenedesmus* was inferior to the control, lipid extraction improved the nutritional properties of this algae significantly.

The feeding trial with *Spirulina* was continued for another 16 days without any decrease in the excellent performance of this algae. Since the amino acid pattern of both algae was comparable, other factors than the proteinaceous portions of the algae must be responsible for the different results.

Contrary to most of the data found in the literature it was possible in this study to successfully utilize algae as the sole source of protein. The high capacity of poultry to excrete uric acid allows elevated algal concentrations in the diet without endangering the animals. The author suggests that the growth-depressing effect of untreated *Scenedesmus* could be due to the high lipid content (18%) in the algal sample.

Most of the authors cited came to the conclusion that a level of 5 to 10% algae in the diet (partly replacing other protein sources) has no deleterious effect on chicken. Prolonged feeding on higher algal concentrations cause an adverse effect on weight gain in proportion to the amount of algae in the diet.

It seems to be no accident that most of the experiments performed with algae on farm

Table 42
WEIGHT GAIN AND FEED UTILIZATION BY PIGS
FED VARYING LEVELS OF ALGAE (MIXTURE OF
***CHLORELLA* AND *SCENEDESMUS*) FOR 46 DAYS[267]**

Amount of algae in diet (%)	0	2.5	5.0	10.0
Weight gain (kg)	35.3	33.8	33.4	34.6
Feed conversion efficiency[a]	3.85	3.76	3.85	3.90

[a] Feed consumed/weight gain.

animals have been conducted on poultry. Judging from the available information it can be assumed that the incorporation of algae in poultry rations offers the most promising prospect for a commercial utilization of algal biomass, except perhaps for fish culture.

Pig Feed

Among the potential animals for which algae can be used as feed, pigs appear to have been the species of choice in some places. Hintz et al.[76,267] were among the first to test the nutritive value of sewage-grown algae with different animals. The algae represented a mixture of *Chlorella* and *Scenedesmus* harvested by centrifugation, and dried either in air or by drum-drying. Algae were incorporated at 2.5, 5.0, and 10% in the diet, substituting soybean and cottonseed meal. Weight gain and feed utilization of pigs in these experiments are listed in Table 42. The pelleted feed was readily consumed and the algae-fed pigs performed as well as the controls.

In a digestion test on boars the apparent digestibility of *Scenedesmus quadricauda*, probably spray-dried, was examined by Hennig et al.[253] by means of the differential method. The authors added 20 to 45% algal material to a diet based on barley. The Net Feed Efficiency (NFE) for the two algal-containing rations was 99 and 68%, respectively. The digestibility was low compared to other values reported in the literature and was perhaps due to a high amount of crude fiber (18%) in the algal samples.

The utilization of algal proteins in hogs was studied by Witt et al.[268] using oven-dried (70 to 80°C) *Scenedesmus* in the first experiment.

The conclusion was that a protein supplement containing 75% algal protein and 25% fish meal protein was suitable for pigs. In a second series of experiments the same authors tested the influence of fish meal, soybean meal, and drum-dried *Scenedesmus* on the growth and the quality of the carcasses of swine.[269]

No significant difference in growth could be found between the different groups (group I fish meal, group II soybean meal, group III 75% soybean and 25% fish meal protein, group IV 25% soybean and 75% algal protein). Fish meal and the mixture of algae plus soybean meal had a significantly positive influence on the ripening process of the meat.

A less common protein carrier in pig rations, alfalfa-leaf meal, was substituted by drum-dried *Scenedesmus* in another study reported from Germany. Here the performance of pigs reared on algal feeds was compared to other animals fed on alfalfa (on an isonitrogen basis) and on casein.[256] Both algae and casein gave relatively high apparent digestibility as compared to alfalfa (Table 43).

In a similar approach Witt and Schröder[270] tested the same algae and fed 2 different diets to pigs (diet I: 77.5% barley and 22.5% farm-mixed concentrate; diet II: 76.5% barley, 21.5% algae, 2% minerals). They found that the daily weight gain in the algae-fed groups was slightly (5%) higher than in the control group. The authors concluded that the algal protein is excellently suitable to be utilized as the *only* protein supplement for fattening pigs.

This conclusion could not be confirmed by others since it was observed that the addition of higher concentrations of algae in pig rations resulted in reduced digestibility and in

Table 43
PROTEIN REQUIREMENT OF GROWING PIGS AND NUTRITIVE VALUE OF *SCENEDESMUS*[256]

	Casein + 2.1% met	Scenedesmus		Alfalfa	
		5.2%	12.5%	13.5%	46.8%
N consumed/day (g)	1.76	2.75	10.0	2.75	7.2
N apparent digestibility (%)	100.9	90.2	74.5	59.9	65.4
N balance (g)	−0.41	+0.15	+5.72	−2.01	+0.47
Biological Value		86.9			38.8
Net Protein Utilization		64.7			38.3

depression of growth. Data reported from Singapore showed that steam-boiled sewage-grown algae can replace about 50% of soybean meal (given at 15.5% in the diet) with little reduction in growth rate and feed efficiency.[271] However, total replacement of soybean meal with algae resulted in a significant reduction in both the weight gain as well as in feed efficiency ration, thereby demonstrating that the nutritive value of the algae was clearly inferior to that of soybean meal. A digestibility trial showed that the algal protein was poorly utilized (coefficient: 27.1). This does not agree with the researchers who found much higher digestibility coefficients. This discrepancy was probably due to the algal species used and to the processing method applied. No pathological changes were observed, however, and the meat was free from any abnormal taste or taint.

The majority of the investigations on pigs were performed with *Chlorella, Scenedesmus*, or mixtures of sewage-grown algae. One of the few studies which employed *Spirulina* was reported by Fevrier et al.[272] who fed this algae to pigs as feed additive in long-term experiments. A first series of experiments was made on early weaned piglets (from 12 to 42 days) with an algal proportion representing 12% of the total proteins in replacement of skim milk or soybean meal. The apparent digestibility decreased slightly from 84.0 of the control diet to 77.9 of the algal diet. In a second study *Spirulina* was fed continuously at 5% to sows during the fist two reproductive cycles. Although the feeding of *Spirulina* did not seem to give rise to particular problems, the authors recommended restricting the incorporation of algae to a level not exceeding 25% of the total dietary protein, especially in young animals.

Recently, two experiments were reported by Yap et al.[273] who evaluated the feasibility of replacing 33% of soy protein in a basal diet with proteins from *Spirulina maxima*, *Arthrospira (Spirulina) platensis*, and *Chlorella* sp. to pigs weaned to a dry diet at 4 to 8 days of age. Animals fed the basal diet up to the 26th day gained weight at a rate not significantly different from those fed with algal diets. There was no sign of diarrhea, loss of appetite, toxicity, or of gross histopathological lesions of the gastrointestinal tract. Terminal blood hemoglobin, serum protein, albumin, and urea concentrations were also similar in all groups in each experiment. The authors suggest that at least 50% of the protein supplied by soybean meal (33% of total) may be safely replaced in weaned pigs by algae.

Another promising report is by Lo,[274] who found enhanced growth of various animals which were fed with algae as a diet supplement. The author concluded that the algae in the feed provide a growth factor that is not available in standard rations and that the algae have some antibacterial properties.

Nearly all the researchers came to the conclusion that microalgal biomass, provided it is processed properly is a feed ingredient of good nutritional quality, well suited in a diet for rearing pigs, in which it may replace conventional protein carriers such as soybean meal or fish meal. In general, no difficulties in acceptability of algae were observed.

Feed for Ruminants

So far, only very limited nutritional evaluations have been performed with ruminants, mainly because of the large amount of algae required for appropriate experiments. The possibility of using wet cakes, which may be sterilized before feeding the cattle without further processing, has not received much attention.

Hintz et al.[76] have tested the nutritive value of sewage-grown algae on lambs, sheep, and cattle. Differently processed (air-dried, drum-dried) mixtures of *Chlorella* sp., *Scenedesmus obliquus*, and *S. quadricauda* were fed to wethers. The rations consisted of 60% algae and 40% hay, alfalfa-oat hay diets serving as control. Whereas the digestibility of crude protein was about 72% in all diets, the digestibility of the nonprotein, nonfat organic matter ("carbohydrate") in the drum-dried algae was significantly lower than that of the hay ration. This was attributed to the lower digestibility of the crude fiber of the algae.

To study the value of algae as source of protein in weanling lambs, cottonseed meal, alfalfa pellets, and alfalfa-algae (10:4) pellets were compared. Under the experimental conditions the alfalfa pellets were generally unsatisfactory, because the lambs would not consume the amount of feed needed to supply nitrogen for maintenance. When sufficient nitrogen was offered in cottonseed meal or algal-enriched alfalfa, the lambs were able to maintain their weight and at higher levels to gain weight.

In a further study algae-containing diets were fed to beef steers. The rations were alfalfa hay, algae and hay (2:8), and algae and hay (4:6), all algae being drum-dried. The animals did not eat the 40% algal pellets readily and therefore the intake of other rations was limited to the daily intake of the 40% algal ration. The addition of 20 to 40% algae to alfalfa hay did not decrease the digestibility of dry matter, organic matter, or crude protein, but 40% algae significantly lowered carbohydrate digestibility from 68 to 52% and was similar to that obtained in the sheep trial. The digestion coefficients for protein were similar to the values obtained for sheep (74%).

Studies were carried out on the usefulness of *Spirulina* as a replacement for groundnut cake (30%) in the rations of infant calves.[54] After a feeding period of 28 weeks the body weight of the experimental animals was reduced by 8% only compared to the control group, indicating that this algae is an ideal substitute for protein-rich oil cakes in the concentrate mixture of infant calf rations.

The utilization of sewage-algae in association with the paper used for their filtration for feeding of sheep were tested by Davis et al.[275] The sewage-grown algae were collected by filtration and the filter paper plus the algae were incorporated with barley (30:70) into sheep rations and were compared with other rations of paper and barley (30:70) and alfalfa and barley (30:70). The sheep utilized algal paper poorly. Its incorporation as roughage in a high grain diet ensured its consumption by the animals but they consumed lesser amounts and digested less nitrogen than could be achieved with an alternate source of roughage such as alfalfa. The sewage-algae used were found to have relatively indigestible cell walls, even for ruminants, and the apparent digestibility of algal nitrogen was lower than that found with other protein concentrates.

Sheep feeding experiments to assess the nutritional value of sewage grown Chlorella were carried out by Hasdai and Ben-Ghedalia.[276] The algae, harvested by flocculation with alum and then drum-dried, substituted 50% of the dietary protein and were fed to young rams, soybean meal serving as control. The digestibilities of dry matter were 69.3 and 79.3% and those of organic matter were 75.3 and 82.2% for the algal diet and the control, respectively. Apparent nitrogen digestibility of the algal diet was 71.3% while the value for the control diet was 83%; the calculated digestibility of the algae was 61.7%. The reasons for these differences could have been the high amount (35.5%) of unabsorbable minerals in the algae.

Due to harvesting by alum-induced flocculation, the *Chlorella* meal contained 5.7% aluminum (AL). Since it was expected that this high concentration would interfere with phosphate (P) absorption, dicalcium phosphate was added to the diet. In ruminants, there is an intensive endogenous P secretion into the stomach which might be much higher than the amount of ingested P. Thus, feeding on Al may cause a depletion of body P reserves. Despite the high concentration of P in the algal diet, the amount of absorbed P was very low, although the plasma P concentration was found to be normal.

These observations indicate that for the feeding of ruminants, aluminum-flocculated algae are not very suitable. Since aluminum may cause P deficiencies in the animals, other methods of harvesting have to be developed in order to obtain an algal biomass devoid of excessive ash.

A digestibility of 65.4% was obtained by Calderon et al.[277] with *Spirulina*, constituting 20% of a complete sheep diet. It was found that this algae could replace all the soybean meal (21%) of rations for fattening sheep or growing calves without significant effect on weight gain or feed conversion efficiency. The lower digestibility of the algae compared to soya might be compensated for by providing a better profile of absorbable amino acids.

One of the very few studies on the utilization of fresh, untreated algae for feeding of ruminants has been reported from Bulgaria. Ganowski et al.[278] fed 1 ℓ of concentrated native *Scenedesmus acutus* (2 to 3 × 10^8 cells/mℓ) daily to calves for a period of 3 weeks. The authors determined the digestibility of the algae and examined effects on the bacterial flora in the rumen of the animals. Only minor differences were observed for the digestibility coefficient between control and experimental animals. Content of hemoglobin, number of erythrocytes, and level of carotenoids in the serum of the experimental animals were increased, while the number and species of infusoria as well as the pH of the gastric acid remained unchanged.

Insect Feeding

Investigations aimed at the development of simple and inexpensive artificial diets for the mass rearing of lepidopterous insects are gaining in importance. In some of these studies the utilization of algae as an ingredient in insect-feed has been tried, *Spirulina* serving as protein source in the larval diet of the bollworm *Heliothis zea* and the tobacco budworm *Heliothis virescens*.[279] In Egypt, investigations were made to evaluate the efficiency of the same algae as a protein supplement to be used in various proportions in diets for rearing the cotton parasites *Spodoptera littoralis* and *Heliothis armigera*. The algae may partly substitute the kidney beans in the standard diet with high efficiency for *Heliothis* and comparatively less efficiency for *Spodoptera*.[280]

There are few reports in the literature on the utilization of algae as proteinaceous matter for the feeding of silkworms (*Bombyx mori* P.). The effect of different concentrations (7.5, 15.0, 20.0, 30.0, and 40.0%) of freeze-dried *Spirulina*, added to the standard diet, was examined by Hou and Chen.[281] The silkworms (1st instar) were able to tolerate the untreated algal powder up to 40% in comparison with cellulose. In contrast, the algal powder washed with methanol inhibited the feeding activity of the silkworm to some extent. Thus, *Spirulina* powder could be incorporated into the silkworm diet without any washing treatment.

One day after feeding, weight gain of the feeding larvae was significantly lower on the diet containing 40% algae than on the other diets. Thereafter, the larval weight gain resulting from the algae-containing diets increased in proportion to the dietary content of the algae and was similar after 15 days to that of the standard diet. Larval survival of insects fed on diets containing 20 to 40% algae was not different from that of larvae fed on the soybean meal standard diet. Mortality increased at lower algal concentrations, being 40% at 15% algae in the feed and 10% at 40% algae.

The feeding behavior of newly hatched silkworm larvae to *Chlorella* was studied by Ogata et al.[282] In their experiments mulberry powder-cellulose-agar jelly was adopted as basal diet and the cellulose in this diet was replaced by *Chlorella*. The number of excreted feces of the larvae was counted. The results showed that the amount of feces on the algal diets was always lower than on the cellulose-containing diet, indicating that the algae had some factors which inhibited larval feeding. This difficulty was removed partly by extracting the algae with methanol. The crude protein fraction caused lower degrees of feeding inhibition which varied considerably between several insect hybrid strains.

These preliminary studies suggest that it might be possible to use certain algae as a protein source for rearing silkworm or other insect larvae. The published data, however, are too scanty to permit final conclusions on the usefulness of algae in this form of application and more investigations are needed to ascertain the beneficial effects.

Aquaculture

Certain species of microalgae are known to have nutritive roles during at least part of the life cycle, for commercially important fish, molluscs, and crustaceans. Although efforts are being made to develop alternative feeds, it appears that aquaculturists will remain dependent upon microalgae for the near future.

Microalgae may be used directly, i.e., unprocessed, to built up short food chains,[283-287] or in form of dried material in pelleted or otherwise processed feed preparations.[288-290] The use of fishmeal as a source of protein in dry compound feeds, as practiced up to now, causes dependence on marine fishery with its increasing problems.

In tropical and subtropical areas with favorable light and temperature conditions for algal growth, the direct feeding of phytoplankton to herbivorous fish may represent a practical form for utilizing algae in aquaculture. This method often is combined with wastewater treatment and water reclamation systems. The feasibility of this concept has been demonstrated in Thailand and Israel with domestic sewage,[288,294] and in Malaysia with rubber wastewater or palm-oil mill sludge discharge.[295,296] The principal of increasing fish production by manuring ponds to stimulate algal growth has been applied in Asia (particularly in China) for centuries,[284,291] but also in Western countries this aspect is gaining more importance.[297,298]

Algae may also be used to produce zooplankton such as *Brachionus* or *Daphnia*, which can be used as live or dehydrated protein supplement in aquaculture,[299-302] Marine algae, either grown in clean or in wastewater, can be used to feed oysters, clams, and copepods.[301,303,304] Microalgae, grown in hypersaline industrial wastewaters serve as food for *Artemia* and *Brachionus*.[305] For all these applications, size, digestibility, and chemical composition of the algae are of importance in view of the specific requirements of the particular consumer.[283,285,306]

Although progress has been made in improving algal cultivation technology, success in producing large quantities of microalgae for use in aquaculture has been marginal for various reasons.[307] Continuous outdoor culturing of algal monocultures has been so far achieved only in very few cases. Controlled indoor cultivation of sufficient amounts of algae to meet food requirements of commercial undertakings in aquaculture is too expensive to be economical. The utilization of natural phytoplanktonic food is restricted to growth periods with favorable environmental conditions. Induced blooming of naturally occurring phytoplankton in outdoor ponds very often results in the growth of mixed cultures dominated by different species. For all these reasons, efforts are being exerted to identify an alternative to living algae.

Comparatively few results have been reported on the utilization of dried algal material, either used solely or incorporated in conventional feed preparations. This is probably due to the limited number of commercial algae production units which are capable of producing larger quantities suitable processed algal material. The most promising application of dried algae seems to be in pisciculture. In this concept, several problems have to be examined in order to establish the feasibility of using algae in fish feed preparation.

Table 44
FINAL WEIGHT, FEED CONVERSION EFFICIENCY, AND SURVIVAL RATE OF CARP AND GRASS CARP FED DIETS CONTAINING DIFFERENT PROPORTIONS OF DRUM-DRIED *SCENEDESMUS*[289]

	Carp			Grass carp		
Amount of algae in diet (%)	Final weight, 12 weeks (g/ fish)	FCE[a]	Mortality (%)	Final weight, 17 weeks (g/ fish)	FCE[a]	Mortality
0	17.08	3.46	1.0	32.12	3.04	Out of 400
10	19.33	3.28	1.0	39.92	2.36	fish, 395
20	25.53	2.72	5.0	49.17	2.04	survived
30	28.27	2.47	2.0	54.71	1.88	
40	34.04	2.28	3.0	52.37	1.80	
50	34.56	2.23	1.0	60.48	1.73	
60	32.48	2.31	9.0	66.48	1.57	
70	12.18	5.43	68.0	85.73	1.42	
80	9.20	6.28	80.0	95.60	1.34	
90	10.34	5.63	87.0	71.77	1.69	

[a] FCE: Feed conversion efficiency = feed consumed/weight gain.

- What is the digestibility of the algal protein as compared to the conventional protein carrier
- Would fish accept pelleted diets with high concentrations of algae
- What is the nutritive value of the algae (growth rate, PER, digestibility, etc.)
- Will residues from the harvesting and processing steps impair the values of the algae and affect the performance of the fish
- Would the algae impart objectionable color, odor, or taste to the fish

Major findings from experiments performed at different algal pilot plants are summarized in what follows to give an impression of the potential of microalgae for rearing fish.

The nutritive value of drum-dried *Scenedesmus* was evaluated in detailed experiments with common carp (*Cyprinus carpia*) and grass carp (*Ctenopharyngodon idella*).[289,308] Grass carps were fed one of the following diets exclusively for a period of 20 weeks: (1) commercial trout feed (containing fish meal); (2) laboratory mixture (LM) (containing 50% whey powder, 40% soybean meal, 10% fat); (3) LM + dried alfalfa meal (68:32); (4) LM + algae (68:32); (5) solely algae. It was found that diet 4 produced the best growth performance. In addition to remarkable variations in weight gain, the appearance of the various groups showed striking differences. While the group fed diets 1 and 4 appeared healthy and homogenous, fish fed on diet 5 showed severe deformations.

Feeding of diets composed of 100% LM and 68% of LM + 32% of algae, yeast, or casein resulted in feed conversion ratios of 3.53, 2.00, 2.41, and 2.00, respectively and PER values of 1.08, 1.54, 1.32, and 1.25, respectively, demonstrating that for some fish algae are nutritionally comparable to casein.[289]

Based on the findings that the feeding value of the algae was best when used as a component rather than as an exclusive feed, experiments were performed with diets, in which the laboratory mixture (diet 2) was blended with increasing algal contents (0 to 90%) and fed to grass carp over a period of 4 months (Table 44). Fish fed with the diet containing 80% algae gained about 3 times as much weight as fish fed an algae-free diet. The diet with 90% algae, however, affected a significant reduction in the weight gain for unclear reasons.

In corresponding experiments the diets were fed to young carp. Low algal diets exerted a similar effect to that observed with grass carp. Algal concentrations higher than 70%, however, resulted in negative effects on weight gain and in increasing mortality. The two species of fish respond differently to algal diets. There was no specific interpretation for the high mortality of the carp, but it was speculated that the nucleic acid content of the algae played a harmful role.

With young grass carp optimal growth was achieved with mixtures of 80% algal powder, whereas for the common carp best results were obtained with mixtures which contained between 30 to 60% of algae. Detailed studies were carried out in Israel on the performance of high-rate algal ponds and the utilization of sewage-grown algal biomass, feeding experiments being carried out with different fish species.[288,309]

Protein digestibility of different algae was determined by feeding carp a diet containing 60% algal or fishmeal protein and 40% starch. The digestibility of algal protein was found to be lower than that of fish meal, but comparable to values obtained for other vegetable proteins (sunflower and cottonseed oil cakes with a digestibility of 74.5 and 75.2%, respectively).

The digestibility of the algal protein is affected by the drying method and is higher in drum-dried than in sun-dried material due to insufficient cell rupture by the latter process (see section on nutritional studies). These differences could be verified by feeding differently processed algae to carp. The variations, however, are not as pronounced as the ones obtained with monogastric animals.

Since fish meal is one of the most expensive components in conventional fish feed formulations, it would be of commercial relevance if this protein source was replaced by algal protein. Thus, suitable trials were carried out on carp with diets, where fish meal was replaced completely by vegetable protein (algae and soybean meal) with increasing concentrations of algae (0, 5, 15, and 25%) and correspondingly decreasing amounts of soybean meal. While the diet containing fish meal resulted in a maximum weight gain (272.5 g), the other diets gave weight gains of 168.5, 192.5, 211.2, and 237 g, respectively. It can be concluded that algal protein is less effective than fish meal in promoting carp growth, but more effective than soya protein. No additional effect was observed for the diets with amino acid supplementation (met, lys), which agrees with other reports indicating that carp cannot utilize free amino acids in the diet.[310]

Because of the common practice of harvesting sewage-grown algae by aluminum-induced flocculation, it has to be questioned whether aluminum residues in the processed algal biomass cause negative effects in fish. In order to examine this possibility fish feeds were mixed with increasing concentrations of $Al_2(SO_4)_3$ (up to 140 g/kg) and $Al(OH)_3$ (up to 176 g g/kg). In spite of these high metal contaminations no serious health problems could be recorded except a slight decrease in the survival rate (from 93% to 83.4%). This may be explained by the neutral to basic pH prevailing in the guts of the carp so that aluminum compounds cannot dissolve in the intestine of the fish and react with other dietary components. High aluminum levels were found in the feces of the fish, but only traces or none at all in the blood and tissue. Even *Tilapia*, which have stomach and gastric acid juices are not affected by the metal, probably due to the fact that in wastewater with its usually high phosphate concentrations, most of the aluminum is present in the form of aluminum phosphate or aluminum hydroxide, insoluble even at lower pH levels.

There is evidence that with certain types of diets, raising the protein content in the feed will lower the fat content in the body to a considerable degree.[311,312] Also, the source of the dietary protein may influence body fat. Feeding of three different feed mixtures with varying levels of protein produced rather similar feed efficiency ratios and identical protein levels in the fish body, but surprisingly different fat levels (Table 45). This observation agrees with earlier findings that fish fed algal-containing diets (32%) always showed very low amount of body fat.

Most of these fish feeding experiments have been performed in tanks under more or less

Table 45
INFLUENCE OF DIFFERENT DIETARY PROTEINS ON
BODY COMPOSITION OF CARP[311]

Protein source	Feed composition (% of dry matter)		Fish composition (g/kg live weight)	
	Protein	Ether extract	Protein	Ether extract
Fish meal	53	10	150	85
Algae/whey/soya	35	12	150	28
Casein/whey/soya	44	11	150	18

Table 46
RESULTS OF POND EXPERIMENTS:
POLYCULTURE EFFECTS OF DIFFERENT
DIETS ON YIELDS OF FISH[288]

	Diet[a]		
	A	B	C
Common carp Yield (kg/ha/120 day)	3187	1922	3767
Silver carp Yield	796.7	680	790
Tilapia Yield	620	603	672
Bighead Yield	116.1	132	168
Feed consumption (kg)	7955	8075	8096
Total yield (kg)	4720	3337	5397
FCE[b]	1.69	2.42	1.5

[a] Diet A: fish meal 15%, soybean meal 5%, wheat 80%; diet B: fish meal 3%, soybean meal 28%, wheat 65%; diet C: fish meal 3%, algae 23%, wheat 72%.
[b] FCE: Feed conversion efficiency = feed consumed/weight gain.

controlled and standardized conditions. The ultimate examination of the suitability of algae as a protein source in fish feed, however, should be made in ponds under natural environmental conditions. Under such circumstances the effect of the diet will depend not only on the relative value of the algae as compared to fish meal or other protein source in promoting growth, but also on the relationship between the standing crop of fish and the production of natural food in the pond.

A few data on corresponding experiments are available from Israel, shown in Table 46.[288] The yield in fish ponds fed with algal diets exceeded that of the control pond by more than 10%, the soya protein diet resulting in lowest yields.

In all experiments the algal diet was readily accepted by the fish and no difference in the survival rate was observed as compared to the control diets; also, the growth rate was similar to that of the control. In promoting growth, algae were found to be more effective than soybean meal but inferior to fish meal. This may be due to lower digestibility and/or to the limited amount of sulfur-containing amino acids and other nutritious components. There was only limited variance in the digestibility among different algal species.

As a preliminary conclusion, the evidence shows that algae meal is probably the only vegetable protein source which can replace fish meal in the diets of Tilapia and common carp, although the relative value of algal protein is smaller than that of fish meal protein.

THERAPEUTIC PROPERTIES OF ALGAE

The earliest reports about the utilization of algae or algal constituents were concerned with pharmaceutical applications. These reports originate from Chinese records of 2700 B.C., where marine algae were recommended for the treatment of several kinds of maladies, and it was Plinius the Elder who praised algae as useful for various applications. On the other hand, Virgil used the phrase "vilior algae", meaning "still more worthless than algae", while describing an absolutely useless item. The pharmaceutical value of algae has been recognized during the last two centuries due to systematic research and improved analytical methods.[313]

While there is a multitude of publications on the therapeutic properties of marine algae (anthelmintics, antihypertensives, antiblood coagulation and antihyperlipemic activities, depression of cholesterol level, antibiotics, antitumor activities, enzyme activators, vitamin sources, etc.), only very few reports are available on similar applications of freshwater microalgae.

The effect of ointments mixed with 20% alcoholic extracts of *Scenedesmus acutus* was tested in Czechoslovakia on persons suffering from trophic and varicose ulcers, burns, nonhealing wounds, or eczema.[314] Out of 109 patients treated with this ointment, 91% were healed and in 7% an improvement could be observed. Out of 112 control patients suffering from the same ailments, but treated with placebo, only 1 person improved. The stimulating effect of the algal ointment on the granulation and epithel formation in the treated patients was attributed to the "natural form of chlorophyll", the carotenoids, and the B-group vitamins present in the algal extract.

Another publication from Czechoslovakia describes the production and pharmaceutical application of *Scenedesmus*-containing preparations for the treatment of skin diseases.[315] More than 650 reports were collected from physicians and veterinary surgeons on the action of the algal preparations. The best results were achieved for the treatment of eczema with children, ulcers cruris, and gynecological diseases.

The administration of *Chlorella* to cases of uncurable wounds has also been reported from Japan.[316] In these studies, however, the algae was given in form of *Chlorella* tablets or *Chlorella* essence tablets (6/day). The preparations showed a tendency toward improvement in granulation of tissues and promotion of epithel formation. Another report from Japan describes the application of *Chlorella* tablets in cases of gastric and duodenal ulcer and gastritis.[317,318] The oral administration of the tablets (2 g of *Chlorella* given for 20 to 90 days) had considerable positive effects in respect to eliminating the subjective symptoms and preventing the decrease of leukocytes. The tablets were clinically observed to have a tendency to promote the intestinal peristaltic.

Quite different aspects were investigated by the Japanese navy, who tested the effect of *Chlorella* on the weight of healthy adults and the morbidity rate from common cold during a voyage lasting 75 days.[319] Administration of algae (2 g/day) had a positive effect on weight gain for persons working mostly under deck in the engine room where no sunlight enters. The morbidity rate of cold was reduced by 30% in the experimental group taking the algal preparation.

Although the findings show promising attributes for algae, it has to be kept in mind that the investigations were initiated by the *Chlorella*-producing industry and thus the independence of the researchers may be questioned.

In the course of nutritional studies on rats it was observed that rats fed *Scenedesmus*-containing diets showed lower cholesterol levels than the control animals. Experiments to establish the hypocholesterolemic effect of *Scenedesmus* and *Spirulina* were reported from India.[320,321] A reference group of three diets, two containing algae at 10% and 15% protein level and one containing casein (10% protein) were used; cholesterol and bile salts were

Table 47
HYPOCHOLESTEROLEMIC EFFECT OF DIETS CONTAINING *SCENEDESMUS* AND *SPIRULINA*[320,321]

Diet	Serum cholesterol (mg/100 mℓ serum)		Total liver cholesterol (mg/100 g fresh liver)
	Total	Free	
Scenedesmus			
Casein (10%)	251.9	75.0	47.6
Alga (10%)	121.2	50.7	39.7
Alga (15%)	94.2	32.7	34.3
Spirulina			
Casein (10%)	256.6	78.0	48.5
Alga (10%)	233.5	65.6	43.8
Alga(15%)	220.0	53.0	36.4
Alga (10%) + 0.3% met	229.0	56.0	40.6

added to the diets to induce hypercholesterolemia. The animals were fed on the different diets for 4 weeks. The serum and liver cholesterol levels after the termination of the experiment are listed in Table 47. The serum cholesterol level was highest in rats fed casein as a protein source. Incorporation of *Scenedesmus* at 10% protein level lowered the serum cholesterol level significantly; the 15% algal protein diet further reduced the concentration. A similar trend was found for the cholesterol level in the liver, which was considerably lower in rats fed on algal diets as compared to those fed on casein. The cholesterol-lowering property of sun-dried *Spirulina* was not as pronounced as that of drum-dried *Scenedesmus*. Since the individual constituents of both algae have not been evaluated, it is difficult to explain the definite reason for the lesser cholesterol-lowering property of *Spirulina*.[322] Similar experiments were reported by Rolle and Pabst[323] who tested the cholesterol-reducing activity of drum-dried *Scenedesmus* on rat experiments. In animals fed for 6 to 8 weeks with standard diet enriched with 3% cholesterol the average concentration of blood plasma cholesterol increased from 2.0 to 3.6 mmol/ℓ. In contrast, the final concentration reached only 2.4 to 2.8 mmol/ℓ in animals receiving the same diet but to which 20% algal powder was added. The plasma cholesterol level in animals receiving only algal diet did not appear to be influenced, the mean value of the test was 2.0 in the beginning and 2.2 mmol/ℓ at the end of the experiment. The level of plasma triglyceride in animals fed on algae with and without the addition of cholesterol was lower than that of the control. Also, the algae-enriched diet prevented an excessive deposition of cholesterol in the liver. The increase in weight of the livers of rats receiving the cholesterol-enriched standard diet and cholesterol-enriched algal diet were 2.0 and 0.8 g, respectively; the cholesterol content of the livers were increased to 600 and 180 mg, respectively.

A study of the effect of hydrophilic and lipophilic *Scenedesmus* fractions as well as the algal residues for their cholesterol-lowering properties, showed that water-extracted algae lowered the plasma cholesterol levels in cholesterol stressed animals.[324] The content of plasma triglyceride in animals receiving different *Scenedesmus* fractions was lowered by 35 to 55% in nearly all groups. The deposition of cholesterol in the liver was reduced up to 50% by untreated algal powder as well as by algal extracts with water or organic solvents. As a matter of fact, the algal residues after extraction showed the best effects. Similar effects have been reported for marine algae.[325]

But for all this evidence, there is no rigorous scientific evidence for therapeutic effects that can be attributed clearly to the action of microalgae. On the other hand, the scarcity of warranted information does not preclude the possibility that certain microalgal species may possess properties which can be of therapeutic value.

ACKNOWLEDGMENT

I am sincerely grateful to Dr. W. Pabst (Kernforschungsanlage Jülich, West Germany, who generously placed his entire collection of literature at my disposal, for his informed direction and helpful criticism. I am also indebted to many other colleagues who provided me with a fund of information.

REFERENCES

1. **Farrar, W. V.,** "Tecuitlatl". A glimpse of Aztec food technology, *Nature (London),* 211, 341, 1966.
2. **Spoehr, H. A. and Milner, H. W.,** The chemical composition of *Chlorella;* effect of environmental conditions, *Plant Physiol.,* 24, 120, 1949.
3. **Kanazawa, T.,** Changes of amino acid composition of *Chlorella* cells during their life cycle, *Plant Cell Physiol.,* 5, 333, 1964.
4. **Kanazawa, T. and Kanazawa, K.,** Changes in composition patterns of basic proteins in *Chlorella* cells during their life cycle, *Plant Cell Physiol.,* 9, 701, 1968.
5. **Matsuka, M., Otsuka, H., and Hase, E.,** Changes in contents of carbohydrate and fatty acid in the cells of *Chlorella* protothecoides during the processes of de- and regeneration of chloroplasts, *Plant Cell Physiol.,* 7, 651, 1966.
6. **Altmann, H., Fetter, F., Cabela, E., and Szilvinyi, A.,** Die Proteinsynthese im Verlauf des Entwicklungszyklus der einzelligen Grünalge *Chlorella pyrenoidosa, Bodenkultur,* 20, 262, 1969.
7. **Mostert, E. S. and Grobbelaar, J. U.,** Protein manipulation of mass cultured algae, University of the Orange Free State Publ., South Africa, Series C, No.3, 86—90.
8. **Tamiya, H.,** Synchronous cultures of algae, *Ann. Rev. Plant Physiol.,* 17, 1, 1966.
9. **Aaronson, S., Berner, T. and Dubinsky, Z.,** Microalgae as a source of chemicals and natural products, in *Algae Biomass,* Shelef, G. and Soeder, C. J., Eds., Elsevier/North-Holland, Amsterdam, 1980, 575.
10. **Durant, N. W. and Jolly, C.,** Green algae, *Chlorella* as a contribution to the food supply of man, *Fish. Ind. Res.,* 5, 67, 1969.
11. **Richardson, B., Orcutt, D. M., Schwertner, H. A., Martinez, C. L., and Wickline, H. E.,** Effects of nitrogen limitation on the growth and composition of unicellular algae in continuous culture, *Appl. Microbiol.,* 18, 245, 1969.
12. **Rhee, G.-Y.,** Effects of N:P atomic ratios and nitrate limitation on algal growth, cell composition, and nitrate uptake, *Limnol. Oceanogr.,* 23, 10, 1978.
13. **Ruppel, H.-G.,** Untersuchungen über die Zusammensetzung von *Chlorella* bei Synchronisation im Licht-Dunkel-Wechsel, *Flora,* 152, 113, 1962.
14. **Lorenzen, H.,** Time measurements in unicellular algae and its influence on productivity, in *Algae Biomass,* Shelef, G. and Soeder, C. J., Eds., Elsevier/North-Holland, Amsterdam, 1980, 411.
15. **Dubinsky, Z., Berner, T., and Aaronson, S.,** Potential of large-scale algal culture for biomass and lipid production in arid lands, *Biotechnol. Bioeng. Symp.,* 8, 51, 1978.
16. **Tipnis, H. P. and Pratt, R.,** Protein and lipid content of *Chlorella vulgaris* in relation to light, *Nature (London),* 188, 1031, 1960.
17. **Becker, E. W. and Venkataraman, L. V.,** Production and utilization of the blue-green alga *Spirulina* in India, *Biomass,* 4, 105, 1984.
18. **Miller, S. A.,** Nutritional factors in single-cell protein, in *Single-Cell Protein,* Mateles, R. I. and Tannenbaum, S. R., Eds., MIT Press, Cambridge, Mass., 1968, 79.
19. **Durand-Chastel, H.,** Production and use of *Spirulina* in Mexico, in *Algae Biomass,* Shelef, G. and Soeder, C. J., Eds., Elsevier/North-Holland, Amsterdam, 1980, 51.
20. **Trubachev, N. I., Gitel'zon I. I., Kalacheva, G. S., Barashkov, V. A., Belyanin, V. N., and Andreeva, R. I.,** Biochemical composition of several blue-green algae and *Chlorella, Prikl. Biokhim. Mikrobiol.,* 12, 196, 1976.

21. **Soeder, C. J.,** Chemical composition of microalgal biomass as compared to some other types of single cell protein (SCP, University of the Orange Free State Publ., South Africa, Series C, No.3, 73—85.

22. **Becker, E. W.,** Biotechnology and exploitation of the green algae *Scenedesmus obliquus* in India, *Biomass* 4, 1, 1984.

23. **Hindak, F. and Pribil, S.,** Chemical composition, protein digestibility and heat of combustion of filamentous green algae, *Biol. Plant. (Praha),* 10, 234, 1968.

24. **Parsons, T. R., Stephens, K., and Strickland, J. D. H.,** On the chemical composition of eleven species of marine phytoplankters, *J. Fish. Res. Board Can.,* 8, 1001, 1961.

25. **Eddy, B. P.,** The suitability of some algae for mass cultivation for food, with special reference to *Dunaliella bioculata, J. Exp. Bot.,* 7, 372, 1956.

26. **Collyer, D. M. and Fogg, G. E.,** Studies on fat accumulation by algae, *J. Exp. Bot.,* 6, 256, 1955.

27. **Ricketts, T. R.,** On the chemical composition of some unicellular algae, *Phytochemistry,* 5, 67, 1966.

28. Protein Advisory Group, PAG ad hoc working group meeting on clinical evaluation and acceptable nucleicacid levels of SCP for human consumption, *PAG Bull.,* 5(3), 17, 1975.

29. **Subbulakshmi, G., Becker, E. W., and Venkataraman, L. V.,** Effect of processing on the nutrient content of the green alga *Scenedesmus acutus, Nutr. Rep. Int.,* 14, 581, 1976.

30. **Gibbs, N. and Duffus, C. M.,** Natural protoplast *Dunaliella* as a source of protein, *Appl. Environ. Microbiol.,* 31, 602, 1976.

31. **Durand-Chastel, H.,** General characteristics of blue-green algae (Cyanobacteria): *Spirulina,* in *CRC handbood of Biosolar Resources,* Zaborsky, O. R., Ed., CRC press, Boca Raton, Fla., 1982, 19.

32. **Boussiba, S. and Richmond, A. E.,** C-Phycocyanin as a storage protein in the blue-green alga *Spirulina platensis, Arch. Microbiol.,* 25, 143, 1980.

33. **Boussiba, S. and Richmond, A. E.,** Isolation and characterization of phycocyanins from the blue-green alga *Spirulina platensis, Arch. Microbiol.,* 120, 155, 1979.

34. **Adrian, J. and Fragne, R.,** Comportement thermique des microorganismes alimentaires: levures et algues spirulines, *Ind. Aliment. Agric.,* 1365, 1975.

35. **Adrian, J.,** Evolution de la lysine des lagues spirulines soumises a des traitments thermiques varies, *Ann. Nutr. Alim.,* 29, 477, 1975.

36. **Carpenter, K. J.,** The estimation of the available lysine in animal protein foods, *Biochem. J.,* 77, 604, 1960.

37. **Lipinsky, E. S. and Litchfield, J. H.,** Algae, bacteria, and yeasts as food or feed, *CRC Crit. Rev. Food Technol.,* 581, 1970.

38. **Mitchell, H. H. and Block, R. J.,** Some relationships between amino acid contents of protein and their nutritive value for the rat, *J. Biol. Chem.,* 163, 599, 1946.

39. **Oser, B. L.,** Method for integrating essential amino acid content in the nutritional evaluation of protein, *J. Am. Diet. Assoc.,* 27, 396, 1951.

40. **Leveille, G. A., Sauberlich, J. W., and Shockley, J. W.,** Protein value and the amino acid deficiencies of various algae for growth of rats and chicks, *J. Nutr.,* 76, 423, 1962.

41. **Diem, K. and Lentner, C.,** *Documenta Geigy, Wissenschaftliche Tabellen,* Georg Thieme Verlag, Stuttgart, 1975.

42. **FAO/WHO,** Energy and protein requirement report of a "Joint FAO/WHO ad hoc Expert Committee", No. 52, 1973.

43. **Paoletti, C., Vincencini, M., Bocci, F., and Materassi, R.,** Composizione biochimica generale delle biomasse de *Spirulina platensis* e *Spirulina maxima,* in *Prospettive della Coltura di Spirulina in Italia,* Materassi, R., Ed., Accademia dei Georgofili, Firenze, 1980, 111.

44. **Clement, G., Giddey, C., and Menzi, R.,** Amino acid composition and nutritive value of the alga *Spirulina maxima, J. Sci. Food Agric.,* 18, 497, 1967.

45. **Walz, O. P.,** Criteria for the evaluation of the protein quality of single cell proteins with regard to the nutrition of monogastric animals and the human nutrition, *Wissensch. Umwelt.,* 4, 268, 1982.

46. **El-Fouly, M. H., Abdalla, F. E., El-Baz, F. K., and Fawzi, A. F. A.,** Mass production of microalgae in the National Research Centre, *Proc. 2nd Egypt. Algae Symp.,* Cairo, 111, 1980.

47. **Priestley, G.,** Algal proteins, in *Food From Waste,* Birch, G. G., Parker, K. J., and Worgan, J. T., Eds., Applied Science Publications, London, 1976, 114.

48. **Lubitz, J. A.,** The protein quality, digestibility and composition of algae, *Chlorella* 71105, *J. Food Sci.,* 28, 229, 1963.

49. **Ben-Amotz, A. and Avron, M.,** Glycerol, β-carotene and dry algal meal production by commercial cultivation of *Dunaliella,* in *Algae Biomass,* Shelef, G. and Soeder, C. J., Eds., Elsevier/North-Holland, Amsterdam, 1980, 603.

50. **Sager, R. and Ramanis, Z.,** A genetic map of non-Mendelian genes in *Chlamydomonas, Proc. Natl. Acad. Sci. U.S.A.,* 65, 593, 1970.

51. **Hudson, B. J. F. and Karis, I. G.,** The lipids of the alga *Spirulina, J. Sci. Food Agric.,* 25, 759, 1974.

52. **Nichols, B. W. and Wood, B. J. B.,** The occurence and biosynthesis of gamma-linolenic acid in a blue-green alga, *Spirulina platensis, Lipids,* 3, 46, 1967.

53. **Kenyon, C. N., Rippka, R., and Stanier, R. Y.,** Fatty acid composition and physiological properties of some filamentous blue-green algae, *Arch. Mikrobiol.,* 83, 216, 1972.

54. **Becker, E. W. and Venkataraman, L. V.,** *Biotechnology and Exploitation of Algae — The Indian Approach,* German Agency for Technical Cooperation. Eschborn, West Germany, 1982, 1.

55. **Podojil, M., Livansky, K., Prokes, B., and Wurst, M.,** Fatty acids in green algae cultivated on a pilot-plant scale, *Folia Microbiol.,* 23, 444, 1978.

56. **Soeder, C. J.,** Verschiedene Single Cell Protein-Typen und ihre biochemischen Eigenschaften. Fette-seifen-anstrichmittel, 81, 352, 1979.

57. **Fried, A., Tietz, A., Ben-Amotz, A., and Eichenberger, W.,** Lipid composition of the halotolerant alga, *Dunaliella bardawil, Biochim. Biophys. Acta,* 713, 419, 1982.

58. **Wood, B. J. B.,** Fatty acids and saponifiable lipids, in *Algal Physiology and Biochemistry,* Stewart, W. D. P., Ed., Blackwell Scientific, Oxford, 1974, 236.

59. **Percival, E. E. and Turvey, J. R.,** Polysaccharides of algae, in *CRC Handbook of Microbiology,* Laskin, A. I. and Lechevalier. H. A., Eds., CRC Press, Cleveland, 1974, 532.

60. **Chuecas, L. and Riley, J. P.,** Component fatty acids of the total lipids of some marine phytoplankton, *J. Mar. Biol. Assoc. U.K.,* 49, 97, 1969.

61. **Shifrin, N. S. and Chisholm, S. W.,** Phytoplankton lipids: environmental influences on production and possible commercial application, in *Algae Biomass,* Shelef, G. and Soeder, C. J., Eds., Elsevier/North-Holland, Amsterdam, 1980, 627.

62. **Materassi, R., Paoletti, C., Balloni, W., and Florenzano, G.,** Some considerations on the production of lipid substances by microalgae and cyanobacteria, in *Algae Biomass,* Shelef, G. and Soeder. C. J., Elsevier/North-Holland, Amsterdam, 1980, 619.

63. **Paoletti, C., Pushparaj, B., Florenzano, G., Capella, P., and Lercker, G.,** Unsaponifiable matter of green and blue-green algal lipids as a factor of biochemical differentiation of their biomasses. I. Total unsaponifiable and hydrobcarbon fraction, *Lipids,* 11, 258, 1976.

64. **Paoletti, C., Pushparaj, B., Florenzano, G., Capella, P., and Lercker, G.,** Unsaponifiable matter of green and blue-green algal lipids as a factor of biochemical differentiation of their biomasses. II. Terpenic alcohol and sterol fractions, *Lipids,* 11, 266, 1976.

65. **Opute, F. I.,** Studies on fat accumulation in *Nitzschia palea, Ann. Bot.,* 38. 889, 1974.

66. **Spoehr, H. A. and Milner, H. W.,** The chemical composition of *Chlorella.* Effect of environmental conditions, *Plant Physiol.,* 24, 120, 1949.

67. **Campbell, J., Stevens, S. E., and Balkwill, D. L.,** Accumulation of poly-β-hydroxybutyrate in *Spirulina platensis, J. Bacteriol.,* 149, 361.

68. **Aaronson, S.,** Effect of incubation temperature on the macromolecular and lipid content of the phyto-flagellate *Ochromonas danica, J. Phycol.,* 9, 111, 1973.

69. **Smith, A. E. and Morris, I.,** Synthesis of lipid during photosynthesis by phytoplankton of the Southern Ocean, *Science,* 207, 197, 1980.

70. **Dickson, G., Galloway, R. A., and Patterson, G. W.,** Environmentally induced changes in the fatty acids of *Chlorella, Plant Physiol. (Lancaster),* 44, 1413, 1969.

71. **Mannino, S. and Benelli, T. G.,** Costituenti minerali di biomasse di *Spirulina maxima,* in *Prospettive della Coltura di Spirulina in Italia,* Materassi, R., Ed., Accademia dei Georgofili. Firenze, 1980, 131.

72. **Kamiya, A. and Miyachi, S.,** General characteristics of green microalgae: *Chlorella,* in *CRC Handbook of Biosolar Resources,* Zaborsky, O. R., Ed., CRC Press, Boca Raton, Fla., 1982, 25.

73. **Robinson, R. K. and Toerien, D. F.,** The algae — a future source of protein in *Development in Food Proteins,* Hudson. B. J. F., Ed., Vol. 1, Applied Science Publications, London, 1982, 289.

74. **Prasomzup, U.,** Effect of different levels of algal diets on rats, *Food,* 11(4), 294, 1979.

75. **Paoletti, C., Materassi, R., and Balloni, W.,** Su un metodo di determinazione rapida degli acidi nucleici delle proteine algali, *Ann. Microbiol.,* 23, 95, 1973.

76. **Hintz, H. F., Heitmann, H., Weird, W. C., Torell, D. T., and Meyer, J. H.,** Nutritive value of algae grown on sewage, *J. Anim. Sci.,* 25, 675, 1966.

77. **Robinson, R. K. and Guzman-Juarez, M.,** The nutritional potential of the algae, *Plant Foods for Man,* 2, 195, 1978.

78. **Combs, G. F.,** Alga (*Chlorella*) as as a source of nutrients for the chick, *Science,* 116, 453, 1952.

79. **Kraut, H. and Rolle, I.,** Vitamin-B$_1$ and B$_2$-Gehalt der Grünalge *Scenedesmus obliquus, Nutr. Dieta,* 10. 183, 1968.

80. **Brown, F., Cuthbertson, W. F. J., and Fogg, G. E.,** Vitamin B$_{12}$ activity of *Chlorella vulgaris* Beij and *Anabaena cylindrica* Lemm, *Nature (London),* 177, 188, 1956.

81. **Fink, H., Herold, E., and Lundin, H.,** Uber den Vitamin B$_{12}$ Gehalt und den Vitamin E-Gehalt der einzelligen Grünalge *Scenedesmus obliquus, Z. Naturforsch.,* 13b, 605, 1958.

411

82. Interdepartmental Committee on Nutrition (1960). Publ. Hlth. Rep. (Wash.), 75, 687.
83. **Anon.,** A blue-green alga as human food source, *Nutr. Rev.,* 26, 182, 1968.
84. **Payer, H. D., Pabst, W., and Runkel, K. H.,** Review of the nutritional and toxicological properties of the green alga *Scenedesmus obliquus* as a single cell protein, in *Algae Biomass,* Shelef, G. and Soeder, C. J., Eds., Elsevier/North-Holland, Amsterdam, 1980, 787.
85. **Cook, B. B.,** Nutritive value of waste-grown algae, *Am. J. Public Health,* 52, 243, 1962.
86. **Fisher, A. W. and Burlew, J. S.,** Nutritional value of microscopic algae, in *Algal Culture: From Laboratory to Pilot Plant,* Publ. no,. 600, Burlew, J. S., Ed., The Carnegie Institution, Washington, D.C., 1953, 303.
87. **Mackie, W. and Preston, R. D.,** Cell wall and intercellular region polysaccharides, in *Algal Physiology and Biochemistry,* Stewart, W. D. P., Ed., Blackwell Scientific, Oxford, 1974, 40.
88. **Percival, E.,** Algal polysaccharides, in Pigman, W. and Horton, D., Eds., *The Carbohydrates: Chemistry and Biochemistry,* Vol 2B, Academic Press, New York, 1970, 537.
89. **Hedenskog, G., Enebo, L., Vendlova, J., and Prokes, B.,** Investigation of some methods for increasing the digestibility in vitro of microalgae, *Biotechnol. Bioeng.,* 11, 37, 1969.
90. **Hedenskog, G. and Mogren, H.,** Some methods for processing of single-cell protein, *Biotechnol. Bioeng.,* 15, 129, 1973.
91. **Becker, E. W., Venkataraman, L. V., and Khanum, P. M.,** Effect of different methods of processing on the protein efficiency ratio of the green alga *Scenedesmus acutus, Nutr. Rep. Int.,* 14, 305, 1976.
92. **Becker, E. W., Venkataraman, L. V., and Khanum, P. M.,** Digestibility coefficient and biological value of the proteins of the alga *Scenedesmus acutus* processed by different methods, *Nutr. Rep. Int.,* 14, 457, 1976.
93. **Pabst, W.,** Die Proteinqualität einiger Mikroalgenarten, ermittelt im Ratten-Bilanzversuch: *Scenedesmus, Coelastrum Uronema, Z. Ernährungswiss.,* 13, 73, 1974.
94. **Omstedt, P. T., von der Decken, A., Hedenskog, G., and Morgen, H.,** Nutritive value of processed *Saccharomyces cerevisiae, Scenedesmus obliquus* and *Spirulina platensis* as measured by protein synthesis in vitro in rat skeletal muscle, *J. Sci. Food Agric.,* 24, 1103, 1973.
95. **Kraut, H., Jekat, F., and Pabst, W.,** Ausnutzungsgrad und biologischer Wert des Proteins der einzelligen Grünalge *Scenedesmus obliquus,* ermittelt im Rattenbilanzversuch, *Nutr. Dieta,* 8, 130, 1966.
96. **Meffert, M. E. and Pabst, W.,** Über die Verwertbarkeit der Substanz von *Scenedesmus obliquus* als Eiweißquelle in Rattenbilanzversuchen, *Nutr. Dieta,* 5, 235, 1963.
97. **Höcht, H., Martin, H. H., and Kandler, O.,** Zur Kenntnis der chemischen Zusammensetzung der Zellwand der Blaualgen, *Z. Pflanzenphysiol.,* 53, 39, 1965.
98. **Drews, G. and Weckesser, J.,** Function, structure and composition of cell walls and external layers, in *The Biology of Cyanobacteria,* Carr, N. G. and Whitton, B. A., Eds., Blackwell Scientific, Oxford, 1982, 333.
99. **Becker, E. W.,** Comparative toxicological studies with algae in India, Peru and Thailand, in *Algae Biomass,* Shelef, G. and Soeder, C. J., Eds., Elsevier/North-Holland, Amsterdam, 1980, 767.
100. **Casu, B., Naggi, A., and Vercellotti, J. R.,** Polisaccaridi di riserva della *Spirulina platensis* estrazione e caratterizzazione, in *Prospettive dell Coltura di Spirulina in Italia,* Materassi, R., Ed., Accademia dei Gergofili Firenzer, 1980, 145.
101. **Olaitan, S. A. and Northcote, D. H.,** Polysaccharides of *Chlorella pyrenoidosa, Biochem. J.,* 82, 509, 1962.
102. **Codd, G. A. and Merrett, M. J.,** Photosynthetic products of division synchronized cultures of *Euglena, Plant Physiol. Lancaster,* 47, 635, 1971.
103. **Liaaen-Jensen, S.,** Carotenoids — a chemosystematic approach, *Pure Appl. Chem.,* 51, 661, 1979.
104. **Walz, O. P., Koch, F., and Brune, H.,** Untersuchungen zu einigen Qualitätsmerkmalen der Grünalge *Scenedesmus acutus* an Schweinen und Küken, *Z. Tierphysiol. Tierernähr. Futtermittelkd.,* 35, 55, 1975.
105. **Ben-Amotz, A., Sussman, I., and Avron, M.,** Glycerol production by *Dunaliella, Experientia,* 38, 49, 1982.
106. **Williams, L. A., Foo, E. L., Foo, A. S., Kühn, I., and Heden, C.-G.,** Solar bioconversion system based on algal glycerol production, *Biotechnol. Bioeng. Symp.,* 8, 115, 1978.
107. **Miki, K., Tajima, O., Matsurra, E., Yamada, K., and Fukimbara, T.,** Isolation and identification of a photodynamic agent of *Chlorella, Nippon Nogeikagaku Kaishi,* 54, 721, 1980.
108. **Uchiyama, M.,** Phaeophorbide, its generation, biological effect and analysis, *Shokuhin Eisei Kenkyu,* 31, 423, 1981.
109. **Al'bitskaya, O. N., Zaitseva, G. N., Rogozhin, S. V., Pakhomova, M. V., Oshanina, N. P., and Voronkova, S. S.,** Characterization of the protein product from the *Spirulina platensis* biomass, *Prikl. Biokhim. Mikrobiol.,* 15, 751, 1979.
110. **Mitsuda, H.,** Utilisation of *Chlorella* for food, *Congr. Food Sci. Technol.,* London, Sept. 1962.
111. **Akeson, W. R. and Stahmann, M. A.,** A pepsin-pancreatin digest index of protein quality evaluation, *J. Nutr.,* 83, 257, 1964.

112. **Tamiya, H.,** Chemical composition and acceptability as food and feed of mass cultured unicellular algae, Cf. Casey, R. P. and Lubitz, J. A., Algae as food for space travel. A review, *Food Technol.,* 17, 1386, 1964.

113. **de Hernandez, J. T. and Shimada, A. S.,** Estudios sobre el valor nutritivo del alga espirulina *(Spirulina maxima),* Arch. Latinoam, Nutr., 28, 196, 1978.

114. **Bock, H. D. and Wünsche, J.,** Möglichkeiten zur Verbesserung von Grünalgenmehl, *Sitzungsber. Dtsch. Akad. Landw. Wiss.,* 16, 113, 1967.

115. **Erchul, B. A. F. and Isenberg, D. L.,** Protein quality of various algal biomasses produced by a water reclamation plant, *J. Nutr.,* 95, 374, 1968.

116. **Yamaguchi, M. M., Hwang, H. D., Kawaguchi, K., and Kandatsu, M.,** Metabolism of ^{15}N-labelled *Chlorella* protein in growing rats with special regard to the nutritive value, *Br. J. Nutr.,* 30, 411, 1973.

117. **Lee, B. H., Picard, G. A., and Goulet, G.,** Effects of processing methods on the nutritive value and digestibility of *Oocystis* alga in rats, *Nutr. Rep. Int.,* 25, 417, 1982.

118. **Fink, H. and Herold, E.,** Über die Eiweißqualität einzelliger Grünalgen und ihre Lebernekrose verhütende Wirkung. III. Mitt. Über den Einfluß des Trocknens auf das diätetische Verhalten der einzelligen Zuchtalge *Scenedesmus obliquus, Hoppe-Seyler's Z. Physiol. Chem.,* 311, 13, 1958.

119. **Mayer, A. M. and Evenari, E.,** The nutritional value of oven-dried *Chlorella, Bull. Res. Counc. Isr.,* 4(4), 401, 1955.

120. **McDowell, M. E., and Leveille, G. A.,** Feeding experiments with algae, *Fed. Proc. Fed. Am. Soc. Exp. Biology,* 22, 1431, 1963.

121. **Campbell, J. A.,** Evaluation of protein in foods for regulating purposes, *J. Agric. Food. Chem.,* 8, 323, 1960.

122. National Academy of Sciences-National Research Council, *Evaluation of Protein Quality,* Publ. No. 1100, NAS/NRC, Washington, D.C., 1963.

123. **Liener, J. E.,** Nutritional value of food protein products, in *Soybean Chemistry and Technology,* Vol. 1. *Proteins,* Smith, A. D. and Circle, S. D., Eds., AVI Publishing, Westport, Conn., 1972.

124. **Bourges, H., Sotomayor, A., Mendoza, E., and Chavez, A.,** Utilization of the alga *Spirulina* as a protein source, *Nutr. Rep. Int.,* 4, 31, 1971.

125. **Pabst, W., Jekat, F., and Rolle, I.,** Die Ausnutzung von Kohlenhydraten, Fett, Phosphor und Stickstoff in walzengetrockneter *Scenedesmus*-Substanz ermittelt im Ratten-Bilanzversuch., *Nutr. Dieta,* 6, 279, 1964.

126. **Bowman, R. O., Middlebrook, J. B., and McDaniel, R. H.,** Nutritional value of algae for mice, *Aerospace Med.,* 33, 589, 1962.

127. **Thananunkul, D., Reungmanipaitoon, S., Prasomsup, U., Pongjiwanich, S., and Klafs, H. J.,** The protein quality of algae produced in Thailand as determined by biological assays with rats, *Food,* 10, 200, 1977.

128. **Mokady, S., Yannai, S., Einav, P., and Berk, Z.,** Nutritional evaluation of the protein of several algae species for broilers, *Arch. Hydrobiol. Beih. Ergebn. Limnol.,* 11, 89, 1978.

129. **van der Decken, A. and Omstedt, P. T.,** Effects of dietary protein and amino acid mixture on protein synthesis *in vitro* in rat liver. *Nutr. Metab.,* 16, 325, 1974.

130. **Klafs, H. J., Reungmanipaitoon, S., and Paramadilok, P.,** The protein quality of algae produced in Thailand as determined by biological assays with rats, *Food,* 9, 6, 1977.

131. **Cheeke, P. R., Gasper, E., Boersma, L., and Oldfield, J. E.,** Nutritional evaluation with rats of algae *(Chlorella)* grown on swine manure, *Nutr. Rep. Int.,* 16, 579, 1977.

132. **Lubitz, J. A.,** Animal nutrition studies with *Chlorella* 71105, in *Biologistics for Space Systems Symposium,* Techn. Doc. Rep. No. AMRL-TDR-62-116, 1962, 331.

133. **Baranowski, J. D., Dominguez, C. A., and Magarelli, P. C.,** Effects of drying on selected qualities of *Spirulina platensis* protein, *J. Agric. Food Chem.,* 32, 1385, 1984.

134. **Chung, P., Pond, W. G., Kingsbury, J. M., Walker, E. F., and Krook, L.,** Production and nutritive value of *Arthrospira platensis,* a spiral blue-green algae grown on swine wastes, *J. Anim. Sci.,* 47, 319, 1978.

135. **Barlow, E. W. R., Boersma, L., Phinney, H. K., and Miner, J. R.,** Algal growth in diluted pig waste. *Agric. Environ.,* 2, 339, 1975.

136. **Mokady, S., Yannai, S., Einav, P., and Berk, Z.,** Algae grown on waste water as a source of protein for young chickens and rats. *Nutr. Rep. Int.,* 19, 383, 1979.

137. **Kosaric, N., Nguyon, H. T., and Bergougnou, M. A.,** Growth of *Spirulina maxima* algae in effluents of secondary waste water treatment plants, *Biotechnol. Bioeng.,* 16, 881, 1974.

138. **Contreras, A., Herbert, D. C., Grubbs, B. G., and Cameron, I. L.,** Blue-green alga, *Spirulina,* as the sole dietary source of protein in sexually maturing rats, *Nutr. Rep. Int.,* 19, 749, 1979.

139. **Cook, B. B., Lau, E. W., and Bailey, E. M.,** The protein quality of waste-grown green algae. I. Quality of protein in mixtures of algae, nonfat powdered milk, and cereals, *J. Nutr.,* 81, 23, 1963.

140. **Narasima, D. L. R., Venkataraman, G. S., Duggal, S. K., and Eggum, O.,** Nutritional quality of the blue-geen alga *Spirulina platensis* Geitler, *J. Sci. Food Agric.,* 33, 456, 1982.

141. **Bock, H. D. and Wünsche, J.**, Untersuchungen über die Proteine-von zwei Grünalgenmehlen, *Jahrb. Tierernahr. Futter.*, 6, 544, 1968.

142. **Bock, H. D. and Wünsche, J.**, Möglichkeiten zur Verbesserung der Proteinqualität von Grünalgenmehl, *Sitzungsber. Dtsch. Akad. Landwirtschaftswissensch.*, 16, 113, 1967.

143. **Fink, H. and Herold, E.**, Über die Eiweißqualität einzelliger Grünalgen und ihre Lebernekrose verhütende Wirkung. I. Mitteilung, *Hoppe-Seyler's Z. Physiol. Chem.*, 305, 182, 1956.

144. **Fink H. and Herold, E.**, Über die Eiweißqualität einzelliger Grünalgen und ihre Lebernekrose verhütende Wirkung, *Hoppe-Seyler's Z. Physiol. Chem.*, 307, 202, 1957.

145. **Fink, H., Schlie, I., and Herold, E.**, Über die Eiweißqualität einzelliger Grünalgen und ihre Beziehungen zur alimentären Lebernekrose der Ratte, *Naturwissenschaften*, 41, 169, 1954.

146. **Pabst, W.**, Ernährungsversuche zur Bestimmung der Proteinqualität von Mikroalgen, in Symp. Mikrob. proteingewinnung, Wagner, F., Ed., Verlag Chemie Weinheim, 1975, 173.

147. **Summers, J. D. and Fisher, H.**, Nutritional requirements of the protein-depleted chicken. II. Effect of different protein depletion regimes on nucleic acid and enzyme activity in the liver, *J. Nutr.*, 76, 187, 1962.

148. **Anusuya Devi, M., Rajasekaran, T., Becker, E. W., and Venkataraman, L. V.**, Serum protein regeneration studies on rats fed on algal diets, *Nutr. Rep. Int.*, 19, 785, 1979.

149. **Rosenthal, O., Fahl, J. C., and Vans, H. M.**, Response of alkaline phosphatase of rat liver to protein depletion and inanition, *J. Biol. Chem.*, 194, 299, 1952.

150. **Hundley, J. M. and Ing, R. B.**, Algae as a source of lysine and threonine in supplementing wheat and bread diet, *Science*, 124, 536, 1956.

151. **Venkataraman, L. V., Becker, E. W., and Khanum, P. M.**, Supplementary value of the proteins of alga *Scenedesmus acutus* to rice, ragi, wheat and peanut proteins, *Nutr. Rep. Int.*, 15, 145, 1977.

152. **Gross, R., Schöneberger, H., Gross, U., and Lorenzen, H.**, The nutritional quality of *Scenedesmus acutus* produced in a semi-industrial plant in Peru, *Ber. Dtsch. Bot. Ger.*, 95, 323, 1982.

153. **Waslien, C., Calloway, D. H., Margen, S., and Costa, F.**, Uric acid levels in men fed algae and yeast as protein sources, *J. Food Sci.*, 35, 294, 1970.

154. **Müller-Wecker, H. and Kofranyi, E.**, Zur Bestimmung der biologischen Wertigkeit von Nahrungsproteinen. XVIII. Einzeller als zusätzliche Nahrungsquelle, *Hoppe-Seyler's Z. Physiol. Chem.*, 354, 1034, 1973.

155. **FAO/WHO/UNICEF**, *Safety of Single Cell Proteins*, Document 2.23/2 of the Meeting at Geneva, Sept. 1969.

156. **Zöllner, N. and Griebsch, A.**, in *Purine Metabolism in Man*, Vol. 41B, Sperling, O., De Vries, A., and Wyngaarden, J. B., Eds., Plenum Press, New York, 1973.

157. **Edozien, J. C., Udo, U. U., Young, V. R., and Scrimshaw, N. S.**, Effects of high levels of yeast feeding on uric acid metabolism of young men, *Nature (London)*, 228, 180, 1970.

158. **Feldheim, W.**, Untersuchungen über die Verwendung von Mikroalgen in der menschlichen Ernäahung. I. Ernührungsversuche mit algenhaltigen Kostformen in Thailand, *Int. J. Vitam. Nutr. Res.*, 42, 600, 1972.

159. **Griebsch, A. and Zöllner, N.**, Harnsäure-Plasmaspiegel und renale Harnsäureausscheidung bei Belastung mit Algen, einer purinreichen Eiweißquelle, *Verh. Dtsch. Ges. Inn. Med.*, 77, 173, 1970.

160. **Gorham, P. R. and Carmichael, W. W.**, Toxic substances from fresh-water algae, *Prog. Water Technol.*, 12, 189, 1980.

161. **Trollope, D. R. and Evans, B.**, Concentrations of copper, iron, lead, nickel, and zinc in freshwater algal blooms, *Environ. Pollut.*, 11, 109, 1976.

162. **Becker, E. W.**, Limitations of heavy metal removal from waste water by means of algae, *Water Res.*, 17, 459, 1982.

163. **Förstner, U. and Wittmann, G. T. W.**, *Metal Pollution in the Aquatic Environment*, Springer-Verlag, Berlin, 1979.

164. **Becker, E. W.**, The legislative background for utilization of microalgae and other types of single cell protein, *Arch. Hydrobiol. Beih. Engebin. Limnol.*, 11, 56, 1978.

165. **Payer, H. D. and Runkel, K. H.**, Environmental pollutants in freshwater algae from open-air mass cultures, *Arch. Hydrobiol. Beih. Ergebn. Limnol.*, 11, 184, 1978.

166. **Payer, H. D., Runker, K. H., Schramel, P., Stengel, E., Bhumiratana, A., and Soeder, C. J.**, Environmental influences on the accumulation of lead, cadmium, mercury, antimony, arsenic, selenium, bromine, and tin in unicellular algae cultivated in Thailand and in Germany, *Chemosphere*, 6, 413, 1976.

167. **Anon.**, Evaluation of certain food additives and the contaminants mercury, lead, and cadmium, *WHO Tech. Rep. Ser.*, 505, 1972.

168. **Anon.**, Normes internationales pour l'eau de boisson, Who (Geneva), 1972.

169. International Union of Pure and Applied Chemistry, Proposed Guidelines for Testing of Single Cell Protein Destined as Major Protein Source for Animal Feed, Tech. Rep. No. 12, IUPAC Secretariat, Oxford, 1974.

170. **Boudene, C., Collas, E., and Jenkins, C.,** Recherche et dosage de divers toxiques minéraux dans les algues spirulines de différentes origines et évaluation de la toxicité a long terme chez la rat d'un lot d'algues spirulines de provenance mexicaine, *Ann. Nutr. Aliment.,* 29, 577, 1975.

171. **Macka, W., Wihlidal, H., Stehlik, G., Washüttl, J., and Bancher, E.,** Uptake of $^{203}Hg^{2-}$ and $^{115m}Cd^{2+}$ by *Chlamydomonas reinhardii, Chemosphere,* 10, 787, 1979.

172. **Geisweid, H. J. and Urbach, W.,** Sorption of cadmium by the green microalgae *Chlorella vulgaris, Ankistrodesmus braunii* and *Eremosphaera viridis, Z. Pflanzenphysiol.,* 109, 127, 1983.

173. **Broda, E.,** The uptake of heavy cationic trace elements by microorganisms, *Ann. Microbiol.,* 22, 93, 1972.

174. **Sakaguchi, T., Tsuji, T., Nakajima, A., and Horikoshi, T.,** Accumulation of cadmium by green microalgae, *Eur. J. Appl. Microbiol. Biotechnol.,* 8, 207, 1979.

175. **Rice, T. R.,** The accumulation and exchange of strontium by marine planktonic algae, *Limnol. Oceanogr.,* 1, 123, 1956.

176. **Horikoshi, T., Nakajima, A., and Sakaguchi, T.,** Uptake of uranium by *Chlorella regularis, Agric. Biol. Chem.,* 43, 617, 1979.

177. **Yannai, S., Mokady, S., Sachs, K., Kantorowitz, B., and Berk, Z.,** Secondary toxicology and contaminants of algae grown on waste water, *Nutr. Rep. Int.,* 19, 391, 1979.

178. **Yannai, S. and Mokady, S.,** Short-term and multi-generation toxicity tests of algae grown in wastewater as a source of protein for several animal species, *Arch. Hydrobiol. Beih. Ergebin. Limnol.,* 20, 173, 1985.

179. **Yannai, S., Mokady, S., Sachs, K., Kantorowitz, B., and Berk, Z.,** Certain contaminants in algae and in animals fed algae-containing diets, and secondary toxicity of the algae, in *Algae Biomass,* Shelef, G. and Soeder, C. J., Eds., Elsevier/North-Holland, Amsterdam, 1980, 757.

180. **Bories, G. and Tulliez, J.,** Determination du 3,4-benzopyrene dans les algues spirulines produites et traitees suivant differents procedes, *Ann. Nutr. Aliment.,* 29, 573, 1975.

181. **Benemann, J., Koopman, B., Weissman, J., Eisenberg, D., and Goebel, R.,** Development of microalgae harvesting and high-rate pond technologies in California, in *Algae Biomass,* Shelef, G. and Soeder, C. J., Eds., Elsevier/North-Holland, Amsterdam, 1980, 457.

182. **Groenweg, J., Klein, B., Mohn, F. H., Runkel, K. H., and Stengel, E.,** First results of outdoor treatment of pig manure with algal bacterial systems, in *Algae Biomass,* Shelef, G. and Soeder, G. C., Eds., Elsevier/North-Holland, Amsterdam, 1980, 225.

183. **Abeliovich, A.,** Factors limiting algal growth in high rate oxidation ponds, in *Algae Biomass,* Shelef, G. and Soeder, C. J., Eds., Elsevier/North-Holland, Amsterdam, 1980, 205.

184. **Grobbelaar, J. U.,** Infections: experiences in miniponds, U. O. F. S. Publ., Series C, No. 3, 116, 1981.

185. **Heussler, P., Castillo, J., and Merino, F.,** Parasite problems in the outdoor cultivation of *Scenedesmus, Arch. Hydrobiol. Beih. Ergebn. Limnol.,* 11, 223, 1978.

186. **Venkataraman, L. V. and Sindhu Kanya, T. C.,** Larval contamination in mass outdoor cultures of *Spirulina platensis, Proc. Ind. Acad. Sci.,* 90, 655, 1981.

187. **Lee, B. I. and Dodd, J. C.,** Production of algae from pig wastewater in high-rate ponds, in *Wastewater Treatment and Resource Recovery. Report of a Workshop on High-Rate Algae Ponds: Singapore,* IDRC-154e, Canada, 1980.

188. **Yoshino, Y., Hirai, Y., Takahashi, H., Tamamoto, N., and Yamazaki, N.,** The chronic intoxication test on *Spirulina* product fed to Wistar rats, *Jpn. J. Nutr.,* 38, 221, 1980.

189. **Faggi, E.,** Caratteristiche microbiologiche delle biomasse di *Spirulina,* in *Prospettive della Coltura di Spirulina in Italia,* Materassi, R., Ed., Accademia dei Georgofili, Firenze, 1980, 127.

190. **Mahadevaswamy, M. and Venkataraman, L. V.,** Microbial load in mass cultures of green algae *Scenedesmus acutus* and its processed powder, *J. Biosci.,* 3, 439, 1981.

191. **Blasco, R. J.,** (1965). Nature and role of bacterial contaminants in mass cultures of thermopohilic *Chlorella pyrenoidosa, Appl. Microbiol.,* 13, 473, 1965.

192. **Cloete, T. E. and Toerin, D. F.,** Bacterial-algal interactions, U.O.F.S. Publ., Series C, No. 3, 110—115.

193. **Waslien, C. J.,** Unusual sources of protein for man, *CRC Crit. Rev. Food Sci. Nutr.,* 6, 77, 1975.

194. **Dabbah, R.,** Protein from microorganisms, *Food Technol.,* 24, 659, 1970.

195. **Lipinsky, E. S. and Litchfield, J. H.,** Single cell protein in perspective, *Food Technol.,* 28, 16, 1974.

196. Protein Advisory Group, Report of the sixth meeting of the ad hoc Working Group on Single Cell Protein, *PAG Bull.,* 6(2), 2, 1976.

197. Protein Advisory Group, PAG Statement on Single Cell Protein. New York, PAG Statement No.4,

198. Protein Advisory Group, PAG guideline 12 on the production of single cell protein for human consumption, *PAG Bull.,* 2(2), 21, 1972.

199. Protein Advisory Group, PAG guideline no. 11 for the sanitary production and use of dry protein foods, *PAG Bull.,* 2(1), 1, 1972.

200. Protein Advisory Group, Report of the third meeting of the ad hoc Working Group on Single Cell Protein, *PAG Bull.,* 3(4), 1, 1973.

201. Protein Advisory Group, (1974). Guideline No. 6 on Preclinical Testing of Novel Sources of Protein. PAG Bulletin 4(3), 17—31.
202. Protein Advisory Group, PAG guideline no. 15 on nutritional and safety aspects of novel protein sources for animal feeding, *PAG Bull.*, 4(3), 7, 1974.
203. Protein Advisory Group, Report on the second meeting of the PAG ad hoc Working Group on Single Cell Protein. Document 3.14/15, 1—31.
204. Protein Advisory Group, PAG Guideline No. 7 for Human Testing of Supplementary Food Mixtures.
205. **Chamorro, G.,** Étude toxicologique de l'algue *Spirulina* plante pilote productrice de protéines (*Spirulina* de Sosa Texcoco S.A.) UF/MEX/78/048, UNIDO/10.387, p. 1—177.
206. **Pabst. W., Payer, H. D., Rolle, I., and Soeder, C. J.,** Multigeneration feeding studies in mice for safety evaluation of the microalga *Scenedesmus acutus.* I. Biological and haematological data, *Food Cosmet. Toxicol.*, 16, 249, 1978.
207. **Venkataraman, L. V., Becker, E. W., Rajasekaran. T., and Mathew, K. R.,** Investigations on toxicology and safety of algal diets in albino rats, *Food Cosmet. Toxicol.*, 16, 271, 1980.
208. **Venkataraman, L. V., Becker, E. W., Khanum, P. M., and Mathew, K. R.,** Short-term feeding of alga *Scenedesmus acutus* processed by different methods. Growth pattern and histopathological studies, *Nutr. Rep. Int.*, 15, 231, 1977.
209. **Merino, F., Moya, R., Rodriguez, A., and Heussler, P.,** Control of the parasite *Aphelidium* sp. in mass cultures of *Scenedesmus obliques* by chemical agent, *Arch. Hydrobiol. Beih. Eryehn. Limnol.*, 20, 115, 1985.
210. **Chaturvedi, D. K., Verma, S. S., and Khare, S. P.,** Studies on the effect of feeding fresh water alga *Scenedesmus* to starting chicks and laying hens, *Proc. Natl. Work. Algal Systems. Indian Soc. Biotechnol.*, 145, 1980.
211. **Bizzi, A., Chiesara, E., Clementi, F., Della Torre, P., Marabini, L., Rizzi, R., and Villa, A.,** Trattamenti prolungati nel ratto con diete contenenti proteine di *Spirulina* aspetti biochimici, morfologici e tossicologici, in *Prospettive dell Coltura di Spirulina in Italia*, Materassi, R., Ed., Accademia dei Georgofili, Firenze, 205, 1980.
212. **Krishnakumari, M. K., Ramesh, H. P., and Venkataraman, L. V.,** Food safety evaluation: acute oral and dermal effects of the algae *Scenedesmus acutus* and *Spirulina platensis* on albino rats, *J. Food Protect.*, 33, 934, 1981.
213. **Herdman, W. A.,** *Copepoda* as an article of food, *Nature (London,*, 44, 273, 1891.
214. **Jørgensen, J. and Convit, J.,** Cultivation of complexes of algae with other fresh-water microorganisms in the tropics, in *Algal Culture: From Laboaratory to Pilot Plant*, Publ. no. 600, Burlew, J. S., Ed., The Carnegie Institution, Washington, D.C., 1964, 190.
215. **Powell, R. C., Nevels, E. M., and McDowell, M. E.,** Algae feeding in humans, *J. Nutr.*, 75, 7, 1961.
216. **Dam, R., Lee, S., Fry, P. C., and Fox, H.,** Utilization of algae as a protein source for humans, *J. Nutr.*, 86, 376, 1965.
217. **Jørgensen, E. G.,** Antibiotic substances from cells and culture solutions of unicellular algae with special reference to some chlorophyll derivates, *Physiol. Plant.*, 15, 530, 1962.
218. **Kondratiev, Y. I., Bychkov, V. P., Ushakov, A. S., Boiko, N. N., Klyushkina, N. S., Abaturova, E. A., Terpilovsky, A. M., Korneeva, N. A., Belyakova, M. I., and Kasatkina, A. G.,** The use of 50 and 100 gm of dry biological bodies containing unicellular algae in food rations of man, *Vopr. Pitan.*, 25, 9, 1966.
219. **Kondratiev, Y. I., Bychkov, V. P., Ushakov, A. S., Boiko, N. N., Klyushkina, N. S., Abaturova, E. A., Terpilovsky, A. M., Korneeva, N. V., Belyakova, M. I., Vorobieva, E. S., Demochkina, N. G., and Kasatkina, A. G.,** The use of 150 gm of dry biological bodies containing unicellular algae in food rations of man, *Vopr. Pitan.*, 25, 14, 1966.
220. **Kofranyi, E. and Jekat, F.,** Zur Bestimmung der biologischen Wertigkeit von Nahrungsproteinen. XII. Die Mischung von Ei mit Reis. Mais, Soja, Algen, *Hoppe-Seyler's Z. Physiol. Chem.*, 348, 84, 1967.
221. **Lee, S. K., Fox, H. M., Kies, C., and Dam, R.,** The supplementary value of algae protein in human diets, *J. Nutr.*, 92, 281, 1967.
222. **Gross, R., Gross, U., Ramirez, A., Cuadra, K., Collazos, C., and Feldheim, W.,** Nutritional tests with green alga *Scenedesmus* with healthy and malnourished children, *Arch. Hydrobiol. Beih. Ergebn. Limnol.*, 11, 161, 1978.
223. **Becker, E. W.,** *Microalgae — Findings of Three Experimental Projects*, Publ. No. 143, German Agency for Technical Cooperation, Exchborn, West Germany, 1983.
224. **Cremer, H. D. and Dürr, C. M.,** *Guidelines for the Performance of Acceptability Studies*, German Agency for Technical Cooperation, Eschborn, West Germany, 1976.
225. **Gross, U. and Gross, R.,** Acceptance and product selection of food fortified with the microalga *Scenedesmus*, *Arch. Hydrobiol. Beih. Ergben. Limnol.*, 11, 174, 1978.
226. **Jacquet, J.,** Utilisations biologiques des Sprirulines, *Bull. Acad. Vet.*, 47, 133, 1974.

227. **Morimura, Y. and Tamiya, N.,** Preliminary experiments in the use of *Chlorella* as human food, *Food Technol.,* 8, 179, 1954.

228. **Tamura, E., Matsuno, N., and Morimoto, K.,** Nutritional studies on *Chlorella,* Report 3. Human experiment on absorption of *Chlorella, Ann. Rep. Natl. Inst. Nutr. Tokyo,* 25, 27, 1957.

229. **Hayami, H. and Shino, K.,** Nutritional studies on decolorized *Chlorella.* II. Studies on the rate of absorption of the decolorized *Chlorella* on adult men, *Ann. Rep. Natl. Inst. Nutr. Tokyo,* 59, 1958.

230. **Mitsuda, H., Yasumoto, K., and Nakamura, H.,** Needs for new protein isolate techniques to utilize *Chlorella* and other unused resources, *Proc. Int. Congr. Nutr. (Hamburg),* 5, 327, 1966.

231. **Streitelmeier, D. and Koch, R. B.,** Modifications of green algae to be used as food, *Nature (London),* 195, 474, 1962.

232. **Hills, C.,** *The Secrets of Spirulina,* University of the Trees Press, Boulder Creek, Calif., 1980.

233. **Gross, U., de Valencia, J., and Ynami, M.,** Some observations about development and introduction of algae with regard to the target groups, *Arch. Hydrobiol. Beih. Engebn. Limnol.,* 20, 191, 1985.

234. **Hernandez, V., Gross, U., and Gross, R.,** Some remarks about a testing program for single cell protein (SCP) as food additives in Peru, on the example of the microalga *Scenedesmus acutus, Arch. Hydrogiol. Beih, Ergebn. Limnol.,* 11, 211, 1978.

235. **Hashimoto, Y. and Tsutsumi, J.,** Dietary photosensitization in animals, *Shokuhin Eisei Shi* 4, 185, 1963.

236. **Tamura, Y., Nishigaki, S., Maki, T., Shimamura, Y., and Naoi, Y.,** Biological and chemical examination on *Chlorella* tablets, *Shokuhin Eisei Kenkyu,* 28, 753, 1978.

237. **Buri, P. and Vitsupakom, S.,** Mass infection of insect larvae in algae mass cultures. VI. Report on the production and utilization of algae as a protein source in Thailand. Kasetsart University, Bangkok, Thailand, 1977.

238. **Jaleel, S. A. and Soeder, C. J.,** Current trends in microalgal cultures as protein source in West Germany, *Indian Food Packer,* 27, 45, 1978.

239. **Burlew, J. S., Ed.,** *Algal Culture: From Laboratory to Pilot Plant,* Publ. no. 600, The Carnegie Institution, Washington, D.C., 1964.

240. **Aaronson, S.,** A new source of single-cell protein and other nutrients, *Arch. Hydrobiol.,* 6(Suppl 41), 108, 1972.

241. **Fowden, L.,** A comparison of the compositions of some algal proteins, *Ann. Bot. N.S.,* 18(71), 257, 1954.

242. **Cabell, C. A. and Earle, I. P.,** Protein quality assay. Comparison of the rat repletion method with other methods of assaying the nutritive value of proteins in cottonseed meals, *Agric. Food. Chem.,* 2, 787, 1954.

243. **Miller, L. E.,** The loss and regeneration of rat liver enzymes related to diet proteins, *J. Biol. Chem.,* 186, 253, 1950.

244. **Murainatsu, K. and Ashida, K.,** Effect of dietary protein level on growth and liver enzyme activities of rats, *J. Nutr.,* 76, 143, 1962.

245. **Bremer, H. J.,** Hazards and problems in the utilization of microalgae for human nutrition, *Arch. Hydrobiol. Beih. Ergebn. Limnol.,* 11, 218, 1978.

246. **Jaya, V. T., Scarino, M. L., and Spadoni, M. A.,** Caratteristiche nutrizionali in vivo di *Spriulina maxima,* in *Prospettive della Coltura di Spirulina in Italia,* Materassi, R., Ed., Accademia dei Georgofili, Firenze, 1980, 195.

247. **Bender, A. E. and Doell, B. H.,** Biological evaluation of proteins. A new aspect, *Br. J. Nutr.,* 11, 140, 1957.

248. **Fink, H. and Herold, E.,** Biological value of algal protein, *Nutr. Rev.,* 16, 22, 1956.

249. **Ciferri, O.,** Spirulina, the edible microorganism, *Microbiol. Rev.,* 47, 551, 1983.

250. **MacIntyre, T. M.,** The digestibility of dried ground seaweed meal by the laying hen, *Can. J. Agric. Sci.,* 35, 168, 1955.

251. **Grau, C. R. and Klein, N. W.,** Sewage-grown algae as a feedstuff for chicks, *Poult. Sci.,* 36, 1046, 1957.

252. **Arakawa, S., Tsurumi, N., Murakami, K., Muto, S., Hoshino, J., and Yagi, T.,** Experimental breeding of white Leghorn with the *Chlorella*-added combined feed, *Jpn. J. Exp. Med.,* 30, 185, 1960.

253. **Mokady, S., Yannai, S., Einav, P., and Berk, Z.,** Protein nutritive value of several microalgae species for young chickens and rats, in *Algae Biomass,* Shelef, G. and Soeder, C. J., Eds., Elsevier/North-Holland, Amsterdam, 1980, 655.

254. **Lipstein, B. and Hurwitz, S.,** The nutritional value of algae for poultry. Dried *Chlorella* in broiler diets, *Br. Poult. Sci.,* 21, 9, 1980.

255. **Lipstein, B., Hurwitz, S., and Bornstein, S.,** The nutritional value of algae for poultry. Dried *Chlorella* in layer diets, *Br. J. Poult. Sci.,* 21, 23, 1980.

256. **Brune, H. and Walz, O. P.,** Studies on some nutritive effects of the green alga *Scenedesmus acutus* with pigs and broilers, *Arch. Hydrobiol. Ergebn. Limnol.,* 11, 79, 1978.

257. **Blum, J. C. and Calet, C.,** Valeur alimentaire des algues spirulines pour la crossance du poulet de chair, *Ann. Nutr. Aliment.,* 29, 651, 1975.

258. **Reddy, V. R., Reddy, C. V., Varadarajulu, P., and Reddy, G. V.,** Utilisation of algae protein *(Scenedesmus acutus)* in chick rations, *Indian Poult. Gaz.,* 62, 67, 1978.

259. **Yoshida, M. and Hoshii, H.,** Nutritive value of a new type of *Chlorella* for poultry feed, *Jpn. Poult. Sci.,* 19, 56, 1982.

260. **Lincoln, P. and Hill, D. T.,** An integrated microalgae system, in *Algae Biomass,* Shelef, G., and Soeder, C. J., Eds., Elsevier/North-Holland, Amsterdam, 1980, 229.

261. **Ngian, M. F. and Thiruchelvam, S.,** A nutritional evaluation of pig wastewater-grown algae, in Wastewater Treatment and Resource Recovery. Report of a Workshop, Singapore, IDRC-154e.

262. **Brune, H.,** Zur Verträglichkeit der Einzelleralgen *Spirulina maxima* und *Scenedesmus acutus* als alleinige Eiweißquelle für Broiler, *Z. Tierphysiol. Tierern. Futtermittelkd.,* 35, 55, 1982.

263. **Hennig, A., Gruhn, K., Kleemann, J., and Hahn, G.,** Prüfung der Verdaulichkeit und der N-Bilanz von *Scenedesmus quadricauda* an Schweinen und Broilern, *Jahrb. Tierernahrg. Futterg.,* 7, 426, 1970.

264. **Sauveur, B., Zybko, A., Colas, B., Garreau, M., and Rocard, J.,** Protéines alimentaires et qualite de lœuf. I. Effet de quelques protéines sur la qualité intern de l'oeuf et les propriétés fontionelles, *Ann. Zootechnol.,* 28, 271, 1979.

265. **Colas, B., Sauvageot, F., Harscoat, J.-P., and Sauveur, B.,** Protéines alimentaires et qualité de l'oeuf. II. Influence de la nature des protéines distribuées aux poules sur les caractéristiques sensorielles de l'oeuf et la teneru en acides aminés libres du jaune, *Ann. Zootechnol.,* 28, 297, 1979.

266. **Rowland, L. O., Lincoln, E. P., Hooge, D. M., and Valentine, T. L.,** Evaluation of high ash algae meals, *SAAS Abstr.,* 1023, 1979.

267. **Hintz, H. F. and Heitman, H.,** Sewage-grown algae as a protein supplement for swine, *Anim. Prod.,* 9, 135, 1967.

268. **Witt, M., Schroeder, J., and Mehden, H.,** Zur Frage der Verwendung von Algeneiweiß in der Schweinemast, *Zuchtungskunde,* 34, 272, 1962.

269. **Witt, M., Schröder, J., and van der Mehden, H.,** Der Einfluß von Fischmehl, Sojaschrot und Süßwasseralgen *(Scenedesmus obliquus)* als Eiweißbeofitter auf die Mastergebnisse und die Schlachtkörperqualitt beim Schewin, *Landwirtschftl. Forsch.,* 20, 62, 1967.

270. **Witt, M. and Schröder, J.,** Die einzellige Süßwasser-Grünalge *Scenedesmus obliquus,* ein vollwertiges Eiweißbeofitter in der Schweinemast, *Landwirtschftl. Forsch.,* 20, 148, 1967.

271. **Lee, B. Y. and Dodd, C.,** Production of algae from pig wastewater in high-rate ponds, in: Report of a workshop in high-rate algae ponds, Singapore, IDRC - 154e.

272. **Fevrier, C. and Sevet, B.,** *Spirulina maxima* in pig feeds, *Ann. Nutr. Aliment.,* 29, 625, 1975.

273. **Yap, T. N., Wu, J. F., Pond, W. G., and Krook, L.,** Feasibility of feeding *Spirulina maxima, Arthrospira platensis* or *Chlorella sp.* to pigs weaned to a dry diet at 4 to 8 days of age, *Nutr. Rep. Int.,* 25, 543, 1982.

274. **Lo, M. C.,** Microbial treatment and utilization of night soil, in Lee, B. Y., Lee, K. W., McGarry, G., and Graham, M., Eds., Wastewater Treatment and Resource Recovery. A report of a workshop on high-rate algal ponds. IDRC-154e, Canada.

275. **Davis, I. F., Sharkey, M. J., and Williams, D.,** Utilization of sewage algae in association with paper in diets of sheep, *Agric. Environ.,* 2, 333, 1975.

276. **Hasdai, A. and Ben-Ghedalia, D.,** Sewage-grown algae as a source of supplementary nitrogen for ruminants, *J. Agric. Sci. (Cambridge),* 97, 533, 1981.

277. **Calderon, J. F., Merino, H., and Barragan, M.,** Valor alimentico del algal espirulina *(Spirulina geitleri)* para ruminants, Tec. Pecu. Mex., 31, 42, 1976.

278. **Ganowski, H., Usunova, K., and Karabashev, G.,** Effect of the microalga *Scenedesmus acutus* on the digestibility of the mast of calves and on some blood parameters, *Anim. Sci.,* 12, 77, 1975.

279. **Guerra, A. A.,** Algae, an economical protein source for insect diets, *Southwest. Entomol.,* 3, 76, 1977.

280. **Salama, H. S. and Sharaby, A. F.,** Efficiency of algae as a protein source in diets for rearing insects, *Z. Angew. Entomol.,* 90, 329, 1980.

281. **Hou, R. F. and Chen, R. S.,** The blue-green alga, *Spirulina platensis,* as a protein source of artificial rearing of *Bombyx mori, Appl. Entomol. Zool.,* 16, 169, 1981.

282. **Ogata, M., Yamada, H., Fukada, T., and Shirota, M.,** The artificial diet for the silkworm *Bombyx mori* L., containing *Chlorella* as a nutritive source. I. Feeding behavior of newly hatched larvae to *Chlorella, J. Sericult. Sci. Jpn.,* 46, 427, 1977.

283. **de Pauw, N. and Pruder, G.,** Use and production of microalgae as food in aquaculture. Presented at World Conference on Aquaculture, Venice,

284. **Bardach, J. E., Ryther, J. H., and McLarney, W. O.,** *Aquaculture. The Farming and Husbandry of Freshwater and Marine Organisms,* Wiley Interscience, New York, 1972.

285. **Ryther, J. H. and Goldman, J. C.,** Microbes as food in mariculture, *Ann. Rev. Microbiol.,* 29, 429, 1975.

286. **De Pauw, N. and von Vaerenbergh, E.**, Microalgal wastewater treatment systems: Potentials and limits. Presented at Conference Phytodepuration and Employment of the Biomass produced. Inst. de Ecologia, Parma, Italy.

287. **dePauw, N., Verboven, J., and Claus, C.**, Large-scale micro-algae production for nursery rearing of marine bivalves, *Aquacult. Eng.*, 2, 27, 1983.

288. **Sandbank, E. and Hepher, B.**, The utilization of microalgae as feed for fish, *Arch. Hydrobiol. Beih. Ergebn. Limnol.*, 11, 108, 1978.

289. **Meske, C. and Pfeffer, E.**, Growth experiments with carp and grass carp, *Arch. Hydrobiol. Beih. Ergebn. Limnol.*, 11, 98, 1978.

290. **Meske, C. and Pfeffer, E.**, Untersuchungen über Mikroalgen, Hefe oder Casein als Komponenten eines fischmehlfreien Trockenfutters für Karpfen, *Z. Tierphysiol. Tierernahr. Futtermittelkd.*, 38, 177, 1977.

291. **Edwards, P.**, A review of recycling organic wastes into fish with emphasis on the tropics, *Aquaculture*, 21, 261, 1980.

292. **Gaigher, I. G. and Cloete, T. E.**, Human health aspects of fish production in wastewater in South Africa, U.O.F.S. Publ. Series C, No.3, 91—95.

293. **Soong, P.**, Production and development of *Chlorella* and *Spirulina* in Taiwan, in *Algae Biomass*, Shelef, G. and Soeder, C. J., Eds., Elsevier/North-Holland, Amsterdam, 1980, 97.

294. **Taiganides, E. P., Chou, K. C., and Lee, B. Y.**, Animal waste management and utilisation in Singapore, *Agricult. Wastes*, 1, 129, 1979.

295. **John, C. K.**, Cultivation of freshwater fish in the effluent pond of rubber processing factories, in: *Proc. Agro-Ind. Wastes Symp.*, Kuala Lumpur, Malaysia, 182—189.

296. **Sivalingam, P. M.**, Recycling of palm-oil mill sludge discharge nutrients through SCP *(Chlorella vulgaris)* culturing, in: Wastewater treatment and resource recovery. Report of a workshop on high-rate algae ponds, Singapore, IDRC-154e.

297. **Hepher, B. and Schroeder, G. L.**, Wastewater utilization in Israel for aquaculture, in *Wastewater Renovation and Reuse*, D'Itri, G., Ed., Marcel Dekker, New York, 1977, 529.

298. **Noriega-Curtis, P.**, Primary productivity and related fish yield in intensely manured fish ponds, *Aquaculture*, 17, 395, 1979.

299. **Schlüter, M. and Groeneweg, J.**, Mass production of freshwater rotifers on liquid wastes. I. The influence of some environmental factors on population growth of *Brachionus rubens* Ehrenberg 1838, *Aquaculture*, 25, 17, 1981.

300. **Groeneweg, J. and Schlüter, M.**, Mass production of freshwater rotifers on liquid wastes. II. Mass production of *Brachionus rubens* Ehrenberg 1838 in the effluent of high-rate algal ponds used for the treatment of piggery waste, *Aquaculture*, 25, 25, 1981.

301. **De Pauw, N., De Leenheer, L., Laureys, P., Morales, J., and Reartes, J.**, Cultures d'algues et d'invertébrés sur déchets agricoles, in *La Pisciculture en Étang*, Billard, R., Ed., I.N.R.A., Paris, 1980, 189.

302. **Dinges, R.**, The availability of *Daphnia* for water quality improvement and as an animal food source, in *Proc. Wastewater Use in the Production of Food and Fiber*, EPA 660/2-74-041, U.S. Environmental Protection Agency, Washington, D.C., 1974, 142.

303. **Dunstan, W. M. and Menzel, D. W.**, Continuing cultures of natural populations of phytoplankton in dilute treated sewage effluent, *Limnol. Oceanogr.*, 16, 623, 1971.

304. **Witt, U., Koske, P. H., Kuhlmann, D., Lenz, J., and Nellen, W.**, Production of *Nannochloris* sp. (Chlorophyceae) in large scale outdoor tanks and its use as a food organism in marine aquaculture, *Aquaculture*, 23, 171, 1981.

305. **Sorgeloos, P.**, High density culturing of the brine shrimp *Artemia salina* L., *Aquaculture*, 1, 385, 1973.

306. **Soeder, C. J.**, The technical production of microalgae and its prospects in marine aquaculture, in *Harvesting Polluted Waters*, Devik O., Ed., Plenum Press, New York, 1976, 11.

307. **Persoone, G. and Claus, C.**, Mass culture of algae: a bottleneck in the nursery culturing of molluscs, in *Algae Biomass*, Shelef, G. and Soeder, C. J., Eds., Elsevier/North-Holland, Amsterdam, 1980, 265.

308. **Meske, C. and Pruss, H. D.**, Mikroalgen als Komponente von fischmehlfreiem Fischfutter, *Fortschr. Tierphysiol. Tierernahr.*, 8, 71, 1977.

309. **Sandbank, E. and Hepher, B.**, Microalgae grown in wastewater as an ingredient in the diet of warmwater fish, in *Algae Biomass*, Shelef, G. and Soeder, C. J., Eds., Elsevier/North-Holland, Amsterdam, 1980, 697.

310. **Abe, H., Masuda, I., Abe Saito, T., Toyoda, T., and Kitamura, S.**, Nutrition of protein in young carp. Nutritive value of amino acids, *Bull. Jpn. Soc. Sci. Fish.*, 36, 407, 1970.

311. **Reimers, J. and Meske, C.**, Der Einfluß eines fischmehlfreien Trockenfutters auf die Körperzusammensetzung von Karpfen, *Fortschr. Tierphysiol. Tierernahr.*, 8, 82, 1977.

312. **Meske, C. and Pfeffer, E.**, Influences of source and level of dietary protein on body composition of carp, *Proc. World Symp. Finfish Nutr. Fishfeed Technol.*, Vol. 2, 394, 1979.

313. **Hoppe, H. A., Levring, T., and Tanaka, Y., Eds.,** *Marine Algae in Pharmaceutical Science,* Walter de Gruyter, Berlin, 1979.

314. **Safar, G.,** Algoterapie, *Prakt. Lek. (Praha),* 55, 641, 1975.

315. **Rydlo, O.,** Verwendung einiger Mikroalgen in der praktischen Pharmiazie, *Pharmazie,* 6, 146, 1973.

316. **Hasuda, S. and Mito, Y.,** Report on experimental administration of *Chlorella* to cases of incurable wounds, *Shinryo and Shinyaku,* 3(3.17), 1, 1966 (translation).

317. **Yamagishi, Y., Toigawa, M., Suzuki, R., Hara, T., and Warita, F.,** Therapy of digestive ulcer by *Chlorella, Nippon Iji Shimpo,* 1997, 1, 1962 (translation).

318. **Saito, T. and Oka, N.,** Clinical application of *Chlorella* preparations, *Shinryo Shinyaku,* 3(3.16) 1966 (translation).

319. **Kashima, Y. and Tanaka, Y.,** On the changes of weights and morbidity rate of common cold for the crew of the training fleet in 1966. Report of the Central Hospital, the Self-Defense Force.

320. **Anusuya Devi, M., Venkataraman, L. V., and Rajasekaran, T.,** Hypocholesterolemic effect of diets containing algae on albino rats, *Nutr. Rep. Int.,* 20, 83, 1979.

321. **Anusuya Devi, M. and Venkataraman, L. V.,** Hypocholesterolic effect of the blue-green algae *Spirulina platensis* in albino rats, *Nutr. Rep. Int.,* 28, 519, 1983.

322. **Mathe, D., Lutton, C., Rautureau, J., Coste, T., Gonffier, E., Silprice, J. C., and Chevallier, F.,** Effects of dietary fibre and salt mixtures on the cholesterol metabolism of rats, *J. Nutr.,* 107, 466, 1977.

323. **Rolle, I. and Pabst, W.,** Über die cholesterinsenkende Wirkung der einzelligen Grünalge *Scenedesmus acutus* 276-3a. I. Wirkung von walzengetrockneter Algensubstanz, *Nutr. Metab.,* 24, 291, 1980.

324. **Rolle, I. and Pabst, W.,** Über die cholesterinsenkende Wirkung der einzelligen Grünalge Scenedesmus acutus 276-3a. II. Wirkung von Algenfraktionen, *Nutr. Metab.,* 24, 302, 1980.

325. **Nisizawa, K.,** Pharmaceutical studies on marine algae in Japan, in *Marine Algae in Pharmaceutical Science,* Hoppe, H. A., Levring, T., and Tanaka, Y., Eds., Walter de Gruyter, Berlin, 1979, 243.

326. **Anusuya Devi, M. and Venkataraman, L. V.,** The effect of algal protein diets on the regeneration of serum and liver proteins in protein depleted rats, *Qual. Plant Foods Hum. Nutr.,* 33, 287, 1983.

327. **Pabst, W.,** Nutritional evaluation of nonsewage microalgae by the rat balance method, *Arch. Hydrobiol. Beih. Ergebn. Limnol.,* 11, 65, 1978.

328. **Brown, A. D. and Borowitzka, L. J.,** Halotolerance of *Dunaliella, Biochem. Physiol. Protozoa,* Vol. 1, Academic Press, New York, 1979, 139.

329. **Ben-Amotz, A. and Avron, M.,** Accumulation of metabolites by halotolerant algae and its industrial potential, *Ann. Rev. Microbiol.,* 37, 95, 1983.

330. **Devi, M. A. and Venkataraman, L. V.,** Supplementary value of the proteins of the blue-green algae *Spirulina platensis* to rice and wheat proteins, *Nutr. Rep. Int.,* 28, 1029, 1983.

PRODUCTS FROM MICROALGAE

Z. Cohen

INTRODUCTION

Production of chemicals by bacterial fermentation has been known for decades and has now developed into an important industry. Macroalgae are harvested for their polysaccharides including agar, alginic acid, and carrageenan, which are extracted commercially on a large scale. Microalgae, on the other hand, have scarcely been exploited by mankind, either as food or as a source for the production of beneficial materials. The soaring prices of fossil oil and its anticipated depletion have caused significant increases in the prices of chemicals and have generated interest in the cultivation of microalgae as raw materials for chemical production.

In general, the main advantages of growing microalgae as a source of chemicals are

1. Algal cultivation is an efficient biological system for use of solar energy to produce organic matter; indeed, algae grow faster than terrestrial (conventional) plants throughout the year, and produce the highest possible annual yield of biomass. In single-celled organisms such as microalgae the entire biomass contains the desired chemicals, in contrast to terrestrial plants, where such chemicals are located in only a part of the plant, i.e., in the fruits, seeds, leaves, or roots.
2. Algae can be grown well in hot desert climates, utilizing sea and/or brackish water, resources which is not suitable for conventional agricultural plants.
3. The life cycle of most algae is completed within several hours, which makes genetic selection and improvements in the species easy and fast.
4. Many species of algae can be induced to produce particularly high concentrations of compounds of commercial interest such as proteins, lipids, starch, glycerol, natural pigments, and biopolymers.

Microalgae are still a relatively untapped resource, but selection of species and genetic manipulation which will lead to a high production of desired chemicals, coupled with improved technological processes should, in the near future, create a new and promising biotechnological field (chemical algology).

The purpose of this chapter is to describe potential microalgal chemicals and discuss their possible commercial or other uses. Special consideration will be given to chemicals that are either unique to microalgae or can be produced by microalgae and exploited commercially.

LIPIDS

Lipids and fatty acids are constituents of all cells, where they serve as a source of energy and metabolites and as membrane components. Chemically, lipids are those biological chemicals that can be extracted with organic solvents. Lipids are generally composed of fatty acids esterified to glycerol, sugars, or bases. However, other forms of lipid or lipoid structures are also found (e.g., hydrocarbons, waxes, sterols and sterol esters, chlorophylls, and carotenes).[1]

The lipids are classified as polar and nonpolar. Most nonpolar lipids in algae are triglycerides, but free fatty acids, wax esters, hydrocarbons, and sterols are also found. Triglycerides have the following general structure:

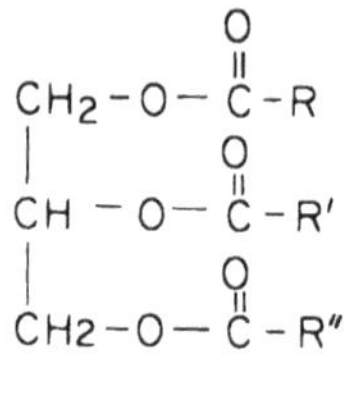

$$CH_2-O-\overset{\overset{O}{\|}}{C}-R$$
$$CH-O-\overset{\overset{O}{\|}}{C}-R'$$
$$CH_2-O-\overset{\overset{O}{\|}}{C}-R''$$

1 Triglyceride R,R′ and R′′ are fatty acid moieties

The polar lipids are essentially glycerides in which one or more of the fatty acids has been replaced with a polar group. These are further classified as phospholipids and glycolipids.

Recently, a new group of polar lipids has been discovered in Chyrsophyceae,[2] Xanthophyceae, Chlorophyceae, and Cyanobacteria.[3] These are the chlorosulfolipids, which are derived from Structure 2:

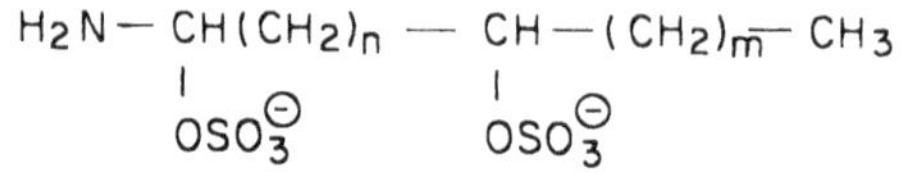

$$H_2N-CH(CH_2)_n-CH-(CH_2)_{\overline{m}}-CH_3$$
$$\qquad\quad | \qquad\qquad\quad |$$
$$\qquad\quad OSO_3^{\ominus} \qquad\quad OSO_3^{\ominus}$$

Diacylglycerol, (*N,N,N,* trimethyl)-homoserine (Structure 3) was reported in six organisms only, five of them algae: *Chlamydomonas reinhardii,*[4] *Volvox Carteri,*[5] *Dunaliella parva,* and *D. tertiolecta*[6] of the Volvocales family, and the phytoflagellate *Ochromonas danica.*[7] Its structure is

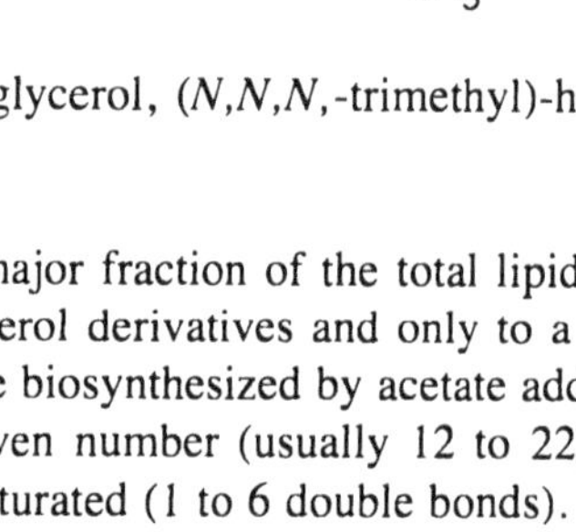

$$CH_2-OA_c$$
$$|$$
$$CH_2-OA_c$$
$$|$$
$$CH_2-O-CH_2-CH_2-CH-\overset{\overset{O}{\|}}{C}\,O^{\ominus}$$
$$\qquad\qquad\qquad\qquad\qquad |$$
$$\qquad\qquad\qquad\quad H_3C-\overset{\oplus}{N}-CH_3$$
$$\qquad\qquad\qquad\qquad\qquad |$$
$$\qquad\qquad\qquad\qquad\quad CH_3$$

3 Diacylglycerol, (*N,N,N,*-trimethyl)-homoserine

Fatty Acids

Fatty acids constitute the major fraction of the total lipids, usually 20 to 40%. Most of it is covalently bound to glycerol derivatives and only to a small extent are they found as free fatty acids. Since they are biosynthesized by acetate addition, practically all fatty acids are straight chains with an even number (usually 12 to 22) of carbon atoms. Algal fatty acids can be saturated or unsaturated (1 to 6 double bonds). The double bonds are virtually always in the *cis* configuration and in polyunsaturated fatty acids (PUFA) the double bonds are allylic (three carbon atoms apart).

Environmental factors play an important role in regulation of the biosynthesis of lipids and fatty acids. However, their effects are sometimes opposite in different algal species (Table 1). Perhaps the most important factor judged by its magnitude, generality, and practical importance, is the nitrogen content of the culture medium. Lipid content tends to decrease for a few days when the medium is deprived of nitrogen and then in many cases there is a sharp rise in the lipid content (Table 2). However, what is more important is the rate of lipid production. Under nitrogen deficiency,[8] overall cell yields and efficiencies were reduced

Table 1
ENVIRONMENTAL EFFECTS ON LIPID
AND PUFA COMPOSITION

	Lipids		PUFA
Light	↑		↑ [11,12]
Temperature	↑	↑ [12]	↓ [11,13-16]
Nitrogen	↑		↑ [17,18]
Nitrogen	↓	↑ [19-20]	↓ [1]
Silicon	↓	↑ [21]	
M_n^{+2}	↑		↑ [22]
Copper tolerant clones		↑ [21]	

Note: Numbers relate to references.

Table 2
LIPID-RICH ALGAE

% Lipid

Alga	N-sufficient	N-starvation	Ref.
Scotiella sp.	48		24
Ochromonas danica	39—71		25
Peridinium cinctum	36		26
Cyrsochromulina polylepsis	48		27
Crysochromulina kappa	33		27
Prymnesium parvum	22—38		27
Chroomonas salina	44		28
Dunaliella primolecta	53.8		29
Dunaliella salina	47.2		30
Botyrococcus braunii	52.9		30
Desert #103 (chlorophyceae)	33.7		30
Radiospahera negevensis	43.0		30
Navicula pelliculosa	22—33		23
Cosmarium laeve	15—33		31
Protosyphon botryoides	37		24
Scenedesmus dimorphus	16—40		31
Nannochloris	20	47.8	21
Ourococcus	27	49.5	21
Nitzschia palea	27.2	39.5	21
Nanochloropsis salina	40.8	72.2	21
Isochrysis sp.	20	47	32
Chlorella pyrenoidosa	14.4	35.8	21
Chlorella vulgaris	33.1	54.2	21

From Aaronson, S., et al., Microalgae as a source of chemicals and natural products, in *Algae Biomass*, Elsevier/North-Holland, Amsterdam, 1982. With permission.

so that the percent lipid yield was not increased. However, Richardson[9] claimed a doubling of the relative lipid content under nitrogen deprivation for *Chlorella sorokiana* and *Oocystis polymorpha* without significant changes in biomass production.

In addition to the environmental and physiological factors that influence the lipid content of microalgae, the yield of lipids is influenced by the method of extraction, including the method for breaking the cells (freezing and thawing, chopping, drying, or direct extraction),

Table 3
MICROALGAE AS A SOURCE OF TRIGLYCERIDES[33]

Algae	Age of culture (days)	Dry biomass (g/ℓ)	Lipid level (% of dry wt)
Characium polymorphum	14	3.72	42
Chlorococcum oleofaciens	17	2.63	44.3
Neochloris oleoabundans	18	1.51	88.8

From Metzger, P., et al., Microalgae as a source of triglycerides, in *Energy from Biomass*, with permission.

and solvents used in extraction. Chloroform-methanol is the most commonly used solvent mixture. However, adding an acid extraction step can increase the lipid yield significantly.[23] The amount of lipids extracted from *Chlorosarinopsis negevensis* rose from 17 to 43% when an acid extraction step was introduced. In *Phaeodactylum tricornutum*, chloroform-methanol extraction yielded 11.6% lipid, while acid extraction provided another 8.3%, for a total of 19.9%.

Examples of outstanding lipid-rich algae are listed in Table 2. Some of these algae are currently being investigated in an effort to exploit algae as a source of fuel oil.

Microalgae have been proposed[33] as an alternative source for edible lipids due to their high triglyceride levels and high productivity (Table 3).

These three species appear to be good oil producers, especially *C. polymorphum*, which grows quickly. Thin layer chromatography (TLC) showed that the oils are essentially triglycerides with less than 1% hydrocarbons. Infrared (IR) and nuclear magnetic resonance (NMR) analysis and determination of the iodine number revealed that these triglycerides are similar to ordinary vegetable oils.

Battelle-Northwest scientists[34] taking advantage of the finding that limited nitrogen leads to increased lipid content at the expense of diminished growth rate, designed a two-stage system in which algal growth was maximized in the first stage and lipid production in the second. In this way lipid yields as high as 80% of the cellular dry weight were reported for various unspecified microalgae. Some of these microalgae were said to produce lipids similar in composition to the commercially important lipids derived from cocoa, corn, linseed, and sperm whale.

Algal lipids can be fractionated to obtain phospholipids, which can be utilized for the production of liposomes. Phospholipids such as PC and PE spontaneously form stable structures in water, in which the polar groups of the lipids are directed toward the water phase and the nonpolar carbon chains toward the inside. Liposomes, which are small bilayered vesicles, can thus be formed. Because of their size and structure, liposomes can fuse with cells. Since various drugs or other active materials can be entrapped within liposomes, they can be utilized as an efficient delivery system for drugs not soluble in water.[35]

Among the various fatty acid constituents of algal lipids, perhaps the most important are the PUFA, and within this group the essential fatty acids (EFA). For humans, these acids are linoleic acid (18:2), γ-linolenic acid (GLA, 18:3-6,9,12), dihomo-γ-linoleic acid (DGLA, 20:3-8,11,14), arachidonic acid (AA, 20:4-5,8,11,14) and eicosapentaenoic acid (EPA, 20:5-5,8,11,14,17). Except for linoleic acid, they are rare in plant and animal sources. Certain algae, however, contain relatively large amounts of EFA (Table 4).

Some of the EFA (DGLA, AA, and EPA) are prostaglandin precursors. Prostaglandins are 20-carbon, oxygenated, unsaturated fatty acid derivatives, which were first isolated from human and sheep prostate glands.

Clinical uses of prostaglandins now being investigated include the induction of labor and abortion, contraception, and the treatment of asthma, bronchitis, ulcers, high blood perssure,

Table 4
OCCURRENCE OF EFA IN ALGAE

Alga	Fatty acid	% of total fatty acids	Total lipids (% of dry wt.)	Ref.
Spirulina platensis	γ-Linolenic acid (18:3)[a]	21	4—9	36
Porphyridium cruentum	Arachidonic acid (20:4)	35	9—14	37
Ochromonas danica	γ-linolenic acid	14	39—71	
	Dihomo γ-linolenic acid (20:3)	5		
	Arachidonic acid	6		
Monodus subterraneous	Eicosapentanoic acid (20:5)	29	11	37

[a] (x:y), x is the carbon number and y is the number of double bonds.

and thrombosis.[38] Unlike hormones, which are produced by specific glands, prostaglandins are synthesized in vivo in low concentrations, released by many tissues, and are not concentrated to any large extent in a particular tissue. It was found that a soybean lipoxygenase,[39] in the presence of readily available reducing agents, converts arachidonic acid to the major prostaglandin PGF$_2$-α. According to Ahern,[40] this would considerably lower the cost of a prostaglandin production process using arachidonic acid as the starting material.

The red alga *Porphyridium cruentum* is the richest known natural source of arachidonic acid. Its AA content at the usual culture temperature (25°C) is usually 36% of the total fatty acids: however, at a lower temperature (16°C) AA constitute as much as 60% of the alga's fatty acids. The production of AA from *Porphyridium* compares favorably with the traditional animal sources. Therefore, the future production of prostaglandins without the need of animal tissues is envisioned. Ahern[10] found that the rate of production of AA from *Porphyridium* is 0.46/ℓ/hr at 32°C and 8000 lux in liquid cultures of 12×10^{-9} cells per liter. These figures led Ahern to estimate a price range of $150 to $1000/kg of AA.

Prostaglandins PGE$_2$ and PGF$_2$ were isolated from the red macroalga *Gracilaria lichenoides* (0.1% of dry weight) after an investigation of the extract's antihypertensive properties.[41] (This is the only mention in the literature of the occurrence of prostaglandins in algae.)

Some of the uses of essential fatty acids for which patents have been applied are (1) antihypertensive;[42] (2) hyperlipidemia treatment;[43,44] (3) treatment of disturbances of consciousness, perception, and movement;[45,46] (4) blood cholesterol reduction;[47] (5) inhibition of blood platelet aggregation;[48] (6) intragastric and antisecretory agents;[49] (7) health food;[50] (8) food additive as imitation chicken flavor;[51] and (9) additives for cosmetics.[52]

EFA affect hyperlipidemia by lowering the levels of plasma lipids (cholesterol and triglycerides). Hyperlipidemia is frequently related to the development of arterosclerosis and coronary heart disease. Huang and Manku[53] have shown that GLA is approximately 170-fold more effective in lowering the plasma cholesterol level than LA, the major constituent of most polyunsaturated oils. GLA is the first product of LA metabolism via the EFA pathway and is formed by the action of delta-6-desaturase (D-6-D). Many factors, such as diabetes, aging, viral infections, *trans* fatty acids, and atopic disorders interfere with D-6-D function.

GLA can be used as a dietary supplement, and the results of tests have been diverse and impressive. Trials with children have shown it is of benefit in treating atopic eczema, while in women it appears to reduce the severity of premenstrual syndrome (PMS) which includes to have a positive effect in heart disease, Parkinson's disease, and multiple sclerosis; however, these reports are preliminary indications at best. Direct provision of GLA could thus have an important role in nutrition. Unfortunately, GLA is rare in foods, being found for practical purposes only in human milk.[53] Oil containing GLA is currently being extracted from seeds of the evening primrose; however, GLA constitutes only 8% of the total fatty acids of this oil.[54] The cyanobacterium *Spirulina platensis* contains more GLA, i.e., 20 to 30% (Table 4) of the total fatty acids (1.25% of its dry weight). *Spirulina* is currently being sold as a health food, primarily for its high protein content. Experiments have shown[55] that it is possible to remove the GLA-containing oil from the algal biomass without destroying the protein. Moreover, this procedure removes the disagreeable odor and taste that the commercial *Spirulina* products suffer.

Radiolabeled fatty acids are an important research tool; some are already available, but at high cost due to difficulties in their preparation. Because of the generally fast growth of microalgae, they can be used to produce fatty acids with uniformly distributed radioisotopes (^{3}H, ^{14}C). The data in Table 5 demonstrate these possibilities. For each fatty acid listed at least one example of an alga containing a relatively high concentration of this fatty acid is shown. For each alga, other major fatty acids are also noted.

Cyanobacteria contain many lipidic compounds that are not found in any other algal class. Most of them are found in one alga, *Lyngbya majuscula*. Some examples are shown in Table 6. For a more detailed presentation the reader is referred to the review of Moore.[67]

Wax

Wax esters are accumulated in aerobically cultured *Euglena gracilis*.[78] After culturing *E. gracilis* heterotrophically for 3 to 5 days to the stationary phase the cells were maintained anaerobically for 2 days and the wax esters, which constituted up to 57.2% of the biomass, were separated (Table 7).

When *Chroomonas salina*, a Cryptomonad, was mass cultured axenically in seawater medium containing 0.25 m glycerol and allowed to autolize, a waxy lipid was produced, 87% of which was wax esters.[28] The major wax ester species had chain lengths of C26 (35%) and C30 (15%) (Table 8).

Sterols

Sterols are apparently components of all algae, although microalgae are not outstanding for their sterol content, which range from 0.001% or less (dry weight basis) in some cyanobacteria to 0.5% or more in some green algae and Chrysophytes. The kinds of sterols found in algae vary. Most species of green algae contain complex mixtures of sterols[79] (Table 9). *Ochromonas danica* contains the highest (1%) concentration of sterols that were found in algae, of which poriferasterol is the main sterol.[80] Chondrilasterol[81] from *Scenedesmus obliquus* (0.2 to 0.3%) or from *Navicula pelliculoss* might be employed as a starting material for the synthesis of pharmacologically important hormones such as cortisone.[82] An unusual steroid has been found in the dinoflagelate *Glenodinium foliaceum*.[83] In addition to the ordinary sterols cholesterol, 24-demethyldinosterol, and dinosterol, 4-methyl gorgostanol *4* has been identified. This sterol is unusual that it has a cyclopropyl ring on its side chain.

Little is known about the conditions favoring synthesis of sterols in algae, but some reports indicate that more sterols are produced under autotrophic than under heterotrophic conditions.[84] (Figure 1.)

Table 5
ALGAL LIPIDS WITH HIGH CONCENTRATION OF CERTAIN FATTY ACIDS

	12:0	14:0	16:0	16:2	16:3	16:4	18:0	18:1	18:2	18:3 α	18:3 γ	18:4	18:5	20:1	20:3	20:4	20:5	22:6	Ref.
Cryptecodynium cohnii	8.0	19.0	20.0					14.0										30.0	55
Skeletonema costatum		32.7	22.4														13.8		56
Hapalosiphon laminusus			53.7	23.8				18.2											57
Phaeodactylum tricornutum			46.9	25.8													8.6		56
Euglena gracilis			15.1		14.4					22.5									17
Ulothrix aequalis			26.0		30.5					36.3									58
Exuviella sp.				10.0														12.0	59
Halosphaera viridis			17.6		44.8														60
Chlorella variegata			13.5		13.4	53.0		16.0											61
Anabaena spiroides			18.0				15.0	12.0	18.0	40.0									62
Spirulina platensis			43.4						12.4		21.4								36
Cryptomonas sp.												44.0				16.0	10.0		63
Prorocentrum minimum			23.1										23.2					24.5	59
Hemiselmis brunescens			12.5									30.5	18.0				13.9		64
Ochromonas danica	13.0		12.0								14.0				5.0	6.0			37
Porphyridium cruentum			23.0						16.0						2.0	36.0	17.0		37
Lauderia borealis		11.7	21.3		12.1												30.3		64
Gonyaulax cattenella			13.9										11.1				11.2	33.9	65

From Pohl, P., Lipids and fatty acids of microalgae, in *Handbook of Biosolar Resources*, Vol. 1, CRC Press, Boca Raton, Fla.

Table 6
SPECIAL LIPIDS OF CYANOBACTERIA

Alga	Type of compound	Structure	No. of known chemicals of each type	Ref.
Lyngbya majuscula	Hydroxy fatty acids		3	68
	Malyngolide (anti-bacterial activity)		1	69
			1	70
	Malyngamides		5	71
	Pukelimides (pyrrolic amides)		7	72
	Majusculamides		2	73
	Lyngbiatoxin	See Table 15		
Oscillatoria tenuis	Geosmin	geosmin		74

Table 6 (continued)
SPECIAL LIPIDS OF CYANOBACTERIA

Alga	Type of compound	Structure	No. of known chemicals of each type	Ref.
Microcystis aeruginosa	β-cyclocytral (monoterpene)	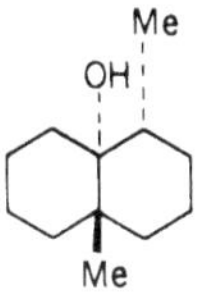		75

β-cyclocytral

Nostoc sp.	Hopene (a triterpene)	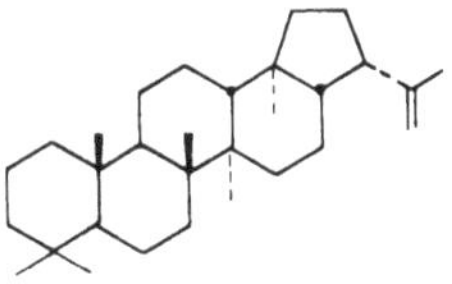		76

Hopene

Nostoc noscurum	Polyhydroxylated derivatives of hopene		4	76
Tolyphothrix conglutinata		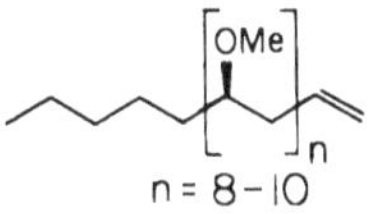		77

n = 8—10

Table 7
FERMENTATIVE PRODUCTION OF WAX ESTERS IN *EUGLENA GRACILIS* GROWN IN VARIOUS MEDIA[78]

Culture medium	Cell yield (dry basis; g/ℓ medium)	Yields of wax esters (g/ℓ medium)
Koren-Hutner	5.05	2.13 (42.2)[a]
Glucose	4.21	2.10 (49.9)
Ethanol	3.06	1.75 (57.2)
Molasses	1.25	0.53 (42.4)

[a] Weight percentage of wax esters with respect to dry weight of cells.

From Inui, H., et al., *Agric. Biol. Chem.*, 47, 2669, 1983. With permission.

Table 8
ALGAL WAX COMPONENTS[77,78]

	No. of carbon atoms				
	12	**13**	**14**	**15**	**16**
Fatty acid (*Euglena*)	10.3	19.5	44.4	12.7	9.2
Fatty alcohol (*Euglena*)	2.7	12.6	46.8	19.3	16.2
Fatty acid (*Chroomonas*)	35		30		14
Fatty alcohol (*Chroomonas*)		53		40	
Total carbon in wax ester	26	27	28	29	30
Euglena	6.1	17.2	36.0	19.0	15.2
Chroomonas	35				15

Table 9
STEROLS IN ALGAE

Sterol	Algae	Ref.
Dinosterol	*Cryptothecodinium cohni*	85
	Gonyaulax tamarensis	86
Cholesterol	*Ochromonas danica*	87
	Chaetocerus simplex calcitrans	88
	Astasia longa	89
Ergosterol	*Haematococcus pluvialis*	90
	Chlorella candida	91
	Cyanidium caldarium	92
22, Dehydrocholesterol	*Porphiridium cruentum*	93
Poriferasterol	*Chlamydomonas reinhardii*	94
	Chlorella ellipsoidea	95
Chondrilasterol	*Euglena gracilis* (green)	96
	Nostoc commune	97
Clionasterol	*Monodus subterraneus*	98
	Tribonema aequale	98
Brassicasterol	*Anabaena cylindrica*	99
Stigmasterol	*Amphora* sp.	100
Ergosterol Δ 8(9)	*Biddulphia aurita*	100
Campesterol Δ 5,22	*Navicula pelliculosa*	100
	Phaeodactylum tricornutum	101
28-Isofucosterol	*Cladophora flexuosa*	102

From Patterson, G. W., Steroids of algae, in *Handbook of Biosolar Resources*, Vol. 1, CRC Press, Boca Raton, Fla.

Hydrocarbons

Tornabene[103] describes patterns of taxonomic importance in the content and structure of hydrocarbons in different algal families. In a variety of marine and freshwater algae including red, green, and brown algae, diatoms, and phytoplankton, *n*-heneicosahexaene (C-21:6) hydrocarbons exist in amounts inversely correlated with the abundance of the long-chain highly unsaturated fatty acids (C-22:6).[104-106] In contrast, nonphotosynthetic diatoms and

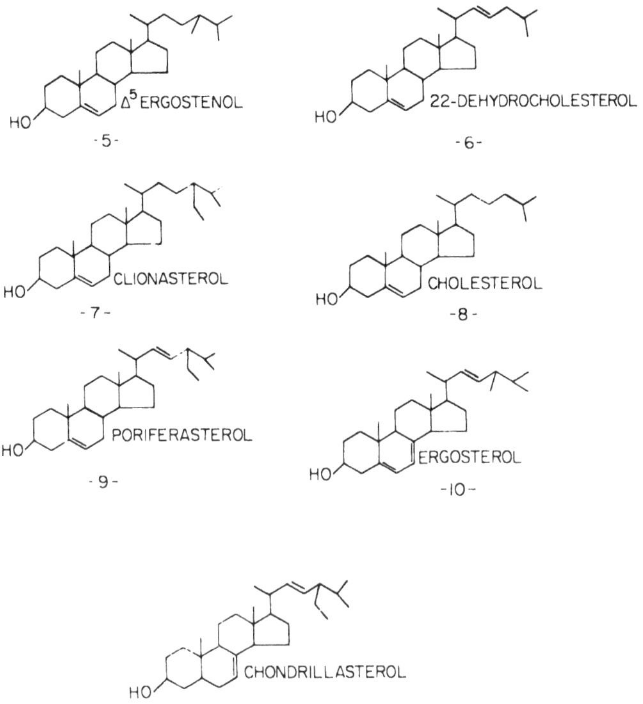

FIGURE 1. Common sterols in microalgae.

dinoflagellates contain traces of aliphatic hydrocarbons but no C-21:6 chains. Other studies[107-108] demonstrated that *n*-pentadecane (C-15) predominates in brown algae and *n*-heptadecane (C-17) in red algae, that olefins predominate in red algae more so than in brown or green algae, and that the C-19 polyunsaturated olefins exist in red, brown, and green algae. The concentrations of these hydrocarbons are generally very low. Mixtures of monoenes, dienes, and trienes (C-15 to C-18) commonly occur in algae, but usually in relatively small quantities. The hydrocarbon of most cyanobacteria consist predominantly of C-17 components, the unique internal methyl branched 7,9-dimethyl hexadecane, and a mixture of 7- and 8-methyl-heptadecane.

Although hydrocarbons have been found only in small amounts in most algae, several algae contain relatively large amounts of hydrocarbons, which led several research groups to explore the possibilities of utilizing the hydrocarbon fraction as fuel oil substitute. The lipid fraction of the green alga *Dunaliella salina* amounts to some 50% of the organic cellular material, and more than 30% of the total lipids consist of acyclic and cyclic hydrocarbons. Carotenes account for 21% of the cell mass, with another 3.5% being saturated and unsaturated C-17 straight chain hydrocarbons and internally branched hydrocarbons identified as 6-methyl hexadecane and 4-methyl octadecane. Temperature, light intensity, and the light spectrum greatly influence the amount and type of lipids synthesized in this alga.[29]

During exponential growth in culture, the alga *Botryococcus braunii* has a hydrocarbon content of about 20%. Under conditions that do not favor growth this organism enters a resting stage where the content of unsaponifiable lipids increases to 80% of the dry weight.[109] The chain length of the hydrocarbons shifts from C27 to C31 during growth to mainly C34 (botryococcene and isobotryococcene) in the resting state. In an alternative resting state hydrocarbon production is very low, yielding less than 1% of the dry weight of the cells. Changes in growth conditions can lead to the production of a variety of hydrocarbons, linear mono- and diunsaturated C29 and C31 (mainly), and branched-chain pentaunsaturated C34 and longer. Bennemann[110] suggests that hydrogenated botrycoccene and isobotrycoccene might also be used as lubricants.

Botryococcus braunii was immobilized on alginate beads.[111] The immobilized cells had a higher than usual photosynthetic productivity. The entrapped culture was active for at least several months. The culture's tendency toward hydrocarbon formation was of the same order of magnitude as that of a free culture. Two drawbacks were mentioned with respect to large scale cultivation of Botryococcus: one is the reported low doubling time (4 to 5 days), but this has recently been reduced significantly (to 36 hr). The second problem lies in the fact that high hydrocarbon concentrations exist only in nongrowing, senescent, and even decaying algae, while at the growing stage the hydrocarbon content is rather poor.

PIGMENTS

Carotenoids

The carotenoids are lipid soluble pigments composed of isoprene units. They can be divided into two main groups: carotenes which include hydrocarbons only and the xanthophylls which are carotene derivatives containing epoxy, hydroxy, ketonic, carboxylic, glycosidic, allenic, or acetylene groups. The carotenoids function both as photoprotectants and, with varying degrees of efficiency, in trapping light energy for photosynthesis. All algae contain carotenoids. Although each species usually contains 5 to 10 different major carotenoids (with a few minor components), the total number of carotenoids in all these classes of organisms is about 60. Some carotenoids occur in most algal classes. These include β-carotene (Structure 12), antheraxanthin (Structure 13), violaxanthin (Structure 14), neoxanthin, and zeaxanthin (Structure 15) (Figure 2). Other carotenoids are restricted to few algal classes. The glycosidic xanthophylls occur only in cyanobacteria, peridinin in Cryptophyceae, and dinoxanthin in Dinophyceae (see Table 10 for a list of algal[113] pigments).

β-Carotene (provitamin A) is a frequent constituent of the carotene mixture in microalgae, but usually less than 1% of the dry weight. However, in *Dunaliella bardawil* β-carotene is accumulated to up to 10% of the dry weight under high light intensity and conditions of limited growth such as high salinity, high temperature, or limited nitrogen supply. β-Carotene globules may protect *D. bardawil* against injury by the high intensity irradiation to which this alga is usually exposed.[114]

Some xanthophylls are metabolized by chickens and fish, and can be used as pigments. Xanthophylls such as lutein and canthaxanthin are used for the pigmentation of chicken skin, tissue, and egg yolks.[115-116] Canthaxanthin is used for intensifying the color of fancy goldfish.[117] Bennemann[110] has stated that metabolizable xanthophylls are valued at about $200/kg as feed additives. Although such xanthophylls were not found in microalgae above 0.5%, other xanthophylls occur in higher concentrations. Diadinoxanthin is reported to constitute 2.4% of the dry weight of *Vacularia virescens*.[118]

An American patent issued to Hoffman-La Roche Inc.[119] describes the process for producing lutein from thermophyllic strains of *Chlorella pyrenoidosa*. These strains were mutagenized by UV irradiation at 254 nm, resulting in the production of new strains capable of an even higher production of lutein, e.g., 1 g of lutein per liter of culture. Lutein has high pigmenting value and low toxicity and is particularly useful in pigmenting poultry, eggs, foodstuffs, feeds, etc.

Chlorophylls

All algae contain one or more types of chlorophyll. Chlorophyll-a is the primary photosynthetic pigment in all algae and is the only chlorophyll in blue green and red algae. Green algae contain chlorophyll-b as well, and chlorophyll-c can be found in many marine algae and fresh water diatoms (Figure 3). Chlorophyll-d has been found in some red algae[120] (Table 10).

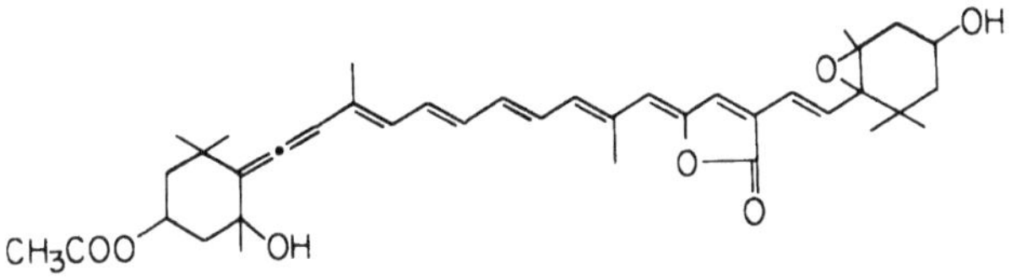

17. Peridinine

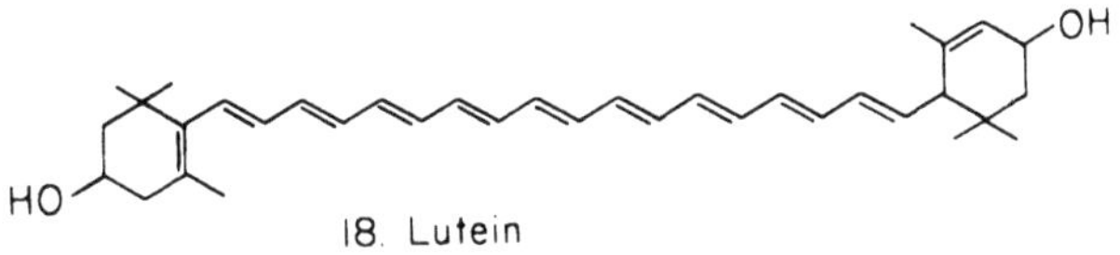

18. Lutein

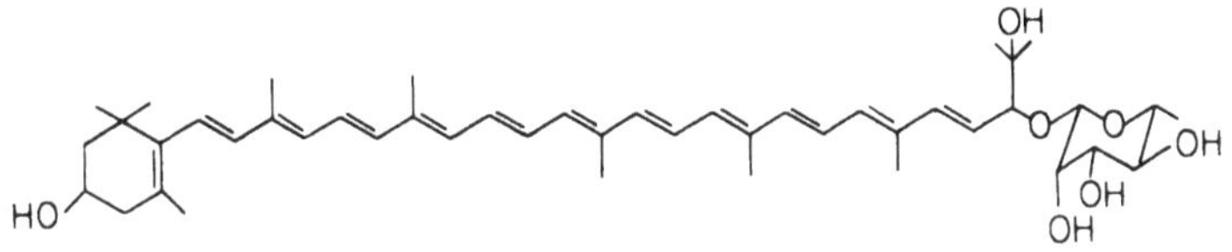

16. Myxoxanthophyll

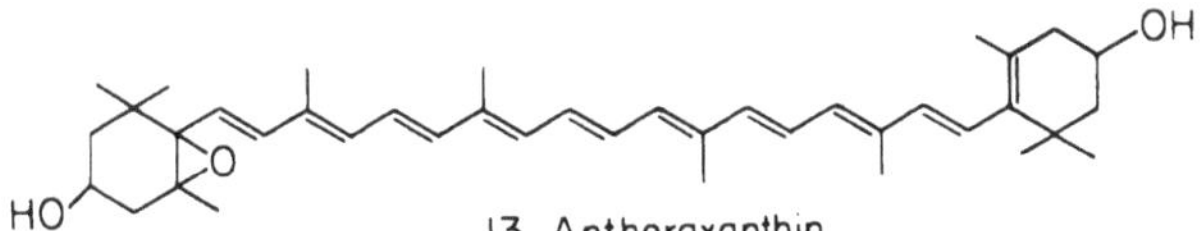

13. Antheraxanthin

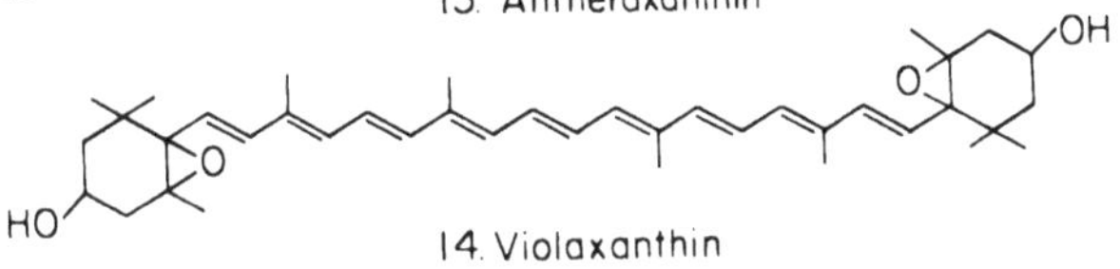

14. Violaxanthin

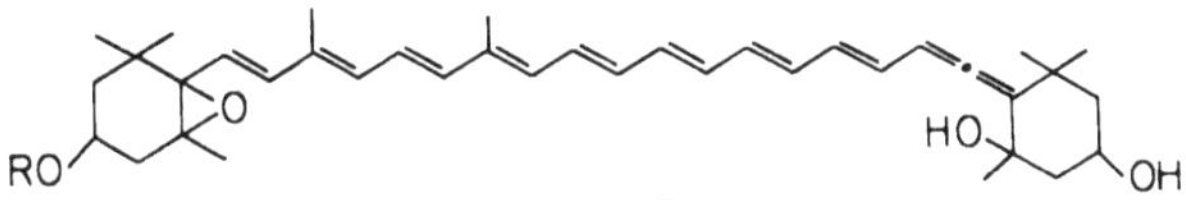

Dinoxanthin R=H
R=Ac

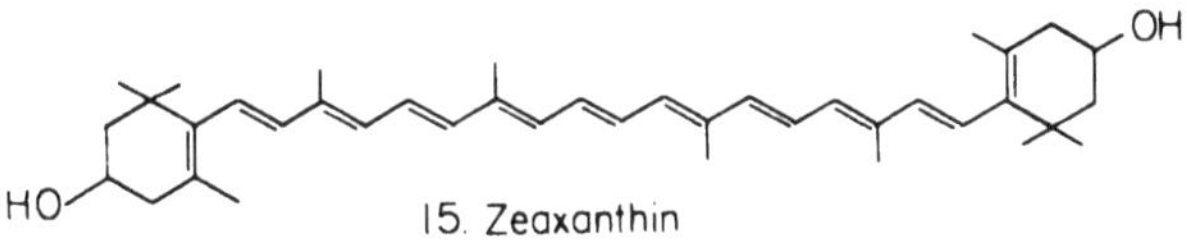

15. Zeaxanthin

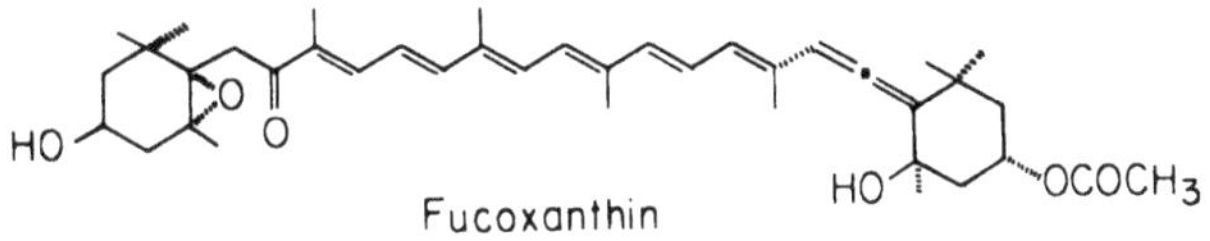

Fucoxanthin

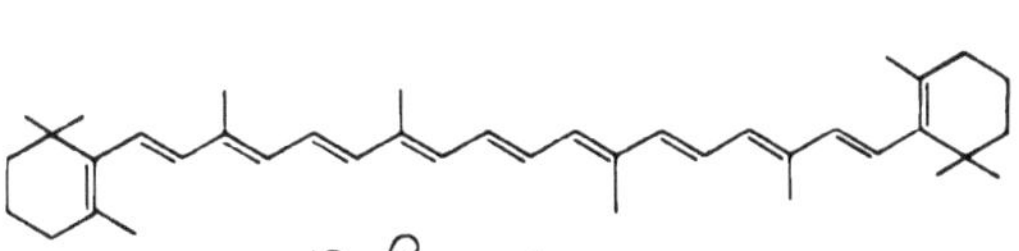

12. β-carotene

FIGURE 2. Common cartenoids in microalgae,

Chlorophylls have traditionally been separated from carotenoids by a series of extractions and chromatographic procedures. However, chlorophylls are very sensitive to heat, light, and oxygen and can be altered rapidly during work-up, leading to a mixture of products that is more difficult to separate than the original. A new method that utilizes reversed phase

Table 10

PHOTOSYNTHETIC PIGMENTS OF MICROALGAE[113]

Class	Chlorophyll	Carotenoids	Phycobilins
Cyanobacteria	a	β-Carotene, *zeaxanthin, echinenone,* canthaxanthin, mutatochrome, antheraxanthin, β-cryptoxanthin, *myxoxanthophyll,* aphanizophyll, oscillaxanthin	C-Phycocyanin C-Phycoerythrin Allophycocyanin
Rhodophyceae	a (+ rarely d)	β-Carotene, *zeaxanthin,* antheraxantin, β-cryptoxanthin, lutein, neoxanthin	R-Phycocyanin R-Phycoerythrin C-Phycocyanin C-Allophycocyanin C-Phycoerythrin
Xanthophyceae	a, c	β-Carotene, *diadinoxanthin,* diatoxanthin, *heteroxanthin, vaucheriaxanthin* ester, neoxanthin, β-cryptoxanthin 5', 6' monoepoxide and 5,6', 5'6'-diepoxide	
Chrysophyceae and Haptophyceae	a, c_1, c_2	β-Carotene, *fucoxanthin,* diatoxanthin, diadinoxanthin, echinenone	
Bacillariophyceae	a, c_1, c_2	β-Carotene, (± α-carotene, + rarely ε-carotene), *fucoxanthin,* diatoxanthin, diadinoxanthin, neoxanthin	
Eustigmatophyceae	a, c	β-Carotene, violaxanthin (± zeaxanthin and antheraxanthin), diatoxanthin, heteroxanthin, vaucheriaxanthin ester, neoxanthin, β-cryptoxanthin 5'6', monoepoxide, and 5,6 5'6'-diepoxide	
Euglenophyceae	a, b	β-Carotene (± γ-carotene), *diadinoxanthin,* diatoxanthin, neoxanthin, β-cryptoxanthin and its 5'6' monoepoxide), echinenone, 3-hydroxy-echinenone, astaxanthin ester	
Chlorophyceae and Prasinophyceae	a, b	β-Carotene, (± α-carotene, rarely γ-carotene and lycopene), *lutein,* violaxanthin, zeaxanthin, antheraxanthin, neoxanthin, β-Cryptoxanthin (or its 5'6'-monoepoxide), lutein-5, 6-epoxide, loroxanthin, pyrenoxanthin, echinenone, c anthaxanthin, 3-hydroxy-echinenone ester, adonirubin eter, adonixanthin ester, crustaxanthin ester, astaxanthin ester, phoenicopterone	
Cryptophyceae	a, c_2	α-Carotene (± β-carotene, rarely ± ε-carotene), *alloxanthin,* crocoxanthin, monoadoxanthin	3 Spectral types of phycoerythrine absorbing at 544 nm, 555 nm and 568 nm 3 Spectral types of phycocyanin absorbing at 615 nm, 630 nm, and 645 nm
Dynophyceae	a, c_2	β-Carotene, *peridinin,* diadinoxanthin, diatoxanthin, dinoxanthin	

Note: The predominant xanthophylls are italicized.

chromatography enables pigment separation within 20 min.[121] The pigment homogenate in 70% methanol is loaded on a "Sep Pak" C-18 cartridge and washed with increasing concentrations of methanol. Methanol, 90% removes oxygenated carotenoids, 100% methanol removes chlorophylls, and carotenes are eluted with acetone. The fractions can then be analyzed by reversed-phase HPLC.

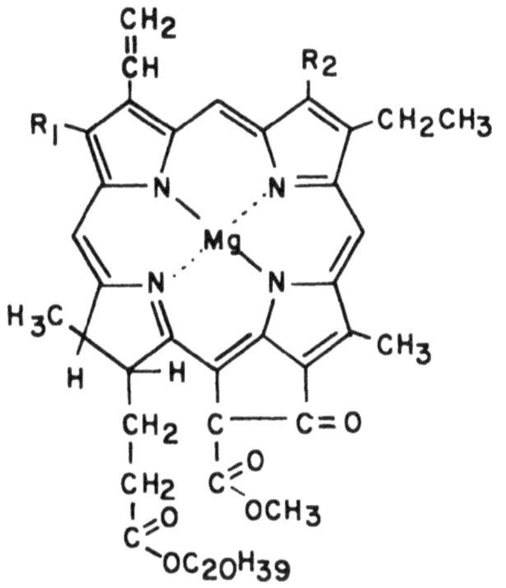

(a) $R_1 = CH_3$ $R_2 = CH_3$

(b) $R_1 = CH_3$ $R_2 = CHO$

19. (a) Chlorophyll a. (b) Chlorophyll b.

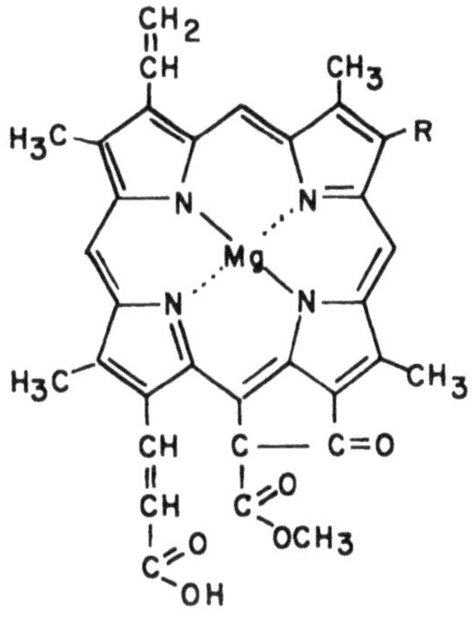

(a) $R = CH_2CH_3$

(b) $R = CH = CH_2$

(a) Chlorophyll a_1

(b) Chlorophyll b_2

FIGURE 3.   Common chlorophylls in microalgae.

A preparation of *Spirulina* chlorophyll in a mixture containing iron oxide and a higher (stearyl and cetyl) alcohol was patented as a strong deodorant. The major deodorization activity was attributed to the chlorophyll in the *Spirulina*.[122]

Phycobilins and Phycobiliproteins

The photosynthetic apparatus of red algae (Rhodophyta), Cryptomonads (Cryptophyta), and of the cyanobacteria contains phycobiliproteins, which are intensely colored proteinaceous accessory pigments. The phycobiliproteins fall into two basic groups: the phycoerythrins, which contain a linear tetrapyrrole pigment, phycoerythrobilin (Figure 4); and the phycocyanins, which have a similar pigment, phycoerythrobilin (Figure 4); and the phycocyanins, which have a similar pigment, phycocyanobilin. Two exceptions to this generalization are R-phycocyanin and phycoerythrocyanin, which possess both chromophores. These linear tetrapyrroles are covalently bound to the polypeptide.

Phycocyanin and phycoerythrin could be utilized as natural pigments for the food, drug, and cosmetics industries to replace the currently used synthetical pigments that are suspected of being carcinogens. Indeed, phycocyanin from *Spirulina* has already been commercialized by Dainippon Ink & Chemicals of Japan under the name of linablue. The product is an odorless nontoxic blue powder with a slight sweetness. When dissolved in water it is a brilliant blue with a faint reddish fluorescence. Its color (max. 618 nm) is between those of blue colors No. 1 (brilliant blue) and No. 2 (indigo carmine). It is stable from pH 4.5 to 8.0, and thermostable up to 60°C, but exhibits poor light stability. Its uses include coloring of candy, ice cream, dairy products, and soft drinks.[124] In another patent Dainippon Ink describe buffer extraction of *Spirulina* and treatment of the extract with an organic solvent, which denatures and precipitates the phycocyanin. The obtained blue pigment is used in eye shadow, eye liner, and lipstick. Since it is not water soluble, it does not run when it is wet by water or sweat. It does not irritate the skin.[125] It is also possible to isolate the blue or red chromophore by enzymatic or acid hydrolysis of the protein to yield a more concentrated

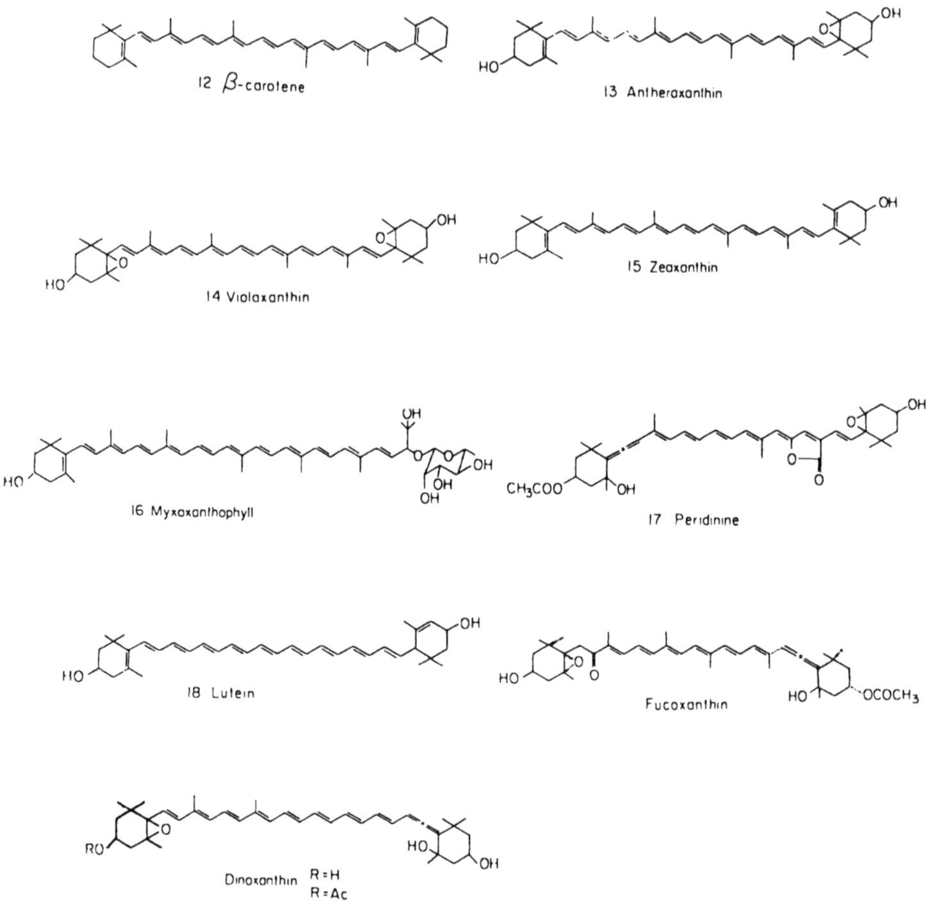

FIGURE 4. Phycobilines in microalgae.

pigment.[126] Chromatographic separation of phycocyanin from other proteins was described by Boussiba,[127] but is too complicated for commercial application.

ENZYMES

Enzyme preparations are required for industrial production of fine chemicals for use in laboratory tests and as reagents. Some algal enzymes can be utilized for such purposes.

ATP Determination

Phosphoglycerate kinase (PGK) has been used for the enzymatic determination of ATP. However, the commercial PGK is not specific for ATP[128] and also converts GTP and ITP at slower rates. PGK from the cyanobacteria *Spirulina platensis* is specific for ATP.[129] With this enzyme it is therefore possible to specifically determine ATP in tissue extracts or in mixtures of nucleotides. GTP and ITP can then be determined by adding PGK from yeast or another source.

Luciferin and luciferase from fireflies are constituents of commercial ATP determination kits. The dinoflagellates *Noctiluca*, *Gonyaulax*, and *Peridinium* also contain bioluminescent systems. The soluble system containing luciferin and luciferase can be extracted with cold water, but the extract does not produce light unless salt is added at a concentration of about

Table 11
RESTRICTION ENZYMES IN
CYANOBACTERIA

Restriction enzyme	Alga
Aca I	*Anabena catanula*
Acy I	*A. cylindrica*
Afa I & II	*A. flosaqua*
Aos I & II	*A. oscillarioides*
Ava I, II, & III	*A. variabilis*
Avr I & II	*A. variabilis*
Mla	*Mastigocladus laminosus*
Mst I & II	*Microcoleus species*

1 M. The luciferase has been obtained in fairly pure form (mol wt 130,000).[130-131] The luciferin has been difficult to purify, for it is susceptible to oxidation. Its molecular weight is probably about 400.[132] Oxygen is necessary for light production,[133] and the in vivo luminiscence spectrum peaks at approximately 474 nm.[134]

Restriction Endonucleases

Restriction endonucleases are part of restriction modification systems in prokaryotic organisms that protect them against foreign genetic information. The restriction modification system consists of an enzyme complex that modifies the host cell DNA, usually by specific methylation, and destroys nonspecifically nonmethylated foreign DNA by endonucleolytic cleavage. Both components of the system, the methylase and the endonuclease, share identical recognition sites.[135]

Restriction endonucleases are now used mainly to cleave a given DNA into characteristic fragments and to splice foreign DNA into a vector molecule. Most restriction enzymes are obtained from bacteria. However, cyanobacteria, which are taxonomically close to bacteria, contain several useful restriction enzymes (Table 11).

Amino Acid Oxidase

Both free and immobilized cells of the algae *Chlorella vulgaris* and *Anacystis nidulans* contain amino acid oxidase (AAO) activity,[137] which is increased by illumination of the cells with red light.

The resulting α-ketoacids are useful as supplements in protein-restricted diets. Both species are photosynthetically active when immobilized, but their AAO activity is low compared to that of the bacteria *Providencia* sp. However, co-immobilization of *C. vulgaris* and Providencia bacteria (1:1) in an illuminated reactor showed a tenfold increase in productivity. The immobilized *Chlorella* cells supply O_2 to the bacteria.

Superoxide Dismutase (SOD)

Free radicals cause damage to proteins, DNA, membranes, and organelles. Recently Fridovich[138] purified SOD which exists in all aerobic organisms tested and which seems to play a major role as a free radical scavenger. This enzyme has become available commercially as a health food and as a therapeutic agent for cancer and other diseases. So far, no evidence has been obtained from controlled experiments to prove its efficacy. However, if its claimed therapeutic characteristics are validated scientifically, then microalgae could prove to be a useful source of this enzyme. It has been purified from *Spirulina* and *Porphyridium*, among others, and since these algae have been proposed as sources of other chemicals (EFA, polysaccharides, pigments), this enzyme could be extracted from them as a by-product.

Table 12
DISTRIBUTION OF HYDROGENASE ACTIVITY AMONG ALGAE[140]

Alga	Ref.	Alga	Ref.
Cyanobacteria		Chlorophyceae	
Anabaena sp.	146	*Ankistrodesmus* spp.	152—154
Anacystis nidulans	146	*Chlamydomonas* spp.	155
Aphanothece halophytica	147	*Chlorella* spp.	148,153
Cyanophora paradoxa	146	*Chlorococcum vacuolatum*	156
Nostoc muscorum	148	*Coelastrum* spp.	146
Oscillatoria limnetica	147	*Crucigenia apiculata*	156
Synechococcus elongatus	149	*Kirchneriella lunaris*	146
Synechocystis sp.	150	*Rhaphidium* spp.	154
Euglenophyceae		*Scenedesmus* sp.	154,155
Euglena spp.	151	*Selenastrum* sp.	153
		Rhodophyceae	
		Porphyridium spp.	150,157

PRODUCTION OF HYDROGEN

Perhaps the most useful class of enzymes are the hydrogen-generating enzymatic systems hydrogenase and nitrogenase (Table 12). Green algae and cyanobacteria form molecular hydrogen.[139] This biophotolysis of water can be achieved by the following two enzymatic processes.[140]

Hydrogenase
This enzyme catalyzes the reaction

$$2H^+ + 2e \rightarrow H_2$$

and is found in both autotrophs and heterotrophs. The reaction is sensitive to CO and independent of ATP.

Nitrogenase
This enzyme transforms molecular nitrogen into ammonia:

$$N_2 + 8e- + 2H^+ \rightarrow 2NH_3 + H_2$$

As can be seen from the reaction, nigrogen fixation is always accompanied by hydrogen production, although most of the reducing potential is used for converting nitrogen to ammonia. However, in the absence of nitrogen, these organisms will evolve hydrogen if given the necessary energy source. Some of them also possess hydrogenase and can recover the hydrogen through this system. The process in nitrogen-fixers is ATP-dependent, insensitive to CO, and irreversible.

All photosynthetic algae can use light energy for photolysis of water. In this process, water is cleaved, yielding molecular oxygen and a reducing agent which, when coupled with hydrogenase or the nitrogenase system, can produce hydrogen exclusively when placed in an atmosphere devoid of nitrogen. The only group of microorganisms that possess two photosynthetic systems like the higher plants as well as the nitrogenase system is the cyanobacteria. These prokaryotic organisms produce hydrogen from water when illuminated, even in the presence of nitrogen. Maximum production, however, occurs in a nitrogen-free atmosphere. Since oxygen inhibits nitrogenase activity, the efficiency of the system might be improved if oxygen could also be excluded.

The practical aspects of producing hydrogen with photosynthetic microorganisms have been thoroughly reviewed by Benemann et al.[141] The most successful biophotolysis system, so far developed, is based on nitrogen-limited cultures of heterocystus cyanobacteria.[142] The ability of those organisms to produce hydrogen and oxygen simultaneously in the laboratory, the demonstration of prolonged production of hydrogen and of roughly stoichiometric amounts of oxygen by nitrogen-limited cultures, and the successful operation of such a biophotolysis system outdoors, make this system the most advanced at present.

In systems using solar energy conversion to produce hydrogen, efficiencies of 0.2% (averaged over several weeks) have been obtained outdoors. Practical systems would require a tenfold higher efficiency. Benemann et al.[141] proposed a general design for a biophotolysis system consisting of vertical, thin-walled glass tubes with an inert gas circulated through the cultures for mixing and for removing hydrogen. They suggested that considerations of gas mass transfer, energy utilization, and economics favor such a system.

Ochiai et al.[143] reported that chlorplast electrodes, upon illumination, could generate an anodic photocurrent with concomitant cleavage of water. However, due to the variable activity and the instability of chloroplast preparations, the use of intact thermophilic algal cells was considered.[144] Living cyanobacteria *(Mastigocladus laminosus)*, immobilized on an optically transparent SnO_2 electrode with calcium alginate, functioned as a photoanode under continuous illumination for periods of time adequate for use in a conventional electrochemical cell. This "living electrode" shows promise as a long-lived photoconverter of solar energy to electrical energy.

Miura and co-workers[145] obtained a patent for hydrogen production in alternating light-dark cycles. The alga *Chlamydomonas reinhardii* was cultured for 12 hr in the light. Cultivation was continued in the dark for another 12 hr, N_2 containing 0.3% O_2 was flushed into the culture, and hydrogen was evolved. The algae were then removed by centrifugation and the supernatant, which contained formic acid formed by the alga, was fed to *Escherichia coli* in which formate hydrogen lyase had been induced. This resulted in the evolution of additional hydrogen.

METHANE PRODUCTION

Solar energy could be exploited for the production of methane through a combined algal-bacterial process. In this process, the algae are mass produced from light and carbon in the first step. In the second step, the algal biomass is then used as a nutrient for anaerobic bacteria that produce methane. The carbon source for the production of algal biomass could be either organic carbon from wastewater (for eukaryotic algae), or carbon dioxide from the atmosphere or from combustion exhaust gases (for both prokaryotic and eukaryotic algae).[158]

Data on the technical feasibility of anaerobic digestion of algal biomass have been reported for many species of algae. The following microscopic algae have been successfully used for the production of methane: mixed culture of *Scenedesmus* spp. and *Chlorella* spp.,[159] mixed culture of *Scenedesmus* spp., *Chlorella* spp., *Euglena* spp., *Oscillatoria* spp., and *Synechocystis* sp.,[160] culture of *Scenedesmus* sp. alone, and together with either *Spirulina* sp., *Euglena* sp., *Micoractinium* sp., *Melosira* sp. or *Oscillatoria* sp., and mixed culture of *Hydrodictyon reticulatum, Cladophora glomerata,*[162] and *Anabaena* sp.[163]

Research conducted by Samson and LeDuy[158] suggested using the cyanobacterium *Spirulina maxima* as the sole catalyst for this process. This species is attractive for such use because it can use atmospheric CO_2 as a carbon source and it is simple to harvest. Furthermore, the fermentability of *S. maxima* seems to be significantly higher than that of other microscopic algae.[161,164]

Certain species of filamentous cyanobacteria have many properties that make then promising as an energy crop: they are readily harvested by simple hydraulic means, grow well,

Table 13
ACCUMULATION OF OSMOTIC METABOLITES IN HALOTOLERANT ALGAE[167]

Organism	Osmotic range (*M* NaCl)	Intracellular osmoregulator	Highest intracellular conc. of osmoregulator (*M*)	Ref.
Cyanobacteria				
Aphanotheca halophytica	1.00—4.00	Polyols, amino acids, carbohydrates	NS	169
Synecoccocus sp.	0.03—1.20	Glycosylglycerol	0.25	170
Chrysophyceae				
Monochrysis luteri	0.15—1.00	Cyclohexanetetrol	0.74	171
Ochromonas malhamensis	0.01—0.15	Isofloridoside	0.30	172
Eustigmatophyceae				
Monallantus salina	0.04—0.60	Mannitol, proline	0.22	173,174
Prasinophyceae				
Asteromonas gracilis	0.50—4.50	Glycerol	5.50	175
Bacillariophyceae				
Cylindrotheca fusiformis	0.57—0.70	Mannose	Not determined	176
Chlorophyceae				
Dunaliella (diff. species)	0.10—5.50	Glycerol	7.00	177
Klebsormidium marinum	0.04—0.27	Sorbitol, proline	1.59	174

Reproduced, with permission, from the *Ann. Rev. Microbiol.*, 37, 1983.

and are the dominant species at high temperatures. Also, they can fix nitrogen, their productivity compares favorably with that of other proposed energy crops, they can adapt to large variations in solar insolation, and they are capable of growing in brackish water.

Keenan[163] studied the use of solar energy to bioproduce methane with the alga *Anabaena flosaqua* and concluded that methane could provide a substantial portion of the fuel requirements for power stations. The overall efficiency of the conversion of solar energy to methane was 1%. He estimated that the energy inputs required for the growth and harvesting of the algal mass would account for only 12% of the fuel value of the methane produced.

OSMOREGULATORS

Osmoregulation by organic metabolites, which has been reviewed by Hellebust[165,166] and more recently by Ben Amotz and Avron[167] enables an alga to grow or survive at a high salt concentration by accumulating an organic solute to equalize the external and internal osmotic pressures (Table 13). There are three main groups of osmoregulators: polyhydric alcohols containing 3 to 6 carbon atoms, e.g., glycerol, mannitol or sorbitol, cyclohexanetetrol; glycosides, e.g., galactoglycerides, floridoside, and isofloridoside; and amino acids, e.g., proline and glutamic acid. A fourth group consists of all of the other osmoregulators.

Ben Amotz and Avron,[167] suggested that osmoregulators protect enzyme activity at high osmotic pressures. Several enzymes are only slightly inhibited by 5 *M* glycerol while being severely inhibited by much lower concentrations of salt. Table 11 shows the variety of chemicals that serve as osmoregulators and the ranges of osmotic pressures these algae can withstand by creating high internal concentrations of osmoregulators. The biosynthetic pathways for these osmoregulators can be utilized for the synthesis of many radiolabeled materials; for example, *Scenedesmus obtusiosculus* was cultivated with $^{11}CO_2$ to yield a ^{11}C glucose.[168] Glucose labelled with the positron-emitting isotope ^{11}C is useful in positron camera studies, e.g., of brain energy metabolism.

The most outstanding example of osmoregulation in algae is *Dunaliella*. This alga can thrive in media containing 0.1 to 8 *M* (saturated) salt solutions by accumulating up to 40%

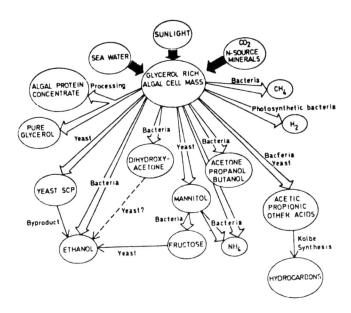

FIGURE 5.   Derivations from glycerol-rich algal cell mass.

of its dry weight as glycerol. These findings have led to several attempts at application. *Dunaliella* is grown in Israel in a pilot plant for production of glycerol and β-carotene. Williams[178] suggested that the glycerol rich biomass could serve as a substrate for a variety of bioconversion processes (Figure 5).

Goldman et al.[179] treated a salt water slurry of *Dunaliella* with several reagents to convert it into benzene soluble materials. Up to 75% of the fixed carbon was claimed to have been converted *in situ* in the presence of benzene during pyrolysis of the algae-salt water slurry.

Research conducted by Ethyl Corporation[180] on *Chlorella* revealed that by manipulation of medium composition and culture depth, the amount of proline accumulated as an osmoregulant could be controlled. Up to 30% proline was obtained. Furthermore, there is no need for expensive harvesting of algae, since when the culture is diluted the cells excrete the proline to the medium, and then once again become enriched to their original level of proline without additional growth when the diluted medium is replaced with high-salt medium.

POLYMERS

Proteins

The great increase in the world's human population during recent decades and man's desire for a better quality of life have made great demands on world food production, especially of protein-rich foods. Microalgae can be utilized as protein sources due to their fast growth rates and high protein contents. The cyanobacteria *Spirulina* contains up to 70% proteins. It has been calculated[181] that about 50,000 kg of protein per hectare could be produced annually. Also, the residue after the extraction of desirable chemicals from most algae is rich in protein and can be utilized as an animal feed.

Polysaccharides

The only group of algal chemicals already utilized to a great extent are the polysaccharides.[182] Agar, alginic acid, and carrageenan are commercially extracted from macroalgae. Carrageenan is a cell wall polysaccharide found in a number of red macroalgae *(Chondrus, Gigartina, Hypnea, and Eucheuma)* from which it is extracted on a large scale. A few of its uses are as an emulsion stabilizer in instant puddings, for improvement of the consistency and smoothness of ice cream, and as an emulsifier in chocolate milk. It is used for similar purposes also in drugs and cosmetics. However, intensive harvesting of these macroalgae and ecological damages have depleted their production, while demand has been constantly on the rise.

The red microalga *Porphyridium* synthesizes and excretes to its medium large amounts of polysaccharide[183] at a rate of up to 50% of the rate of biomass production.[184] This sulfated polysaccharide is of special interest since it resembles carrageenan in its structure and properties (viscosity and emulsifying power). A culture of *Porphyridium* can reach a high density in a relatively short time and release significant amounts of the polysaccharide to the growth medium.[185] The size of the polysaccharide capsule varies with the growth conditions. The cell coats are thinnest during the logarithmic phase of growth and thickest during the stationary phase. When the rate of dissolution of the polysaccharide from the cell surface was measured as a function of its production, it was found that in the stationary growth phase the rate of synthesis is higher than the dissolution of the polysaccharide in the external medium. This enables the creation of a capsule, resulting in the cells reaching their maximum size at this stage. It seems that the thickness of the capsule is affected by the rate of its synthesis, its degree of solubility, and the cell surface area.

During growth in liquid medium, the viscosity of the medium increases rapidly because of the release of polysaccharide from the cell surface. The polysaccharide dissolved in the growth medium is identical to that extracted from the cell surface. Equal amounts of polysaccharide are excreted in the light and in the dark. However, other environmental conditions affect the amount of polysaccharide produced; e.g., under nitrogen starvation the content is the highest.[186,187]

The polysaccharide of *Porphyridium* is composed mainly of xylose, glucose, and galactose and of four minor constituents (guluronic acid, galacturonic acid, a pentose, and an anhydrohexose). Sulfate constitutes 7.6% of the weight of the polymer, but it is not known to which sugar it is attached.[188]

Porphyridium cruentum was immobilized[189] on a spongy form of polyurethane. The culture was irradiated with artificial light and CO_2 and a mineral solution were fed into it. After several days $NaNO_3$ was eliminated from the medium and the viscosity of the medium then increased sharply. After 17 months of continuous circulation of medium the cells were still producing high concentrations of polysaccharides (4 g/ℓ).

The polysaccharide obtained from *Porphyridium aerugineum* was utilized, according to an American patent,[190] for petroleum recovery. The crude oil was displaced from underground sand formations with an aqueous driving fluid thickened with *Porphyridium* polysaccharide. It was claimed that the amounts of petroleum recovered were equal to or greater than the amounts obtained when Kelzan®, a commercial biopolymer, was used even though the concentration of Kelzan® was higher (1500 vs. 1100 ppm).

Starch

A four-membered consortium consisting of a *Chlorella*-like green alga and three heterotrophic bacteria, *Alcaligenes* sp., *Flavobacter* sp., and *Serratia* sp., was selected from a natural source because of its ability to grow at 37°C and to produce starch photosynthetically.[191] It was suggested that the green algae created an ecological niche for the heterotrophic bacteria, probably by excreting carbon and nitrogen compounds. The bacteria consumed these compounds immediately and excluded any possible invasion by undesirable heterotrophic organisms.

With a thermophilic strain of *Chlorella* (when the medium was diluted slowly), the starch content reached approximately 50% of the total dry weight.[192]

Poly-β-Hydroxybutyric Acid (PHB)

Another polymer of potential importance is poly-β-hydroxybutyric acid (PHB). This is a naturally occurring polymer found in a large number of photosynthetic and nonphotosynthetic bacteria. It is a carbon and energy reserve in prokaryotes,[193] like starch in eukaryotic plants. PHB is optically active[194] and chemical, spectral, and synthetic studies[195] support a linear head-to-tail linkage of levorotatory R-β-hydroxybutyric acid units. PHB[196] can be used as a thermoplastic. It takes glass-fiber filling very well. It is also biodegradable in soil. Its properties are quite similar to those of polypropylene homopolymer. PHB has been identified in the cyanobacterium *Spirulina platensis*.[197] PHB accumulated during exponential growth to 6% of the total dry weight and then decreased during the stationary phase.

- 22 -

PHARMACEUTICALS

Antibiotics

Many algae, mainly macroalgae, produce antibiotics. The simplest antibiotic found was acrylic acid (Structure 23), which was first found in *Phaeocystis poucht*.[198,199] It was claimed that it could be produced from its thetine precursor (Structure 24) either in vivo by enzymatic splitting or in vitro by mild alkaline treatment. Acrylic acid inhibits the growth of Gram-positive microorganisms and, to a lesser extent, Gram-negative organisms.[199] Acrylic acid or its precursor was identified in many other algae (a list can be found in Glombitza's review[200]).

Complex sulfur compounds (Structures 25 and 26) were discovered[201] in the green alga *Chara globularis*. These sulfur compounds apparently help the alga to predominate over the other algal flora it encounters.

Phenols, which are known for their antimicrobial activity, are quite frequent in macroalgae. Some phenols were found also in microalgae (Table 14).

Unidentified fatty acids or their photooxidation products are mentioned as being responsible for the antibiotic activity of chlorelin from *Chlorella*[207] and of extracts from *Ochromonas danica*,[208] *Stichococcus mirabilis*,[209] *Protosyphon botyroides*, and *Chlamidomonas reinhardii*.[210] In the diatom *Astrionella japponica*, Pesando[211] found that the fatty acid responsible for the antibiotic activity was eicosapentaenoic acid (20:5).

Scenedesmus obliquus was used in postoperative care of the coagulation surface.[212] An extract of the diatom *Asterionella notata* has an antiviral and antifungal effect.[213] This alga

Table 14
DISTRIBUTION OF PHENOLIC SUBSTANCES

Species	Phenolic substances	Ref.
Cyanobacteria		
Lyngbya majuscula		202

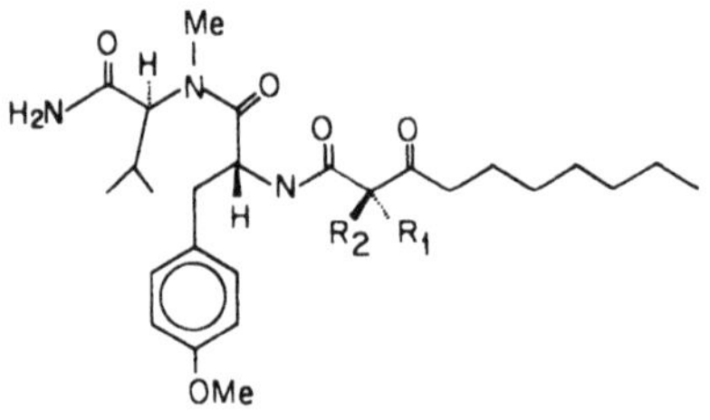

$$R_1 = H, R_2 = Me$$
$$R_1 = Me, R_2 = H$$

L. gracilis		203
Debromo aplysiatoxin (Table 15)		
Oscillatoria nigrovidis & *Schizothrix calciola*		204
Oscillatoxin A, Aplysiatoxins (Table 15)		
Calothrix brevissima		205

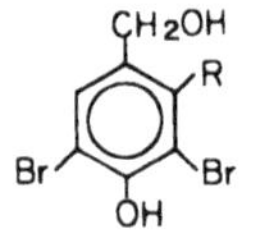

$$R = H$$
$$R = Br$$

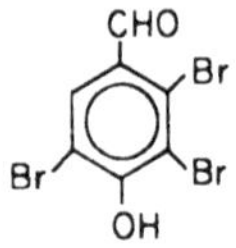

| Chlorophyta | | |
| *Nitella hookeri* | | 206 |

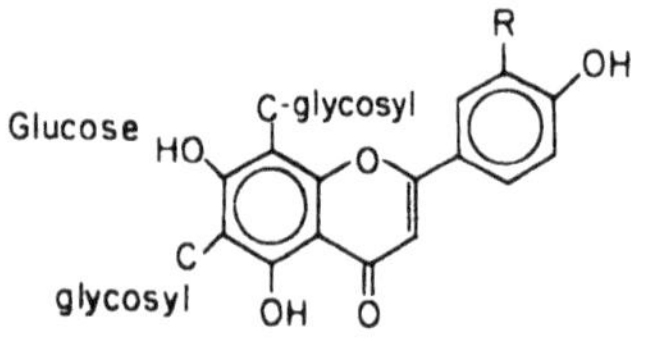

$$R = H$$
$$R = OH$$

and four other diatoms, *Chaetocerus lauderi, C. pseudcurvisteus, C. socialis,* and *Fragilaria pinnata,* have significant antifungal activity, the highest among a total of 48 algae.[214]

Toxins

Algal toxins have been regarded more as a threat to public health than as valuable chemicals. However, once their structure and effects are understood, some could serve as drugs, similar to tetrodotoxin of the puffer fish, which is used as a vasodepressor.[215] Major algal toxins, their structures and activities are presented in Table 15.

CONCLUSION

A special consideration should be given to three algae that show the best potential for practical exploitation. Due to the variety of valuable chemicals they contain, a concerted separation scheme could be envisioned to increase economic feasibility.

- *Dunaliela: Dunaliella bardawil* contains large amounts of β-carotene and glycerol.
- *Spirulina: Spirulina* strains contain a relatively high percentage (about 1%) of the rare GLA. Also valuable are the blue pigment phycocyanin and the carotenoid zeaxantin.
- *Porphyridium:* This alga is one of the very few sources of AA and is also a rich source of a carragenan (like polysaccharide).

The residue after the extraction and separation of the chemicals from all three algae is protein-rich and could be used as feed.

LIST OF ABBREVIATIONS

AA —	Arachidonic acid (20:4)
DGLA —	Dihomo-γ-linolenic acid
EFA —	Essential fatty acids
EPA —	Eicosapentaenoic acid (20:5)
GLA —	γ-Linolenic acid (18:3)
LA —	Linoleic acid (18:2)
PGK —	Phosphoglycerate kinase
PHB —	Polyhydroxybutryic acid
PUFA —	Polyunsaturated fatty acids

Table 15
ALGAL TOXINS

Alga	Structure	Description	Effect	Ref.
Cyanobacteria *Anabaena flos-aquae*	Anatoxin-a	Anatoxin a. A water soluble alkaloid	A potent, postsynoptic depolarizing neuromuscular blocking agent, causes death by respiratory arrest	215—216
Aphanizomenon flos-aquae		A mixture of toxins, one of which is Saxitoxin	Saxitoxin is the major toxin associated with poisonous shellfish and clams	217
Microcystis aeuruginosa		Microcystine Polypeptide containing equimolar amounts of L-methionine, L-tyrosine, D-alanine, D-glutamic acid erythromethylaspartic acid and methylamine	LD_{50} in mice 0.056 mg/kg	217—219
Lyngbya majuscula	Debromoaplysiatoxin Lyngbyatoxin A	Displays activity against P-388 lymphocytic mouse leukemia	Causes swimmer's itch	220
			Has antileukemic activity (both are effective only at or near the chronic level); both		

Oscillatoria nigroviridis and
Schizothrix calciola

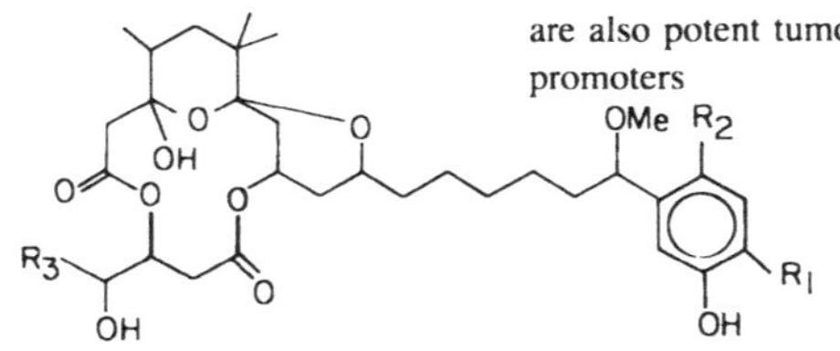

are also potent tumor promoters

204

Oscillatoxin A	$R_1 = R_2 = R_3 = H$
Bromooscillatoxin A	$R_1 = R_3 = H$, $R_2 = Br$
Dibromooscillatoxin A	$R_1 = R_2 = Br$, $R_3 = H$
Bromoaplysiatoxin	$R_1 = R_2 = Br$, $R_3 = Me$
Debromoaplysiatoxin	$R_1 = R_2 = H$, $R_3 = Me$

Dinoflagellates

Peridinium polonicum

An alkaloid with structure similar to 12-methoxybogamine

Ichthyotoxic LD_{min} 2.5 mg/kg in mice

223

Amphidinium kleosii and *A. rhynchocephalum*

Choline sulfate and choline esters

Ichthyotoxic

223

Prymnesium parvum

Glycolipid mol wt 23,000, 1,800

Ichthyotoxic LD_{50} for mice 1.4 mg/kg

223

Gonyulax tamarensis

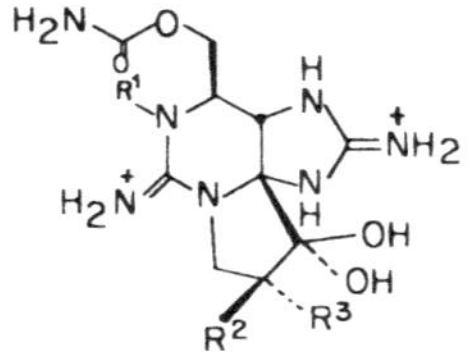

Alkaloids

Paralytic shellfish poisoning neurotoxin; inhibits sodium passage through axonal membranes

223

Saxitoxin	R^1, R^2, $R^3 = H$
Neosaxitoxin	$R^1 = OH$; R^2, $R^3 = H$
Gonyautoxin II	R^1, $R^2 = H$; $R^3 = OH$
Gonyautoxin III	R^1, $R^3 = H$; $R^2 = OH$

Gymnodinium breve

Two toxins of unknown structure

LD_{50} (in mice) 0.25 mg/kg
Ichthyotoxic

224

REFERENCES

1. **Pohl, P.**, Lipids and fatty acids of microalgae, in *Handbook of Biosolar Resources*, Vol. I, Part 1, Zaborsky, O. R., Ed., CRC Press, Boca Raton, Fla., 1982, 383.

2. **Haines, T. H., Pousada, M., Stern, B., and Mayers, G. L.**, Microbial sulpholipids: (R)-13-Chlorol-(R)-14-docosanediol disculphate and polychlorosulpholipids in *Ochromonas danica*, *Biochem. J.*, 113, 565, 1969.

3. **Mercer, E. I. and Davies, C. L.**, Chlorosulpholipids in algae, *Phytochemistry*, 14, 1545, 1975.

4. **Eichenberger, W. and Boschetti, A.**, *FEBS Lett.*, 88, 201, 1978.

5. **Moseley, K. R. and Thompson, G. A., Jr.**, *Plant Physiol.*, 65, 260, 1980.

6. **Evans, R. W., Kates, M., Ginzburg, M., and Ginzburg, B. Z.**, Lipid composition of halotolerant algae, *Dunaliella parva lerche* and *Dunaliella tertiolecta*, *Biochim. Biophys. Acta*, 712, 186, 1982.

7. **Brown, A. E. and Elovson, J.**, *Biochemistry*, 13, 3476, 1974.

8. **Thomas, W. H., Seibert, D. L. R., Alden, M., Neurin, A., and Elderidge, P.**, Yields, Photosynthetic Efficiencies and Proximate Composition of Dense Marine Microalgal Cultures. I. Introduction and Phacodactylum tricornutum, *Biomass*, 5, 181, 1984.

9. **Richardson, B.**, *Appl. Microbiol.*, 18, 235, 1969.

10. **Ahern, T. J., Katoh, S., and Sada, E.**, Arachidonic acid production by the red alga *Porphyridium cruentum*, *Biotechnol. Bioeng.*, 25, 1057, 1983.

11. **Nichols, B. W.**, Light induced changes in the lipids of *Chlorella vulgaris*, *Biochim. Biophys. Acta*, 106, 274, 1965.

12. **Aaronson, S.**, Effect of incubation temperature on the macromolecular and lipid content of the phytoflagellate *Ochromonas danica*, *J. Phycol.*, 9, 111, 1973.

13. **Klienschmidt, M. G. and McMahon, V. A.**, Effect of growth temperature on the lipid composition of *Cyanidium caldarium*. I. Class separation of lipids, *Plant Physiol.*, 46, 286, 1970.

14. **Klienschmidt, M. G. and McMahon, V. A.**, Effect of composition of Cyanidium of growth temperature on the lipid caldarium, *Plant Physiol.*, 46, 290, 1970.

15. **Blee, E. and Schantz, P.**, Biosynthesis of galactolipids in *Euglena gracilis*. II. Changes in fatty acid composition of galactolipids during chloroplast development, *Plant Sci. Lett.*, 13, 257, 1978.

16. **Ackman, R. G., Tocher, C. S., and MacLachlan, J.**, Marine phytoplankter fatty acids, *J. Fish. Res. Board Can.*, 225, 1603, 1968.

17. **Pohl, P. and Wagner, H.**, Control of fatty acid and lipid biosynthesis in *Euglena gracilis* by ammonia, light and DCMU, *Z. Naturforsch. Teil B.*, 53, 1972.

18. **Pohl, P., Passig, T., and Wagner, H.**, Über den Einfluss von anorganischem Stickstoff-Gehalt in der Nahrlosung auf die Fettsaure-Biosynthese in Grunalgen, *Phytochemistry*, 10, 1505, 1971.

19. **Sphoer, H. A. and Milner, H. W.**, *Plant Physiol.*, 24, 120, 1949.

20. **Ketchum, B. H. and Redfield, A. C.**, *J. Cell. Comp. Physiol.*, 33, 281, 1949.

21. **Shifrin, N. S. and Chisholm, S. W.**, Phytoplankton lipids: interspecific differences and effects of nitrate, silicate and light-dark cycles, *J. Phycol.*, 17(4), 374, 1981.

22. **Constantopoulos, G.**, Lipid metabolism of manganese-deficient algae. I. Effect of manganese deficiency on their greening and the lipid composition of *Euglena gracilis*, *Plant Physiol.*, 45, 76, 1970.

23. **Dubinsky, Z. and Aaronson, S.**, Increase of lipid yields from some algae by acid extraction, *Phytochemistry*, 18, 51, 1979.

24. **Collyer, D. M. and Fogg, G. E.**, Studies on fat accumulation by algae, *J. Exp. Bot.*, 6, 256, 1955.

25. **Aaronson, S.**, A new source of single-cell protein and other nutrients, *Arch. Hydrobiol.*, 6(Suppl 41), 108, 1972.

26. **Aaronson, S., Berner, T., and Dubinsky, Z.**, Microalgae as a source of chemicals and natural products, in *Algae Biomass*, Shelef, G. and Soeder, C. J., Eds., Elsevier/North-Holland, Amsterdam, 1980.

27. **Ricketts, T. R.**, On the chemical composition of some unicellular algae, *Phytochemistry*, 5, 67, 1966.

28. **Antia, N. J., Lee, R. F., Nevenzel, C., and Cheng, J. Y.**, Wax ester production by the marine cryptomonad *Chroomonas salina* grown photoheterotrophically on glycerol, *J. Protozool.*, 21, 768, 1974.

29. **Tornabene, T. G., Holzer, G., and Peterson, S. L.**, *Biochem. Biophys. Res. Commun.*, 96, 1349, 1980.

30. **Dubinsky, Z., Berner, T., and Aaronson, S.**, Potential of large scale algal culture for biomass and lipid production in arid lands, *Biotechnol. Bioeng. Symp.*, No. 8, 51, 1978.

31. **Paoletti, C., Pushparaj, B., Florenzano, G., Capella, P., and Lercker, G.**, Unsaponifiable matter of green and blue-green algal lipids as a factor of biochemical differentiation of their biomasses. I. Total unsaponifiable and hydrocarbon fraction, *Lipids*, 11, 258, 1976.

32. **Kaplan, D. and Cohen, Z.**, Unpublished results, 1984.

33. **Metzger, P., Descouls, N., Casadevall, E., and Coute, A.**, Microalgae as a source of triglycerides, *Comm. Eur. Communities*, p. 339, 1983.

34. **Anon.**, Micro-algae groomed for future chemical production, *Chem. Eng.*, p. 67, 1980.

35. **Pohl, P. and Zurheide, F.,** Fatty acids and lipids of marine algae and the control of their biosynthesis by environmental factors, in *Marine Algae in Pharmaceutical Science,* Hope, H. A., Levring, T., and Tanaka, Y., Eds., Walter de Gruyter, Berlin, 1979, 65.

36. **Nichols, B. W. and Wood, B. J. B.,** The occurence and biosynthesis of gamma-linolenic acid in a blue-green alga, *Sirulina platensis, Lipids,* 3, 46, 1968.

37. **Nichols, B. W. and Appleby, R. S.,** The distribution and biosynthesis of arachidonic acid in algae, *Phytochemistry,* 8, 1907, 1969.

38. **Srivastava, K. C.,** *S. Afr. J. Sci.,* 76, 170, 1980.

39. **Bild, G. S., Bhat, S. G., Ramadoss, C. S., and Axelrod, B.,** *J. Biol. Chem.,* 253, 21, 1978.

40. **Ahern, T. J., Katoh, S., and Sada E.,** Prostaglandin synthesis from arachidonic acid by immobilized ram seminal microsomes, *Biotechnol. Bioeng.,* 25(3), 881, 1983.

41. **Gregson, R. P., Marwood, J. F, and Quinn, R. J.,** The occurrence of prostaglandins PGE-2 and PGE-2alpha in a plant — the red alga *Gracilaria Lichenoides, Tetrahedron Lett.,* p. 4505, 1979.

42. **Yuki Gosei Kogyo Co., Ltd.,** Unsaturated fatty acids from salmon milt, *Jpn. Kokai Tokkyo Koho,* 67, 398, 1981.

43. **Baran, J. S. and Liang, C.,** Antihyperlipidemic fatty acids, *S. African Patent 76 04,303,* July 20, 1977.

44. **Baran, J. S. and Liang, C.,** Antihyperlipidemic fatty acids, *Canadian Patent 1,067,509,* Dec. 4, 1979.

45. **Nakajima, T., Rikimi, K., Soda, Y., Kashima, K., Miyamoto, A., and Soejima, Y.,** Therapeutic agent for treating disturbances of consciousness, perception, and movement, *Belgium Patent 875, 227,* Oct. 1, 1979.

46. **Nakajima, T., Rikimi, K., Soda, Y., Kashima, K., Miyamoto, A., and Soejima, Y.,** Use of phosphatidyl compounds in pharmaceutical compositions against attention-, consciousness- and motion disturbances, *German Patent 2, 907, 778,* Aug. 30, 1979.

47. **Spritz, M.,** *J. Clin. Invest.,* 48, 78, 1969.

48. **Liang, C. D., Baran, J. S., Miller, J. E., Steinman, D. H., and Saunders, R. N.,** Synthesis and biological activities of arachidonic and alpha- and beta-alkyl-substituted arachidonic acid derivatives, *Adv. Prostaglandin Thromboxane Res.,* 6, 235, 1980.

49. **Liang, C.,** Phenyl 5Z, 8Z, 11Z, 14Z-eicosatetranoate and congeners, *U.S. Patent 4, 181, 670,* Jan. 1, 1980.

50. **Nippon Suisan Kaisha, Ltd, and Nissui Seiyaku Co., Ltd.,** Health food additives from marine oils and fats, *Japan Kokai Tokkyo Koho JP Patent 82, 86, 254,* May 29, 1982.

51. **Unilever, N. V.,** Dienals, *Netherland Appl. Patent 296, 925,* May 25, 1965.

52. **Williams, J.,** Additive containing arachidonic acid or gamma-linolenic acid for food or cosmetics, *German Patent 2, 352, 797,* Oct. 3, 1973.

53. **Huang, Y. S. and Manku, M. S.,** Triacylglycerol composition of evening primrose oil, *J. Am. Oil Chem. Soc.,* 60, 748, 1983.

53a. **Horrobin, D. F.,** Essential fatty acids: a review, in *Clinical Uses of Essential Fatty Acids,* Eden Press, London, 1982.

54. **Wolf, R. B., Kleiman, R., and England, R. E.,** New sources of gamma-linolenic acid, *J. Am. Oil Chem. Soc.,* 60(11), 1858, 1983.

55. **Cohen, Z.,** Unpublished results.

56. **Harrington, G. W. and Holz, G. G., Jr.,** The monoenoic and docosahexaenoic fatty acids of a hetero-trophic dinoflagellate, *Biochim. Biophys. Acta,* 164, 137, 1968.

57. **Ackman, R. G., Tocher, C. S., and McLachlan, J.,** Marine phytoplankter fatty acids, *J. Fish. Res. Board Can.,* 25, 1603, 1968.

58. **Holton, R. W., Blecker, H. H., and Stevens, T. S.,** Fatty acids in blue-green algae: possible relation to phylogenetic position, *Science,* 160, 545, 1968.

59. **Jamieson, G. R. and Reid, E. H.,** The fatty acid composition of *Ulothrix aequalis* lipids, *Phytochemistry,* 15, 795, 1976.

60. **Harrington, G. W., Beach, D. H., Dunham, J., and Holz, G. G., Jr.,** The polyunsaturated fatty acids of marine dinoflagellates, *J. Protozool.,* 17, 213, 1970.

61. **Ackman, R. G., Addision, R. F., Hooper, S. N., and Prakash, A.,** *Halosphaera viridis:* fatty acid composition and taxonomical relationships, *Fish. Res. Board Can.,* 27, 251, 1970.

62. **Glasl, H. and Pohl, P.,** Fettsauren und lipide in farblosen algen, *Z. Naturforsch. Teil C,* 29, 399, 1974.

63. **Kenyon, C. N., Rippka, R., and Stanier, R. Y.,** Fatty acid composition and physiological properties of some filamentous blue-green algae, *Arch. Mikrobiol.,* 83, 216, 1972.

64. **Beach, D. H., Harrington, G. W., and Holz, G. G., Jr.,** The polyunsaturated fatty acids of marine and freshwater cryptomonads, *J. Protozool.,* 17, 501, 1970.

65. **Chuecas, L. and Riley, J. P.,** Component fatty acids of the total lipids of some marine phytoplankton, *J. Mar. Biol. Assoc. U.K.,* 49, 97, 1969.

66. **Joseph, J. D.,** Identification of 3, 6, 9, 12, 15-octadecapentaenoic acid in laboratory cultured photosynthetic dinoflagellates, *Lipids,* 10, 395, 1975.

67. **Moore, R. E.,** Constituents of blue-green algae, in *Marine Natural Products Chemical and Biological Perspectives,* Vol. 4, Scheuer, P. J., Ed., Academic Press, New York, 1981, 1.

68. **Loui, M. S. M. and Moore, R. E.,** *Phytochemistry,* 18, 1733, 1979.

69. **Starr, T. J., Dieg, E. F., Church, K. K. and Allen, M. B.,** *Tex. Rep. Biol. Med.,* 20, 271, 1962.

70. **Marner, F. -J. and Moore, R. E.,** *Phytochemistry,* 17, 553, 1978.

71. **Mynderse, J. S. and Moore, R. E.,** *J. Org. Chem.,* 43, 4359, 1978.

72. **Simmons, C. J., Marner, F. -J., Cardellina, J. H., Moore, R. E., and Seff, K.,** *Tetrahedron Lett.,* p. 2003, 1979.

73. **Moore, R. E.,** in *Marine Natural Products: Chemical and Biological Perspectives,* Vol. 1, Scheuer, P. J., Ed., Academic Press, New York, 1978, 43.

74. **Safferman, R. S., Rosen, A. A., Mashni, C. I., and Morris, M. E.,** *Environ. Sci. Technol.,* 1, 429, 1967.

75. **Juttner, F.,** *Z. Naturforsch.,* 31c, 491, 1976.

76. **DeRosa, M., Gambacorta, A., Minale, L., and Bullock, J. D.,** *Chem. Commun.,* p. 619, 1971.

77. **Mynderse, J. S. and Moore, R. E.,** *Phytochemistry,* 18, 1181, 1979.

78. **Inui, H., Miyatake, K., Nakano, Y., and Kitaoka, S.,** Production and compositions of wax esters by fermentation of *Euglena gracilis, Agric. Biol. Chem.,* 47, 2669, 1983.

79. **Patterson, G. W.,** Steroids of algae, in *CRC Handbook of Biosolar Research,* Vol. 1, Part 1, Zaborsky, O. R., Ed., CRC Press, Boca Raton, Fla., 1982.

80. **Gershengorn, M. C., Smith, A. R. H., Goulston, G., Goad, L. J., Goodwin, T. W., and Haines, T, H.,** The sterols of *Ochromonas danica* and *Ochromonas malhamensis, Biochemistry,* 7, 1698, 1968.

81. **Bergman, F. and Feeney, R. J.,** Sterols in algae: the occurrence of chondrillasterol in *Scenedesmus obliquus, J. Org. Chem.,* 15, 812, 1950.

82. **Hoppe, H. A.,** Marine algae and their products and constituents in pharmacy, in *Marine Algae in Pharmaceutical Science,* Hoppe, H. A., Levring, T., and Tanaka, Y., Eds., Walter de Gruyter, Berlin, 1979.

83. **Alam, M., Martin, G. E., and Ray, S. M.,** Dinoflagellate sterols. II. Isolation and structure of 4-methylgorgostanol from the dinoflagellate *Glenodinium foliaceum, J. Org. Chem.,* 44, 4466, 1979.

84. **Wright, D. C.,** Fatty Acid and Sterol Compositions of Six Green Algae Grown Autotrophically, Photoheterotrophically, and Heterotrophically, M.S. thesis, University of Maryland, College Park, 1978.

85. **Withers, N. W., Tuttle, R. C., Holz, G. G., Beach, D. H., Goad, L. J., and Goodwin, T. W.,** Dehydrodinosterol, dinosterone, and related steroids of a non-photosynthetic dinoflagellate *Cryptothecodium cohnii, Phytochemistry,* 17, 1987, 1978.

86. **Shimizu, Y., Alam, M., and Kobayashi, A.,** Dinosterol, the major sterol with a unique side chain in the toxic dinoflagellate, *Gonyaulax tamarensis, J. Am. Chem. Soc.,* 98, 1059, 1976.

87. **Collins, R. P. and Kalnins, K.,** Sterols produced by *Synura petersenii* (Chrysophyta), *Comp. Biochem. Physiol.,* 30, 779, 1969.

88. **Boutry, J. L., Barbier, M., and Ricard, M.,** La diatomee *Chaetoceros simplex* calcitrans Paulson et son environment. II. Effets de la luminiere et des irradiations ultraviolet sur la production primaire et la biosynthese des sterols, *J. Exp. Mar. Biol. Ecol.,* 21, 69, 1976.

89. **Rohmer, M. and Brandt, R. D.,** Les sterols et leurs precursors chez Astasia longa prengsheim, *Eur. J. Biochem.,* 36, 446, 1973.

90. **Williams, B. L., Goodwin, T. W., and Ryley, J. F.,** The sterol content of some protozoa, *J. Protozool.,* 13, 227, 1966.

91. **Patterson, G. W.,** Sterols of *Chlorella.* III. Species containing ergosterol, *Comp. Biochem. Physiol.,* 3, 391, 1969.

92. **Seckbach, J. and Ikan, R.,** Sterols and chloroplast structure of *Cyanidium caldarium, Plant Physiol.,* 49, 457, 1972.

93. **Beastall, G. H., Rees, H. H., and Goodwin, T. W.,** Sterols in *Porphyridium cruentum, Tetrahedron Lett.,* p. 4935, 1971.

94. **Bard, M. and Wilson, K. J.,** Isolation of sterol mutants in *Chlamydomonas reinhardi:* chromatographic analyses, *Lipids,* 13, 533, 1978.

95. **Patterson, G. W. and Kraus, R. W.,** Sterols of *Chlorella.* I. The naturally occurring sterols of *Chlorella vulgaris, Chlorella ellipsoidea, Plant Cell Physiol.,* 6, 211, 1965.

96. **Brandt, R. D., Pryce, R. J., Anding, C., and Ourisson, G.,** Sterol biosynthesis in *Euglena gracilis Z.* (Comparative study of free and found sterols in light and dark-grown *Euglena gracilis Z.*), *Eur. J. Biochem.,* 17, 344, 1970.

97. **Paoletti, C., Pushparaj, B., Florenzano, G., Capella, P., and Lerker, G.,** Unsaponifiable matter of green and blue-green algae lipids as a factor of biochemical differentiation of their biomasses. II. Terpenic alcohol and sterol fractions, *Lipids,* 11, 266, 1976.

98. **Mercer, E. I., London, R. A., Kent, I. S. A., and Taylor, A. J.,** Sterols esters and fatty acids of *Botrydium granulatum, Tribonema aequale* and *Monodus subterraneus, Phytochemistry,* 13, 845, 1974.

99. **Teshima, S. and Kanazawa, A.,** Occurrence of Sterols in the blue-green alga, *Anabaena cylindrica, Bull. Jpn. Soc. Sci. Fish.*, 38, 1197, 1972.

100. **Orcutt, D. M. and Patterson, G. W.,** Sterol, fatty acid and elemental composition of diatoms grown in chemically defined media, *Comp. Biochem. Physiol.*, 50B, 579, 1975.

101. **Rubenstein, I. and Goad, L. J.,** Occurrence of (24S)-24-methyl-cholest-5, 22-E-dien-3beta-ol in the diatom *Phaeodactylum tricornutum, Phytochemistry*, 13, 485, 1974.

102. **Dickson, L. G., Patterson, G. W., and Knights, B. A.,** Distribution of sterols in marine chlorophyceae, *Proc. Int. Seaweed Symp.*, 9, 413, 1978.

103. **Tornabene, T. G.,** Microorganisms as hydrocarbon producers, *Experientia*, 38, 43, 1982.

104. **Lee, R. F. and Loeblich, A. R.,** *Phytochemistry*, 10, 593, 1971.

105. **Caccamese, S. and Rinehart, K. L., Jr.,** *Experentia*, 34, 1129, 1978.

106. **Younglood, W. W. and Blumer, M.,** *Mar. Biol.*, 21, 163, 1973.

107. **Clark, R. C. and Blumer, M.,** *Limnol. Oceanogr.*, 12, 79, 1967.

108. **Blumer, M., Guillard, and Chase, T.,** *Mar. Biol.*, 8, 183, 1971.

109. **Bachofen, R.,** The production of hydrocarbons by *Botryococcus braunii, Experentia*, 38, 47, 1982.

110. **Benemann, J. R. and Weissman, J. C.,** Chemicals from microalgae, in *Bioconversion Systems*, Wise, D. L., Ed., CRC Press, Boca Raton, Fla., 1984, 59.

111. **Bailliez, C., Largeau, C., Casadevall, E., and Berkaloff, C.,** Effects de Ímmobilisation en gel d'alginate sur l'algue *Botrycoccus braunii C. R. Acad. Sci. Paris Ser. 3:*, 1983, 199.

112. **Davies, B. H.,** Carotenoids of algae and higher plants, in *CRC Handbook of Biosolar Resources*, Vol. I, Part I, Zaborsky, O. R., Ed., CRC Press, Boca Raton, Fla., 1982.

113. **Bisalputra, T.,** Plastids, in *Algal Physiology and Biochemistry*, Stewart, W. D. P., Ed., Blackwell Scientific, Oxford, 1974, 124.

114. **Ben-Amotz, A., Katz, A., and Avron, M.,** Accumulation of beta-carotene in halotolerant algae: purification and characterization of best-carotene-rich globules from *Dunaliella bardawil (chlorophycea), J. Phycol.*, 18, 539, 1982.

115. **Marusich, W. L. and Bauernfeind, J. C.** Oxycarotenoids in poultry pigmentation. I. Yolk studies, *Poult. Sci.*, 49, 1555, 1970.

116. **Marusich, W. L. and Bauernfeind, J. C.,** Oxycarotenoids in poultry pigmentation. II. Broiler studies, *Poult. Sci.*, 49, 1566, 1970.

117. **Matsuno, T., Nagata, S., Iwahashi, M., Koike, T., and Okada, M.,** Intensification of color of fancy red carp with zeaxanthin and myxoxanthophyll, major carotenoid constituents of *Spirulina, Nippon Suisan Gakkaishi*, 45(5), 627, 1979.

118. **Chapman, D. J. and Haxo, F. T.,** *J. Phycol.*, 2, 89, 1966.

119. **Farrow, W. M. and Tabenkin, B.,** Process for the Preparation of Lutein, U.S. Patent 3,280,502, 1966.

120. **Holt, A. S.,** Further evidence of the relation between 2-desvinyl-2-formyl-chlorophyll-a and chlorophyll-d, *Can. J. Bot.*, 39, 327, 1961.

121. **Eskins, K. and Dutton, H. J.,** Sample preparation for high-performance liquid chromatography of higher plant pigments, *Anal. Chem.*, 51, 1885, 1979.

122. **Yamaguchi, T.,** Strong Deodorant, Japanese patent 79 JP — 116131, 1981.

123. **Gysi, J. R. and Chapman, D. J.,** Phycobilins and phicobiliproteins of algae, in *CRC Handbook of Biosolar Resources*, Vol. 1, Part 1, Zaborsky, O. R., Ed., CRC Press, Boca Raton, Fla., 1982, 83.

124. Dainippon Ink and Chemicals - Phycocynin leaflet.

125. Dainippon Ink and Chemicals Inc., Production of Highly Purified Alcoholophilic Phycocyanin, Japanese Patent 80,77,890, 1980.

126. Dainippon Ink and Chemical Inc., Cosmetics Containing Water Soluble Phycocyanim, Japanese Patent, 79 - 138755, 1981.

127. **Boussiba, S. and Richmond, A. E.,** Isolation and characterization of phycocyanins from the blue-green alga *Spirulina platensis, Arch. Microbiol.*, 120, 155, 1979.

128. **Bergmeyer, H. U.,** Methoden der enzymatischen analyse, *Verlag Chemie*, Weinheim Bergstrasse, 1974, 2142.

129. **Krietsch, W. G. and Kuntz, G. W. K.,** Specific enzymatic determination of adenosine triphosphate, *Anal. Biochem.*, 90, 829, 1978.

130. **Krieger, N. and Hastings, J. W.,** *Science*, 161, 586, 1968.

131. **Krieger, N., Njus, D., and Hastings, J. W.,** *Biochemistry*, 13, 2871, 1974.

132. **Hastings, J. W.,** *Annu. Rev. Biochem.*, 37, 597, 1968.

133. **Hastings, J. W. and Sweeney, B. M.,** *J. Cell. Comp. Physiol.*, 49, 209, 1957.

134. **Schmitter, R. E., Njus, D., Sulzman, F. M., Gooch, V. D., and Hastings, J. W.,** *J. Cell. Physiol.*, 87, 123, 1976.

135. **Puhler, A. and Heumann, W.,** Genetic engineering, in *Biotechnology*, Vol. 1, Rehm, H. J. and Reed, G., Eds., Verlag Chemie, Weinheim, Deerfield Beach, Fla.

136. **Mayer, H.,** Restriction endonucluses, *Biochemica*, 1981.

137. **Wikstrom, P., Szwajcer, E., Brodelius, P., Nilsson, K. N., and Mosbach, K.,** Formation of alpha-keto acids from amino acids using immobilized bacteria and algae, *Biotech. Lett.,* 4, 153, 1982.
138. **Fridovich, I.,** *Accounts Chem. Res.,* 5, 321, 1972.
139. **Spruit, C. J. P.,** Photoreduction and anaerobiosis, in *Physiology and Biochemistry of Algae,* Lewin, R. A., Ed., Academic Press, New York, 1962, 47.
140. **Kumazawa, S. and Mitsui, A.,** Hydrogen metabolism of photosynthetic bacteria and algae, in *CRC Handbook of Biosolar Resources,* Vol. 1, Part 1, Zaborsky, O. R., Ed., CRC Press, Boca Raton, Fla., 1982.
141. **Benemann, J. R., Miyamoto, K., and Hallenbeck, P. C.,** Bioengineering aspects of biophotolysis, *Enzyme Microb. Technol.,* 2, 103, 1980.
142. **Benemann, J. R. and Weare, N. M.,** *Science,* 184, 174, 1974.
143. **Ochiai, H., Shibata, H., Fujishima, A., and Honda, K.,** *Agric. Biol. Chem.,* 43, 881, 1979.
144. **Ochiai, H., Shibata, H., Sawa, Y., and Katoh, T.,** "Living electrode" as a long-lived photoconverter for biophotolysis of water, *Proc. Natl. Acad. Sci. U.S.A.,* 77(5), 2442, 1980.
145. **Miura, Y., Miyamoto, K., and Kiyohito, Y.,** Producing hydrogen by alga in alternating light-dark cycle, *Eur. Pat Appl. EP 77,014,* 16 pp., 1983.
146. **Eisbrenner, B., Distler, E., Floener, L., and Bothe, H.,** The occurrence of the hydrogenase in some blue-green algae, *Arch. Microbiol.,* 118, 117, 1978.
147. **Belkin, S. and Padan, E.,** Hydrogen metabolism in the facultative anoxygenic cyanobacteria (blue-green algae) *Oscillatoria limnetica* and *Aphanothece halophytica, Arch. Microbiol.,* 116, 109, 1978.
148. **Ward, M. A.,** Whole cell and cell-free hydogenases of algae, *Phytochemistry,* 9, 259, 1970.
149. **Frenkel, A., Gaffron, H., and Battley, E. H.,** Photosynthesis and photoreduction by the blue-green alga, *Synechococcus elongatus,* Nag, *Biol. Bull. Woods Hole, Mass.,* 99, 157, 1950.
150. **Frenkel, A. W. and Rieger, C.,** Photoreduction in algae, *Nature (London),* 167, 1030, 195.
151. **Hartman, H. and Krasna, A. I.,** Studies on the Adoption of hydrogenase in *Scenedesmus, J. Biol. Chem.,* 238, 749, 1963.
152. **Kessler, E.,** Reduction of nitrite with molecular hydrogen in algae containing hydrogenase, *Arch. Biochem. Biophys.,* 62, 241, 1956.
153. **Bishop, N. I., Frick, M., and Jones, L. W.,** Photohydrogen production in green algae: water serves as the primary substrate for hydrogen and oxygen production, in *Biological Solar Energy Conversion,* Mitsui, A., Miyachi, S., San Pietro, A., and Tamura, S., Eds., Academic Press, New York, 1977.
154. **Kessler, E. and Czygan, F.-C.,** Physiologische und biochemische beitrage zur taxonomie der *Gattungen Ankistrodesmus* und *Scenedesmus.* I. Hydrogenase, sekundar-carotinoide unde gelatine verflusngung, *Arch. Mikrobiol.,* 55, 320, 1967.
155. **Healey, F. P.,** Hydrogen evolution by several algae, *Planta,* 91, 220, 1970.
156. **Kessler, E.,** Hydrogenase, Photoreduction and anaerobic growth, in *Algal Physiology and Biochemistry,* Stewart, W. D. P., Ed., Blackwell Scientific, Oxford, 1974, 456.
157. **Ben-Amotz, A., Erbes, D. L., Riederer-Henderson, M. A., Peavey, D. B., and Gibbs, M.,** H-2 metabolism in photosynthetic organisms. I. Dark H-2 evolution and uptake by algae and mosses, *Plant Physiol.,* 56, 72, 1975.
158. **Samson, R. and LeDuy, A.,** Biogas production from anaerobic digestion of *Spirulina maxima* algal biomass, *Biotechnol. Bioeng.,* 24, 1919, 1982.
159. **Golueke, C. G., Oswald, W. J., and Gotaas, H. B.,** *Appl. Microbiol.,* 5, 47, 1957.
160. **Golueke, C. G. and Oswald, W. J.,** *Appl. Microbiol.,* 7, 219, 1959.
161. **Uziel, M., Oswald, W. J., and Golueke, C. G.,** Solar Energy Fixation and Conversion with Algal Bacterial Systems, U.S. National Science Foundation Rep. No. NSF-RA-N-74-195, NSF, Washington, D.C., 1974.
162. **Binot, R., Martin, D., Nyns, E. J., and Naveau, H.,** Digestion anaerobie d'algues cultivees dan les eaux de refroidissement industrielles, in *Proc. Heliosynthese Aquaculture Semin.,* Martigues, France, Sept. 20, 1978.
163. **Keenan, J. D.,** Bioconversion of solar energy to methane, *Energy,* 2, 365, 1977.
164. **Uziel, M.,** Solar Energy Fixation and Conversion with Algal Bacterial Systems, Ph.D. thesis, University of California, Berkeley, 1978.
165. **Hellebust, J. A.,** *Ann. Rev. Plant Physiol.,* 27, 485, 1976.
166. **Hellebust, J. A.,** *Can. J. Bot.,* 54, 1735, 1976.
167. **Ben-Amotz, A. and Avron, M.,** Accumulation of metabolities by halotolerant algae and its industrial potential, *Ann. Rev. Microbiol.,* 37, 95, 1983.
168. **Ehrin, E., Ston-Elander, S., and Nilsson, L. G.,** Photosynthetic preparation of c-labelled glucose. Cultivation of the green algae used in the procedure, *Acta Pharm. Suec.,* 20, 73, 1983.
169. **Yopp, J. H., Miller, D. M., and Tindall, D. R.,** in *Energetics and Structure of Halophilic Microorganisms,* Caplan, S. R. and Ginzburg, M., Eds., Elsevier, Amsterdam, 1978, 619.
170. **Borowitzka, L. J.,** *Hydrobiologia,* 81, 33, 1981.

171. **Craigie, J. S.,** *J. Fish Res. Board Can.,* 26, 2959, 1969.
172. **Kauss, H.,** *Nature (London),* 214, 1129, 1967.
173. **Lee, R. E.,** *J. Protozool.,* 25, 163, 1978.
174. **Brown, L. M. and Hellebust, J. A.,** *J. Phycol.,* 16, 265, 1980.
175. **Ben-Amotz, A. and Grunwald, T.,** *Plant Physiol.,* 67, 613, 1981.
176. **Paul, J. S.,** *J. Phycol.,* 15, 280, 1979.
177. **Ben-Amotz, A., Katz, A., and Avron, M.,** *J. Phycol.,* 18, 529, 1982.
178. **Williams, L. A., Foo, E. L., Foo, A. S., Kuhn, I., and Heden, C.-G.,** Solar bioconversion systems based on algal glycerol production, *Biotechnology and Bioengineering Symposium No. 8,* John Wiley & Sons, New York, 1978, 115.
179. **Goldman, Y., Garti, N., Sasson, Y., Ginzburg, B., and Bloch, M. R.,** Conversion of halophilic algae into extractable oils, *Fuel,* 29, 181, 1980.
180. **Leavitt, R.,** *Algal Biomass Workshop,* Boulder, Colorado, March 20, 1984.
181. **Bhattacharjee, J. K.,** Micro-organisms as potential sources of food, *Adv. Appl. Microbiol.,* 13, 139, 1970.
182. **Ramus, J.,** Cell surface polysaccharides of the red alga *Porphyridium,* in *Biogenesis of Plant Cell Wall Polysaccharides,* Loewus, F., Ed., Academic Press, New York, 1973, 333.
183. **Golueke, L. G. and Oswald, W. J.,** *Appl. Microbiol.,* 10, 102, 1962.
184. **Gudin, C. and Thomas, D.,** *C. R. Acad. Sci. Paris Ser. 3:* 293, 35, 1981.
185. **Arad, S.,** Personal communication.
186. **Ramus, J. and Robins, D. M.,** *J. Phycol.,* 11, 70, 1975.
187. **Ramus, J.,** Algae Biopolymer Production, U.S. Patent No. 4,236,349, 1980.
188. **Percival, E. and Foyle, R. A. J.,** The extracellular polysaccharides of porphyridium cruentum and prophyridium aerugineum, *Carbohydrate Res.,* 72, 165, 1979.
189. **Gudin, C., Thepenier, C., Thomas, D., and Chaumont, D.,** Solar and CO-2 biotechnology production of exocellular biomass by biophotoreaction of immobilized cells, *Solar World Forum,* Brighton, England, 1981.
190. **Savins, J. G.,** Oil Recovery Process Employing Thickened Aqueous Driving Fluid, U.S. Patent 4,079,544, 1978.
191. **Lee, Y. and Pirt, S. J.,** Interactions between an alga and three bacterial species in a consortium selected for photosynthetic biomass and starch production, *J. Chem. Technol. Biotechnol.,* 31, 295, 1981.
192. **Pirt, M. W. and Pirt, S. J.,** Photosynthetic production of biomass and starch by *Chlorella* in a chemostat culture, *J. Appl. Chem. Biotechnol.,* 27, 643, 1977.
193. **Dawes, E. A. and Gibbons, D. W.,** *Bacteriol. Rev.,* 28, 126, 1964.
194. **Baptist, J. N. and Werber, F. X.,** *SPE Trans.,* 4, 245, 1964.
195. **Shelton, J. R., Lendo, J. B., and Ayostini, D. E.,** *Polymer Lett.,* 9, 173, 1971.
196. **King, P.,** Biotechnology an industrial view, *J. Chem. Technol. Biotechnol.,* 32, 2, 1982.
197. **Campbell, J., Stevens, E., Jr., and Balkwill, D. L.,** Accumulation of poly-beta-hydroxybutryate in *Spirulina plantensis, J. Bacteriol.,* 749, 361, 1982.
198. **Sieburth, J. M.,** Acrylic acid, and "antibiotic" principle in phaeocystis blooms in antarctic waters, *Science,* 132, 676, 1960.
199. **Sieburth, J. M.,** Antibiotic properties of acrylic acid, a factor in the gastrointestinal antibiosis of polar marine animals, *J. Bacteriol.,* 82, 72, 1961.
200. **Glombitza, K. -W.,** Antibiotics from algae, in *Marine Algae in Pharmaceutical Science,* Hoppe, H. A., Levring, T., and Tanaka, Y., Eds., Walter de Gruyter, Berlin, 1979, 304.
201. **Anthoni, U., Christophersen, C., Ogard-Madsen, J., Wium-Andersen, S., and Jacobsen, N.,** Biologically active sulphur compounds from the green alga *Chara globularis, Phytochemistry,* 19, 1228, 1980.
202. **Marner, F.-J., Moore, R. E., Hirotsu, K., and Clardy, J.,** *J. Org. Chem.,* 42, 2815, 1977.
203. **Mynderse, J. S., Moore, R. E., Kashiwagi, M., and Norton, T. R.,** *Science,* 196, 538, 1977.
204. **Mynderse, J. S. and Moore, R. E.,** *J. Org. Chem.,* 43, 2301, 1978.
205. **Pedersen, M. and DaSilva, E. J.,** Simple brominated phenols in the blue-green alga *Calothrix brevissima West., Plant,* 115, 83, 1973.
206. **Markham, K. R. and Porter, L. J.,** *Phytochemistry,* 8, 1777, 1969.
207. **Pratt, R., Daniels, T. C., Eiler, J. B., Gunnison, J. B., Kumler, W., D., Oneto, J. F., Strait, L. A., Spoehr, H. A., Hardin, G. J., Milner, H. W., Smith, J. H. C., and Strain, H. H.,** Chlorellin, an antibacterial substance from *Chlorella, Science,* 99, 351, 1944.
208. **Aaronson, S. and Bensky, B.,** Effect of aging of a cell population on lipids and during resistance of *Ochroleuca danica, J. Protozool.,* 14, 76, 1967.
209. **Harder, R. and Oppermann, A.,** Ober antibiotische stoffe bei den Grunalgen *Stichococcus* und *Protosiphon botryoides, Arch. Mikrobiol.,* 19, 398, 1953.
210. **Proctor, V. W.,** Studies of algal antibiosis using *Haematococcus* and *Chlamydomonas, Limnol. Oceanogr.,* 1, 125, 1956.

211. **Pesando, D.,** Etude chimique et structurale d'une substance lipidique antibiotique produite par une diatomee marine: *Asterionella japonica, Rev. Intern. Oceanogr. Med.,* 25, 49, 1972.
212. **Fabik A.,** Postoperative care of the coagulation surface following diathermocoagulation using the weed *Scenedesmus obliquus, Cesk. Gynek.,* 35, 4, 1970.
213. **Gauthier, M. S.,** Activite antibacterienne d'une diatomee marine, *Asterionella notata* (Gnum), *Rev. Intern. Oceanogr. Med.,* 25, 103, 1969.
214. **Pesando, D., Gnassia-Barelli, M., and Gueho, E.,** Antifungal properties of some marine planktonic algae, in *Marine Algae in Pharmaceutical Science,* Hoppe, H. A., Levring, T., and Tanaka, Y., Eds., Walter de Gruyter, Berlin, 1979.
215. **Blankenship, J. E.,** *Perspect. Biol. Med.,* 19, 509, 1976.
216. **Carmichael, W. W., Biggs, D. F., and Gorham, P. R.,** *Science,* 187, 542, 1975.
217. **Carmichael, W. W., Biggs, D. F. and Peterson, M. A.,** *Toxicon,* 17, 229, 1979.
218. **Gorham, P. R. and Carmichael, W. W.,** Phycotoxins from blue-green algae, *Pure Appl. Chem.,* 52, 165, 1979.
219. **Bishop, C. T., Anet. E. F. L. J., and Gorham, P. R.,** *Can. J. Biochem. Physiol.,* 37, 453, 1959.
220. **Gregson, R. P. and Lohr, R. R.,** Isolation of peptide hepatotoxins from the blue-green alga *Microcycstis aeruginosa, Comp. Biochem. Physiol.,* 74c,(no. 2), 413, 1983.
221. **Mynderse, J. S., Moore, R. E., Kashiwagi, M., and Norton, T. R.,** *Science,* 196, 538, 1977.
222. **Cardelina, J. H., Marner, F.-J., and Moore, R. E.,** *Science,* 204, 193, 1979c.
223. **Collins, M.,** Algal toxins, *Microbiol. Rev.,* 42, 725, 1978.
224. **Shimizu, Y.,** Dinoflagellate toxins, in *Marine Natural Products, Chemical and Biological Perspectives,* Vol. 1, Scheuer, P. J., Ed., Academic Press, New York, 1978, 1.
225. **Alam, M., Trieff, N. M., Ray, S. M., and Hudson, J. E.,** *J. Pharm. Sci.,* 64, 865, 1975.

BLUE-GREEN ALGAE AS BIOFERTILIZER

L. V. Venkataraman

INTRODUCTION

The biggest constraint in the agriculture of developing countries is the nonavailability of chemical fertilizers at reasonable prices. Nitrogenous fertilizers in particular have been recognized as the limiting factor in food production.[1] By 1985, the anticipated demand for fertilizers in the developing countries will be about 85 million mt of which about 30% constitute nitrogen fertilizers.[2]

Rice has been grown for centuries in tropical Asia without significant inputs of chemical fertilizers and harvests have remained at 1 to 2 t/ha.[3] Rice crop in India yield on an average 1.6 t/ha annually and occupies 40 million ha which is 37% of that utilized for the production of cereals.[4,5] Watanabe et al.[6] recently summarized data from Japan, Thailand, Indonesia, and the Philippines and reported that a single rice crop can take from 37 to 113 kg N/ha from sources other than mineral nitrogen. Nitrogen-fixation is clearly evident in the results of many long-term fertility experiments carried out for several decades at Japanese experimental stations by Yamaguchi.[7] The principle agents of biological nitrogen fixation are (1) blue-green algae (BGA), (2) the water fern, *Azolla* in association with BGA, (3) nonsymbiotic in vitro nitrogen fixing bacteria associated with the roots of plants, and (4) nonsymbiotic bacteria in the bulk of anaerobic soil.

HISTORICAL

The concept of utilizing BGA as fertilizer for fixing nitrogen in paddy fields was developed by De[8] in India, who recognized that field fertility was maintained by the proliferation of cyanobacteria in the soil. This observation was later confirmed by Singh[9] and Watanabe[10]

A large variety of algal species which are capable of fixing atmospheric nitrogen have been identified. Li[11] listed 32 genera and 152 species of BGA capable of fixing nitrogen (Table 1) classified into major groups as follows:

1. Forms with heterocysts which can fix nitrogen aerobically such as *Nostoc, Anabaena, Oscillatoria, Cylindrospermun, Mastigocladus, Tolypothrix*
2. Unicellular or colonial algae fixing nitrogen aerobically like *Gloeocopsa*
3. Nonheterocystous filamentous forms which can fix nitrogen only in the absence of oxygen and in the presence of nitrogen and carbon dioxide (micro-aerophilic), for example, *Oscillatoria* and *Phormidium*
4. Nonheterocystous filamentous forms which can fix nitrogen aerobically like *Trichodesmium*

Watanabe[12] studied soil samples collected from different countries in Southeast Asian regions, India, and Africa[13] and observed that nigrogen-fixing BGA are not present in all environments. Venkataraman[14] found that only 33% of 2213 soil samples collected in India and analyzed for algal flora contained BGA. Forms such as *Nostoc, Anabaena, Calothrix, Aulosira*, and *Plectonema* are ubiquitous to Indian soils, while *Halosiphon, Scytonema*, and *Clyindrospermum* have localized distribution.

Physiology of N_2 Fixation

Among the microorganisms which reduce dinitrogen to ammonia, the cyanobacteria play

Table 1
NITROGEN-FIXING CYANOBACTERIA[11]

Group	Genus	Aerobic sp.	Anaerobic/mi-croaerobic sp.
Unicellular forms	Aphanothece	1	1
	Gloeocapsa	1	1
	Gloeothece	4	4
	Synechococcus	0	3
	Chlorogloeopsis	1	0
	Dermocarpa	0	2
	Xenococcus	0	1
	Myxosarcina	0	1
	Chroococcidiopsis	0	8
	Pleurocapsa	0	7
Nonheterocystous fi-lamentous forms	Oscillatoria	0	5
	Pseudoanabaena	0	4
	Lyngbya-Plectonema-Phormidium	0	16
	Microcoleus	1	1
	Trichodesmus	1	1
Heterocystous fila-mentous forms	Anabaena	34	34
	Anabaenopsis	2	2
	Aulosira	1	1
	Calothrix	12	12
	Cylindrospermum	7	7
	Fischerella	2	2
	Gloeotrichia	4	4
	Hapalosiphon	3	3
	Mastigocladus	1	1
	Nostoc	16	16
	Scytonema	5	5
	Stigonema	1	1
	Tolypothrix	6	6
	Westiella	1	1
	Westiellopsis	1	1

a major role since they fix N_2 both in the free living state and in symbiosis with other organisms. The best known examples of nitrogen-fixing cyanobacteria are filamentous heterocystous species, e.g., *Anabaena* and *Nostoc* which differentiate vegetative cells to heterocysts only when grown in the absence of combined nitrogen. The heterocysts, specialized for aerobic fixation of nitrogen, are the site of the enzyme complex nitrogenase which catalyzes the conversion of dinitrogen to ammonia. In some species they may also function in perennation as location of filament breakage.

There are structural and metabolic differences between heterocysts and vegetative cells which may enhance the effectiveness of heterocysts as sites for nitrogen fixation. The most conspicuous structural difference is the formation of a thick envelope by the heterocysts, which comprises an inner laminate layer consisting of glycoplipids, a central homogenous layer, and an outer fibrous layer. It is assumed that the inner glycolipid layer normally reduces the rate of diffusion of oxygen into the heterocyst.

Heterocysts, compared with vegetative cells, look pallid like vegetative cells, however, they contain chlorophyll-a and cartenoids. Nitrogenase is a complex enzyme which consists of two iron-sulfur proteins, which individually have no enzymatic activity. In the combined form they catalyze the reduction of various substrates, viz., N_2, N^-_3, HCN, C_2H_2, etc.

The larger enzyme protein contains 18 to 36 Fe and 1 to 2 Mo atoms. This Fe-Mo component binds the substrates by catalyzing their reduction when supplied with reductants from the second nitrogenase protein. The reduction of N_2 to ammonia requires ATP which possibly is involved in the reduction of the substrate at the Fe-Mo protein. It is assumed that the substrate is bound to a subunit (Mo cofactor) which consists of one molecule of Mo and an unknown amount of Fe. Both the nitrogenase components are oxygen sensitive, especially the Fe-protein complex.[15] Heterocysts lack the water-splitting photosystem II (PS II) which implies that these cells do not generate oxygen which otherwise would inactivate nitrogenase. Hence, heterocysts are dependent on the supply of carbohydrates from the vegetative cells. The light-stimulated substrate reduction in the heterocysts can be explained by the supply of ATP provided by cyclic photophosphorylation.

The immediate electron carrier to nitrogenase is ferredoxin. Since ferredoxin is continuously transferring electrons to the nitrogenase complex, it must be reduced by an electron donor, which most probably is NADPH. The transfer of electrons to nitrogenase occurs either in the dark or in a PS I dependent reaction. An alternative donor may also be pyruvate.

In N_2-fixing microorganisms, including cyanobacteria, most of N_2-derived nitrogen is incorporated into glutamine by glutamine synthetase and then metabolized by glutamate synthetase.[16] The heterocysts, however, metabolize N_2 or NH^+_4 only to glutamine, glutamate synthatase being seemingly restricted to vegetative cells. This implies that the glutamine, found in the heterocysts, has to move into the vegetative cells in order to undergo amide transfer.

SUCCESSION OF CYANOBACTERIA OR BLUE-GREEN ALGAE (BGA) IN RICE FIELDS

The ecosystem of rice fields is well suited for the growth of BGA since it provides favorable conditions of light, water, temperature, and nutrients. This may explain the higher density of BGA in rice fields than in soils under other agricultural crops. Studies in India have shown seasonal periodicity and succession of algae in unfertilized rice fields as shown in Table 2.

The following two stages in the development of the algal flora have been suggested by Roger and Kulasooriya:[18] (1) At the beginning of the cultivation cycle, the soils of paddy fields are characterized by low pH, high light intensity, and high level of CO_2 which are more suited for the growth of chlorophycease than for blue-green algae. (2) The rice cultivation cycle is marked by lower light intensity and an increase in pH which are better suited for BGA growth. There are no reports on the dominance of BGA at the beginning of rice cultivation. The nature of algal biomass in relation to rice fields in Sengal have been reported by Roger et al.[19,20] (Table 3).

FACTORS AFFECTING ALGAL GROWTH IN PADDY FIELDS

Physical Factors

Growth of BGA in soils is influenced by environmental factors. Light is a major environmental factor which affects the seasonal fluctuations of phytoplankton in paddy fields. Phototrophic microalgae are restricted to the photic zone and are usually located in the upper 0.5 cm of the soil. In the paddy fields, BGA occur especially as surface scum, as a bloom, or as crust-forming aggregates at the soil-water interface. In fields, the light screening effect of the crop canopy appears to cause a rapid decrease in light reaching the water, in direct proportion to the height of the rice plant. Plants, 30 cm high absorb about 50% of the light and 60 cm high plants cut off 90%. Light may have a selective effect on the composition of the algal flora. In general, green algae can withstand more direct intense

Table 2
SEASONAL OCCURRENCE OF INDIGENOUS NITROGEN-FIXING CYANOBACTERIA IN AN UNFERTILIZED RICE FIELD[17]

Month	Species
January	*Nostoc muscorum* (C)
February	*Nostoc muscorum* (C)
March	*Nostoc muscorum* (C), *N. punctiforme* (A), *Anabaena variabilis* (C), *A. ambigua* (C)
April	*Nostoc muscorum* (A), *N. punctiforme* (A), *Anabaena variabilis* (A), *A. ambigua* (C)
May	*Nostoc muscorum* (A), *N. punctiforme* (A), *Anabaena variabilis* (A), *A. ambigua* (C)
June	*Nostoc muscorum* (A), *N. punctiforme* (A), *Anabaena variabilis* (A), *A. ambigua* (A), *Aulosira fertilissima* (R)
July	All of the above (A)
August	All of the above (A), *Cylindrospermum muscicola* (C)
September	All of the above (A), *Westiella* sp. (C)
October	All of the above (A)
November	*Nostoc punctiforme* (C), *Westiella* sp. (C)
December	*Nostoc muscorum* (C)

Note: (A) Abundant, (C) common, (R) rare.

Table 3
ALGAL BIOMASS COMPOSITION IN RELATION TO RICE DEVELOPMENT IN 40 PADDY FIELDS IN SENEGAL[19-20]

Stages of rice development	Algal forms	Dominant flora (% of biomass) mean	N$_2$-fixing algae (% of biomass) mean
Tillering	Diatoms, unicellular green algae	73 (49—99)	2 (0.1—4.0)
Panicle initiation	Filamentous green algae, nonheterocystous blue-green algae	89 (86—93)	3 (0.1—9.0)
Heading to maturity, weak plant cover	Filamentous green algae, nonheterocystous blue-green algae	70 (62—91)	8 (0.2—14)
Heading to maturity, dense plant cover	Blue-green algae	71 (19—99)	38 (13—99)

light compared to BGA of which only a limited number of species such as *Cylindrospermum* or *Aulosira* can thrive under high light intensities.[21,22] The favorable temperature range for the growth of most BGA is about 30 to 35°C and both field and pot experiments indicated a setback to their growth during the cold season.[23]

Soil algae in general, and BGA in particular, have a high capacity to withstand desiccation. In paddy fields, where the dry period is relatively short, N_2-fixing algae comprise only 30% of the algal flora; but where the dry season lasts about 8 months, spores and heterocystous BGA constitute more than 95% of the flora at the end of the dry period, homocystous BGA being present primarily because of their introduction by irrigation water. In paddy soils of Japan, algae are reported to be more abundant when the soil is waterlogged throughout the year than during rice growth. Forms such as *Nostoc, Cylindrospermum,* etc. seem to develop profusely when soil water decreases.[21,24] Wind invariably disturbs the natural spread of BGA and may either disperse them in the field or bring the blooms to accumulate in a corner of the rice field.

Biological Factors

Certain bacteria, fungi, and viruses are known to be pathogenic to BGA. Considering the widespread occurrence of algal viruses and the specific host ranges of the individual strains, it is possible that in natural situations cyanophages may be important in determining algal successions and disappearance. Several forms of cyanophages have been isolated in India which may have deleterious effects on the propagation of BGA.[25]

Many algae release substances that inhibit their own growth or that of other species, or both. Some BGA have been reported to excrete ascorbic acid or urea, and it has been suggested that this directly accelerates the growth and development of the rice plants. Alternatively, these indirectly participate in the process of nitrogen fixation and nitrate reduction, influencing the growth of rice indirectly.

Invertebrates such as cladocerans, copepods, ostracods, mosquito larvae, snails, etc. are common grazers of algae in rice fields. The development of such populations prevents the establishment of algal growth and may wipe out algal blooms within a short period of time.[26] Evidence of preferential grazing among algae has been presented by Watanabe et al.[26] who showed that unicellular green algae are excellent feed for daphnids while filamentous N_2-fixing BGA were fed less effectively.

Soil Factors

pH is certainly the most important soil factor determining the composition of algal flora. BGA thrive on neutral to alkaline soils in a pH range between 5.5 and 8.5 as has been shown in soil culture experiments with low soil nitrogen.[19] Soils of slightly alkaline reaction fixed more nitrogen than those with low pH. Thus, addition of lime to elevate the pH of acidic soil favored algal growth. In general, observations in India indicate that most soil BGA grow best in neutral or near neutral range pH (6.5 to 7.5).[28] Physicochemical properties of the soil, such as soil aggregation or the redox potential, have a direct effect on the growth of BGA.[29]

Agronomic Practices

The growth of BGA is influenced by various agronomic practices such as tillage which has different effects in different seasons. Irrigation practices also have an effect on the growth of algae. In Australian rice fields, when the flow of irrigation water was reduced and its turbidity decreased, an improved growth of BGA was noticed (Table 4).[30] The growth of BGA is influenced by the quality and quantity of the fertilizers as well as by the application techniques. Nitrogen (N) and phosphorous (P) fertilization exerted a positive effect on the growth of total algal biomass, but growth of N_2-fixing bacteria is favored in the absence of nitrogenous fertilizer.[19,31] Studies have been conducted in Senegal on algal flora of 30 rice fields differing in geographic location, stages of growth, and fertilizer treatment.[20] Qualitative and quantitative effects on the growth of BGA have been associated with the application of fertilizers. *Aphanothece* exhibited luxurious growth on treatment with lime, and *Rivularia*

Table 4
EFFECT OF ALGAE UNDER DIFFERENT
LEVELS OF IRRIGATION AND NITROGEN
FERTILIZER ON THE GRAIN YIELD OF PADDY[30]

| | | Yield (kg/ha) | |
| | Nitrogen level | Without | |
Irrigation level	(kg/ha)	algae	With algae
A	0	3,477	4,295
	30	3,993	4,518
	60	5,236	5,491
	90	5,752	5,826
B	0	3,045	3,333
	30	3,624	3,950
	60	4,426	4,553
	90	4,959	4,663
C	0	3,346	3,748
	30	4,033	4,423
	60	4,553	4,798
	90	5,128	5,099

Note: A = Submergence to saturation during all stages; B = saturation to 0.3 bar tension during all stages; C = submergence to saturation from tillering to dough stage and saturation to 0.3 bar tension during the remaining period.

aquatica and *Plectonema boryanum* grew well on treatment with potassium and phosphorous, respectively.[31]

Under nitrogen-deficient conditions, N_2-fixing BGA are greatly favored by the lack of competitiveness from other algae and may develop profusely if other environmental factors are not limiting. Okuda and Yamaguchi[28] observed algal flora in soils treated with different nitrogen fertilizers and they reported nitrogen fixing forms to become abundant only in the unfertilized control.

The requirement of phosphorous for the optimal growth of algae differs considerably among various species. BGA assimilate more P than required and store it in the form of polyphosphate which is probably utilized under P-deficient conditions.[32] *Anabaena* and *Tolypothrix* have been found to fix more nitrogen in the presence of phosphate.[33]

Because of its role of nitrate reductase and nitrogenase, molybdenum (Mo) is required by all algae fixing nitrogen. The minimum concentration for optimum growth, about 200 ppb, is often available in paddy soils, particularly with waterlogged conditions when the soil pH decreases.[34] Mo is likely to be a limiting factor for nitrogen-fixing algae particularly during the drying out of the paddy fields when algal nitrogen fixing is often most active. To overcome this deficiency an addition of sodium molybdate (0.25 kg/ha) to improve algal growth has been recommended.[35] The cost of its addition is low.

Other elements such as K, Na, S, Co, Zn, Cu, etc., are required for optimal growth of algae, but their ecological implications as limiting factors or as factors affecting the composition of algal communities in rice fields have not been demonstrated. Effects of organic manure on BGA and phototrophic nitrogen fixation seem to be variable and both favorable and unfavorable effects on growth have been reported.[19,36-38] In India, addition of green manure stimulated the growth of a large number of BGA. Surface application of straw has also been reported to have a stimulating effect on N-fixing BGA. Sugar factory effluent used sometimes in irrigation has been reported to preferentially enhance the growth of *Oscillatoria*, *Spirulina*, and *Lyngbye*. This treatment, however, may cause difficulties due to bacterial contamination.[39]

A variety of pesticides are used for pest control, and their effect on soil microorganisms in general, and on BGA in particular, have only recently received attention. Many BGA strains were found to have a high degree of resistance to many pesticides. While most of the strains of *Anabaena* could tolerate 100 ppm Ceresan M (N-ethylmercuri p-toluene sulfonilide), a few were sensitive even to a concentration of 0.1 ppm. *Tolypothrix tenus* and *Aulosira fertilissima* tolerate high concentrations of Ceresan, Dithane, and Delapon.[40,41] *Aulosira*, however, is sensitive to substituted urea compounds such as Cotoron and Diuron, as well as to the triazine herbicide Propazine. Growth stimulation of BGA was observed with Lindane, Parathion, Endrin, Diazinone, and BHC at a concentration of 10 ppm.

The growth of four species of BGA and their capacities to fix nitrogen were significantly reduced by γ-BHC and endosulfan.[42] The specific effect of these insecticides on singular metabolic processes have also been reported.[43]

Herbicides in general seem to have limiting effects on the growth and nitrogen fixation of algae. A stimulatory effect of herbicides on the growth and N_2 fixation of *Tolypothrix tenius* and *Calothrix brevissima* has been reported.[44]

In general, little is known about the biochemical interaction between pesticides and BGA. The algae seem to be more resistant than other algal species to pesticides and most BGA are capable of tolerating the level of pesticides recommended for application in the field. Insecticides are generally less toxic to algae than other pesticides and may have a secondary beneficial effect of suppressing grazer population.

AVAILABILITY OF THE FIXED NITROGEN TO THE RICE PLANT

The process involved in the transfer of fixed nitrogen to the rice plant is not fully understood. It is assumed generally that the nutrients fixed by the BGA are made available to the crop through exudation, autolysis, and microbial decomposition.

Definite proof of nitrogen transfer from algae to higher plants has been obtained by Mayland and McIntosh[44] and Stewart,[45] using the isotope 15N. Renaut et al.[46] showed that part of the labelled molecular nitrogen assimilated by *Westiellopsis prolifica*, a common BGA in India paddy fields, is tranferred to rice plants. Specific rhizosphere-algae associations have been reported.[47]

A variety of compounds ranging from sugars and amino acids to growth regulators are known to be excreted copiously by BGA during their exponential growth phase.[48,49] Some biologically potent substances liberated by the BGA have been found to be similar to gibberellins.[23,50] Also, Vitamin B_{12} is liberated into the medium by *Cylindrospermem muscicola* and *Tolypothrix tenuis*.[51]

Exudation is not restricted to nitrogen-fixing BGA; other species such as *Oscillatoria* are known to liberate a variety of compounds into the medium. Some other effects of BGA on paddy fields, e.g., increasing phosphorous availability, decreasing sulfide injury, preventing the growth of weeds, etc., also contribute to their overall beneficial effect.

Under field conditions, only part of the exuded fixed nitrogen is available to the rice plant, some being either reincorporated by the microflora or volatilized. The susceptibility of BGA to decomposition and the amount of nutrients released depend on the physiological stage of the algae, the biodegradability of the algal cell, and the type of associated microflora. Some algae are decomposed within a few days, while others withstand microbial digestion for several weeks. It was demonstrated that a strain of *Bacillus subtilis* is able to degrade several nitrogen-fixing BGA very rapidly, thereby converting about 40% of the algal nitrogen into ammonia.[52]

In most of the cases the composition of the algal biomass is associated with soil desiccation at the end of the rice cultivation period. This has frequently resulted in an increase of the soil fertility and a residual effect on the standing rice crop. In a 5-year experiment De and

Sulaiman[52] found that during the first 3 years the yield of rice was not affected by the presence of BGA. In the following years, however, the yields in the presence of the algae were much higher than those in their absence. Soils with algal growth showed a considerable increase in nitrogen, while there was a loss of nitrogen in soils without algae. Similar results were reported by Watanabe,[53] who performed field experiments with *Tolypothrix tenuis* over several years. It was found that in the first year only 30% of the algae were decomposed, the rest remained as residual soil nitrogen. This slow release of fixed nitrogen seemed to be the reason for continued increase in yield of rice in successive years.

Algal-coated compost granules were prepared and tested for their performance under field conditions in India. The results indicate the practicality of this method. If this were to be taken as a commercial proposition, the algal granules would be required at 50 kg/ha.

In a cropping season of about 3 months, algae may reduce the pH of soil from 9.5 to 8.5. The physical properties of soils may also be greatly improved, peptides, polysaccharides, and lipids extracted by the algae improve soil aggregation thus affecting better water and air movement in the soil. The algae which are successful in sodic and saline soils are *Schizothrix* sp., *Porphyrosiphone notarisii*, *Scytonema* sp., and *Gloeocapsa* sp.

Venkataraman[55] and associates have worked on the effect of BGA in terms of energy subsidy to rice crop. When no chemical nitrogen is supplied to the soil, the additional energy to the crop due to the activity of BGA was reported to be 2.9×10^6 kcal/ha.

The energy compensation by the use of BGA at different levels of added urea N is about 30%. The energy contribution by BGA is impressive and may be up to 2.68×10^6 kcal/ha in terms of rice yield. As a technological assessment, for example, to obtain a grain yield of 3562 kg/ha the energy subsidy required in the form of urea-N would be 1.62×10^6 kcal/ha. The same yield of grain could be obtained by supplementing BGA and lowering the energy input to 1.08×10^6 kcal/ha.

It was shown that even if half the rice grown in India is treated with BGA, 8.55×10^{12} kcal as chemical nitrogen per season could be saved.

ALGALIZATION TECHNOLOGY

The starter culture of the inoculum is normally a soil-based mixture of *Aulosira*, *Tylopothrix*, *Scytonema*, *Nostoc*, *Anabaena*, and *Plectonema*. The idea of using a mixture of algae is to offset the ecological or edaphic dangers to any one particular strain in a given area. The only shortcoming of this system is that when several forms are used as inocula, the ultimate proportion of individual strains in the soil tends to be unpredictable. Under field conditions, this variation is less important, since the establishment of any efficient form will be satisfactory.[56] Different methods, four of algal multiplcation have been suggested: (1) trough or tank method, (2) pit method, (3) field method, and (4) nursery plus algal multiplication/propagation method.

The standard conditions for the propagation of BGA are shown in Table 5 and Figures 1 and 2. In the trough method, 8 to 10 kg soil mixed with 200 g of superphosphate and 10 g of BHC are placed in water-filled iron trays or cement tanks. Dried algal culture (400 g) is spinkled on the surface of the standing water and exposed to the sun. In this way, a single harvest of algae will normally yield about 1.5 to 2.0 kg of material. The process can be repeated until the entire soil in the trough is exhausted, which normally takes about 4 to 5 harvests.

The principle of the pit method is the same as the trough or tank method, the advantage being that the farmer can lay this unit in a corner of his field or courtyard. For this, shallow pits are dug in the ground and layered with polyethelene sheets to hold the water. Additions of sawdust or rice husk to the standing water has been found to enhance the production of algal biomass production.[2] Based on observations of such units in many villages in India, the cost of production has been estimated to be about $.50 (U.S.) for 10 kg algal material.[30,57]

Table 5
DETAILS FOR PREPARATION OF BGA FOR
APPLICATION TO RICE FIELDS

Attribute	Condition
Culture basin	Shallow trays or tanks made of galvanized iron or brick and mortar (2 × 1 × 0.25m)
Nutrients	Depending on the evaporation rate, the basins are filled with the following mixture: 5 kg soil, 100 g superphosphate, 1 g sodium molybdate suspended in 100 ℓ water; up to a depth of 5—15 cm.
Starter culture	400 g of soil-based mixture of *Aulospira, Tolypothrix, Scytonema, Nostoc, Anabaena,* and *Plectonema,* supplied by agricultural centers.
Aeration	Medium to be kept in open air after sprinkling of the starter culture on the surface of the standing water
pH	7.0—7.5; if acidic, to be corrected with lime
Light	Open sunlight
Retention time	7—8 days or until water dries up completely
Contamination control	To prevent mosquito breeding and other insects, Folidol (1 ppm), Parathion (7.5 ppm), Carbofuron (3% granules) or any other insecticide is added
Yield	1.5—2 kg of material/basin

FIGURE 1. Rural oriented open air method for producing BGA, showing the dried algal flasks ready for harvesting.

When commercial fertilizer nitrogen is to be used, the dose can be reduced by one third if properly supplemented with algae. The sun-dried algal material can be stored for long periods in a dry state without any loss in their viability. The dried algal material should be broadcast over the standing water in the field at the rate of 8 to 10 kg (soil based material from the production units) per hectare, 1 week after transplantation. Addition of excess algal

FIGURE 2.   BGA multiplication unit set up in villages of Tamil Nadu, India, consisting of cement tanks for raising algal inoculum.

material is not harmful for it will accelerate the multiplcation and establishment of the inoculum in the field. Algae should be applied for at least 3 consecutive seasons.

Farmers can also produce algae for their own requirements when they raise paddy nurseries. If 320 m² are allotted to the nursery, an additional 40 m² can be attached for algal production. By the time the seedlings are ready for transplantation, an amount of 15 to 20 kg algal material, sufficient for 3 to 4 acres of paddy field, will also become available.

The desirable characteristics of BGA strains suitable for field inoculation have been summerized by Stewart et al.[33] as follows: "Strains selected for use in the field should be fast growing and capable of fixing N under aerobic, microaerobic and anaerobic conditions. They should also be able to grow photoautotrophically and chemoheterotrophically and store endogenous carbohydrate reserves. They should evolve little H_2 and liberate nitrogen in excess of their requirements for optimum growth. Cyanobacteria differ in whether they liberate extracellular NH_4 and on inhibition of glutamine synthetase. The way in which glutamine synthetase of cyanobacteria is regulated could be of importance in strain selection".

NITROGEN FIXATION IN SYMBIOSES INVOLVING BGA: *AZOLLA*

There is a great variety of symbiotic associations formed between BGA and other plants or even animals. In some of these cases, the BGA are of a proven nitrogen-fixing species, and in many instances there is direct evidence of nitrogen fixation by the intact symbiotic system, e.g., by the lichens *Collema* spp., *Liptoqium lichenoides, Peltigera* spp., *Stereocaulon* sp., and *Lichina* spp. by liverworts and algal symbionts such as *Blasia* and *Cavicularia* and by the water fern *Azolla*. Physiological studies showed that the metabolism of nitrogen-fixing BGA is greatly modified by symbiosis. The *Nostoc* content of *Peltigera cania* was estimated as 2.7% of the total thallus nitrogen. The nitrogenase activity of the symbiotic *Nostoc* on this basis appeared to be 2 or 3 times that of free-living alga.

Distributed all over the world, the water fern *Azolla* occurs in warm temperatures and tropical regions and has long been used by Chinese farmers as fertilizer. The importance of this as a source of nitrogen and green manure in rice fields is of interest in recent time.

Reports are available of its present use in China, India, Vietnam, Indonesia, the Philippines, and the U.S. Variations of *Azolla* strains occur in different areas: *A. pinnata* is most commonly used in Southeast Asia, *A. imbricate* is used in Vietnam, and *A. mexicana* and *A. feliculoides* are used in the U.S.

Azolla assimilates atmospheric nitrogen in association with the nitrogen-fixing BGA *Anabaena azolla,* which lives in the cavities of the fern in the upper lobes as symbiont.[58] There are six *Azolla* species: *A. feliculoides, A. caroliniana, A. mexicana, A. microphylla, A. pinnata,* and *A. nilotica,* all of which are associated with nitrogen-fixing BGA.

The use of *Azolla* as fertilizer depends on its ability to grow rapidly with a doubling period of 1 to 3 days under optimum conditions and on its capacity to fix large quantities of nitrogen. Its high organic content also benefits the soil in which it grows. Studies have shown that the increase in rice yield by one crop of *Azolla* is equivalent to that obtained by applying 30 kg of nitrogen in the form of urea.[59]

The optimum physiological and environmental conditions for *Azolla* vary with strains. The red *A. feliculoides* used in China and *A. mexicana* can withstand low temperatures. *A. pinnata,* commonly found in India has a temperature optimum of 20 to 30°C and the growth is inhibited above 30 and below 15°C. The optimum pH for the growth in soil is 5.7, while that of the isolated alga is about 7.5 to 8.5.

Azolla may either be grown in rotation with rice, preferring sheltered and shaded habitats, and ploughed in prior to rice planting, or as in China, where it is grown together with rice in alternate rows.[60-63] It is inoculated at the rate of 0.6 to 1 kg/m². About 10 t to wet material or 1 ton of dry material is recommended for 1 ha of land. In areas where availability of water limits the maintenance of *Azolla* nurseries, the fern can be introduced into the transplanted field at the rate of 200 to 1000 kg/ha, which will grow and cover the field in 20 to 40 days. The water in the fields is drained off and the *Azolla* is incorporated into the soil manually or with the help of a rotary hoe.

An essential requirement for *Azolla* is added phosphate and the maintenance of 5 to 10 cm of standing water in the field. Growth of inoculum has to be started 1 month before planting to obtain adequate material. For this purpose, the bunded fields are ploughed, levelled, irrigated, and then fresh *Azolla* is spread over the standing water along with 4 to 8 kg of P_2O_5.

The positive effects of *Azolla* as a fertilizer for rice has been well documented by several investigators[64-66] (Table 6). When twice inoculated, *Azolla* affected a significant increase in the yield of rice equivalent to that obtained by incorporating 25 kg nitrogen fertilizer per hectare. A thick layer of *Azolla* provides 30 to 35 kg N/ha for rice crops per season (Table 7).[66]

For all its merits, the technology of *Azolla* production has certain limitations. In the northern states of India where rice is grown during the Kharif season, *Azolla* is to be multiplied during the middle of May to the first week of June in order to make it effective for use in the wet season. Unfortunately, the growth of *Azolla* is adversely affected by high temperatures. Therefore, its multiplication in open fields during hot summer months is rather difficult in large parts of India and in order to circumvent these obstacles one has to select temperature and light tolerant strains.

The floating *Azolla* has the tendency to drift in the fields due to wind. Therefore, it is necessary to divide larger fields into smaller ones which will prevent functional and operational problems in rural areas. Also, the heavy growth of *Azolla* at early stages might hamper the seedling growth and tiller production. Above all, *Azolla* is essentially a weed and its uncontrolled spread in irrigation channels and tanks may carry negative implications.

EFFECTS OF BGA ON RICE GRAIN YIELD

The annual reports of the All India Coordinated Project on Algae in India as well as many

Table 6
GRAIN YIELD OF RICE (KG/HA) AS INFLUENCED BY
BGA AND *AZOLLA* WITH AND WITHOUT ADDITION OF
NITROGEN FERTILIZER[17]

Treatment	Location					
	A	B[a]	C	D	E[b]	F
Control	3,171	1,021	4,856	3,233	828	5,620
25 kg N/ha	3,588	1,200	5,236	4,067	1,269	6,239
50kg N/ha	4,028	1,401	5,279	4,300	1,564	6,624
75kg N/ha	3,426	1,537	5,659	5,400	2,018	7,137
Azolla (5 t/ha)	3,449	1,158	5,025	3,433	1,431	5,855
BGA (10 kg/ha)	3,542	1,149	4,867	4,000	1,338	6,026
25 kg N/ha + *Azolla*	3,704	1,390	5,195	4,200	1,711	6,944
25 kg N/ha + BGA	4,005	1,341	4,856	5,400	1,572	7,008

[a] Crop effects by gallmidge.
[b] Delayed transplantation.

Table 7
EFFECT OF *AZOLLA* ON RICE CROP (COIMBATORE, INDIA)

Treatment	Spacing of seedlings (cm × cm)	Grain yield (kg/ha)	Increase over control (kg/ha)
No N (control)		2,747	—
60 kg N/ha	20 × 20	3,811	1,064
60 kg N/ha	10 × 40	3,755	1,008
Azolla 2 t/ha + 30 kg N/ha	20 × 20	3,455	708
Azolla 2 t/ha + 30 kg N/ha	10 × 20	3,566	819
Azolla as green manure, dual crop	20 × 20	3,710	963
Azolla as green manure, dual crop	10 × 40	3,110	363

others[18,66-69] have recorded extensive data to document the positive role of BGA application of 20 to 30 kg N/ha per cropping season. If commercial nitrogen fertilizers are used, the dose can be reduced by 25% by supplementing it with algae. Even with the application of high levels of nitrogen fertilizer, the yield per unit input can be increased through an algal complement (Table 8).

Similar results were reported from China. In 1976, 32,700 kg of fresh *Anabaena* strains were produced, enough to inoculate 67 ha and to develop into 500 tons in the late paddy fields. The increase in grain yield was about 10 to 15%.[71] In 1978, a total of 20,000 ha rice paddy fields were inoculated with nitrogen-fixing algae. The effect of algal biofertilizer on the rice plants could be seen from the increases of effective tillering and average number of crop grains of each tiller. From the data obtained in experimental paddy fields, the average increase in grain yield was 450 kg/ha, when the biomass of fresh algae was 7500 kg/ha.

When BGA were inoculated into the rice seedling nursery beds, sturdy rice seedlings were obtained. The dry weight of seedling was increased and the root system was well developed. This resulted in the increases of both plant growth and grain yield (about 10%).

Sizable portions of land the world over are unusable for agriculture either because of high alkali content or salinity. Scientists in India have worked out a practical procedure of reclaiming such lands with BGA.[71] Kaushik et al.[72] have shown that selected species of BGA can thrive well in such soils, where no vegetation can grow. About 25 strains of BGA

Table 8
EFFECT OF BGA APPLICATION
(10 kg/ha) ON RICE YIELD[30]

Treatment	Yield (kg/ha)
Addition of N (kg/ha)	
0	3,750
0 + BGA	3,942
20	4,038
20 + BGA	4,519
30	4,326
30 + BGA	4,807
40	4,519
40 + BGA	4,807
50	4,615
50 + BGA	5,000
Addition of N:P:K (kg/ha)	
0:25:25	3,476
0:25:25 + BGA	3,918
25:25:25	3,902
25:25:25 + BGA	4,528
50:50:50	4,732
50:50:50 + BGA	5,384

collected from all over India are available for field application. Thomas[73] observed a decrease of 25 to 30% in salinity due to repeated cultivation of *Anabaena* in saline-rich soil.

PROSPECTS

The practical goal at present is to train farmers in different areas to produce algal material under the technical supervision of the proper agencies and help them in marketing the produce for large-scale cultivation. Under rural reconstruction programs, village-level cooperatives can be organized for the production and marketing of BGA. Concerted efforts should also be made to create a consumer demand through field level workers and mass media. Promotional information regarding hoarding the gains of BGA application to rice in the rural areas of India is shown in Figure 3.

The application of BGA as an algal biofertilizer to improve the agriculture economy should be supported by a strong and well-organized extension program. This necessitates generation of trained personnel who can disseminate the know-how to farmers. About 5 years ago, the blue-green algal technology was taken to the farm fields in several states of India and the response has been very encouraging.

Large-scale algal inoculum production at state seed farms and farmer training centers have already been initiated by the Agricultural Departments of the Indian states of Tamil Nadu and Uttar Pradesh. Several Asian countries such as China, Burma, Vietnam, Thailand, and the Philippines have also begun work to popularize this practice.

The ultimate success of BGA technology will depend on the cooperation between the researchers, extension personnel, and the ultimate users — the farmers. The present trend and interest seem to show that the application of BGA in rice cultivation is already well on the way to successful exploitation.

As a general conclusion, it can be stated that BGA play a beneficial role in the fertility of paddy fields. Since the ecology of the algae and their mode of action with respect to the rice plant are still not sufficiently understood, long-term field experiments are needed to determine the mechanisms of algal succession, the limiting factors for the growth of the algae, and the agricultural practices favoring the occurrence and growth of nitrogen-fixing BGA.

FIGURE 3. Promotional hoarding (in local language) on the gains of BGA application to rice, displayed in rural areas of Tamil Nadu, India.

ACKNOWLEDGMENT

I wish to express my gratitude to several researchers in India and abroad working in the field of algal biofertilizer who readily responded to my request for reprints and other valuable material needed for the preparation of this manuscript. These include G. S. Venkataraman, S. K. Goyal, and B. D. Koushik of IARI, New Delhi, J. Thomas of BARC, Bombay, S. Kannaiyan of Coimbatore, S. P. Singh of Varanasi, and P. R. Roger of IRRI, Los Banos, and F. Majid of BCSIR, Dacca.

I also wish to thank my colleagues at the CFTRI, Mysore M. Anusuya Devi and T. Shanmugasundaram, for helping in the preparation of the paper, the late B. A. Sathyatarayana Rao and K. M. Dastur for editing the manuscript, and C. V. Venkatesh for secretarial help.

REFERENCES

1. **Venkataraman, G. S.,** Bluegreen algae for rice production, *Soil Bull. FAO,* 46, 1981.
2. **Venkataraman, G. S.,** Algal biofertilizers. Potential and problems, in *Proc. Natl. Work. Alg. Syst. Madras, India,* Sechadre, C. V. Ed., Indian Society of Biotechnology, IIT, New Delhi, 1980, 1.
3. **Surajit, K. and De Dutta,** *Rice Production Principles and Practices,* John Wiley & Sons, New York, 1981.
4. **Venkataraman, G. S.,** *Algal Biofertilizer for Rice,* Indian Agricultural Research Institute, New Delhi, 1980.
5. **Venkataraman, G. S.,** Algal inoculation of rice fields, in *Nitrogen and Rice,* Watanabe, I., Ed., International Rice Research Institute, Los Banos, Philippines, 1980, 311.
6. **Watanabe, I., Lee, K. K., Alimagno, B. V., Sato, D. C., Del Rosario, D. C., and De Guzman, M. R.,** Biological N_2 fixation in paddy field studies by *in situ* acatylene reduction assays, *IRRI Res. Pap. Ser.,* 3, 1, 1977.
7. **Yamaguchi, M.,** Biological nitrogen fixation in flooded rice fields, in *Nitrogen and Rice,* Watanabe, I., Ed., International Rice Research Institute, Los Banos, Philippines, 193, 1980.

8. **De, P. K.**, The role of bluegreen algae in nitrogen fixation in rice fields, *Proc. R. Soc. London Ser. B:*, 127, 121, 1939.
9. **Singh, R. N.**, The fixation of elementary nitrogen by some of the commonest bluegreen algae from the paddy field soils of the United Provinces and Bihar, *Indian J. Agric. Sci.*, 2, 743, 1942.
10. **Watanabe, A., Nishigaki, S., and Konishi, O.**, Effect of nitrogen fixing bluegreen algae on the growth of rice plants, *Nature (London)*, 168, 748, 1951.
11. **Li, S.**, Nitrogen-fixing blue-green algae, a source of biofertilizer, in *Proc. Joint China-U.S. Phycology Symp.*, 479, 1981.
12. **Watanabe, A.**, Distribution of nitrogen fixing blue-green algae in various areas of South and East Asia, *J. Gen. Appl. Microbiol.*, 5, 21, 1959.
13. **Watanabe, A. and Yamamoto, Y.**, Algal nitrogen fixation in the tropics, *Plant Soil*, (Special ed.) 403, 1971.
14. **Venkataraman, G. S.**, The role of bluegreen algae in tropical rice cultivation, in *Nitrogen Fixation by Free-Living Microorganisms*, Stewart, W. D. P., Ed., Cambridge University Press, England, 1975, 207.
15. **Bothe, H.**, Nitrogen fixation, in *Botanical Monographs, Vol. 19, The Biology of Cyanobacteria*, Carr, N. G. and Whitton, B. A., Eds., Blackwell Scientific, Oxford, 1982, 87.
16. **Wolk, C. P.**, Heterocysts, in *Botanical Monographs, Vol. 19, The Biology of Cyanobacteria*, Carr, N. G. and Whitton, B. A., Eds., Blackwell Scientific, Oxford, 1982, 97.
17. **Roger, P. A. and Kulasooriya, S. A.**, *Blue-Green Algae and Rice*, International Rice Research Institute, Los Banos, Philippines, 1980.
18. **Roger, P. A. and Reynaud, P. A.**, Ecology of bluegreen algae in paddy fields, in *Nitrogen and Rice*, Watanabe I., Ed., International Rice Research Institute, Los Banos, Philippines, 289, 1979.
19. **Roger, P. and Reynaud, P.**, La biomasse algale dans les rizieres du Senegal: importance relative des Cyanophycees fixatrice de N_2, *Rev. Ecol. Biol. Soil*, 14, 519, 530.
20. **Traore, T. M., Roger, P. A., Reynaud, P. A., and Sasson, A.**, Etude de la fixation de N_2 par les cyanobacteries dans une riziere du Mali, *Cahiers Orstom Ser. Biol.*, 13, 181, 1978.
21. **Singh, P. K.**, Algal inoculation and its growth in waterlogged rice fields, *Phycos*, 15, 5, 1976.
22. **Gupta, A. B. and Shukla, A. C.**, Effect of algal extracts of *Phormidium sp.* on growth and development of rice seedlings, *Hydrobiologia*, 34, 77, 1969.
23. **Ciferri, R.**, Le alghe delle acque dolci pavesi e vervellesi e le loro associazioni neele, risaie, *Riso*, 12, 30, 1963.
24. **Singh, P. K.**, Occurrence and distribution of cyanophages in ponds, sewage and rice fields, *Arch. Mikrobiol.*, 89, 169, 1973.
25. **Venkataraman, G. S.**, The role of blue-green algae in agriculture, *Sci. Cult.*, 27, 9, 1961.
26. **Watanabe, A., Ito, R., and Sasa, T.**, Microalgae as a source of nutrients for daphnids, *J. Gen. Appl. Microbiol.*, 1, 137, 1955.
27. **Prasad, B. N., Mehrotra, R. K., and Singh, Y.**, On pH tolerance of some bluegreen algae, *Acta Bot. India*, 6, 130, 1978.
28. **Okuda, A. and Yamaguchi, M.**, Algae and atmospheric nitrogen fixation in paddy soils. II. Relation between the growth of blue-green algae and physical or chemical properties of soil and effect of soil treatments and inoculation on the nitrogen fixation, *Mem. Res. Inst. Food Sci.*, 4, 1, 1952.
29. **Venkataraman, G. S., Ed.**, All India Coordinated Project on Algae Annual Report, Vol. 3, Department of Science and Technology New Delhi, India, 1978.
30. **Srinivasan, S.**, Blue-green algae for paddy, *Aduthurai Rep.*, 1, 29, 1977.
31. **Mahapatra, I. C., Mudholkar, N. J., and Patnaik, H.**, Effect of N, P and K fertilizers on the prevalence of algae under field conditions, *Indian J. Agron.*, 16, 19, 1971.
32. **Arora, S. K.**, Effect of nitrogen compounds on the nitrogen metabolization during the growth of blue-green algae, *Riso*, 19, 217, 1970.
33. **Stewart, W. D. P., Rowel, P., Ladha, J. K., and Sampaio, M. J. A.**, Blue-green algae (Cyanobacteria) — some aspects related to their role as source of fixed nitrogen in paddy soils, in *Nitrogen and Rice*, Watanabe, I., Ed., International Rice Research Institute, Los Banos, Philippines, 1979, 263.
34. **Subramanyan, R., Manna, G. B., and Patnaik, S.**, Preliminary observations on the interaction of different rice types to inoculation of blue-green algae in relation to rice culture, *Proc. Indian Acad. Sci.*, 62B, 171, 1965.
35. **Kikuchi, E., Furusaka, C., and Kurihara, Y.**, Survey of the fauna and flora in the water and soils of paddy fields. The Reports of the Institute for Agricultural Research, Tohoku University, 26, 35, 1975.
36. **Marathe, K. V.**, A study of the effects of green manures on the subterranean algal flora of paddy field soils, *J. Biol. Sci.*, 7, 1, 1964.
37. **Lehri, L. K. and Mehrotra, C. L.**, Note on the effect of artificial inoculation of *Aulosira fertilissima* in rice crops, *Ind. J. Agric. Sci.*, 40, 65, 1970.
38. **Jutono**, Blue-green algae in rice soils of Jogjakarta, Central Java, *Soil Biol. Biochem.*, 5, 91, 1973.

39. **Gangawane, L. V. and Saler, R. S.,** Tolerance of certain fungicides by nitrogen-fixing blue-green algae, *Curr. Sci.,* 8, 306, 1979.

40. **Wang, L., Liu, F., Ye, J., Cui, S., Li, S., Zeng, J., You, J., and Shi, Y.,** Effect of some fungicides and insecticides on the growth of nitrogen-fixing blue-green algae, *Acta Hydrobiol. Sin.,* 4, 461, 1959.

41. **Watanabe, A.,** The blue-green algae as the nitrogen fixers, *Int. Cong. Microbiol.,* Moscow, 66, Symp. C 77, 1967.

42. **Raghava Reddy, H. R.,** Studies on the Influence of pH, Phosphate and Pesticides on Nitrogen Fixation and Radio Carbon Assimilation by Two Blue-Green Algae, M.Sc. thesis, University of the Agricultural Sciences, Bangalore, India, 1976.

43. **Ibrahim, A. A.,** Effect of certain herbicides on growth of nitrogen-fixing algae and rice plants, *Symp. Biol. Hung.,* 11, 445, 1972.

44. **Mayland, H. F. and McIntosh, T. H.,** Availability of biologically fixed N^{15} to higher plants. *Nature (London),* 209, 421, 1966.

45. **Stewart, W. D. P.,** Transfer of biologically fixed nitrogen in a sand dune slack region, *Nature (London),* 214, 603, 1967.

46. **Renaut, J., Sasson, A., Pearson, H. W., and Stewart, W. D. P.,** Nitrogen fixing algae in Morocco, in *Nitrogen Fixation by Free-living Microorganisms,* Stewart, W. D. P., Ed., Cambridge Free-living Microorganisms, Stewart, W. D. P., Ed., Cambridge University Press, England, 1975, 229.

47. **Watanabe, A.,** Collection and cultivation of nitrogen-fixing blue-green algae and their effect on the growth and crop yield of rice plants, *Stud. Rokugawa Inst. Tokyo,* 9, 162, 1961.

48. **Venkataraman, G. S. and Neelakanthan, S.,** Effect of cellular constituents of the nitrogen-fixing blue-green algae *Cylindrospermum muscicola* on the root growth of the rice seedlings, *J. Gen. Appl. Microbiol.,* 13, 53, 1967.

49. **Singh, V. P. and Trehan, T.,** Effect of extracellular products of *Aulosira fertilissima* on the growth of rice seedlings, *Plant Soil,* 38, 457, 1973.

50. **Okuda, A. and Yamaguchi, M.,** Nitrogen fixing microorganisms in paddy soils. VI. Vitamin B_{12} activity in nitrogen-fixing blue-green algae, *Soil Plant Food (Tokyo),* 6, 76, 1960.

51. **Watanabe, A. and Kiyohara, T.,** Decomposition of blue-green algae as affected by the action of soil bacteria, *J. Gen. Appl. Microbiol.,* 5, 175, 1960.

52. **De, K. P. and Sulaiman, M.,** The influence of algal growth in the rice fields on the yield of crops, *Indian J. Agric. Sci.,* 20, 327, 1950.

53. **Watanabe, A.,** On the effect of the atmospheric nitrogen-fixing blue-green algae on the yield of rice, *Bot. Mag.,* 69, 530, 1956.

54. **Venkataraman, G. S.,** Blue-green algae as a biological nitrogen input in rice cultivation, in *Proc. Natl. Symp. Nitrogen Assimilation and Crop Productivity,* Hissar, India, 132, 1977.

55. **Venkataraman, G. S.,** *Algal Biofertilizers and Rice Cultivation,* Today and Tomorrows Printers, Faridabad, India, 1972.

56. **Venkataraman, G. S., Ed.,** All India Coordinated Project on Algae Annual Report, Vol. 2, Department of Science and Technology New Delhi, India, 1977.

57. **Moore, A.,** *Azolla* biology and agronomic significance, *Bot. Rev.,* 35, 17, 1969.

58. **Kananaiyan, S.,** Azolla in rice. Training on multiplication and use of *Azolla* biofertilizer for rice production, Tamil Nadu Agric. Univ. Coimbatore, India, 1982.

59. **Lumpskin, T. A. and Plucknett, D. L.,** *Azolla* botany, physiology and use as green manure, *Econ. Bot.,* 34, 111, 1980.

60. **Lin, C.,** Use of *Azolla* in rice production in China, in *Nitrogen and Rice,* International Rice Research Institute, Los Banos, Philippines, 1971, 375.

61. **Tran, Q. T. and Dao, T. T.,** *Azolla* as a green manure. Reitranise studies, *Agric. Prob.,* 4, 119, 1975.

62. **Venkataraman, A.,** Propagation of *Azolla* in China, *Azolla* a Biofertilizer, Tamil Nadu Agric. Univ. Coimbatore, India, 1980.

63. **Singh, P. K.,** *Azolla* plants as fertilizer and feed. Subject matter training cum discussion on use of *Azolla*, Mandya, India, 1980.

64. **Watanabe, A.,** *Azolla* utilization in rice culture, *Int. Rice Res. New Lett.,* 2, 3, 1977.

65. **Govindarajan, K., Kananaiyan, S., and Ramachandran, M.,** Effect of *Azolla* inoculation on rice, in *Proc. 20th Ann. Microbiol. Conf.* Hissur, India, 60, 1979.

66. **Venkataraman, G. S., Ed.,** All India Coordinated Project on Algae Annual Report Vol. 1, Department of Science and Technology, New Delhi, India, 1976.

67. **Venkataraman, G. S., Ed.,** All India Coordinated Project on Algae Annual Report, Vol. 3, Department of Science and Technology, New Delhi, India, 1979.

68. **Venkataraman, G. S., Ed.,** All India Coordinated Project on Algae Annual Report, Vol. 4, Department of Science and Technology, New Delhi, India, 1980.

69. **Venkataraman, G. S., Ed.,** All India Coordinated Project on Algae, Annual Report, Vol. 5, Department of Science and Technology, New Delhi, India, 1981.

70. **Li, S.,** Dinitrogen-fixing blue-green algae and their role in crop yield of rice, in *GIAM VI, Global Impacts of Applied Microbiology*, Academic Press, London, 1981, 287.

71. **Singh, R. N.,** Reclamation of ''Usar'' lands in India through blue-green algae, *Nature (London)*, 165, 325, 1950.

72. **Kaushik, B. D., Krishnamurthy, G. S. R., and Venkataraman, G. S.,** Influence of blue-green algae on saline-alkaline soils, *Sci. Cult.*, 47, 169, 1981.

73. **Thomas, J.,** Isotopes in biological nitrogen fixation, paper presented at Int. Atomic Energy Agency, Proc. Adv. Group Meet., FAO/IAEA, Vienna, 1977.

74. **Kananaiyan, S.,** Studies on the effect of Azolla application of rice crop, in *Proc. Int. Rice Res. conf.*, International Rice Research Institute, Los Banos, Philippines, 1983, 1.

ECONOMIC ASPECTS OF THE MANAGEMENT OF ALGAL PRODUCTION

Yacov Tsur and Eithan Hochman

INTRODUCTION

This chapter concerns the economic aspects associated with the development of the biotechnology of algal mass culture. Economic considerations have a major role in the chain of events that lead from the first research ideas and lab experiments up to the development of a self-sustained industry. Throughout the process there is an interaction between economic factors, such as prices of inputs and outputs as well as cost of research, and the various biological factors.

On the macroeconomic level the process includes the introduction of a new agro-technology for an agricultural crop. In our case, algal biomass production may represent an appropriate technology to support settlement in arid zones (see Preface). The climatic and environmental conditions combined with poor quality (saline) of water may bestow an advantage to such an industry, serving as a classic example of the induced innovations theory developed by Binswanger and Ruttan.[1] This theory claims that research directions in different locations are influenced by the specific constraints, the economic value of the various resources being determined by their relative scarcity.

The carriers of technological changes can be divided into three groups: the individual growers, the researchers, and the state and public institutions. While the first two supply the most important elements of human factors, the last group mainly supplies the budget and resources, as well as extension services. The Schumpeterian system emphasizes the importance of leadership.[2] Schumpeter placed great emphasis on the function of the entrepreneur whose initiative is the main factor in introducing the new technologies. He perceived three stages: (1) invention, (2) innovation, and (3) imitation, which correspond to basic, applicative, and commercial research and development of the product. Economic considerations, i.e., the pursuit of profits by the entrepreneurs motivate all three stages.

On the micro level the importance of the economic factors becomes crucial. The conventional approach of cost-benefit analysis is now a known method used in feasibility studies of new products. The calculations are however, quite often erroneous and misleading. A common mistake in such an analysis is the confusion between the use of average quantities vs. marginal quantities. Thus, to base the decision of how much algae to produce on the average cost per kilogram of algae may be wrong. The cost of producing the last kilogram is the relevant information that determines the quantity produced. Another error which is of particular importance in our case stems from ignoring the role of time in economic analysis. As we shall see, time enters both the production and the decision processes. While the conventional production function concept is static and deals with the transformation of inputs into outputs, in our case growth processes of the algal biomass are a part of the production process and require, therefore, a different approach to production analysis. Also, the fact that management and production of algal biomass are carried over time imposes a dynamic framework on the decision-making process, a subject which recently received attention.[3-5]

This work focuses mainly on management aspects of the current production of algal biomass. Thus, it provides the potential producer with the tools required to make economically sound decisions, e.g., whether to enter the industry, what algal product to produce and how to determine the preferred production policy.

We shall not discuss issues such as the impact of aggregate demand and supply on algal product prices, specific investment cost structures and their interaction with current management decisions, and welfare economic aspects of introducing a new industry into a developing region.

In the following section, a model is constructed in order to derive optimal management of a growing inventory with particular reference to algal biomass production. In the section following that, an empirical application is discussed.

OPTIMAL DECISION RULES

Although the main concern is with algal biomass, many of the ideas presented in this chapter are quite general and can be applied to any other form of renewable resource such as fish (in ponds or sea), woods, or livestock inventory.

Consider an inventory of biomass, denoted by $B(t)$, which grows over time at a rate $dB/dt(^{def}\dot{B})$, in which 3 sets of factors affect growth:[6,7] (1) ecological and environmental factors such as the intensity of solar irradiance and diurnal and seasonal fluctuations in temperature; (2) managerial factors such as the quality of growth facilities, supplementary materials that accelerate growth, the amount of energy applied in mixing the number of replacements of the growth medium, placement of cover on the pond in winter, etc.; (3) maintenance of the population density (B).

The terms "growth process" and "production process" are used interchangeably. A given set of ecological and environmental conditions are denoted as the production environment. A given set of managerial factors which are of a long run nature and involve fixed cost investment (e.g., type of ponds) is denoted by production technology, whereas short run managerial factors are termed production policy. A given set of production environment, technology, and policy is denoted by production regime. In a given production regime growth is a function of the population density and can be written implicitly as

$$\dot{B} = G(B) \tag{1}$$

The growth function $G(B)$ is generally assumed to be bell-shaped, reflecting the fact that as B increases some growing factors become limited, thereby supressing growth. A well-known example of $G(B)$ is the logistic growth curve

$$\dot{B} = G(B) = \rho B(1 - B/M) \tag{2}$$

where ρ is a growth parameter and M is the maximum population density that allows positive growth (Figure 1).

For a given initial population density B_o, Equation 1 provides the state of the system at any time instant t. Given the population density level, $B(t)$, the grower has to make the following decision: to harvest one (marginal) biomass unit now or to let it grow and harvest next period. The chosen policy is the one that provides the higher gains. Let P indicate the price of the product and assume, for a while, costless production, harvesting, and processing. If the decision is to harvest now, the gain equals P. If harvesting is postponed until next period, the discounted gains equal

$$P(1 + \partial G/\partial B)/(1 + r) \tag{3}$$

where r is the interest rate. This is because the left in unit of biomass is responsible for the growth of $\partial G/\partial B$ out of the total growth at that period.

Having found that gains involved for each policy the optimal decision should be made as follows:

$$\text{decision} = \begin{cases} \text{harvest now} & \text{if } P \geq P(1 + \partial G/\partial B)/(1 + r) \\ \text{harvest next period} & \text{otherwise} \end{cases}$$

or equivalently

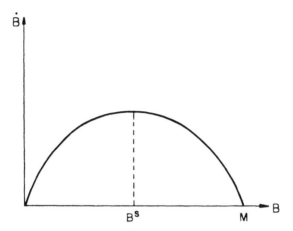

FIGURE 1. Logistic growth curve. $\dot{B} = dB/dt$, B = algal population density; B^s = maximum sustainable growth; M = maximum population density in which positive growth occurs.

$$\text{decision} = \begin{cases} \text{harvest now} & \text{if } r \geq \partial G/\partial B \\ \text{harvest next period} & \text{otherwise} \end{cases}$$

Of course, the grower faces the same decision problem at the next period and so on. Hence the two-period solution, in fact, characterizes the complete solution which can be described as follows: at the beginning of the growing process B is small and any reasonable growth process (e.g., the logistic) admits large $G'(b)$ ($^{\text{def}}\partial G/\partial B$). Hence, $G'(B) > r$ and the optimal decision is not to harvest. As time progresses, population density increases and growth factors become limiting, suppressing growth until at some time instant t^r

$$\partial G/\partial B = r \tag{4}$$

From there on the optimal harvesting decision is obviously to keep population density at a level that satisfies the condition of Equation 4.

Figure 2 describes the above situation. At population level B^s growth is maximal. Hence B^s is generally referred to as Maximum Sustainable Growth (MSG) level. At that level $\partial \dot{B}/\partial B = O$ and the B vs. time curve passes through an inflection point. With interest rate equal to r_1 optimal biomass level (i.e., the level of B that equate r_1 to $\partial G/\partial B$) is B^{r_2}. In the same manner B^{r_2} is optimal under r_2. Note that for any positive interest rate the optimal biomass level, B^r, is smaller than the MSG level B^s. Only at zero interest rate will B^s be economically optimal. Note also that if any growth factor changes over time independent of biomass level, then the optimal biomass level will change over time as well. To see this, assume $\dot{B} = G(B,t) = F(B)g(t)$ where $g(t)$ reflects factors that suppress growth as a result of aging independent of biomass level B. Hence $0 < g(t) < 1$ and $g(t)$ decreases with time. Optimality condition Equation 4 becomes

$$F'(B) = r/g(t) \tag{5}$$

With fixed interest rate r, $r/g(t)$ increases over time, hence optimal level must be decreased in order to increase $F'(B)$ and keep the equality above valid.

Let us now relax the "costless" assumption and see how the introduction of cost affects optimal behavior. Two general types of cost exist: fixed and operating costs. The first is assumed given (in the short run) and does not affect operating behavior. Operating costs in algal production involves direct production cost (such as supplementing materials, inducing

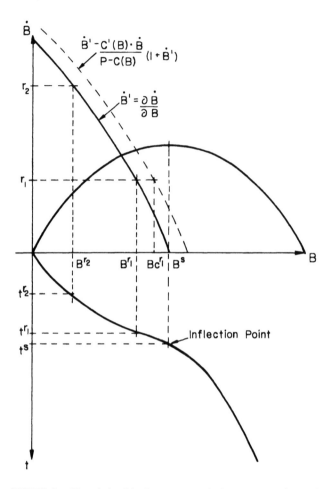

FIGURE 2.   The relationships between growth, interest rate, price, and cost in determining management decisions. B, $\dot{B}$, B^s as in Figure 1; r = interest rate; B^r = optimum population density corresponding to interest rate r; $B_c{}^r$ = B^r, including operating cost; t = time; t^s = time required to reach B^s; t^r = time required to reach B^r; C = unit operating cost.

water flow), harvesting cost, and drying and processing cost. Different algal products (e.g., human health food, animal food) entail different production, harvesting, drying, and processing methods and hence a different structure of operating cost and different product price as well. Let us postpone the discussion of these issues to the next section, where an empirical example will be given. At this point we prefer to speak generally and let C(B) indicate unit operating cost. The two-period decision scenario depicted above is slightly changed. The gains from harvesting one (marginal) unit of biomass now is P − C(B). The gains from waiting and harvesting next period is comprised of two parts: the left in biomass unit have produced G′(B) during that period which amount to a value of [P − C(B)] (1 + G′(B)); the change in cost due to total growth at that period is approximately C′(B).G(B) (C′(B)^{def}C/B).

The discounted gains from waiting is then given by:

$$1/(1 + r) [P − C(B)] (1 + G'(B)) − C'(B)G(B)$$

Optimal behavior requires that both gains be equal. This provides the following optimality conditions:

$$G'(B) - \frac{C'(B)G(B)}{P - C(B)} = r \qquad (6)$$

If $C'(B) = 0$, i.e., if unit operating costs are independent of biomass level, then Equation 6 reduces to Equation 4, the costless case. If $C'(B) < 0$ then the left hand side of Equation 6 increased and $G'(B)$ must decrease in order to keep the equation valid, which implies (under $G''(B) < 0$) increasing optimal biomass level. In other words, if unit operating costs decrease with biomass level, then the incentive to leave a biomass unit another period is enhanced since it involves, in addition to growth, the effect on reduced operating costs. The same idea works in the opposite direction if $C'(B) > 0$. A rigorous treatment of the optimization problem is given in the Appendix.

The above problem of managing renewable inventory is similar to the classical capital investment problem in which some production process involves capital input, K. Assume no deterioration and let $F(K)$ denote the net productivity of capital which is the rate of return from a capital stock of size K. The investor wishes to decide whether to invest and increase K by an additional (marginal) unit. An investment decision will provide a return of $F'(K)$ for each dollar invested in buying the additional (marginal) unit. On the other hand the investor could put the money in the bank and receive r, the interest rate, as a return. Obviously, the decision is to invest as long as $F'(K) > r$. As the investment takes place, K increase and, generally, its efficiency decreases (each additional unit of K contributes less than the previous one to the production process) that is

$$F''(K) < 0 \qquad (F''(K) = \partial^2 F/\partial K^2) \qquad (7)$$

In economic terms we assume decreasing marginal productivity of K. Eventually for some level of K

$$F'(K) = r \qquad (8)$$

and at that point it does not pay to invest any more and the capital stock level is determined by this condition.

Returning to our renewable biomass managing problem, the optimality condition (Equation 6) identifies three basic components:

1. The biological growth process (Equation 1,which in economic terms is the production process.
2. The cost structure of (a) harvesting and drying technologies, and (b) production technologies. Each of these costs involves fixed cost (infrastructure, ponds, harvesting facilities, etc.) and variable costs (nutrients, labor, energy, etc.).
3. Market considerations as represented by P, the product price, and by prices of inputs to the production process.

The economic feasibility of producing a given biomass is a consequence of a combination of these three components. The first two are highly affected by managerial and human factors and we shall dwell on these issues. The third component is exogenous to the system assuming competitive markets.

The next section is devoted to a discussion of these components in relation to algal biomass production accompanied by an empirical example.

AN EMPIRICAL APPLICATION

A model for the optimal management of algal biomass production was developed in the second section and the formal presentation is given in the Appendix.

In this section we simulate an application of this model to the case of *Spirulina* production. We selected the data used in this application from reports concerning the production of *Spirulina* in the Negev desert. (In the simulation we are grateful to have permission to use a dynamic optimization algorithm developed at the World Bank from Alex Mirhause.[11]

As pointed out in the preceding section the structure of equations and parameters of the model can be classified into three main components: (1) growth (production) equations and their parameters; (2) cost functions and their parameters; and (3) inputs and outputs markets.

Growth and production of the algal biomass are depicted by Equation 16 in the Appendix and in our applicaton had the specific form

$$\Delta B_t = \rho(1 - (B_t/M))B_t - h_t \tag{9}$$

where ρ is the growth parameter, B_t is the level of biomass at time t, h_t is the harvesting at time t, and $\Delta B_t = B_t - B_{t-1}$ is the additional biomass growth minus harvesting at time t.

Thus, the biomass at time t is determined by the level of biomass at time $t-1$, plus the growth at time t and minus the amount harvested.

The growth function consists of a linear coefficient and an asymptotic level M, and were estimated from a logistic growth function at the labs of the Desert Research Institute.[6] The amount harvested is controlled by the grower and is a decision-variable to be determined by the optimization rules.

These growth and production functions are heavily influenced by the human factor. The grower's abilities can be classified into two groups; one includes the agro-technical abilities to produce the maximal possible amount of algae and the second relates to his managerial and economic abilities to manage the operations with minimum costs and maximum profit.

The managerial abilities play an important role in the cost and revenue components which constitute the second and third groups of equations. In our application they are represented in the profit function (see also Equation 15 in the Appendix):

$$V_N = \sum_{t=1}^{N} [P - C(B_t)] \cdot h_t \cdot (1/(1 + r))^t \tag{10}$$

where P is the price per kg of harvested *Spirulina*, $C(B_t)$ is the cost per kg of harvested *Spirulina*, r is the interest, and V_N is the present value of the sum of profit for a planning horizon of N periods.

Since both the price and the cost per unit of *Spirulina* do not depend on the quantity sold, it is possible to separately calculate the operating profit per kilogram,* and then obtain the total operating profit per period by multiplying by the amount harvested. The sum over the N periods of the discounted value of operating profit per period yields the quantity V_N.

Now, the unit cost equation is represented in our case by $C(B_t)$, which is equal to

$$\frac{\text{harvesting costs per m}^3}{B_t} = \frac{C}{B_t}$$

plus production and drying costs per kilogram, indicated by d.

Thus, we have decreasing marginal costs with respect to the level of algal biomass. As noted above the costs depend, among other factors, on the managerial ability of the producer.

---

* Operating profit measures the profit before deducting the fixed costs.

They also depend on the destination and final use of the algae. Thus, if the *Spirulina* is produced for human consumption as health products, the costs will be much higher than if the final use is for fish or animal food. The obvious reasons will be higher standards of quality as well as required government health regulations, e.g., the U.S. Food and Drug Administration. If the *Spirulina* is used to feed fish nearby, drying and overall production costs can be reduced drastically by leaving a high percentage of water in the produce.

The price received from sales for human consumption is many times higher than for feeding fish. This brings us to another component, the demand for the final product. In our application this factor is represented by the price P in Equation 12, since we assume the producer to have only a small share of the market and is hence unable to influence the price. If the producer has a large share of the market, and therefore exercises an oligopolistic or a monopolistic power, he may be facing a negative sloped demand curve, i.e., D(p) with [D(p)]/p < 0, instead of a fixed price p for all levels of quantities sold. However, though it may make the calculations more complicated, the nature of solution will not change.

Up to this stage, the procedure to calculate the operating profits over time was described. As in the static case of the conventional cost-benefit analysis fixed costs of investments must be taken into account. Since these are incurred at the initial state of operation in a fixed amount, they only determine the decision whether to start production or not. If the discounted net value of operation is greater (smaller), than the fixed costs production will take place (cease).

The level of fixed costs is determined by the investment in the ponds, i.e., land scrapping, insulation of floor and walls, equipment needed for turbulation of water, as well as other required facilities for harvesting, drying, and producing the final product. Again, the final use of the product will influence the nature of investments, e.g., covering the ponds may be an important requirement for the quality of the final product in the case of human consumption.

Two basic types of products are considered in the application: (1) expensive or high standard or quality product (e.g., for human consumption), and (2) cheap product (e.g., for fish feeding). These two basic models were evaluated for various changes in parameters.

The empirical set of data was used to reproduce Equations 11 and 12 in the following equations and this basic model was evaluated for various changes in parameters. The corresponding basic equations are a human consumption:

$$0.27(1 - (B_t/1.6))B_t = \Delta B_t \tag{11}$$

and

$$\sum_{t=1}^{N} (16 - C(B_t))h_t(i/(1 + 0.15))^t = V_N \text{ per unit of product} \tag{12}$$

where

$$C(B_t) = 0.03/B_t + 5$$

An optimal solution is fully characterized by the trajectory of optimal biomass density over time. Since we assume growth, in a given production regime to be dependent on biomass level only (no aging effects), the optimal density level reaches a steady-state level (see discussion in the second section). The optimal harvesting trajectory is chosen in order to keep density level at its optimal trajectory. If harvesting is allowed to take negative values (i.e., it is possible to add biomass after the process starts), then optimal density level can be reached instantaneously. We assume this is the case with algal culture. Optimal solution therefore contains optimal steady-state density level (denoted by b_{ss}) and its corresponding

Table 1
OPTIMAL PROFIT AND STEADY-STATE LEVELS OF OPTIMAL DENSITY AND OPTIMAL HARVESTING: THE CASE OF A QUALITY PRODUCT

Parameter changes	Optimal profit[a]	B_{ss} optimal	h_{ss} optimal
Basic	51.92	0.796	0.108
M = 0.2	37.99	0.795	0.080
M = 0.35	67.86	0.798	0.140
d = 10	28.09	0.798	0.108
d = 1	71.02	0.796	0.108

[a]   ($)/m³.

Table 2
THE "INEXPENSIVE PRODUCT" CASE

Parameter changes	Optimal profit[a]	B_{ss} optimal	h_{ss} density
Basic	12.44	0.842	0.319
C = 0.03	13.45	0.816	0.320
C = 0.15	11.16	0.879	0.317
M = 5; C = 0.02; d = 0.05	8.46	0.810	0.200
M = 0.4; C = 0.02; d = 0.05	6.59	0.815	0.160

[a]   ($)/m³.

optimal harvesting (denoted by h_{ss}), which is the growth at the density level b_{ss}. Table 1 presents optimal solutions and their corresponding profits for the quality product for different levels of the growth parameter ρ and the cost parameter d. From this table we can conclude:

1.   An increase in the rate of growth, ρ, *cet. par.* increases the steady-state level of biomass B_{ss}, and the amount of harvesting at steady state, h_{ss}.
2.   An increase in [P − C(B)], *cet. par.*, leaves B_{ss} and h_{ss} unchanged, but increases optimal profits.

Thus, optimal behavior and profit were found to be very sensitive to changes in the growth function, i.e., the production function and this on the human factors and extension services discussed above. In the previous example we considered a case of a quality algal product. *Spirulina* pills marketed as health food, with relatively high price (e.g., p = 16) and high unit costs of production (e.g., d = 5). For the sake of comparison we consider a case of a less refined product, that can be used to feed fish and this will receive a lower price, but will require lower costs of production having to meet lower quality requirements. In such a case to have a profitable production a higher rate of growth is required, and we assumed for the sake of demonstration that such a rate of growth is attainable. Thus, the example of a "cheap product" is depicted by the basic Equations 13 and 14 and the various parametric changes by Table 2.

$$0.8(1 - (B_t/1.6))B_t = \Delta B_t \tag{13}$$

$$\sum_{t=1}^{N} (1 - C(B_t))h_t(1/(1 + r))^t = V_N \qquad (14)$$

where

$$C(B_t) = 0.08/B_t + 0.05$$

and r is the daily equivalent of 10% annual interest rate.

APPENDIX

A Mathematical Formulation of the Optimization Problem
The (discounted) value of the algal biomass is

$$PV = \int_0^\infty e^{-\delta t}(P - C(B))h(t)dt \qquad (15)$$

where δ is the (continuous) discount rate, P is price of a unit biomass, C(B) is unit production cost (include also harvesting, drying, processing and packing), h(t) is biomass harvested at time t. (We have assumed for a while an infinite planning horizon.)

The growth process is given by

$$dB/dt = G(B,t,U(t)) - h(t) \qquad (16)$$

where U(t) represents all factors other than B that affect growth and which may be controlled by the grower. We assume the environment and technology are given so h(t) indicates production policy.

The optimization problem can be stated as:

$$\underset{U(t),h(t)}{\text{Max}} \quad PV \quad \text{subject to the growth Equation 16} \qquad (17)$$

Assume first that U(t) is given so that Equation 17 is maximized over h(t) alone. Since the state Equation 16 is linear in h(t), the control variable, it is possible to directly apply the Euler condition (see "A Historical Outline of Applied Algology", Chapter 2) and to get the following optimality condition:

$$G'(B) - \frac{C'(B)G(B)}{P - C(B)} = \delta \qquad (18)$$

Equation 18 is with complete analogy to Equation 6.

In many situations it is possible, due to biological or economical limitations, to grow only part of the year, say T days. Assuming identical conditions over the years the optimal solution for 1 year is also optimal for all other years. The problem can be stated as

$$\text{Max} \int_0^T e^{-\delta t}[P - C(B)]h(t)dt + e^{-T}[P - C_T]B_T \qquad (19)$$

subject to Equation 16, where C_T is unit cost of impulse harvesting, i.e., emptying the pond. Note that T, the cycle length, can enter the maximization argument if the determination of optimal cycle length is required.

In solving Equation 19 we define the Hamiltonian

$$H = e^{-\delta t}[P - C(B)]h(t) + \lambda(t)[G(B) - h(t)] \qquad (20)$$

where $\lambda(t)$ is an auxialiary variable with an important econmic interpretation.[8] Pontryagin's Maximum Principle[9] states that h(t) is determined so as to maximize H for each time period $0 < t < T$. For time T, the transversality condition[9] states

$$\lambda(T) = \partial/\partial B\{e^{\delta T}[P - C_T]B_T \quad (21)$$

By rewriting Equation 20 as follows

$$H = e^{-} {}^{|}[P - C(B)] - (t) h(t) + (t)G(B)$$

it is easy to verify that the maximum principle implies for $0 < t < T$

$$\text{Optimal harvesting policy (t)} = \begin{cases} h_{min} \text{ if (t)} < 0 \\ h^*(t) \text{ if (t)} = 0 \\ h_{max} \text{ if (t)} > 0 \end{cases} \quad (22)$$

where h_{min} (h_{max}) is the lower (upper) bound of $h(h^*(t)$ is the harvesting level that satisfies Equation 18 and $\delta(t) \stackrel{def}{} e^{-\delta t}[P - C(B)] - \lambda(t)$. The transversality conditions imply for time T

$$\lambda(T) = e^{-\delta T}(P - C_T) \quad (23)$$

Along the singular solution $\lambda(t) = 0$, i.e., $\lambda(t) = e^{-|}[P - C(B)]$, hence Equation 23 will be effective only if $C(B(T)) \neq C_T$ (i.e., only if the unit cost of impulse harvesting differs from the unit harvesting cost), which is not the case in algaculture.

As noted earlier, T can be an argument of the maximization in order to determine optimal cycle length. This can be done by a numerical search method.

Up to now we have assumed the vector of other control variables U(t) to be fixed. This vector includes variables such as quantities and nutrient content of supplemented fed materials, water flow rate, indoor or outdoor pond, etc. The subset of variables of U which are truly continuous and time dependent should be treated in the maximization process the same as h(t). This involves the generalization to multiple control theory,[9] which in principle creates no problem, but for practical purposes is intractable. Those variables of U which are time independent can be treated in the maximization problem via standard numerical methods.[10]

LIST OF SYMBOLS

t	Time index
$B_t = B(t)$	Biomass population density at time t
$\dot{B} = dB/dt$	Rate of change of B over time
ΔB_t	Discrete change in B between (t-1) to t
G(B)	Growth rate
ρ	Growth parameter
M	Maximum population density that allows positive growth
P	Price of algal product
r	Interest rate
B^s	Biomass population density at which growth is maximum
K	Capital
C	Unit cost
δ	Continuous discount rate

h(t)	Biomass harvested at time t
U	Control vector
T	Cycle length
H	The Hamiltonian
$\lambda(t)$	Auxiliary (shadow price) variable
h_{min}	Lower bound of h(t)
h_{max}	Upper bound of h(t)
h*(t)	Optimal harvesting policy
PV	Present value of the stream of future profits
V_N	Present value of sum of profits for planning horizon of N periods
b_{ss}	Steady-state level for optimal biomass density
h_{ss}	Optimal harvesting level
d	Unit drying and processing costs

REFERENCES

1. **Binswanger, H. P. and Ruttan, V. W.,** *Induced Innovation,* John Hopkins University Press, Baltimore, 1978 (see chap. 1 to 5).
2. **Schumpeter, J. A.,** The Analysis of Economic Change, Reprinted in *Readings in Business Cycle Theory,* The Blakiston Co., Philadelphia, 1944.
3. **Clarck, C. W.,** *Mathematical Bioeconomics. The Optimal Management of Renewable Resources,* Wiley Interscience, New York, 1976.
4. **Rausser, G. C. and Hochman, E.,** *Dynamic Agricultural Systems: Economic Prediction and Control,* North Holland, New York, 1979.
5. **Hochman, E. and Lee, I. M.,** *Optimal Decisions in the Broiler Producing Firm: A Problem of Growing Inventory,* Giannini Foundation Monogr. No. 29, University of California, Berkeley, 1972.
6. **Richmond, A., Vonshak, A., and Arad, S.,** Environmental limitations in outdoor production of algal biomass, in *Algal Biomass,* Shelef, G. and Soeder, C. J., Ed., Elsevier/North Holland, Amsterdam, 1980, 65.
7. **Vonshak, A., Abeliovich, A., Boussiba, S., Arad, S., and Richmond, A.,** Production of *Spirulina* biomass: effects of environmental factors and population density. *Biomass,* 2, 175, 1982.
8. **Dorfman, R.,** An economic interpretation of optimal control theory, *AER V LIX, N.* 5, 817, 1969.
9. **Pontryagin, L. S., Boltyanskii, V. G., Gamkrelioze, R. V., and Mishenko, E. F.,** *The Mathematical Theory of Optimal Processes,* Interscience, New York, 1962.
10. **Talpaz, H. and Tsur, Y.,** Optimizing aquaculture management of a single-species fish population, *Agric. Syst.,* 9(2), 127, 1982.
11. **Mirhause, A.,** Personal communication, 1981.

FUTURE PROSPECTS

A. Richmond

The culture of algae for industrial purposes is a novel biotechnology, naturally prompting the question of whether it will ever become an important means of production of food, feeds, and chemicals. The author believes there is solid evidence that the utilization of algae for various economic purposes will gradually develop into an important industrial endeavor, particularly in warm, sunny regions where the water available is unsuitable for the production of conventional agricultural crops.

A crucial question in analyzing the future prospects of algaculture concerns the market potential for algal products. No comprehensive answers are available yet, but certain points have become clear. The nutritional value of various algae as animal feeds and as food for humans has been substantiated by hundreds of research studies. In addition, scores of studies have described various therapeutic effects of microalgae. There is a growing list of chemicals that can be extracted from microalgae for commercial purposes. Most notably are glycerol and β-carotene from *Dunaliella*, sulfonated carrageenan-like polysaccharides from *Porphyridium* sp. and phycocyanin from *Spirulina* which seem well on their way to commercial production. Other products, such as linoleic acid from *Spirulina* and arachidonic acid from *Porphyridium* represent potential products from microalgae. Most important in this respect is the fact that while tens of thousands of algal species have been taxonomically defined, only a small fraction have been surveyed for their possible economic potential. Yet, it seems possible to produce from these microorganisms nearly all the products that are currently obtained from conventional crops. Such crops do not usually produce continuously year-round, and most importantly, require a supply of sweetwater, a resource which is becoming scarce in arid and semi-arid lands all over the world. Being unicellular and having a short life cycle, microalgae in general and cyanobacteria in particular represent a most suitable material for genetic manipulation. Transfer of genes for the synthesis of specific molecules should eventually allow large-scale production of many products that are presently obtained rather inefficiently from terrestial plants or from heterotrophic microorganisms cultured in expensive reactors.

It seems clear that the pressing issue in commercial algaculture is not so much the market potential of algae products as it is the cost of their production. The current prices of commonly used animal feeds, i.e., various grains, and of high-protein plants such as soybean for human consumption, are approximately one-tenth the cost of production of algae in the small industrial plants in operation today.

What, then, are the chances for microalgae in the competition with conventional crops for land, water, and capital? There are several indications that the cost of production of microalgae, which today confines them to the health food market, will gradually but consistently decline, in parallel with continuous expansion of the market. In general, the history of agriculture and industry provides numerous examples for the natural course of development of new fields of endeavor. Repeatedly, be it with rubber trees or wheat, experience has shown that with time, productivity increases and the cost per unit production decreases by an order of magnitude or more.

The scale of production has the most decisive effect on the cost of unit-product. A cost-benefit analysis for the production of *Scenedesmus* in Peru[1] bears this out. With a cultivation area of 2500 m^2, the total production cost of 1 kg of drum-dried *Scenedesmus* harvested by centrifugation was $7.50. With a fourfold increase in cultivation area with carbon feeding by self-produced CO_2 and harvesting by flotation, the production cost per kilogram of dry matter was reduced about 50%. A further tenfold increase of the cultivation area to 100,000 m² decreased the calculated production cost to $1.17/kg. The capital investment was calculated as $2,851,000, depreciation; maintenance, $309,000; and variable costs, $703,000. The annual output rate was considered to be approximately 1000 tons, i.e., 100 tons/ha/year or approximately 27 g dry wt m^{-2}/day.

It is almost self-evident that due to the substantial capital investment needed per unit pond area, the cost of production is greatly dependent on the yield per area. In the above example, a rather high output rate was assumed and this was a major reason for the relatively low cost calculated for large-scale production. In reality, to date only about 30 to 50% of these outputs have actually been obtained over a long-term production period, making the actual cost per unit product being 5 to 10 times higher. Nevertheless, yields greater than 10 kg/m^{-2}/year are possible and have been obtained for periods lasting a few weeks, (see "Outdoor Mass Cultures of Microalgae"). Theoretically, the photosynthetic efficiency can be approximately 8% where the temperature does not limit growth, which implies a tremendous output rate, i.e., about 200 tons of dry matter/ha/year. However, even a consistent output one third the theoretical maximum will significantly reduce the current cost of production, which ranges between $5 to $10 (U.S.)/kg of drum- or spray-dried algal powder.

The current low yields of 15 to 30 tons/ha/year that are usually obtained in outdoor mass cultures reflect the present state of ignorance in large-scale production of microalgae. It seems appropriate to compare the present know-how in algaculture with that thousands of years ago at the beginning of agriculture, when the cultivated plant species and the cultivation methods used only reflected man's preliminary selection of species and rudimentary skills in cultivating them. Even a superficial view of the methods and machinery currently used for commercial production of microalgae would reveal that the industry is still in the initial phase of trial and error, adapting machines and materials from other industries. Every phase of this biotechnology may be significantly improved.

One example concerns pond-lining, which represents a major investment cost. Lining made of PVC with a long durability is being developed today. If lining could be safely used for two or three decades, the capital investment for pond construction would be significantly curtailed.

The cost of nutrients, which comprises some 30% of the running cost of production, could also be reduced. A major contribution to this end would be the provision of an inexpensive source of CO_2. Industrial waste would be one such source, being suitable, however, only where industry is in close proximity to the production site. Another possibility with a wider application is to extract CO_2 by burning a 10:1 mix of calcium carbonate and crude oil (using the heat produced for dehydrating the product). Recently, tremendous reservoirs of CO_2 have been discovered in the southwestern U.S. Tapping such resources for algaculture should be considered. In addition, with algae such as *Spirulina*, which are grown on an alkaline medium with a pH of 10.5 a substantial amount of carbon required for an average output may be obtained from the CO_2 in the air. The use of species selected for growth in highly alkaline pH would reduce the cost of carbon nutrition.

The most expensive nutritional input except carbon is nitrogen. An attainable target with this nutrient would be the use of nitrogen-fixing microorganisms in conjuction or in symbiosis with the cultured algal species, thus reducing or even eliminating the need for a nitrogen fertilizer.

It can be expected that improved methods for harvesting the algal mass will become available in the future. To aid in harvesting unicellular microalgae, simple flocculation techniques, based on physical methods such as electro-flocculation and flotation, will have to be developed. Such means should replace the cumbersome and costly chemical methods commonly used to affect flocculation. New, edible chemical flocculants, as well as novel techniques for screening and filtration could be expected to become significantly improved for the harvesting of microalgae.

Expenses on power would be cut if the conventional electric grid could be dispensed with, and photovoltaic power (in the not too distant future) could be used for stirring, pumping, filtration, and even drying of the algal mass. In principle, photovoltaic power can be used most effectively for algaculture, since power is needed mainly during the day and a good parallel exists between peak production of solar power and peak power demand by the algal system. Storage capacity of power, which at the present hinders the economics of photovoltaic power, must be minimal when photovoltaic power is used in algaculture.

Natural energies could be also used for dehydration of the harvested algal mass. In arid, sun-rich areas (the natural sites for algaculture) air is relatively dry, being suitable for rapid drying processes. Pretreatment of the algal mass with solar steam will be advantageous in sun-rich areas once solar technologies for steam production, which are being developed today, are put into general use.

As long as the commercial aspects of microalgal culture were not recognized, there were no opportunities to relate technically to problems associated with large-scale operation. There is every reason to assume that with the establishment and gradual expansion of a microalgal industry, this biotechnology will continuously improve, becoming increasingly more specialized and more efficient, with immediate effects on reduction of the cost of production. For example, a breakthrough in commercial algaculture resulted from the increased demand for some microalgal species as health foods, which prompted establishment of large-scale algaculture in several countries. The experience gained in the present production plants will have far-reaching effects on advancing this novel biotechnology. Also, the know-how obtained in large scale commercial production of health foods will be instrumental in the development of other uses for microalgae, e.g., supplements for human food and various special chemicals. A significant cut in the cost of algal production could be expected when the market is further expanded, e.g., through demand for special feed supplements for animals, particularly fish and other aquatic organisms, but also for poultry, hogs, and ruminants.

One of the most important advantages of algaculture which will have a decisive effect on its future prospects is the ability of many algal species to thrive in saline water, unsuitable for the production of most agricultural crops. When experience and understanding in this biotechnology becomes such that production is greatly simplified and its cost significantly reduced, algaculture in brackish water will have clear economic advantages in many lands. Somewhat paradoxically perhaps, it is in the hot, dry, usually poor lands of the world which often have untapped sea and brackish water resources, and where only meager conventional agriculture now exists, that algaculture may become a highly productive venture.

REFERENCE

1. **Becker, W.,** Personal communication.

INDEX

I